Biogeochemistry: Science and Applications

Biogeochemistry: Science and Applications

Editor: Karolina Jensen

RCALLISTO
REFERENCE

www.callistoreference.com

Callisto Reference,
118-35 Queens Blvd., Suite 400,
Forest Hills, NY 11375, USA

Visit us on the World Wide Web at:
www.callistoreference.com

ISBN: 978-1-63239-823-9 (Hardback)

Cataloging-in-publication Data

Biogeochemistry : science and applications / edited by Karolina Jensen.
   p. cm.
Includes bibliographical references and index.
ISBN 978-1-63239-823-9
   1. Biogeochemistry. 2. Biochemistry. 3. Geochemistry. I. Jensen, Karolina.
QH343.7 .B56 2017
574.5222--dc23

Table of Contents

Preface

Biogeochemistry is an upcoming field of science that has undergone rapid development over the past few decades. It involves the study of the composition of the environment and the various chemical, geological and other processes related to it. This field of study also examines the cycles of chemical elements like carbon and nitrogen. There has been rapid progress in this field and its applications are finding their way across multiple industries. The aim of this book is to present researches that have transformed this discipline and aided its advancement. This book includes contributions of experts and scientists which will provide innovative insights into this field. It brings forth some of the most innovative concepts and elucidates unexplored aspects of biogeochemistry. The objective of this text is to give a general view of the different areas of biogeochemistry and its applications.

This book was inspired by the evolution of our times; to answer the curiosity of inquisitive minds. Many developments have occurred across the globe in the recent past which has transformed the progress in the field.

This book was developed from a mere concept to drafts to chapters and finally compiled together as a complete text to benefit the readers across all nations. To ensure the quality of the content we instilled two significant steps in our procedure. The first was to appoint an editorial team that would verify the data and statistics provided in the book and also select the most appropriate and valuable contributions from the plentiful contributions we received from authors worldwide. The next step was to appoint an expert of the topic as the Editor-in-Chief, who would head the project and finally make the necessary amendments and modifications to make the text reader-friendly. I was then commissioned to examine all the material to present the topics in the most comprehensible and productive format.

I would like to take this opportunity to thank all the contributing authors who were supportive enough to contribute their time and knowledge to this project. I also wish to convey my regards to my family who have been extremely supportive during the entire project.

Editor

Self-Organising Maps and Correlation Analysis as a Tool to Explore Patterns in Excitation-Emission Matrix Data Sets and to Discriminate Dissolved Organic Matter Fluorescence Components

Elisabet Ejarque-Gonzalez*, Andrea Butturini

Departament d'Ecologia, Facultat de Biologia, Universitat de Barcelona, Barcelona, Catalunya, Spain

Abstract

Dissolved organic matter (DOM) is a complex mixture of organic compounds, ubiquitous in marine and freshwater systems. Fluorescence spectroscopy, by means of Excitation-Emission Matrices (EEM), has become an indispensable tool to study DOM sources, transport and fate in aquatic ecosystems. However the statistical treatment of large and heterogeneous EEM data sets still represents an important challenge for biogeochemists. Recently, Self-Organising Maps (SOM) has been proposed as a tool to explore patterns in large EEM data sets. SOM is a pattern recognition method which clusterizes and reduces the dimensionality of input EEMs without relying on any assumption about the data structure. In this paper, we show how SOM, coupled with a correlation analysis of the component planes, can be used both to explore patterns among samples, as well as to identify individual fluorescence components. We analysed a large and heterogeneous EEM data set, including samples from a river catchment collected under a range of hydrological conditions, along a 60-km downstream gradient, and under the influence of different degrees of anthropogenic impact. According to our results, chemical industry effluents appeared to have unique and distinctive spectral characteristics. On the other hand, river samples collected under flash flood conditions showed homogeneous EEM shapes. The correlation analysis of the component planes suggested the presence of four fluorescence components, consistent with DOM components previously described in the literature. A remarkable strength of this methodology was that outlier samples appeared naturally integrated in the analysis. We conclude that SOM coupled with a correlation analysis procedure is a promising tool for studying large and heterogeneous EEM data sets.

Editor: Matthias Dehmer, UMIT, Austria

Funding: This research was funded by the Spanish Ministry of Economy and Competitiveness (CGL2011-30151-C02-02). Elisabet Ejarque-Gonzalez's research was in part supported by an FPU doctoral scholarship from the MEC (AP2008-03431). Both authors are members of the GRACCIE consortium. The funders had no role in study design, data collection and analysis, decision to publish, or preparation of the manuscript.

Competing Interests: The authors have declared that no competing interests exist.

* E-mail: elisabet.ejarque@ub.edu

Introduction

Excitation-Emission Matrices (EEMs) are three-dimensional fluorescence data that provide information about the composition of fluorescent chemical mixtures. They constitute optical landscapes that extend over the dimensions of excitation and emission wavelengths {λex–λem}, and where fluorophores appear in the form of peaks. In the field of marine and freshwater biochemistry, EEMs have been used for the study of dissolved organic matter (DOM), being a comprehensive analytical technique with which to characterise a highly complex mixture of organic compounds [1–3]. Indeed, EEMs have served to advance scientific knowledge about the ecology and biogeochemistry of DOM in aquatic systems [1,2]. Most importantly, they have contributed to evidence that some fractions of DOM are highly reactive organic molecules that are involved in numerous ecosystem processes, such as bacterial uptake [4–6], metal binding [7,8], photoreactivity [9–11] and light attenuation [12]. Overall these findings suggest the major involvement of DOM in the global carbon cycle [13,14].

Despite the great potential for EEMs to increase knowledge about DOM behaviour in the environment, their interpretation and statistical treatment remain a challenge [15]. The spectral shapes of EEMs are complex mixtures of multiple and overlapping independent fluorescence phenomena, caused by the wide range of organic molecules contained in DOM. As only about 25% of these molecules have been identified [16], there is a lack of chemical standards to be used to separate the signal of bulk DOM into its individual components. For that reason, there is a need to develop pattern recognition methods capable of detecting and isolating the signal of different fluorescing moieties in the absence of any previous knowledge about the composition of DOM in a given sample.

A well-suited tool to satisfy these needs are Self-Organising Maps (SOM). SOM is an artificial neural network algorithm that mirrors the biological brain function [17]. Due to its unsupervised self-learning capacity, it is capable of recognizing patterns in complex data sets without following any assumptions about the data structure. Although it has been increasingly used within analytical chemistry in recent years [18] it has not been until

A

B

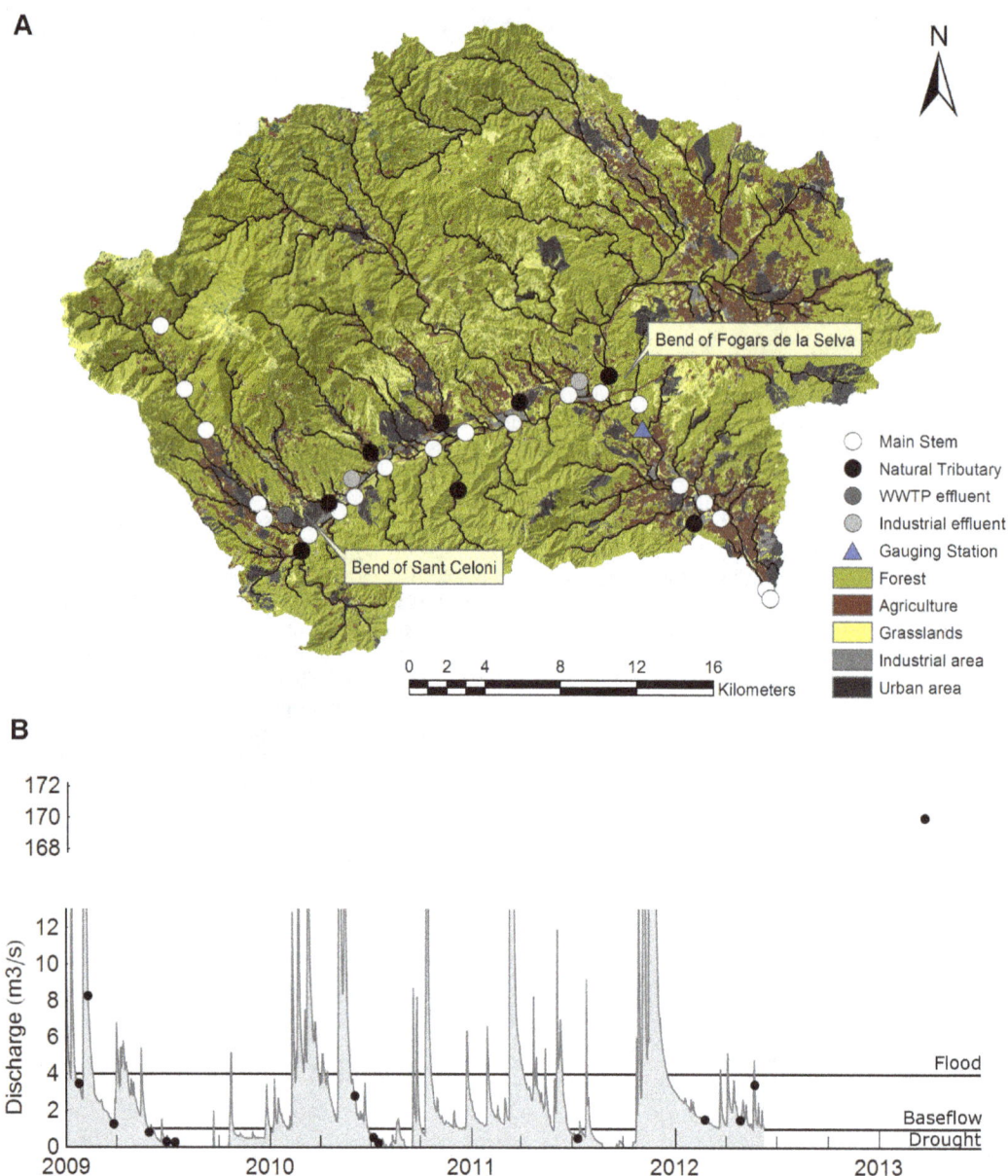

Figure 1. Experimental setting of the data set. A) Study site within the catchment from which the samples were collected. The river was operationally divided into three reaches: the "headwaters", the "middle reaches" and the "lowland". The divisions between segments correspond to the two big bends of Sant Celoni and Fogars de la Selva. B) Hydrogram contextualising the 15 sampling dates. Discharge data were recorded in the gauging station at Fogars de la Selva. Sampling dates were operationally divided into "flood" ($Q > 4$ m$^3 \cdot$s^{-1}), "baseflow" ($4 > Q > 1$ m$^3 \cdot$s^{-1}) and "drought" ($Q < 1$ m$^3 \cdot$s^{-1}) categories. As continuous monitoring was interrupted, the discharge on the last sampling date (2013/06/03) was measured individually on that date. All discharge data were provided by the Catalan Water Authority (Agència Catalana de l'Aigua, [24]).

recently that SOM has been used to analyse EEM data sets [19,20], and the potential for SOM to equate or even outperform other state-of-the-art EEM data treatment methods like partial least-squares regression (PLS), principal components analysis (PCA) and parallel factor analysis (PARAFAC) has been highlighted [15,18,21,22]. The map space produced by SOM offers multiple possibilities for the graphical representation of the output, allowing to unveil patterns among samples (best matching unit and unified distance matrices), as well as to explore what variables (wavelength coordinates in the case of EEM data sets) are the most influent in creating the sample patterns (component planes) [18]. However, pattern recognition at the variable level has

remained at a qualitative stage, and the specific need to isolate independent fluorophores has not been covered.

Furthermore, previous analyses of EEM data sets with SOM were performed on data from engineered systems, where the diversity of fluorophores was essentially homogeneous among the samples [19,20]. However, EEM data sets collected in natural water systems are subject to contain a wide diversity of spectral shapes, due to the multiple environmental factors that influence DOM quality [23]. In this case, data pattern interpretation may become more challenging, as the presence of outliers may alter the stability of the SOM output, and hence its reliability.

In this context, this study aims at expanding the evidences that SOM is a suitable tool for the study of EEM data sets. Specifically, we focus on two aspects. On the one hand, we aim to further test the performance of SOM when a high heterogeneity of spectral shapes is contained within the data set. We address this point by assessing the stability of the quantization and the neighbourhood relations of the SOM output under a leave-one-out cross-validation approach. On the other hand, we search for independent fluorophores by extending SOM with a correlation analysis of component planes. This constitutes a novel approach to discriminate areas of the EEM (i.e. groups of wavelength coordinates) representing different fluorophores.

Materials and Methods

Ethics statement

Some of the sampling sites included in this study were located in the protected areas of the Parc Natural del Montseny and Parc del Montnegre-Corredor, both under the authority of the Diputació de Barcelona. No specific permission was required to conduct the fieldwork. We confirm that our study did not involve any endangered or protected species.

Data set

Our EEM data set included 270 samples from a Mediterranean river catchment called La Tordera (865 km^2), situated to the north-west of Barcelona, Catalunya. The sampling strategy was designed in order to assess the influence of space and hydrology on the EEM spectral shapes. Accordingly, in order to characterise the longitudinal dimension, water samples were collected at 20 sites along the main stem (60 km long). The sites were operationally categorised into three main reaches, referred to as "*headwaters*", "*middle reaches*" and "*lowland*", divided by the bends of Sant Celoni and Fogars de la Selva (Figure 1A). Each of these three river reaches has distinctive properties. The "*headwaters*" section corresponds to a forested catchment area with accentuated slopes and incipient human pressure, the "*middle reaches*" are characterised by intensive anthropogenic activity, receiving both diffuse inputs from urban activities and point source effluents of waste water treatment plants (WWTPs) and industries; and finally the "*lowland*" corresponds to a shallow and meandering geomorphology with a lower density of direct anthropogenic effluents. Eleven influent waters were also sampled upstream from the confluence with the main stem. Some of them correspond to natural tributaries with varying degrees of anthropogenic impact, whereas others correspond to WWTPs or effluents from chemical industries.

The seasonal hydrological variability was captured by sampling on 15 different dates during which a wide range of hydrological conditions was encountered: from flash floods to severe summer droughts (Figure 1B). In this case, samples were also operationally defined according to three categories: "*flood*" corresponds to discharges higher than 4 m$^3 \cdot$s^{-1}, "*drought*" to discharges lower than 1 m$^3 \cdot$s^{-1}, and "*baseflow*" to flows between 1 and 4 m$^3 \cdot$s^{-1}. We used discharge data from the gauging station of Fogars de la Selva, provided by the Agència Catalana de l'Aigua (Catalan Water Authority, [24]), as a reference.

Due to the wide variety of drained land cover, water sources and hydrological conditions included in the sampling design, the final EEM data set was expected to include a wide variety of spectral shapes.

Field and laboratory procedures

Samples were collected in acid-rinsed glass bottles, and were kept refrigerated in the dark until arrival at the laboratory. Next, samples were filtered with 0.22-μm-pore nylon membranes and kept refrigerated until their spectral analysis, which was conducted within the next two days. Fluorescence analyses were performed using a Shimadzu RF-5301 PC spectrofluorometer equipped with a xenon lamp and a light-source compensation system (S/R mode). For every EEM, 21 synchronous scans were collected at 1-nm increments both in emission and in excitation. During each scan, excitation was measured over a wavelength range of 230 nm<λex<410 nm. Initial emission wavelengths ranged from 310 nm to 530 nm, at intervals of 10 nm. The bandwidth used for both excitation and emission was 5 nm. Spectra were acquired with a 1-cm quartz cell.

Absorption spectra were measured for fluorescence inner filter correction purposes using a Shimadzu UV-Visible UV1700 Pharma Spec spectrophotometer. Data were collected in double beam mode with wavelength scanned from 200 to 800 nm and with milliQ water as the blank. The slit width was set to 1 nm.

Raw EEM data were corrected and normalised to allow for inter study comparison following the steps described by Goletz et al. [25]. Spectral corrections were applied to both emission and excitation measurements to correct for wavelength-dependent inefficiencies of the detection system. An excitation correction function was determined using Rhodamine B as a quantum counter [26], whereas for emission a correction file was obtained by comparing the reference spectra of quinine sulphate and tryptophan provided by the National Institute of Standards and Technology (NIST) according to the procedure described by Gardecki and Maroncelli [27]. Next, data were normalised by the area under the Raman peak of a deionised water sample at λex = 350 nm and λem = {371–428} nm [28]. Inner filter effects were corrected by comparing absorbance measurements according to Lackowicz [26], as described by Larsson et al. [29]. Finally, a blank EEM of deionised water, measured on the same day of analysis and having undergone the same correction and normalisation procedures, was subtracted from every EEM sample.

Optical indices calculation

Specific Ultra-Violet Absorbance (SUVA), as a surrogate measurement for DOC aromaticity, was measured as the Napierian absorption coefficient at λ_{abs} = 254 nm normalised by DOC concentration [30]. DOC concentration was determined by oxidative combustion and infrared analysis using a Shimadzu TOC Analyser TOC-V$_{CSH}$.

The Humification Index (HIX), indicator of the humification degree of humic substances, was calculated as the ratio between the area under {λex$_{254}$, λem$_{(435–480)}$} and the area under {λex$_{254}$, λem$_{(330–345)}$}, as described by Zsolnay [31]. Finally, the Fluorescence Index (FI) [32,33], indicator of the allochthonous vs autochthonous origin of DOM, was calculated as the fluorescence intensity at {λex, λem} = {370,470} nm divided by that at {λex, λem} = {370,520} nm.

Self-organising maps

Self-Organising Maps (SOM) – also known as Kohonen maps – are a special type of two-layered artificial neural network (ANN). ANNs are mathematical models mirrored in the functioning of the biological nervous system, which have the ability to learn the patterns of input features and predict an output. They consist of an adaptive system of interconnected neurons – or processing units – that change their structure during a learning phase. In this phase, weight vectors (called prototype vectors or, in this context, prototype EEMs) that lie in the connections between neurons are adjusted to minimize the overall error of the network prediction [34].

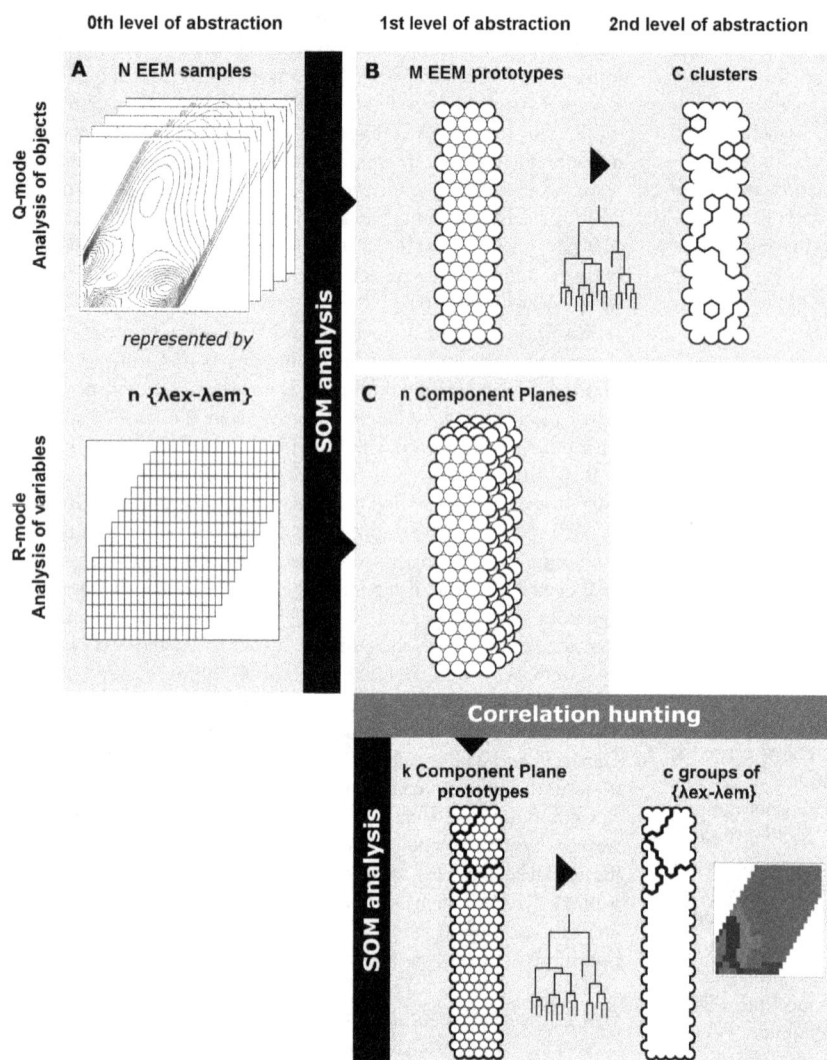

Figure 2. Summary of the methodology applied in this study. A) N initial samples are reduced to M prototype EEMs by SOM analysis. B) EEM prototypes are clustered to facilitate exploration of the relationships between the sample EEMs. C) SOM is performed on the correlation matrix of the component planes of the Q-mode SOM analysis. The output corresponds to an aggregation of highly correlated wavelength coordinates in a single neuron unit. D) Neuron units are clustered in order to find groups of highly correlated wavelength coordinates. E) Wavelength coordinate clusters are displayed in an EEM optical space in order to evaluate their biogeochemical meaning. Adapted and extended from Vesanto and Alhoniemi [38].

By the end of the learning process, the EEM samples have been assigned to their best matching unit (BMU), that is, the unit that has the most similar prototype EEM. Thus, the outcome of the SOM will be a grid in which each unit will contain a prototype EEM whose spectral properties vary gradually but unevenly across the grid, according to the characteristics of the input data. By projecting the original EEMs on their BMU in the SOM grid, sample patterns can be explored.

According to Cattell [35], this analysis can be considered as an analysis in the Q mode, as it consists of a comparison between objects [36]. It can be seen as an exercise involving reduction of the dimensionality, in which samples become distributed over a two-dimensional grid, as well as a classification process, whereby samples become grouped into discrete units [37]. Moreover, in order to facilitate visual inspection of the distribution of the samples across the SOM grid, the analysis can be complemented with a clustering analysis of the neural EEM prototypes [38].

Correlation analysis and the determination of EEM fluorescence components

In the SOM grid, it is possible to represent the intensity of a given wavelength coordinate of the prototype EEMs throughout the different neurons using a colour scale. This kind of visualisation is called a component plane [17], and shows how the fluorescence magnitude on a given coordinate varies from neuron to neuron over the SOM grid. Two highly correlated wavelength coordinates will therefore produce two similar component planes [39,40]. When the number of variables in the data set is low, it is possible to visually compare the patterns among component planes and detect which ones are positively, negatively or not correlated [41,42]. However, this becomes an unfeasible task when dealing with high-dimensional data, as is the case of EEMs (in our case, defined by 366 λex–λem coordinates). Barreto-Sanz and Perez-Uribe [39] proposed a methodology to simplify this task by projecting the correlations between the component planes on a new SOM grid. This new projection

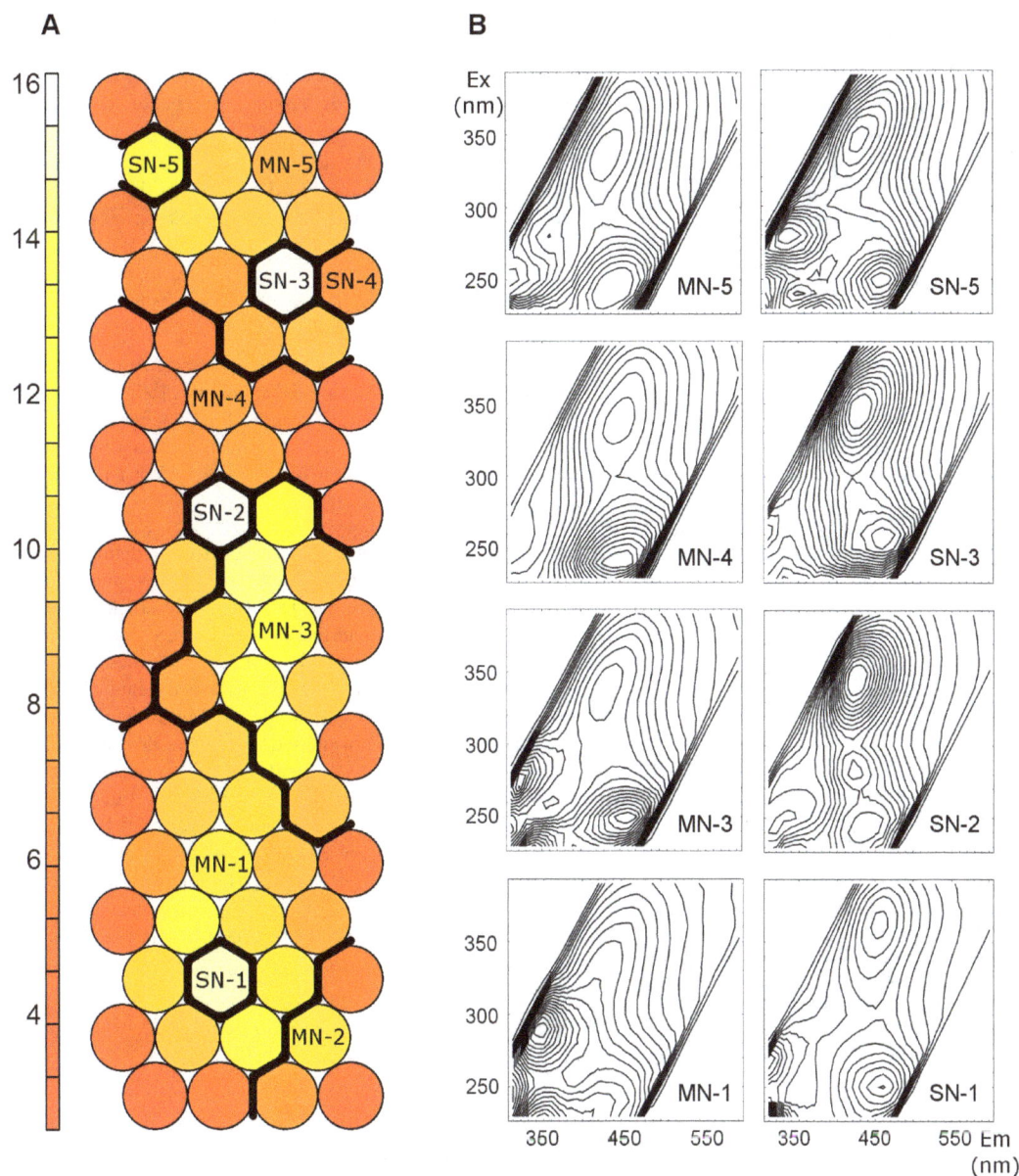

Figure 3. Clustering of the U-matrix of the SOM analysis in the Q-mode. A) Ten regions were defined in the SOM grid (black solid lines), based on hierarchical clustering of the U-matrix. B) EEM prototypes representing the main SOM regions.

groups highly correlated variables into the same neuron, and moderately correlated variables into nearby neurons. At this point, a hierarchical clustering analysis can be used to determine a consistent number of groups of {λex–λem} coordinates, each of which can be considered as a different fluorescence component. As in this case the analysis involves exploring dependences between the descriptors, it can be considered as an R-mode SOM analysis [35,36].

Computations

SOM analysis was conducted using the Kohonen package for R [37]. The successive steps undertaken in our computations are conceptualised in the flow diagram shown in Figure 2. EEMs were pre-processed by normalising their fluorescence intensity by their maximum, in order remove effects of changes in concentration and focus specifically on qualitative variations [43]. The input

matrix for the SOM analysis in the Q-mode contained 270 linearized EEMs with fluorescence data from 366 λex–λem coordinate pairs (Figure 2A). The output layer was an hexagonal grid (Figure 2B). Its size was chosen to be the largest size that ensured stability of the quantization error [44]. In addition, dimensions were set to preserve the proportions of the two highest eigenvalues of the covariance matrix of the input data [19,45–47]. During the training phase, the learning rate decreased linearly from 0.05 to 0.01. The initial neighbourhood size included two-thirds of all distances of the map units, and decreased linearly during the first third of the iterations. After that, only the winning unit was being adapted. In order to emphasise dissimilarities between the neurons of the SOM grid, a hierarchical cluster analysis with complete linkage was performed using the Lance-Williams update formula [48].

A

B

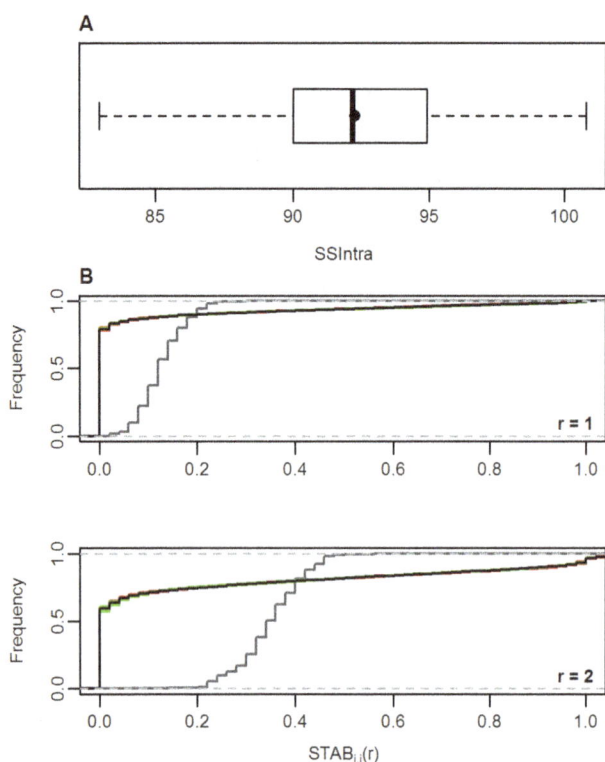

Figure 4. Outlier sensitivity test. A) Quantization stability: variation of the average SSIntra among 270 LOO subsets. The black dot indicates the mean. The absence of outlier values of CV(SSIntra) and the similar mean and median should be noted. B) Stability of neighbourhood relations: Histograms of the stabilities over all pairs of observations. In red, histograms of the LOO subsets in which the left-out sample was assigned to a single-neuron cluster. In green, histograms of the remaining LOO subsets. In black: histogram of the whole data set. It should be noted that there is hardly any difference between them. In grey, theoretical histogram of a randomly distributed map, following a binomial distribution defined according to de Bodt et al. [44]. This demonstrates that the SOM results are organised in a far from random distribution.

The influence of outliers on the performance of SOM was assessed by evaluating the quality of the SOM output in a series of leave-one-out (LOO) sample subsets. As measures of output quality, we used the SOM reliability criteria described by de Bodt et al. [44], which tested the stability of both the quantization and the topology of the SOM model. The stability of the quantization was assessed using the intra-class sum of squares (SSIntra) statistic, which is the sum of the squared distances between the observed data and their corresponding neural centroid. On the other hand, the stability of the neighbourhood relations was inspected by computing the histograms of all pairwise neighbourhood stabilities of a given LOO subset. SSIntra and neighbourhood stabilities were computed as described in de Bodt et al. [44]. For every LOO subset, the statistics were averaged over 50 runs of the SOM analysis, in order to minimise the variability of the output due to random initialisation of the reference vectors [49].

In parallel, 366 component planes were obtained from the SOM analysis (Figure 2C), one for each $\{\lambda ex-\lambda em\}$ coordinate that defined our original EEMs. In order to discriminate the number of fluorescence components within the samples, a correlation analysis was performed, based on the steps defined by Barreto-Sanz and Pérez-Uribe [39]. These steps included:

- Transformation of the component planes into normalised vectors.
- Calculation of the Pearson's correlation between each pair of vectors, obtaining a covariance matrix of dimensions (366×366).
- Computation of a SOM analysis of this covariance matrix, hereafter referred to as the SOM analysis in the R-mode. In this grid, neurons grouped highly correlated $\{\lambda em-\lambda em\}$ coordinates.
- Clustering of the U-matrix with a hierarchical cluster analysis with complete linkage using the Lance-Williams update formula [48].
- The optimal number of groups (i.e. fluorescence components) was determined by inspecting the silhouettes [50] of a range of partitions, from two to nine groups. The best partition had a high average $s_{(i)}$, and the fewest objects with a negative $s_{(i)}$, where $s_{(i)}$ is a measurement of how well object i matches its assigned cluster.

Eventually, the correlation analysis led to the definition of a number of EEM regions containing uncorrelated fluorescence phenomena and hence, assumed to reflect different fluorescence components. Next, the components in every sample were quantified as area-normalised fluorescence volumes, following the Fluorescence Regional Integration described Chen et al. [51].

Finally, the fluorescence components found by correlation analysis, and expressed as normalised volumes as described above, were evaluated as descriptors of the data set by performing a non-metric multidimensional scaling (NMDS). The analysis was performed using the vegan package for R [52], and Bray-Curtis dissimilarities. Each variable was centred and scaled to a mean of 0 and a standard deviation of 1. In addition, the relationship between the fluorescence components and the optical indices of HIX, SUVA and FI was tested with a vector fit analysis within the NMDS ordination.

Results

SOM codebooks

The output of the SOM analysis trained on the 270-sample data set is summarised in Figure 3. The unified distance matrix (frequently referred to as U-matrix, Figure 3A) represents the distances between the EEM prototypes of neighbouring neurons using a colour scale [53]. This kind of visualisation is the most frequently used method to explore dissimilarity and clustering patterns in the SOM grid [17].

In our results, inter-neighbouring distances were clearly uneven across the SOM grid, indicating the presence of dissimilarity patterns. Low distances dominated in the upper-middle part of the U-matrix, whereas high dissimilarities were observed in the central region of the lower part of the SOM grid. In order to further emphasize and differentiate regions with higher similarities between neurons, a 10-cluster division was applied to the U-matrix (Figure 3A). It should be noted here that the partitioning of the U-matrix was used only for visualisation purposes. Some neurons had such a high dissimilarity to their neighbouring neurons (lowest values in the U-matrix) that they formed stand-alone clusters by themselves (hereafter referred to as SN-1 to SN-5, where SN stands for single neuron). The rest of the grid was partitioned into five multi-neuron zones (hereafter referred to as MN-1 to MN-5). The nomenclature specified in Figure 3 will be used hereafter to facilitate description of the distribution of samples throughout the SOM grid in order to explore relationships between samples.

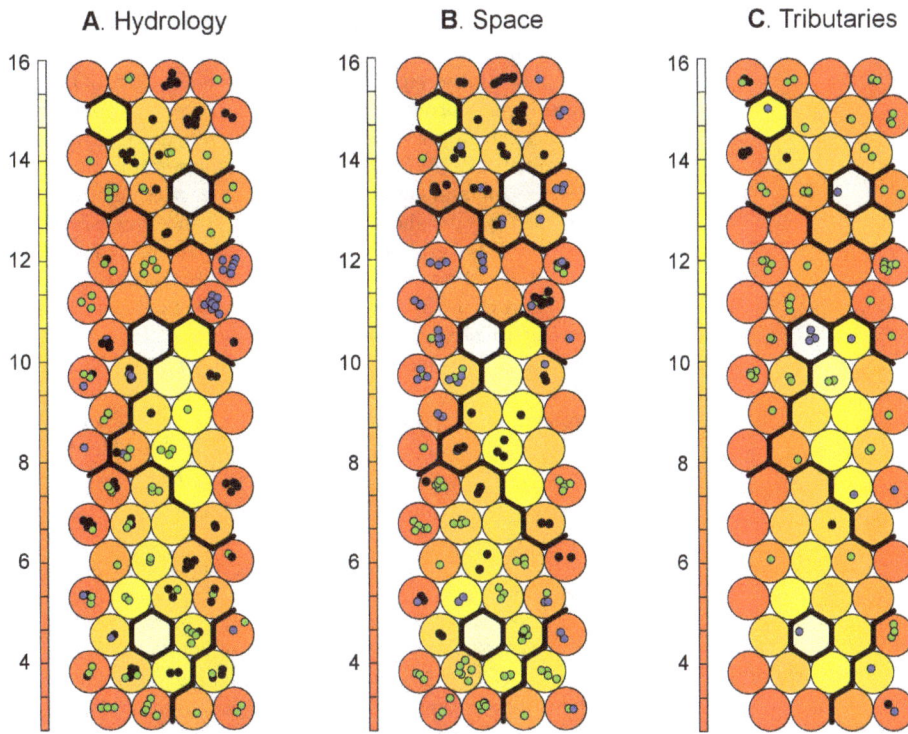

Figure 5. Projection of space, discharge, and type of tributary onto the U-matrix. Neuron colour scale indicates, for every neuron, the sum of the euclidean distances to all its immediate neighbours. Samples are projected on the SOM grid and coloured according to A) hydrology: blue represents flood conditions, black represents base flow, and green drought; B) space: blue corresponds to headwater samples, black middle reaches samples, and green are the lowland samples; C) types of tributary: blue are industrial, black are WWTP, and green are natural tributaries.

Outlier sensitivity analysis

The outlier sensitivity test showed that the presence of a few samples with very distinctive and infrequent spectral shapes (especially those assigned to single-neuron clusters) did not affect the SOM outcome in a meaningful way. The SSIntra computed for the 270 LOO subsets followed a Gaussian distribution without any outlier values (Figure 4A). Moreover, the mean was almost identical to the median (92.27 and 92.17, respectively), further indicating that none of the LOO subsets exhibited a statistically relevant differentiated quantization structure.

The histograms of neighbourhood stability showed that at a radius of one and two neurons, the neighbourhood relations remained almost the same irrespective of the sample left out by the LOO subsets (Figure 4B). This demonstrates that the topology of the SOM output is preserved in the presence of specific outlier samples. Furthermore, all the histograms of the LOO subsets are clearly different from the theoretical histogram of a randomly organised map (Figure 4B). This indicates that in every SOM analysis, corresponding to different LOO subsets, the samples are meaningfully organised in the SOM grid, in a far from random distribution [44].

Table 1. Characteristics of the silhouettes of a range of hierarchical partitionings of the R-mode SOM grid.

# groups	$\bar{S}$	S_{min}	S_{max}	$n_{(S<0)}$
2	0.56	−0.74	0.86	17
3	0.57	−0.50	0.80	13
4	0.54	−0.29	0.74	9
5	0.48	−0.43	0.72	13
6	0.48	−0.44	0.70	8
7	0.41	−0.23	0.70	7
8	0.42	−0.23	0.70	7
9	0.35	−0.33	0.70	16

The silhouettes analysis [5] corresponds to the calculation of $s_{(i)}$ for every object in the data set, where $s_{(i)}$ is a measurement of how well object i matches its assigned cluster. $\bar{S}$ corresponds to the average $s_{(i)}$, S_{min} to the minimum $s_{(i)}$, S_{max} to the maximum $s_{(i)}$ and $n_{(S<0)}$ to the number of objects that have a negative $s_{(i)}$. Values of S near one indicate that the object is very well clustered, whereas negative S indicates that the object might be assigned to the wrong group.

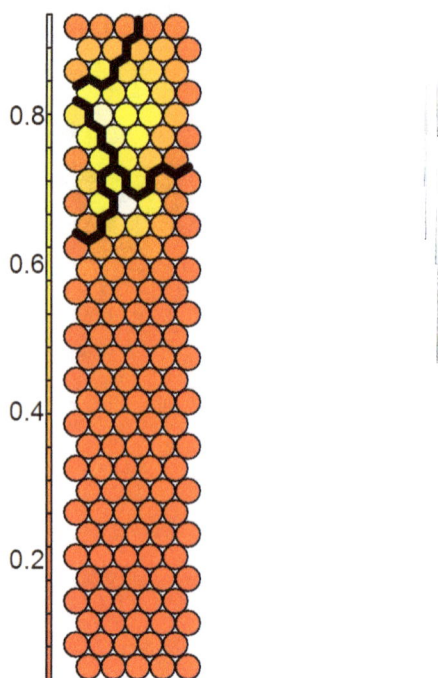

Figure 6. Clustering of the U-matrix of the SOM analysis in the R-mode. Every cluster groups highly correlated wavelength coordinates, representing different fluorescence components.

Sample projection

The samples in our data set were collected along a longitudinal downstream gradient, and under a variety of hydrological conditions. In order to test the influence of space and hydrology on the distributions of EEM spectral shapes, samples were projected onto the SOM grid, and coloured according to their sampling location ("headwaters", "middle reaches" and "lowland" categories) and hydrology ("flood", "baseflow" and "drought" categories, Figure 5).

In terms of hydrology (Figure 5A), samples collected during flood conditions were grouped into three main neurons, all situated in region MN-4. However, baseflow and drought samples were distributed across the grid. In the case of space (Figure 5B), the three categories appeared in different parts of the SOM grid. Headwater samples appeared mainly in region MN-4, samples from the middle reaches in regions MN-3 and MN-5, and those from the lowland mainly in region MN-1. Specifically, the neurons in region MN-4, which contained samples from middle reaches or the lowland, were the very same neurons that corresponded to the flood category in the hydrological projection. This combination of a single category for hydrology (flood) and multi category for space (whole length of the river) in a single neuron suggests a homogenisation effect on the spectral shape of EEMs over the whole length of the river under flood conditions.

Tributaries are presented separately in Figure 5C, coloured according to their origin: riverine, sewage-treated or industrial. It is noteworthy that single-neuron clusters contained exclusively industrial effluents, indicating that these sources produce DOM spectral shapes that are dissimilar with respect to the DOM from riverine and sewage-treated water. In contrast, WWTP samples appeared mainly in region MN-5, and natural tributaries were spread over the whole grid, but mainly in regions MN-4 and MN-5, those also associated with headwaters and middle reach sampling locations.

Determination of fluorescence components

The U-matrix of the SOM analysis in the R mode is shown in Figure 6. It can be seen that the bottom half of the SOM grid contains highly correlated wavelength coordinates, expressed by the homogeneous dark red-coloured neurons that indicate short distances between them. In the top part, there is a central light-coloured region and darker neurons in the margins, indicating the presence of greater heterogeneity among these units. Hence, overall the SOM grid contains a high number of neurons with highly correlated wavelength coordinates, and in contrast, a small set of neurons with larger dissimilarities between them, thus containing a higher diversity of fluorescence signals.

Next, the hierarchical clustering and silhouette analysis of the SOM units showed that four clusters was the best number of fluorescence components, as it exhibited the optimal combination of the minimal number of presumably misplaced samples ($n_{(S<0)}$) and the highest average silhouette ($\bar{S}$), (Table 1).

The four groups of wavelength coordinates (hereafter referred to as C1 to C4) are represented on the excitation-emission space in Figure 7. It can be seen that they appear spatially grouped in the optical plane and, moreover, that they overlap regions previously related to specific DOM fluorophores in the literature (Table 2). C4 corresponds to the V region of Chen et al. [51] and broadly to peak C of Coble [54], which were associated with humic-like substances. This component has been detected in a wide range of aquatic environments but mainly in waters draining forested catchments [2], and hence, represents an indicator of terrestrially derived DOM [54]. In the same emission range, but at the lowest excitation wavelengths, component C3 is apparent. Similarly to C4, it has also been associated with humic-like components of terrestrial origin but with a higher molecular weight and more freshly released character [2,55]. In the region of the EEM with the lowest emissions are two spots centred at $\lambda ex/\lambda em = 230/330$ nm and $270/310$ nm (C1), similarly to the coordinates of maximal fluorescence of tyrosine [56]. Hence, components appearing at these wavelengths have been attributed to peptide material resembling or containing tyrosine, indicating the presence of autochthonous microbially derived DOM [57]. Finally, C2 covers an area surrounding the previous protein-like spots, overlapping the region occupied by tryptophan [56]. This component has also been reported to reflect microbial activity, and has been used as an indicator of anthropogenic DOM inputs [58–60].

SOM fluorescence components as descriptors of the data set

Finally, we evaluated the capacity of these four fluorescence components to describe patterns in our data set as new independent variables by performing a NMDS. The results are shown in Figure 8. For the sake of simplicity in exploring the distribution of the samples in the NMDS space, panels A and B include only the main stem sites, whereas panel C includes only the tributary sites. However, it should be noted that all three figures come from the same analysis, and therefore the loadings of the variables (i.e. the fluorescence components C1 to C4) and the vector fit analysis of the optical indices is the same in the three panels.

In summary, the first axis separates the humic-like components C3 and C4 (negative side) from the protein-like components C1 and C2 (positive side). HIX and FI are oriented, respectively, in the negative and the positive directions of the first axis with a high level of significance (p<0.001). This reinforces our interpretation of the components, such that C1 and C2 are related to microbially

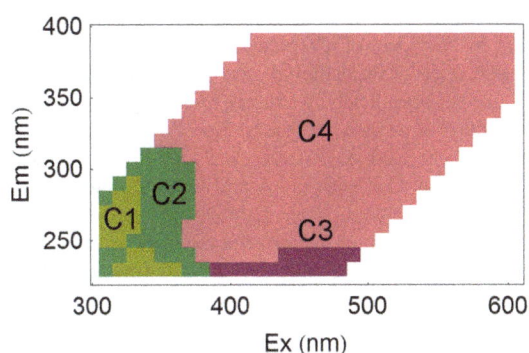

Figure 7. Localisation of the fluorescence components. Representation of the four groups of wavelength coordinates determined by correlation analysis on the excitation-emission space.

derived components, whereas C3 and C4 are related to terrestrially derived components. The second axis separates C1 and C3 from C2 and C4, suggesting further differentiation within the protein- and the humic-like groups of components. SUVA appears directed towards C3, however with a weaker level of significance ($p<0.05$). It is noteworthy, though, that SUVA and HIX appear perpendicular, showing independency from one another, even though they have previously been found to characterise a similar aspect of DOM [61].

According to our sampling design, we checked the role of hydrology and space in this new ordination based on fluorescence composition. In panel A, objects are coloured according to the discharge category under which they were sampled. The samples collected during flood conditions appear clearly aligned between the region of C3 and C4 and that of component C2. Samples from baseflow and drought conditions appear more broadly distributed throughout the whole NMDS plane. Drought samples seem to be more dispersed and occupy the negative secondary axis, which is not directly associated with any fluorescence component or optical index.

In space, the most important segregation occurs on the second axis. The sites from the lowland appear on the negative side, whereas those from the headwaters and the middle reaches are found on the positive side. Furthermore, headwater samples appear slightly more concentrated in the region between C3 and C4, similarly to the situation for flood samples in panel A.

Finally, panel C shows the tributary sites, which comprise a mixture of natural and anthropogenic water types. This figure shows a very clear pattern, consisting of an aggregation of industrial and WWTP effluents near component C2. This suggests a relationship between C2 and anthropogenically derived DOM.

Discussion

SOM coupled with a correlation analysis offers a flexible tool that enables, in the first stage, a similarity-based classification of EEMs and, in the second stage, a reduction of the dimensionality by grouping highly correlated {λex–λem} coordinates (Figure 2). Hence the methodology consists of two main parts: first, an analysis of the objects (i.e. sample EEMs) and second, an analysis of the variables (i.e. wavelength coordinates). In essence, the analysis of the objects is an exercise of classification of the samples, based on their spectral similarities; whereas the analysis of the variables reduces the dimensionality by grouping those coordinates that are highly correlated. This correlation analysis has meaningful biogeochemical implications, as each group of correlated wavelength pairs is assumed to be an independent fluorescent component, with consistent distributions in the λex–λem space according to the literature [54,62].

As a classification system, SOM has the advantage that it shows a low degree of dependency on the frequency at which a sample (or a spectral shape) is represented in the data set. By means of an outlier sensitivity test, the SOM quantization and topological structure was found to be robust to the presence of outlier samples. Accordingly, a single sample with unique and distinctive features can be classified on its own without affecting the classification of the other samples. In this way, outliers are not a distorting element, but a result integrated into the whole output. In our data set, this was exemplified by the neurons SN-1, SN-3 and SN-5, each of which represented only one sample. Specifically, they represented industrial effluents, which had very different spectral shapes with respect to the river water samples. This robustness to outliers provides the advantage that a data set can be analysed irrespective of its heterogeneity. This circumvents the main limitation of other currently used and well-established methods for EEM data treatment, like PCA, PLS or PARAFAC, which are highly sensitive to the presence of outliers [63–65] as they largely depend on least-squares solutions [18]. In least squares methods, the overall model is adjusted to include a better fit of an outlier, even if it results in a lower overall fit [66]. However, in SOM every sample only modifies its BMU and its neighbourhood, resulting in a less apparent influence of the presence of an outlier on the whole model outcome.

Furthermore, this classification stage leads not only to the grouping of samples with a high degree of similarity in terms of spectral shapes, but also to the generation of a reduced number of EEM prototypes (Figure 2, 0th to 1st level of abstraction). This reduced data set contains all the initial diversity of spectral shapes, but with the relative frequencies more evenly distributed. For instance, in our work, one EEM prototype could represent either a large number of samples that were very similar to one another (e.g. 13 headwater samples in a single neuron in SOM region MN-4,

Table 2. Wavelength coordinate boundaries of the fluorescence components.

Component	Correspondence with		Approximate boundaries	
	Coble 1996 [54]	Parlanti 2000 [62]	λex (nm)	λem (nm)
C1	B	γ	250–280 and 230–240	310–330 and 320–360
C2	T	δ	240–300	340–370
C3	A	α'	230–240	>370
C4	C	α	>250	>400

Summary of the location of the fluorescence components determined by correlation analysis and correspondence with previous components described in the literature.

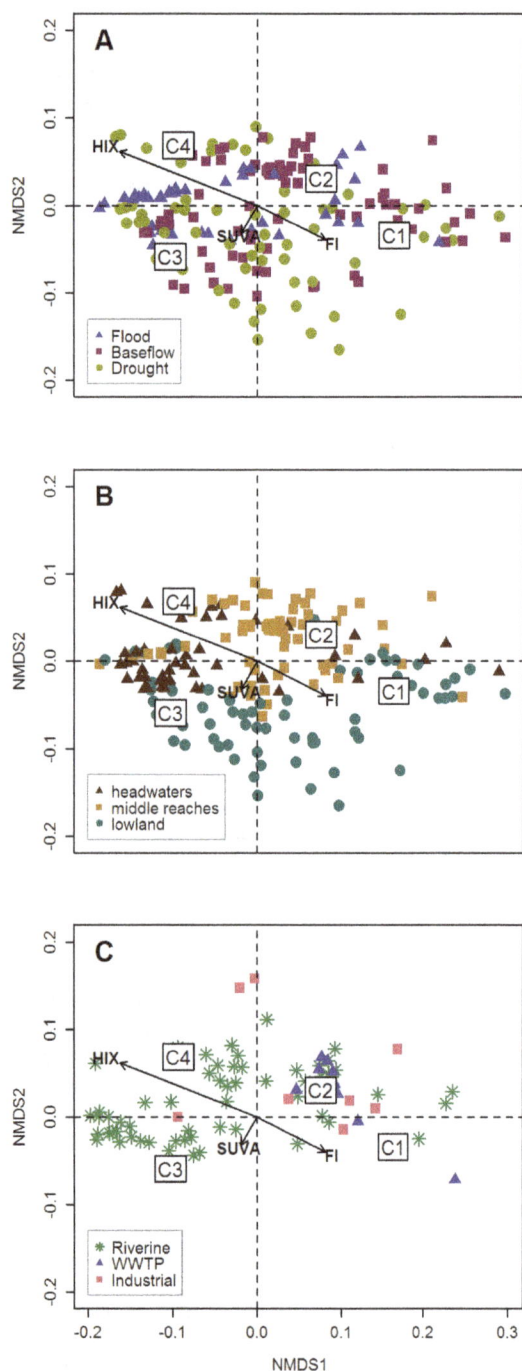

Figure 8. Multivariate analysis of our data set based on the four fluorescence components determined by SOM analysis. A non-metric multidimensional scaling was complemented with a vector fit analysis with the optical indices HIX, SUVA and FI. A) Main stem sites are coloured according to their discharge category. B) Main stem sites are coloured according to their downstream distance. C) Tributary sites are represented according to their source type.

we distinguished four areas in the EEM that were highly correlated (Figure 2, 1st to 2nd level of abstraction). Our four components had consistent properties in relation to previous descriptions in the literature (Table 2). Specifically, we distinguished two protein-like components, one of which appeared specifically related to anthropogenically derived DOM, as well as two humic-like components that coincided with the A and C areas described by Coble [54].

This methodology for detecting fluorescence components represents a novel statistical approach. In the procedure, the partitioning of the SOM grid represents a key step where the final decision is taken about the number of fluorescence components present in the data set. This step requires particular attention. Specifically, there are several clustering techniques that could be used to classify the neurons in a SOM grid. It has been reported that SOMs create clusters similar to those created by hierarchical clustering [38,67]. Indeed, we computed a hierarchical clustering with complete linkage using the Lance-Williams update formula, and our clusters were consistent with the (dis)similarity patterns of the U-matrix (Figures 3 and 6). However, in SOM grids of higher resolution (i.e., number of neurons) the U-matrix can present more complex patterns of clustering and subclustering. In this case, the results of a hierarchical clustering analysis may not follow the results of the U-matrix very closely [68]. As a better approximation, computation of Vellido's algorithm and the use of the U-matrix neural neighbourhood distances as a cluster distance function have been proposed [39,68] as, in this case, the neighbourhood conditions become explicit in the analysis and the output fits better with the results of the U-matrix. Hence, future studies should test the performance of different clustering techniques when larger data sets – and hence, larger SOM grids – are concerned.

Finally, after the regionalisation of EEMs into four fluorescence components, we quantified their contribution in every sample using the FRI technique originally described by Chen et al. [51]. This technique has been widely applied to track changes in DOM composition [69–71]. It has the advantage that it integrates the whole shape of the EEM region and accounts for the fluorescence provided by shoulders and other spectral features that would be omitted if only the maximal value of the region was taken into account. However, it has recently been pointed out that the numerical method used for integration can have important consequences for the accuracy of the results. Specifically, the Riemann summation method proposed by Chen et al. [51] and used in this paper may result in the underestimation of the protein-like fractions, and in the overestimation of humic-like fractions [72]. In order to minimise this bias, future studies may consider the use of other methods, such as the composite trapezoidal rule or the composite Simpson's rule [72].

Despite the main focus being on the methodology, some biogeochemically meaningful information arose throughout the study. Hydrology and downstream distance were found to be relevant shapers of DOM spectral properties. Floods exhibited differentiated patterns with respect to baseflow and drought conditions. Floods appeared to have a homogenisation effect on EEM spectral characteristics, with a gradual shift downstream between the presence of humic-like components with high HIX and SUVA. This indicates the prevalence of terrestrial humic-like material along the whole length of the river that rapidly transfers to the coastal system with little chance of being transformed [73]. The presence of C2 with high FI indicates some impact of industrial and WWTP effluents during downstream transport [58,74]. Outside flood conditions, samples collected from the headwaters, the middle reaches and the lowland could be

Figure 5B), or just a single sample with very unique properties (e.g. an industrial effluent in SN-1, SN-3 or SN-5, Figure 5C). This re-weighting effect of the representativeness within the data set allows for an analysis of correlations among variables (i.e. λex–λem coordinates) that can detect fluorophores that were initially represented at only low levels. Indeed, in our correlation analysis,

distinguished from each other. They exhibited successively lower HIX and higher FI values from the headwaters to the lowland. This indicated a shift from terrestrial-like characteristics to an autochthonously generated DOM character during downstream transport. Furthermore, industrial effluents exhibited unique and distinctive properties with respect to the rest of the data set.

In summary, our results open a new viewpoint to the statistical treatment of EEMs. Thanks to its robustness to the presence of outliers, SOM can be applied to EEM data sets including both high- and low-represented spectral shapes. This may have important practical implications especially for the study of the biogeochemical behaviour of DOM in natural systems, as sampling designs will be less restricted to the requirements of the statistical treatment, and more adaptable to research needs.

Conclusions

In this paper, the use of SOM in combination with a correlation analysis has been presented as a powerful method to deal with large and complex EEM data sets. Specifically, our findings indicate that:

- SOM analysis coupled with a correlation analysis as described by Barreto-Sanz and Perez-Uribe [39] allows an analysis both at the object and at the variable level. Hence, it serves not only to explore the differences in fluorescence properties between samples, as shown by Bieroza et al. [19,20], but also helps to identify particular fluorescence components, as shown herein.

- It is robust to the presence of outlier samples. That is, samples with very distinct features are discerned while having little effect on the ordination and classification of the other samples. This distinct property makes it possible to work with heterogeneous data sets.

- The correlation analysis performed on the SOM EEM prototypes has an enhanced capacity to detect fluorophores that are represented at only low levels in the original EEM data set.

Therefore, we conclude that SOM analysis coupled with a correlation analysis of the component planes expands the toolbox of the fluorescence DOM researchers by enabling the analysis of complex and heterogeneous EEM data sets. This may open new possibilities for advancing our understanding of DOM character and biogeochemical behaviour.

Acknowledgments

We would like to thank Mark Maroncelli (Department of Chemistry, Penn State University, USA) for providing advice on fluorescence spectral corrections. We are also grateful to Jose Eduardo Serrão and two anonymous reviewers for their contributions during the review process.

Author Contributions

Conceived and designed the experiments: EEG AB. Analyzed the data: EEG AB. Wrote the paper: EEG AB.

References

1. Hudson N, Baker A, Reynolds D (2007) Fluorescence analysis of dissolved organic matter in natural, waste and polluted waters - A review. River Res Appl 23: 631–649. doi:10.1002/rra.1005.
2. Fellman JB, Hood E, Spencer RGM (2010) Fluorescence spectroscopy opens new windows into dissolved organic matter dynamics in freshwater ecosystems: A review. Limnol Oceanogr 55: 2452–2462. doi:10.4319/lo.2010.55.6.2452.
3. Nebbioso A, Piccolo A (2013) Molecular characterization of dissolved organic matter (DOM): a critical review. Anal Bioanal Chem 405: 109–124. doi:10.1007/s00216-012-6363-2.
4. Azam F, Fenchel T, Field JG, Gray JS, Meyerreil LA, et al. (1983) The Ecological Role of Water-Column Microbes in the Sea. Mar Ecol Ser 10: 257–263. doi:10.3354/meps010257.
5. Findlay S (2010) Stream microbial ecology. J North Am Benthol Soc 29: 170–181. doi:10.1899/09-023.1.
6. Cory RM, Kaplan LA (2012) Biological lability of streamwater fluorescent dissolved organic matter. Limnol Oceanogr 57: 1347–1360. doi:10.4319/lo.2012.57.5.1347.
7. Elkins KM, Nelson DJ (2002) Spectroscopic approaches to the study of the interaction of aluminum with humic substances. Coord Chem Rev 228: 205–225. doi:10.1016/S0010-8545(02)00040-1.
8. Brooks ML, McKnight DM, Clements WH (2007) Photochemical control of copper complexation by dissolved organic matter in Rocky Mountain streams, Colorado. Limnol Oceanogr 52: 766–779.
9. Bertilsson S, Tranvik LJ (2000) Photochemical transformation of dissolved organic matter in lakes. Limnol Oceanogr 45: 753–762.
10. Mostofa KMG, Yoshioka T, Konohira E, Tanoue E (2007) Photodegradation of fluorescent dissolved organic matter in river waters. Geochem J 41: 323–331.
11. Osburn CL, Retamal L, Vincent WF (2009) Photoreactivity of chromophoric dissolved organic matter transported by the Mackenzie River to the Beaufort Sea. Mar Chem 115: 10–20. doi:10.1016/j.marchem.2009.05.003.
12. Foden J, Sivyer DB, Mills DK, Devlin MJ (2008) Spatial and temporal distribution of chromophoric dissolved organic matter (CDOM) fluorescence and its contribution to light attenuation in UK waterbodies. Estuar Coast Shelf Sci 79: 707–717. doi:10.1016/j.ecss.2008.06.015.
13. Cole JJ, Prairie YT, Caraco NF, McDowell WH, Tranvik LJ, et al. (2007) Plumbing the global carbon cycle: Integrating inland waters into the terrestrial carbon budget RID B-9108-2008 RID E-9767-2010 RID B-4951-2011. Ecosystems 10: 171–184. doi:10.1007/s10021-006-9013-8.
14. Tranvik LJ, Downing JA, Cotner JB, Loiselle SA, Striegl RG, et al. (2009) Lakes and reservoirs as regulators of carbon cycling and climate. Limnol Oceanogr 54: 2298–2314. doi:10.4319/lo.2009.54.6_part_2.2298.
15. Bieroza M, Baker A, Bridgeman J (2011) Classification and calibration of organic matter fluorescence data with multiway analysis methods and artificial neural networks: an operational tool for improved drinking water treatment. Environmetrics 22: 256–270. doi:10.1002/env.1045.
16. Benner R (2002) Chemical composition and reactivity. In: Hansell D, Carlson C, editors. Biogeochemistry of Marine Dissolved Organic Matter. New York. pp. 59–90.
17. Kohonen T (2001) Self-Organizing Maps. 3rd editio. Springer Berlin Heidelberg.
18. Brereton RG (2012) Self organising maps for visualising and modelling. Chem Cent J 6 Suppl 2: S1. Available: http://www.pubmedcentral.nih.gov/articlerender.fcgi?artid=3395104&tool=pmcentrez&rendertype=abstract. Accessed 2014 April 9.
19. Bieroza M, Baker A, Bridgeman J (2009) Exploratory analysis of excitation-emission matrix fluorescence spectra with self-organizing maps as a basis for determination of organic matter removal efficiency at water treatment works. J Geophys Res 114: G00F07–G00F07. doi:10.1029/2009JG000940.
20. Bieroza M, Baker A, Bridgeman J (2012) Exploratory analysis of excitation-emission matrix fluorescence spectra with self-organizing maps: A tutorial. Educ Chem Eng 7: e22–e31. Available: http://www.sciencedirect.com/science/article/pii/S1749772811000157.
21. Lloyd GR, Brereton RG, Duncan JC (2008) Self Organising Maps for distinguishing polymer groups using thermal response curves obtained by dynamic mechanical analysis. Analyst 133: 1046–1059. Available: http://www.ncbi.nlm.nih.gov/pubmed/18645646. Accessed 16 April 2014.
22. Bieroza M, Baker A, Bridgeman J (2012) New data mining and calibration approaches to the assessment of water treatment efficiency. Adv Eng Softw 44: 126–135. doi:10.1016/j.advengsoft.2011.05.031.
23. Jaffe R, McKnight D, Maie N, Cory R, McDowell WH, et al. (2008) Spatial and temporal variations in DOM composition in ecosystems: The importance of long-term monitoring of optical properties RID C-2277-2009 RID E-9767-2010. J Geophys Res 113: G04032–G04032. doi:10.1029/2008JG000683.
24. ACA (2013) Consulta de dades - Xarxes de control. Agència Catalana l'Aigua, General Catalunya. Available: aca-web.gencat.cat/aca/.
25. Goletz C, Wagner M, Gruebel A, Schmidt W, Korf N, et al. (2011) Standardization of fluorescence excitation-emission-matrices in aquatic milieu. Talanta 85: 650–656. doi:10.1016/j.talanta.2011.04.045.
26. Lakowicz JR (2006) Principles of Fluorescence Spectroscopy. 3rd ed. Springer. Available: http://www.springer.com/chemistry/analytical+chemistry/book/978-0-387-31278-1.
27. Gardecki JA, Maroncelli M (1998) Set of secondary emission standards for calibration of the spectral responsivity in emission spectroscopy. Appl Spectrosc 52: 1179–1189. doi:10.1366/0003702981945192.
28. Lawaetz AJ, Stedmon CA (2009) Fluorescence Intensity Calibration Using the Raman Scatter Peak of Water RID B-5841-2008. Appl Spectrosc 63: 936–940.

29. Larsson T, Wedborg M, Turner D (2007) Correction of inner-filter effect in fluorescence excitation-emission matrix spectrometry using Raman scatter RID B-2620-2010 RID A-7870-2010. Anal Chim Acta 583: 357–363. doi:10.1016/j.aca.2006.09.067.

30. Weishaar JL, Aiken GR, Bergamaschi BA, Fram MS, Fujii R, et al. (2003) Evaluation of specific ultraviolet absorbance as an indicator of the chemical composition and reactivity of dissolved organic carbon. Environ Sci Technol 37: 4702–4708. doi:10.1021/es030360x.

31. Zsolnay A, Baigar E, Jimenez M, Steinweg B, Saccomandi F (1999) Differentiating with fluorescence spectroscopy the sources of dissolved organic matter in soils subjected to drying. Chemosphere 38: 45–50. doi:10.1016/S0045-6535(98)00166-0.

32. McKnight DM, Boyer EW, Westerhoff PK, Doran PT, Kulbe T, et al. (2001) Spectrofluorometric characterization of dissolved organic matter for indication of precursor organic material and aromaticity. Limnol Oceanogr 46: 38–48.

33. Cory RM, Miller MP, McKnight DM, Guerard JJ, Miller PL (2010) Effect of instrument-specific response on the analysis of fulvic acid fluorescence spectra. Limnol Oceanogr 8: 67–78.

34. Kohonen T (1998) The self-organizing map. Neurocomputing 21: 1–6. doi:10.1016/S0925-2312(98)00030-7.

35. Cattell RB (1952) Factor Analysis: An Introduction and Manual for the Psychologist and Social Scientist. New York, USA: Harper.

36. Legendre P, Legendre L (1998) Numerical Ecology. 2nd Englis. Elsevier.

37. Wehrens R, Buydens LMC (2007) Self- and super-organizing maps in R: The kohonen package. J Stat Softw 21: 1–19.

38. Vesanto J, Alhoniemi E (2000) Clustering of the self-organizing map. IEEE Trans Neural Networks 11: 586–600. doi:10.1109/72.846731.

39. Barreto-Sanz MA, Perez-Uribe A (2007) Improving the correlation hunting in a large quantity of SOM - Component planes classification of agro-ecological variables related with productivity in the sugar cane culture. Lect NOTES Comput Sci 4669: 379–388.

40. Vesanto J (1999) SOM-based data visualization methods. Intell Data Anal 3: 111–126. Available: http://linkinghub.elsevier.com/retrieve/pii/S108846X9900013X.

41. Çinar Ö, Merdun H (2008) Application of an unsupervised artificial neural network technique to multivariant surface water quality data. Ecol Res 24: 163–173. Available: http://link.springer.com/10.1007/s11284-008-0495-z. Accessed 9 April 2014.

42. Mat-Desa W, Ismail D, NicDaeid N (2011) Classification and Source Determination of Medium Petroleum Distillates by Chemometric and Artificial Neural Networks: A Self Organizing Feature Approach. Anal Chem 83: 7745–7754. Available: http://pubs.acs.org/doi/abs/10.1021/ac202315y. Accessed 2014 April 9.

43. Boehme JR, Coble PG (2000) Characterization of colored dissolved organic matter using high-energy laser fragmentation. Environ Sci Technol 34: 3283–3290. doi:10.1021/es9911263.

44. De Bodt E, Cottrell M, Verleysen M (2002) Statistical tools to assess the reliability of self-organizing maps. Neural Netw 15: 967–978. Available: http://www.ncbi.nlm.nih.gov/pubmed/12416687.

45. Vesanto J (2000) Neural Network Tool for Data Mining: SOM Toolbox. Available: http://cda.psych.uiuc.edu/martinez/edatoolbox/Docs/toolmet2000.pdf.

46. Park Y-S, Tison J, Lek S, Giraudel J-L, Coste M, et al. (2006) Application of a self-organizing map to select representative species in multivariate analysis: A case study determining diatom distribution patterns across France. Ecol Inform 1: 247–257. Available: http://linkinghub.elsevier.com/retrieve/pii/S1574954106000525. Accessed 2014 January 30.

47. Céréghino R, Park Y-S (2009) Review of the Self-Organizing Map (SOM) approach in water resources: Commentary. Environ Model Softw 24: 945–947. Available: http://linkinghub.elsevier.com/retrieve/pii/S1364815209000188. Accessed 2014 January 27.

48. Lance GN, Williams WT (1967) A General Theory of Classificatory Sorting Strategies.1. Hierarchical Systems. Comput J 9: 373–&.

49. Cottrell M, Bodt E De, Verleysen M (2001) A Statistical Tool to Assess the Reliability of Self- Organizing Maps. Advances in Self-Organising Maps. Lincoln (United Kingdom): Springer Verlag. pp. 7–14.

50. Rousseeuw PJ (1987) Silhouettes: A graphical aid to the interpretation and validation of cluster analysis. J Comput Appl Math 20: 53–65. Available: http://www.sciencedirect.com/science/article/pii/0377042787901257.

51. Chen W, Westerhoff P, Leenheer JA, Booksh K (2003) Fluorescence excitation - Emission matrix regional integration to quantify spectra for dissolved organic matter. Environ Sci Technol 37: 5701–5710. doi:10.1021/es034354c.

52. Oksanen J, Blanchet FG, Kindt R;, Legendre P, Minchin PR;, et al. (2012) vegan: Community Ecology Package. Available: http://cran.r-project.org/package=vegan.

53. Ultsch A, Siemon HP (1990) Kohonen's Self Organizing Feature Maps for Exploratory Data Analysis. Proceedings of International Neural Network Conference (INNC'90). Dordrecht, The Netherlands: Kluwer. pp. 305–308.

54. Coble PG (1996) Characterization of marine and terrestrial DOM in seawater using excitation emission matrix spectroscopy. Mar Chem 51: 325–346. doi:10.1016/0304-4203(95)00062-3.

55. Huguet A, Vacher L, Relexans S, Saubusse S, Froidefond JM, et al. (2009) Properties of fluorescent dissolved organic matter in the Gironde Estuary. Org Geochem 40: 706–719. doi:10.1016/j.orggeochem.2009.03.002.

56. Yamashita Y, Tanoue E (2003) Chemical characterization of protein-like fluorophores in DOM in relation to aromatic amino acids. Mar Chem 82: 255–271. doi:10.1016/S0304-4203(03)00073-2.

57. Cammack WKL, Kalff J, Prairie YT, Smith EM (2004) Fluorescent dissolved organic matter in lakes: Relationships with heterotrophic metabolism. Limnol Oceanogr 49: 2034–2045.

58. Baker A (2001) Fluorescence excitation-emission matrix characterization of some sewage-impacted rivers. Environ Sci Technol 35: 948–953. doi:10.1021/es000177t.

59. Henderson RK, Baker A, Murphy KR, Hambly A, Stuetz RM, et al. (2009) Fluorescence as a potential monitoring tool for recycled water systems: A review. Water Res 43: 863–881. doi:10.1016/j.watres.2008.11.027.

60. Borisover M, Laor Y, Saadi I, Lado M, Bukhanovsky N (2011) Tracing Organic Footprints from Industrial Effluent Discharge in Recalcitrant Riverine Chromophoric Dissolved Organic Matter. Water Air Soil Pollut 222: 255–269. doi:10.1007/s11270-011-0821-x.

61. Hur J, Kim G (2009) Comparison of the heterogeneity within bulk sediment humic substances from a stream and reservoir via selected operational descriptors. Chemosphere 75: 483–490. doi:10.1016/j.chemosphere.2008.12.056.

62. Parlanti E, Worz K, Geoffroy L, Lamotte M (2000) Dissolved organic matter fluorescence spectroscopy as a tool to estimate biological activity in a coastal zone submitted to anthropogenic inputs. Org Geochem 31: 1765–1781. doi:10.1016/S0146-6380(00)00124-8.

63. Engelen S, Hubert M (2011) Detecting outlying samples in a parallel factor analysis model. Anal Chim Acta 705: 155–165. Available: http://www.ncbi.nlm.nih.gov/pubmed/21962358. Accessed 16 April 2014.

64. Bro R, Vidal M (2011) EEMizer: Automated modeling of fluorescence EEM data. Chemom Intell Lab Syst 106: 86–92. Available: http://linkinghub.elsevier.com/retrieve/pii/S0169743910001152. Accessed 2014 March 19.

65. Stedmon CA, Bro R (2008) Characterizing dissolved organic matter fluorescence with parallel factor analysis: a tutorial. Limnol Oceanogr 6: 572–579.

66. Quinn GP, Keough MJ (2010) Experimental Design and Data Analysis for Biologists. 10th ed. New York, USA: Cambridge University Press.

67. Oja M, Nikkilä J, Törönen P, Wong G, Castrén E, et al. (2006) Exploratory Clustering of Gene Expression Profiles of Mutated Yeast Strains. In: Zhang W, Shmulevich I, editors. Computational and Statistical Approaches to Genomics. Springer US. pp. 61–74. Available: http://dx.doi.org/10.1007/0-387-26288-1_5.

68. Vesanto J, Sulkava M (2002) Distance matrix based clustering of the Self-Organizing Map. Lect NOTES Comput Sci 2415: 951–956.

69. Wang Z, Wu Z, Tang S (2009) Characterization of dissolved organic matter in a submerged membrane bioreactor by using three-dimensional excitation and emission matrix fluorescence spectroscopy. Water Res 43: 1533–1540. Available: http://www.ncbi.nlm.nih.gov/pubmed/19138782. Accessed 28 January 2014.

70. Marhuenda-Egea FC, Martínez-Sabater E, Jordá J, Moral R, Bustamante M a, et al. (2007) Dissolved organic matter fractions formed during composting of winery and distillery residues: evaluation of the process by fluorescence excitation-emission matrix. Chemosphere 68: 301–309. Available: http://www.ncbi.nlm.nih.gov/pubmed/17292449. Accessed 28 January 2014.

71. Shao Z-H, He P-J, Zhang D-Q, Shao L-M (2009) Characterization of water-extractable organic matter during the biostabilization of municipal solid waste. J Hazard Mater 164: 1191–1197. Available: http://www.ncbi.nlm.nih.gov/pubmed/18963454. Accessed 28 January 2014.

72. Zhou J, Wang J-J, Baudon A, Chow AT (2013) Improved Fluorescence Excitation-Emission Matrix Regional Integration to Quantify Spectra for Fluorescent Dissolved Organic Matter. J Environ Qual 42: 925–930. doi:10.2134/jeq2012.0460.

73. Battin TJ, Kaplan LA, Findlay S, Hopkinson CS, Marti E, et al. (2008) Biophysical controls on organic carbon fluxes in fluvial networks. Nat Geosci 1: 95–100. doi:10.1038/ngeo101.

74. Baker A, Spencer RGM (2004) Characterization of dissolved organic matter from source to sea using fluorescence and absorbance spectroscopy RID A-6298-2011. Sci Total Environ 333: 217–232. doi:10.1016/j.scitotenv.2004.04.013.

2

Complex Effects of Ecosystem Engineer Loss on Benthic Ecosystem Response to Detrital Macroalgae

Francesca Rossi[1]*, **Britta Gribsholt**[2], **Frederic Gazeau**[3,4], **Valentina Di Santo**[5], **Jack J. Middelburg**[2,6]

1 Laboratoire Ecologie des Systèmes marins côtiers, Université Montpellier 2, Montpellier, France, 2 Department of Ecosystems, Royal Netherlands Institute for Sea Research, Yerseke, the Netherlands, 3 Centre National de la Recherche Scientifique-Institut National des Sciences de l'Univers, Laboratoire d' Oceanographie de Villefranche, Villefranche-sur-mer, France, 4 Université Pierre et Marie Curie, Observatoire Océanologique de Villefranche, Villefranche-sur-mer, France, 5 Department of Biology, Boston University, Boston, Massachusetts, United States of America, 6 Department of Earth Sciences – Geochemistry, Faculty of Geosciences, Utrecht University, Utrecht, The Netherlands

Abstract

Ecosystem engineers change abiotic conditions, community assembly and ecosystem functioning. Consequently, their loss may modify thresholds of ecosystem response to disturbance and undermine ecosystem stability. This study investigates how loss of the bioturbating lugworm *Arenicola marina* modifies the response to macroalgal detrital enrichment of sediment biogeochemical properties, microphytobenthos and macrofauna assemblages. A field manipulative experiment was done on an intertidal sandflat (Oosterschelde estuary, The Netherlands). Lugworms were deliberately excluded from 1 × m sediment plots and different amounts of detrital *Ulva* (0, 200 or 600 g Wet Weight) were added twice. Sediment biogeochemistry changes were evaluated through benthic respiration, sediment organic carbon content and porewater inorganic carbon as well as detrital macroalgae remaining in the sediment one month after enrichment. Microalgal biomass and macrofauna composition were measured at the same time. Macroalgal carbon mineralization and transfer to the benthic consumers were also investigated during decomposition at low enrichment level (200 g WW). The interaction between lugworm exclusion and detrital enrichment did not modify sediment organic carbon or benthic respiration. Weak but significant changes were instead found for porewater inorganic carbon and microalgal biomass. Lugworm exclusion caused an increase of porewater carbon and a decrease of microalgal biomass, while detrital enrichment drove these values back to values typical of lugworm-dominated sediments. Lugworm exclusion also decreased the amount of macroalgae remaining into the sediment and accelerated detrital carbon mineralization and CO_2 release to the water column. Eventually, the interaction between lugworm exclusion and detrital enrichment affected macrofauna abundance and diversity, which collapsed at high level of enrichment only when the lugworms were present. This study reveals that in nature the role of this ecosystem engineer may be variable and sometimes have no or even negative effects on stability, conversely to what it should be expected based on current research knowledge.

Editor: Andrew Davies, Bangor University, United Kingdom

Funding: This work was funded by a PIONIER grant of the Netherlands Organization for Scientific Research to JJM. The funders had no role in study design, data analysis, decision to publish, or preparation of the manuscript.

Competing Interests: The authors have declared that no competing interest exist.

* E-mail: francesca.rossi@univ-montp2

Introduction

It is widely recognized that species loss may affect ecosystem stability and, in turn, have important social and ecological consequences [1–3]. Overall, when more species are present in a community, ecosystem functions are more stable in time and increase their resistance or resilience against disturbance, due to an increased variety of life strategies, functional traits and responses to environmental disturbance [4–6]. Different species contributing to similar ecosystem functions can occur under different environmental conditions and ensure that functions are performed even when some other species are displaced [4,5,7].

Empirical evidence has shown that sometimes species loss can have complex, often idiosyncratic effects on the stability of ecosystem functions and on both their resistance and resilience against disturbance [8,9]. This is especially true for the marine coastal ecosystem, where species identity, functional traits and environmental context may rule ecosystem functions and their

response to disturbances [10–12]. One or few species can be key contributors to ecosystem functions to the extent that their loss can outweigh the effect of species richness on ecosystem functions, resilience or resistance [10,13,14]. Such overwhelming effect can, however, vary with the environmental context and change with increasing ecological complexity [15]. It seems thus evident that understanding the role of species loss in marine systems might require a large body of knowledge of the functional role of individual species and of the consequences of losing particular species for ecosystem functions and for their response to disturbance.

Ecosystem engineer species strongly modify resource availability and environmental conditions, which in turn can alter communities and ecosystem functions [16,17]. These species are thus ideal candidates to test hypotheses on how local extinction of key contributors to ecosystem functions can modulate the response of ecosystems to disturbance. In marine sediments and terrestrial soils, ecosystem engineers are often species that contribute to particle mixing and solute penetration into the sediment through the process of bioturbation, which includes sediment reworking and burrow ventilation [18]. These bioturbating engineer species can be great contributors to diagenetic reactions, to the recycling of sedimentary organic matter, and to benthic community structure and biodiversity [18–21]. They also have the potential to strongly modify the response to disturbance of benthic communities and functions [22]. For instance bioturbation by *Marenzellaria viridis* has been found to decrease hypoxic crises in the Baltic Sea [23]. It has also been found that in soil, reworking activity of dominant contributors to bioturbation, such as earthworms, modulates the response of plant communities to invasions [24]. Surprisingly, however, ecosystem engineer loss has been poorly considered in studies investigating ecosystem response to disturbance.

In marine coastal systems, a large portion of seaweeds and seagrasses is not consumed by herbivores but returns to the environment as decaying organic matter [25]. In shallow-water coastal habitats, detrital seaweeds and seagrasses can supply important amounts of organic matter to benthic ecosystems. They represent an important source of disturbance because during decomposition they can alter sediment biogeochemistry and, in turn, benthic communities of consumers. Despite detrital seaweed and seagrass supply is often a natural event, proliferation and blooms of seaweeds as a consequence of eutrophication can dramatically increase detrital seaweed biomass and exacerbate the effects on benthic ecosystems [26–29]. Biomass decomposition can induce hypoxia followed by the release of reduced toxic compounds such as sulphides, which are toxic for several benthic species and can therefore strongly reduce species abundance and diversity [26,27,29–33]. Detrital decomposition can also provide food, directly for detritivores or indirectly, by stimulating bacterial metabolism, thereby altering the benthic food web [28,34]. These effects strongly depend on the amount of detrital inputs and on its decomposition patterns, which may strongly vary as a consequence of benthic invertebrate composition and identity [35].

Bioturbating fauna can modulate the recycling of detrital macroalgae enhancing organic matter mineralization and microbial activity [36–38]. Some burrowing bioturbating species, such as certain crabs or polychaetes, can also incorporate detritus into their burrow wall lining and represent important sinks for detrital macroalgae [39]. In addition, often bioturbating infauna are also detritivores that can consume directly detritus and contribute to its recycling in the food web. In a laboratory experiment, for instance, it was found that nitrogen and carbon fluxes at the sediment-water interface changed following both the addition of the detrital green macroalgae *Chaetomorpha linum* and the presence of the bioturbating polychaete *Nereis diversicolor*. This polychaete ingested macroalgal fragments, enhanced microbial metabolism and contributed to diffuse solutes in the interstitial sediment and at the sediment-water interface [36].

One common well-recognized bioturbating species typical of temperate waters is the burrowing lugworm *Arenicola marina*. This worm lives head down in J-shaped burrows where it ingests sediment and defecates on the surface. Its activity has been recognized to destabilize sediment and counteract the effect of microphytobenthos, increase sediment microhabitats and alter sediment biogeochemistry, by enhancing solute penetration into sediment depth and re-distributing organic material along the vertical profile [40]. Consequently, this species may alter other fauna and vegetation [36,37,41]. This study reports on how the loss of this bioturbating lugworm may alter the response to detrital macroalgae of intertidal benthic macrofauna and sediment biogeochemistry. It was hypothesized that the deliberate exclusion

of lugworms would increase the magnitude of changes caused by the disturbance of macroalgal detritus for (i) carbon availability and benthic respiration; (ii) microalgal biomass and (iii) macrofauna assemblages. Recycling of the organic carbon derived from detrital decomposition was also investigated to explore some of the mechanisms that might regulate the effect of *A. marina* on benthic responses to macroalgal detritus.

Materials and Methods

Study Site and Experimental Design

Two manipulative experiments were done on an intertidal flat of the Oosterschelde estuary (The Netherlands) dominated by the bioturbating lugworm *Arenicola marina*, during summer 2006. The study intertidal flat, named Slikken van Viane (51° 37′ 00″ N, 4° 01′ 00″ E), was a sheltered area situated on the upper intertidal close to a salt marsh. Sediment grain-size was 65% fine sand and 35% very fine sand (Rossi F., unpublished data). The Oosterschelde estuary is a national park designated as zone VI by the International Union for Conservation of Nature (IUCN). This IUCN classification has among its objectives the facilitation of scientific research and environmental monitoring, mainly related to the conservation and sustainable use of natural resources (http://www.iucn.org). No specific permits were therefore required for performing these scientific experiments, which did not involve endangered or protected species. The first experiment tested the effect of *A. marina* exclusion on the response of macrofauna assemblages and sediment biogeochemistry to different amounts of detrital macroalgae. The second experiment measured the changes in detrital recycling following *A. marina* exclusion. For the first experiment (Fig. 1), all lugworms were excluded from 15 randomly chosen sediment plots of 1×1 m, some meters apart, by burying mosquito nets to the depth of approximately 7 cm (hereafter exclusion A− treatment), following the method proposed by [41] in a more extensive area. A procedural control for the effect of the net and of the burial was done for other 15 sediment plots of 1×1 m by burying mosquito nets with 10×10 cm holes (hereafter procedural control A+ treatment).

One month after lugworm exclusion, the sediment plots were organically-enriched by addition of detrital *Ulva* spp. Fresh *Ulva* thalli were previously collected in the Oosterschelde estuary and washed to remove animals and sediment. Thalli were then weighted and frozen in separate plastic bags. Freezing is

▪ Lugworm excluded plots (A-)
☐ Procedural control plots (A+)

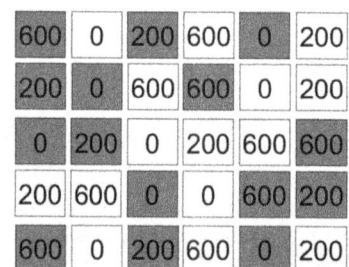

600	0	200	600	0	200
200	0	600	600	0	200
0	200	0	200	600	600
200	600	0	0	600	200
600	0	200	600	0	200

Figure 1. Schematic experimental design for experiment 1. Each rectangle corresponds to 1 m² plot. Numbers refers to the detrital enrichment. (0: no addition; 200:200 g WW were added twice at a 1 week interval; 600:600 g WW were added twice with a 1 week interval). Distance among plots are not in scale.

commonly used in experiments that aims at testing effects of detrital macroalgae and decomposition patterns. This treatment is necessary in order to create detritus and ensure that *Ulva* is dead when buried since this seaweed can survive for very long periods in the dark [30]. The thalli were left defrosting at air temperature and then added to the top 1 cm of the sediment by gently hand-churning the detritus with the sediment. Enrichment was done twice, on July 3 and July 10, 2006. Each time either 200 or 600 g of wet weighted *Ulva* (hereafter treatments 200 and 600, respectively) were added to 10 exclusion (A−) and 10 procedural control (A−) plots. These amounts corresponded to 40 and 120 g dry weight. Other 10 plots (5 A− and 5 A+) were not enriched (hereafter treatment 0), but the sediment was hand-churned to control for the manipulation. There was no procedural control for the physical presence of detritus as previous experiments had demonstrated no effects [28–30]. The amount of *Ulva* spp. was based on the available literature to cause medium and high levels of disturbance (e.g. [27,30,32,33]).

The second experiment was run at the same time of experiment 1. Two additional procedural control plots for burial and 2 exclusion plots were enriched with 200 g ^{13}C-labelled *Ulva* twice, resembling the treatment "200". These four plots were used to trace the fate of the carbon released from detrital *Ulva*. The labeling of *Ulva* was done in the laboratory. Leaves were washed and carefully cleaned to eliminate epiphytes and animals. Then, thalli were transferred to aquaria containing filtered seawater supplemented with NaH^{13}CO$_3$ (99% ^{13}C) to the concentration of 0.2 g L^{-1}. The aquaria were oxygenated and covered with transparent PVC to limit ^{13}C bacterial respiration.

Field Sampling

For the first experiment, sampling was done on July 30 and August 1, 2006, one month after detrital enrichment. This timing was chosen because previous studies suggested that effects of algal mats on macrofauna occur within 2 to 10 weeks from the deposition [31–33,42] and that decomposition of *Ulva* occurs within a month (e.g. [43,44]).

Sediment samples were taken for organic carbon (hereafter OC), porewater total inorganic carbon (DIC = H2CO$_3$+ HCO$_3^-$+CO$_3^{2-}$), microalgal biomass (chlorophyll *a*) and macrofauna species composition, abundance and diversity. The sediment for chlorophyll *a* and OC was collected using a cut syringe of

3 mm inner diameter (i.d.). For chlorophyll *a*, the sediment was collected to 1 cm depth, kept in the dark and preserved at −80°C. The sediment for OC was collected to the depth of 6 cm, sectioned into 0–1, 1–2 and 2–6 cm depths and preserved freeze-dried. *Ulva* detritus was removed from this sediment under the stereo-microscope. The sediment for porewater DIC extraction was collected with cores of 5 cm i.d. and sectioned into 0–1, 1–2 and 2–6 cm depth intervals at the lowest and highest levels of organic enrichment (0 and 600) in both the control and the exclusion treatments. Two cores were collected from each plot and then pooled to reach the amount of water needed for the analyses. Porewater was extracted by centrifugation (1344 g, 15 min) and the supernatant was filtered on Millipore 45 μm. The extracted water was stored in glass vials, sealed, immediately poisoned with saturated HgCl$_2$ to stop bacterial activity and analyzed within 3 d. Only 3 out of 5 randomly-chosen replicate plots were sampled for OC and TCO$_2$.

The sediment for macrofauna was collected using 11 cm i.d. PVC core. This sediment was sieved (0.5 mm mesh) and macrofauna fixed in 4% buffered formol. Before sampling, all plots were photographed to estimate the abundance of *A. marina* using visible mounds. In the laboratory, all plots were independently examined by four researchers to reduce estimate bias. Five randomly-chosen natural plots were also photographed and then sampled using 11 cm i.d. cores to test for the effect of the procedural control A+ on the natural abundance of *A. marina* and of the other benthic macrofauna. No differences were detected between natural and procedural control treatments (see Results section) and this latter was used for evaluating the effect of detrital enrichment and lugworm exclusion.

Sediment cores were also used to measure fluxes of inorganic carbon (TCO$_2$) and oxygen (O$_2$) at the sediment-water interface. Intact sediment cores were collected by pushing Plexiglas cores (11 cm i.d.) into the sediment to a depth of 20 cm at each of three replicate plots for the lowest and the highest levels of organic enrichment (0 and 600) in both the control and the exclusion treatments. The net was carefully cut around the cores to allow the core penetration into the sediment. Cores were sealed with rubber stoppers and dug out of the sediment. All cores were brought to a climate controlled room within 2 h, and kept in the dark at *in situ* temperature (10°C) before further handling. The same day, the cores were incubated in the dark for 12 h. Incubation in the dark

Arenicola-excluded sediment   Procedural control sediment
(A-)   (A+)

Figure 2. Comparison of sediment surface appearance after *Arenicola marina* exclusion at low tide. A− represent the sediment after 1 month from the burial of net to exclude burrowing species. A+ is the procedural control treatment (see Materials and Methods for details).

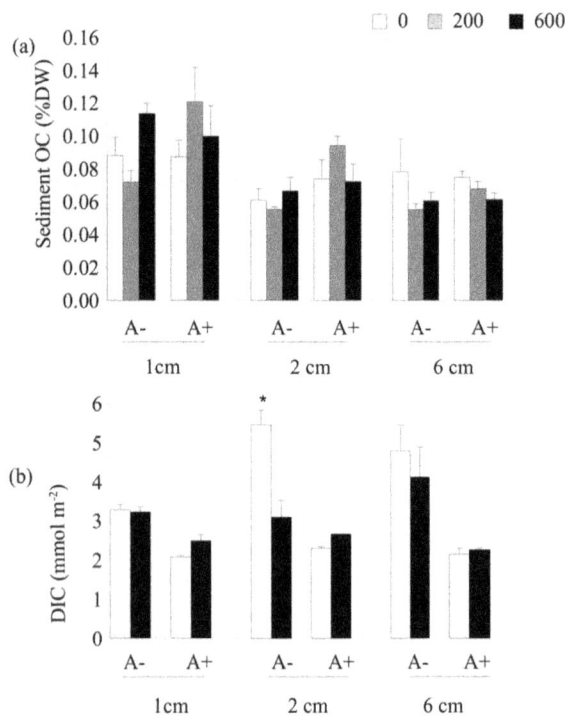

Figure 3. Mean (+1SE) for (a) bulk sediment organic carbon (OC) and (b) porewater dissolved inorganic carbon (DIC) at different sediment depths (*1 cm:* surface sediment to 1 cm depth; *2 cm:* from the depth of 1 to 2 cm; *6 cm:* from the depth of 2 to 6 cm). In the graphs, A−: *A. marina* excluded; **A+:** procedural control for burial; the legend indicate the levels of detrital addition (0: no addition; 200:200 g WW were added twice at a 1 week interval; 600:600 g WW were added twice with a 1 week interval). Data were sampled one month after detrital addition. Asterisk indicates significant differences.

did not allow to gather additional information concerning the (auto)trophic state of the sediment, which would need a light incubations of other sediment cores. Our system was however limited to carry 12 cores at a time and no further analyses could be done. Prior to incubations, cores were inundated with well-oxygenated Oosterschelde water collected the same day ($[NH_4^+] = 20$ µmol L⁻, $[NO_3^-] = 110$ µmol L^{-1}, salinity = 22) and left to acclimatize in the dark for 6–8 h. During incubation, the overlying water-column (25–30 cm) was continuously stirred with a magnetic stirrer, maintaining continuous water circulation at a rate well below the resuspension limit. Dissolved O_2 never decreased below 65% of saturation. Water samples for TCO_2 and O_2 were collected in glass vials and sealed avoiding air bubbles. All samples were immediately poisoned with saturated $HgCl_2$ to stop bacterial activity and then analyzed within 3 d.

For experiment 2, sediment samples were taken from the four plots with labelled *Ulva*. Sampling was done on four occasions, 6, 8, 15 and 17 days after detrital enrichment. Each time, one plexiglass intact sediment core (11 cm i.d.) was collected to 20 cm depth, following the modality used to sample the cores for estimating benthic respiration. To reduce plot damage during core collection, the central part of the plot (20 cm from the edges) was divided into 4 quadrants of 30×30 cm. Each sampling occasion, a sediment core was taken from a different quadrant. An area of 15×15 cm was thus damages each time, corresponding to 2.25% of the entire plot area. The sampled sediment was replaced with plexiglass tubes to avoid any collapse and to leave the remaining

Table 1. Two-way analyses of variance for sediment organic carbon (OC) and porewater DIC at different sediment depths one month after detrital enrichment.

Sediment OC	df	0–1 cm MS	F	1–2 cm MS	F	2–6 cm MS	F
Exclusion = E	1	0.67×10^{-3}	0.40	1.60×10^{-3}	3.80	0.09×10^{-3}	0.52
Detrital enrichment = D	2	0.61×10^{-3}	0.96	0.09×10^{-3}	0.40	0.62×10^{-3}	2.22
E×D	2	1.7×10^{-3}	2.64	0.42×10^{-3}	1.90	0.17×10^{-3}	0.62
Residual	12	0.63×10^{-3}		0.22×10^{-3}		0.28×10^{-3}	
Porewater DIC							
Exclusion = E	1	1.89	17.50	5.06	1.03	10.15	4.80
Detrital enrichment = D	1	0.06	2.00	1.28	5.12	0.15	0.28
E×D	1	0.11	3.43	4.93	**19.71**	0.30	0.57*
Residual	8	0.03		1.00		2.11	

Significant values (P<0.05) are in bold. Data were not-transformed ($P_{C-cochran}$ >0.05). Asterisk indicates pooling of interaction when P>0.25.

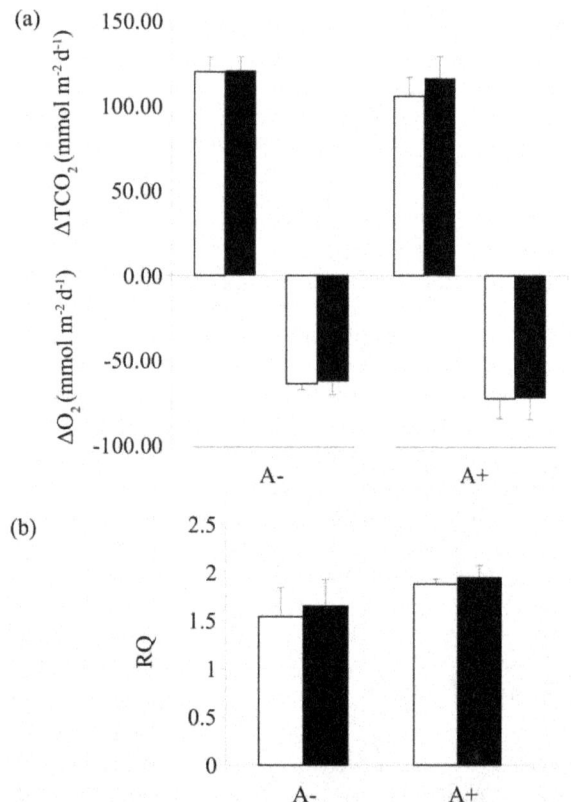

Figure 4. Mean (+1SE) for (a) benthic oxygen (O_2) consumption and total carbon dioxide (TCO_2) released during 12 hours dark incubation; (b) Respiratory quotient (RQ). In the graphs, A−: *A. marina* excluded; **A+:** procedural control for burial; the legend indicate the levels of detrital addition (0: no addition; 200:200 g WW were added twice at a 1 week interval; 600:600 g WW were added twice with a 1 week interval). Data were sampled one month after detrital addition.

Table 2. Two-way analyses of variance for sediment-water exchanges of O_2 and TCO_2 and for the benthic respiratory coefficient RQ.

		O_2		TCO_2		RQ	
	df	MS	F	MS	F	MS	F
Exclusion = E	1	$8.8 \ 10^{-6}$	2.38	252.08	0.78	0.29	2.39
Detrital enrichment = D	1	$8.0 \ 10^{-9}$	0.00	102.08	0.31	0.02	0.18
E×D	1	$4.38 \ 1^{-7}$	0.11*	70.08	0.22*	0.00	0.01*
Residual	8	$4.10 \ 1^{-6}$		325.08		0.14	

Significant values (P<0.05) are in bold. Data were not-transformed ($P_{C\text{-cochran}}$ >0.05), except for O_2, which was arc-tangent transformed to homogenise the residual variances. Asterisk indicates pooling of interaction when P>0.25.

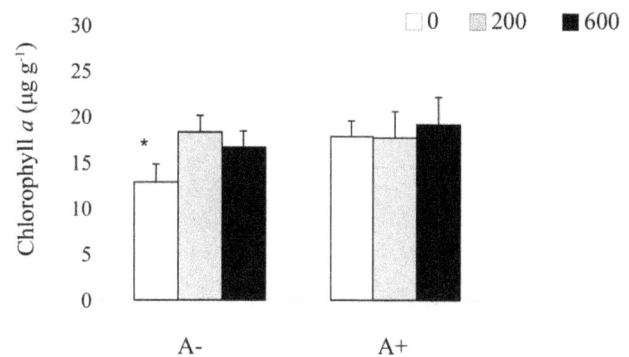

Figure 5. Mean (+1SE) for chlorophyll *a* concentration at the surface sediment (1 cm depth). In the graphs, A−: *A. marina* excluded; A+: procedural control for burial; the legend indicate the levels of detrital addition (0: no addition; 200:200 g WW were added twice at a 1 week interval; 600:600 g WW were added twice with a 1 week interval). Data were sampled one month after detrital addition. Asterisk indicates significant differences.

part of the plots undisturbed. The cores were handled like the cores used for evaluating oxygen and carbon exchange at the sediment-water interface for the experiment 1. However, before incubating for wet respiration and determining release of $T^{13}CO_2$ in the water column, the labelled cores were incubated for dry fluxes during 6 h. At the beginning and at the end of the incubation, air samples were collected using syringes and pumped into tubes containing soda lime to trap CO_2 and measure the $^{13}CO_2$ derived from *Ulva* decomposition. A LICOR automated CO_2 analyzer was used to quantify CO_2 release every hour. Sediment-water exchange rate and macrofauna data were generated as for experiment 1 (see "Analytical measures" section).

Following flux incubations for both experiments, the cores were sieved at 0.5 mm to collect the macrofauna. No lugworm was included in the cores, meaning that any observed pattern represented the result of sediment and macrofauna alteration due to *Arenicola*, but not the direct activity of the lugworm during incubation.

Analytical Measures

In the laboratory, macrofauna were identified to the species level, counted and dried at 60°C for 48 h to obtain dry weight biomass to be used for stable isotope analyses relative to the experiment 2. The detrital *Ulva* remaining in the samples was also collected, dried and weighed. Chlorophyll *a* was extracted in darkness for 24 h at 0–4°C using a 90% acetone solution. The sediment was homogenized and sub-samples of about 1 g dry weight were taken. Pigments were measured spectrophotometrically before and after 0.1 N HCl acidification. Pre-combusted (450°C for 3 h) sediment was used as a blank. Measurements for bulk sediment OC were made using a Perkin-Elmer CHN element analyzer on freeze-dried material, after carbonate removal with HCl.

Flux rates of TCO_2 and O_2 between the sediment and the overlying water were estimated as difference between the initial and the final TCO_2 or O_2 concentrations during dark incubation, assuming constant solute exchange with time. The oxygen dissolved into the water samples was analyzed using the standard method of Winkler titration, while water TCO_2 and porewater DIC were determined by the flow injection/diffusion cell technique, which is based on the diffusion of CO_2 through a hydrophobic membrane into a flow of deionized water, thus generating a gradient of conductivity proportional to the concentration of CO_2.

Water TCO_2 (wet fluxes), gaseous CO_2 (dry flux), bulk sediment and macrofauna collected in the labelled plots were analyzed for carbon isotope composition ($\delta^{13}C$). The gaseous $^{13}CO_2$ (dry flux) captured using soda lime was analyzed after acidification. The ^{13}C isotope signature for TCO_2 dry and wet fluxes was analyzed using a GC column coupled to a Finningan delta S mass spectrometer IRMS. The carbon isotopic composition for sediment, residual detritus and animals was determined using a Fisons elemental analyzer coupled to a Finningan delta S mass spectrometer. ^{13}C to ^{12}C ratio was referred to Vienna PDB and expressed in units of ‰. Reproducibility of the measurements is better than 0.2‰ [45]. The excess of the heavy isotope of carbon (above background) was expressed as total uptake (I) in milligrams of ^{13}C. I was calculated as the product of excess ^{13}C (E) and carbon or biomass (C):

$$I = E \times C$$

$$E = F_{control} - F_{sample}$$

where F indicates the fraction of isotope added and it is calculated as:

$$F = R/R + 1$$

where R can be calculated from the measured $\delta^{13}C$ using $R_{reference}$ of 0.0112372:

$$\delta^{13}C = \left[R_{sample} / R_{reference} - 1 \right] \times 10^3$$

Statistical Analyses

For experiment 1, data were analyzed with two-way orthogonal mixed model of analysis of variance (ANOVA), with exclusion (2 levels: procedural control A+ and exclusion of *A. marina*, A−) and detrital enrichment (3 or 2 levels: 0, 200 and 600) as fixed and random factors, respectively. Detrital enrichment was considered a random factor because the amount of detritus to add was chosen within a range of values to simulate conditions of low and high

Table 3. Two-way analyses of variance for chlorophyll *a*, macrofauna abundance and number (N) of species.

	df	Chlorophyll a		Abundance		N of species	
		MS	F	MS	F	MS	F
Exclusion = E	1	72.23	1.27	0.48	0.69	0.03	0.00
Detrital enrichment = D	2	53.74	**4.02**	0.29	0.56	30.00	**7.11**
E×D	2	56.80	**4.25**	2.34	**4.81**	17.73	**4.21**
Residual	24	13.37		0.49		4.22	

Significant values (*P*<0.05) are in bold. Data were sqrt-transformed for macrofauna abundance.

enrichment, but no specific hypothesis concerned the chosen amounts. A significant interaction between exclusion and detrital enrichment was expected if the exclusion of *A. marina* changed the response to detrital enrichment. These analyses were done for the total abundance of macrofauna individuals, number of macrofauna species, chlorophyll *a*, sediment OC, porewater DIC, water O_2 and TCO_2 exchange rate. *C-cochran* test was used to test for residual variance homogeneity before running the analyses. When this test was significant (*P*<0.05), data were transformed. Pooling was done when *P*>0.25. *A posteriori* multiple comparison SNK test was run when interaction term was significant.

The variables collected for evaluating the recycling of detrital ^{13}C (experiment 2) were analyzed with repeated measure ANOVA, with time (6, 8, 15 and 17 sampling days from the end of detrital addition) as the repeated measure random factor and *A. marina* exclusion as fixed factor. Plots (2 levels) was nested in exclusion. The interaction of time with the exclusion treatment was measured over the interaction time×plot. When significant interaction was found, the main effect exclusion was not interpreted. Assumption about the correlation among the different observations of the repeated measure factor (time) was tested with Mauchley's sphericity test.

Decomposition rate of macroalgal detritus was estimated by fitting the first-order model:

$$B_t = B_0 e^{kt}$$

Where B_t is the g of detrital biomass at time t (days), B_0 is the initial biomass, at time 0, at the moment of the second detrital addition and k is the specific decomposition constant. The two detrital additions were done at a distance of one week, which may greatly affect the estimates of B_0. We therefore considered the estimates of residual *Ulva* as the sum of two decomposition curves, with the same constant k but different decomposing time:

$$B_t = B_{(0-7)} e^{k(t+7)} + B_{(0)} e^{kt}$$

The curves of ^{13}C benthic incorporation (including loss due to respiration) were fitted to the logistic curve:

$$I_t = Y/(1 + q e^{-rt})$$

$$q = (Y - I_0)/I_0$$

Where Y is the ^{13}C loss from macroalgal decomposition and r is the capacity of benthos to incorporate ^{13}C.

Results

After the exclusion of *A. marina*, no burrows were visible in the excluded plots (Fig. 2) and the number of burrows in the procedural control (treatment A+) was comparable to that of natural sediment (mean±SE, n = 5:8.72±1.77 and 10.33±1.17 burrows m^{-2} for the natural and the procedural control, respectively). Overall, there were no differences in macrofauna species composition between the natural sediment and the procedural control plots. Macrofauna abundance (mean±SE, n = 5) was 294±42 and 220±53 individual m^2, whereas the number of species was 10±0.2 and 10±0.5 for A+ and natural sediment, respectively. Therefore, A+ could be used as a control for the effect of lugworm exclusion and detrital macroalgae.

Overall, in the natural sediments and in the A+ plots the gastropod *Hydrobia ulvae* was numerically dominant, representing 90% of macrofauna abundance. The remaining 10% was made up by oligochaetes (3%), deposit feeder polychaetes (3%) such as spionids, nereids, capitellids and cirratulids and bivalves (3%), mainly *Macoma balthica* and *Cerastoderma edule*.

Carbon Availability and Benthic Respiration

At the time of sampling, sediment organic carbon (OC) did not vary following detrital enrichment, the exclusion of *A. marina* or their interactions (Table 1, Fig. 3a; ANOVA, *P*>0.05). Macroalgal debris was still present in the sediment where detrital macroalgae were added at the highest enrichment level (treatment 600). The biomass of this macroalgal debris was smaller in the lugworm-excluded sediment than in procedural controls (mean±SE: lugworm exclusion, treatment A−: 2.3±1.4 gDW m^{-2}; A+: 12.0±6.6 gDW m^{-2}; ANOVA: $F_{1,8} = 6.19$, *P*=0.038).

There was a significant effect of the interaction between lugworm exclusion and detrital enrichment on porewater DIC to the depth of 1–2 cm (Table 1; Fig. 3b). In the non-enriched sediment, porewater DIC concentration was higher in the lugworm exclusion than in the control plots (SNK test at *P*=0.05: non-enriched plots, treatment 0: A−>A+). Moreover, in the lugworm-exclusion plots, porewater DIC was lower in the enriched than in the non-enriched plots (SNK test at *P*=0.05: *Arenicola* exclusion, A−: 0>600). Values in detrital-enriched, lugworm-excluded sediment were similar to those measured for the A+ treatment (Fig. 3b). A similar pattern was observed for superficial (0–1 cm) and deeper sediment (2–6 cm in Fig. 3b), but no significant changes were measured.

Both *A. marina* exclusion and detrital enrichment or their interaction did not significantly change sediment-water exchanges of TCO_2 and O_2 nor the RQ respiration coefficient (TCO_2:O_2 ratio) (Table 2; Fig. 4).

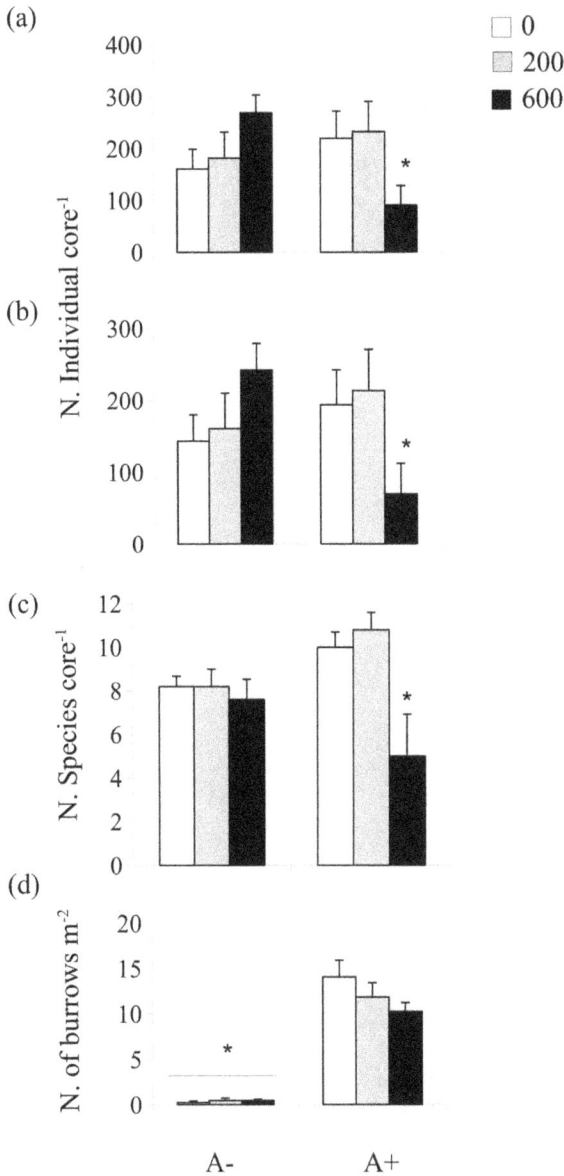

Figure 6. Mean (+1SE) for (a) total number of macrofauna individuals, (b) abundance of _Hydrobia ulvae_, (c) number of macrofauna species and (d) numbers of _Arenicola marina_ burrows in the plots where the mosquito net was buried (A−) and in the procedural control plots (A+) at the end of the experiment, e.g; 1 month from the second addition of detrital _Ulva_ (0: no addition; 200:200 g WW were added twice at a 1 week interval; 600:600 g WW were added twice with a 1 week interval). Data were sampled one month after detrital addition. Asterisk indicates significant differences.

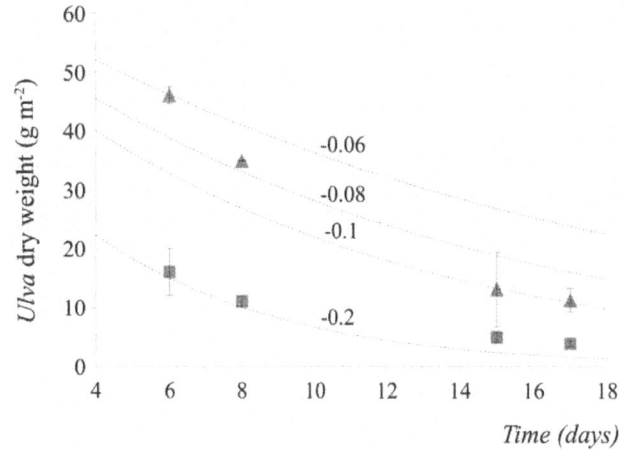

Figure 7. Decomposing curves for _Ulva_ detritus in lugworm excluded (treatment A−; square symbols in the graphs) or procedural control sediments (treatment A+; triangle symbols in the graph). Symbols indicate mean (± SE) for macroalgal biomass and are relative to experiments 2.

A significant interaction between _A. marina_ exclusion and detrital enrichment was also found for macrofauna abundance and number of species (Table 3). Abundance was lowered by the highest level of detrital enrichment in the plots where _A. marina_ was present (Fig. 6a; SNK test: A+: 0 = 200>600). Overall, both the dominant gastropod _Hydrobia ulvae_ and all other remaining macrofauna decreased in abundance, especially nereids, spionids and bivalves (ANOVA: $F_{2, 24} = 4.26$, $P = 0.026$ for _H. ulvae_, Fig. 6b; $F_{2, 24} = 4.65$, $P = 0.019$ for remaining taxa, data were square-root transformed; SNK tests for both: A+: 0 = 200>600). Number of species showed a similar pattern, decreasing in response to the highest level of enrichment in the A+ plots (Fi. 6C; SNK test: A+: 0 = 200>600). Detrital enrichment seemed to not affect the abundance of the lugworms as the number of _Arenicola_ burrows remained relatively stable at different levels of detrital enrichment (2 way ANOVA: $F_{2, 12} = 0.88$, $P = 0.34$, Fig. 6d).

The Fate of Detrital Carbon

There were differences in the rate of _Ulva_ decomposition following _A. marina_ exclusion. The decomposition constant k was −0.2 for the detritus added to the _A. marina_ excluded sediment (A− plots in Fig. 7). Interestingly, the A+ plots did not fit well the exponential curve using the same constant k through the decomposition. Rather, at the beginning of decomposition, they fitted k values between −0.06 and −0.08, whereas at the end they fitted a K constant of −0.1 (Fig. 7). There was a significant interaction between exclusion treatment and time after detrital enrichment for the flux rate of $T^{13}CO_2$ from the sediment to the water column (Table 4). More $T^{13}CO_2$ was released by the sediment in the exclusion treatment within 6 days from burial (Fig. 8a). Respiration rates were (mean ± SE) 0.84±0.25 and 0.12±0.03 mg ^{13}C m^{-2} day^{-1} for dry and wet fluxes, respectively.

Part of the ^{13}C remained in the benthos and it was incorporated in the bulk sediment and in the macrofauna. The amount incorporated was variable among plots, but it did not significantly vary between exclusion treatments or time (Table 4; Fig. 8 b–c). _H. ulvae_ incorporated the largest amount of ^{13}C among all macrofauna taxa (mean ± SE: 3.2±0.8 and 1.2±0.4 mg ^{13}C m^{-2}, for _H. ulvae_ and remaining fauna, respectively).

Microalgal Biomass and Macrofauna

There was a significant interaction between _A. marina_ exclusion and detrital enrichment for microalgal biomass (Table 3; Fig. 5). Lugworm exclusion had a negative effect on microalgal biomass in the non-enriched plots (SNK test: 0: A+>A−). In the lugworm-excluded sediment, detrital enrichment enhanced microalgal biomass (SNK test: A−: 0<200 = 600). Values were similar to those measured for the A+ control plots (treatments 200 and 600 in Fig. 5).

Table 4. Repeated measure analyses of variance and sphericity tests for the ^{13}C incorporation in macrofauna and bulk sediment, and for the $^{13}CO_2$ released from the benthos (benthic respiration) during detrital decomposition at an intermediate level of detrital enrichment (200 gWW plot^{-1} added twice).

	df	Macrofauna MS	F	Bulk sediment MS	F	Benthic respiration MS	F
Exclusion = E	1	$5.97\ 10^{-7}$	0.12	$4.27\ 10^{-4}$	0.22	$1.52\ 10^{-5}$	**9.73**
Time = T	3	$3.13\ 10^{-7}$	0.60	$8.78\ 10^{-4}$	1.02	$1.84\ 10^{-5}$	**14.45**
Residual (Plots)	2	$4.96\ 10^{-6}$		$1.96\ 10^{-3}$		$1.56\ 10^{-6}$	
T×E	3	$6.21\ 10^{-7}$	0.12	$8.79\ 10^{-4}$	1.02	$8.98\ 10^{-6}$	**7.07**
Residual (T×Plots)	6	$5.21\ 10^{-6}$		$8.59\ 10^{-4}$		$1.27\ 10^{-6}$	
Sphericity test (χ^2) n = 6		7.80		4.83		8.06	

Significant values (P<0.05) are in bold. Sphericity test was not significant.

Discussion

Bioturbation can enhance solute penetration in deeper sediments, their exchange between the sediment and the overlying water and re-distribute organic material along the vertical profile [36,37,41]. As such, the loss of species that are important contributors to bioturbation may greatly change the response to detrital decomposition of sediment biogeochemistry. In laboratory experiments, for instance, *A. marina* and the polychaete *Nereis diversicolor* were found to increase turnover of carbon and nutrients during detrital macroalgal decomposition [19,36]. The exclusion of *A. marina* from intertidal sediments was thus expected to modify the biogeochemical response to macroalgal detrital enrichment. Conversely, at the end of decomposition (1 month after the second enrichment) benthic respiration (O_2 and TCO_2 sediment-water exchanges) and sediment organic carbon content did not change as an effect of lugworm exclusion.

Ulva amount and time of sampling (one month after enrichment) allowed decomposition and release of a large quantity of organic matter, which could be mineralized or accumulated in the sediment, thereby modifying sediment organic carbon, inorganic carbon dissolved in the interstitial water and fluxes at the sediment-water interface. *Ulva* halftime decomposition rate is within 15 days [30]. By considering the range of k constant calculated for this study (between −0.06 and −0.2), at least 85% *Ulva* would be decomposed at the time of sampling for both enrichment levels. If one considers that dried sediment weighs about 1600 g l^{-1} and *Ulva* organic carbon content is around 40% of dried tissues [34], this decomposed detritus corresponds to organic carbon concentrations ranging between 0.15 and 1.2% sediment dry weight for the top first cm, roughly 2–10 times the concentration of organic carbon found in natural sediments for low and high enrichment, respectively (see the non-enriched sediment in Fig. 2a). The lack of sediment carbon accumulation and changes in TCO_2 fluxes at the sediment-water interface indicated that the benthic system was able to remove the carbon derived from *in situ* decomposition and that bioturbation did not play a central role in detrital carbon recycling.

It is important to re-iterate that fluxes were measured during 12 h dark incubation of sediment cores without lugworms, which indicated the overall indirect effect of lugworms on detrital benthic mineralization, but did not include any direct effects of lugworms due to burrow ventilation during the period of incubation. The importance of ventilation was instead assessed in the field by

measuring porewater DIC, which increased after lugworm removal, indicating a decreased solute exchange due to *A. marina* loss. Interestingly, these changes were variable along the vertical sediment profile and were offset by the addition of detrital macroalgae, which supported the idea that fast mineralization of detrital carbon in these sediments is only in part due to bioturbation. The complexity of the relationships between species, functions and response to disturbance can increase with increasing ecological realism [15,46]. Our experiments were done *in situ* and they were designed to test effects of *A. marina* under natural environmental conditions. Detrital macroalgae were deliberately added to the sediment, without using artificial nets that could limit detrital loss. This method exposed decomposing detritus to tides and waves, especially when detritus remained at the sediment surface. The role of bioturbation on sediment biogeochemistry has been mainly assessed through laboratory studies. The few field studies on the role of *Arenicola* on porewater, fluxes and organic matter decomposition have revealed that its role may vary and decrease considerably in nature [19,37,47]. By comparing field to laboratory experiments, it was found, for instance, that solute exchange was more variable in the field because it was not only ruled by bioturbation but also by advection related to waves and tides and by differences in the reactivity of sediment organic matter related to the complexity of biological communities of primary producers and consumers [47]. This is particularly true when experiments are done in permeable sediment, where *A. marina* is often a dominant species [48]. In these sediments, hydrodynamics can enhance advective flow from the porewater to the overlying water column, rule the exchange of solutes and organic matter at the sediment-water interface and overwhelm the biological effect of bioturbation [49]. Although our experiments were done in a relatively sheltered sand flat, as indicated by the dominance of fine-grained sand, numerous ripple marks were visible at low tide (Fig. 2) and they probably indicated strong hydrodynamics generated by waves and tides. In these sediments, hydrodynamics could accelerate the exchange of solutes produced during decomposition, the removal of macroalgal detritus from the sediment and override the effect of bioturbation on the response of sediment biogeochemistry to detrital macroalgae.

The decomposition rate of *Ulva* in these sediments was fast, as indicated by the estimated k constants (k = −0.2 and −0.1 for A− and A+, respectively; Fig. 7). These values were at the limit or even exceeded the range of values typical of *Ulva* detritus, which varies between −0.04 and −0.1 [34]. More importantly, the estimated k

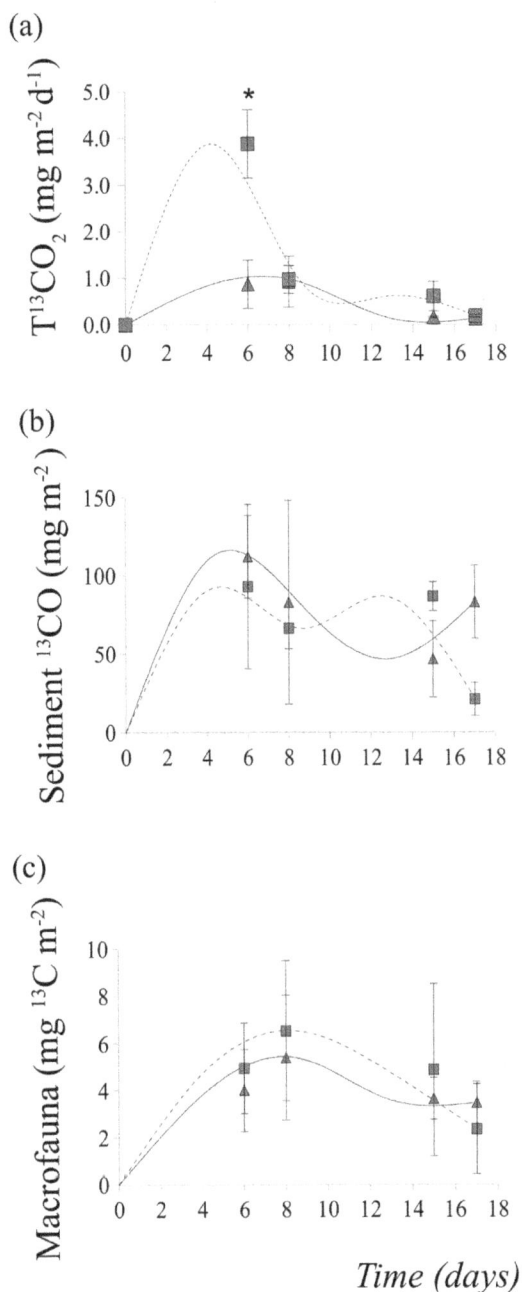

Figure 8. Mean ± SE for (a) $T^{13}CO_2$ released from the sediment to the water column (sum of dry and wet fluxes); (b) amount of ^{13}C organic carbon in the bulk sediment (averaged over depth profile) and (c) amount of ^{13}C carbon incorporated by macrofauna. Data are fitted to the LOESS smoothing with an interval of 4 and 1 polynomial degree. Values are averaged over the two plots where A. marina was excluded (treatment A−; square symbols in the graphs) and the two procedural control plots (treatment A+; triangle symbols in the graph). Data refers to the experiment 2. Asterisk indicates significant differences.

constant for the lugworm-exclusion sediment was twice that of the control plots. In addition, the biomass of macroalgal debris was, on average, 6 times larger in controls than in lugworm-excluded sediments (2 and 12 g DW per plot^{-1} for A+ and A−, respectively). This may bring to the conclusion that despite hydrodynamics could play an important role in determining

benthic response to detrital macroalgae, A. marina could represent an important sink for detritus and it could modify *in situ* decomposition rate, as found for other burrowing bioturbators [39,50]. These species can mechanically bury macroalgae in their funnels, thereby decreasing detritus availability to surface sediment and slowing down decomposition rate, while maintaining a pool of organic matter in the sediment. Buried detritus and slowed decomposition rate could, in turn, modify mineralization processes. In this study, a peak in $T^{13}CO_2$ release to the water column was identified at an early stage of decomposition in A. marina-excluded sediment plots, suggesting that the loss of A. marina accelerated the mineralization of detrital Ulva at an early stage of decomposition. This explanation of increased mineralization following A. marina loss was supported by the finding that porewater DIC decreased in response to detrital addition in A. marina excluded sediments. In addition, simultaneously to this decrease, macroalgal biomass increased. Probably, without the mechanical action of burying macroalgal fragments by *Arenicola*, detrital macroalgae remaining on sediment surface and decomposing faster released nutrients that enhanced bacterial metabolism and their carbon consumption. Evidence for increased carbon consumption during biomass decomposition has been found for plant litter in freshwater ecosystems [51,52]. In addition, benthic microalgae are often regulated by bottom-up processes [53] and they could greatly benefit from nutrient release, thereby contributing to recycling the mineralized detrital carbon. Interestingly, detrital enrichment and lugworms had similar effects on microalgal biomass. Probably, both lugworm bioirrigation and detrital decomposition can provide nutrients necessary to regulate microalgal growth [19,28,37,47].

A. marina is believed to influence the assembly rules and the distribution of macrofauna and, as such, their patterns of response to disturbances by altering the sedimentary habitat (e. g. [37,41]). Our results showed a negative effect of A. marina on macrofauna response to detrital addition since both macrofauna abundance and diversity decreased following the highest level of enrichment in presence of lugworms. The increased detrital burial by A. marina could play a central role in determining this response. Patches of detrital macroalgae in sediments can govern the spatial and temporal variability of macrofauna patterns of distribution. The presence of detritus may decrease numbers of individuals and diversity because of hypoxia during decomposition. Once detritus has decomposed, fast recolonizing taxa can promptly occupy the previously disturbed patches and benefit from increased food supply [29]. Detrital burial due to lugworm activity probably prolonged hypoxic conditions and slowed down macrofauna recovery. The decrease in abundance common to all taxa and the loss of diversity could corroborate this explanation as anoxic conditions related to detrital decomposition deplete most taxa abundance and diversity [29]. Moreover, burial could decrease detrital availability to surface consumers and therefore prevent their colonization, once hypoxia ceases. Macroalgal supply to sediment surface can represent an important source of food for surface grazers and detritivores, such as the gastropod Hydrobia ulvae [34,54]. Part of detrital carbon was transferred to macrofauna, especially to H. ulvae (experiment 2, Fig. 7c), which was the numerically dominant species in these area. Its important role for detrital carbon transport to the food web was found previously in the surrounding area [34]. This species is highly mobile and it crawls on sediment surface for feeding. Its local abundance might thus vary rapidly following food availability on surface sediment.

In summary, our field experiment showed that A. marina can be an important sink for detrital Ulva. However, conversely to what it should be expected based on current research knowledge, this species has moderate, sometimes negative effects on the response to detrital enrichment of sediment biogeochemistry and macrofauna. *A. marina* loss causes the sedimentary ecosystem to become a faster recycler of detrital macroalgae, enhancing detrital loss, carbon consumption and mineralization. This may bring to the conclusion that the loss of this species might be unimportant or somewhat beneficial for certain intertidal ecosystems. However, fast recycling and detrital dispersal could undermine carbon storage capacity of benthos and, in turn, the stability of whole estuarine ecosystem facing organic enrichment and eutrophication.

These findings clearly demonstrate that under natural conditions, the local extinction of an ecosystem engineer may have complex effects on the ecological stability of marine sediment biogeochemistry and benthic communities of consumers. They also highlight the importance of performing experiments under natural conditions if we are to understand and predict the response of ecosystems to perturbation and species loss.

Acknowledgments

The authors wish to thank the technicians for their invaluable help in the chemical analyses and E. Weerman, JJ Jansen and F Montserrat and numerous other students for their help in the field. S. Thrush is kindly thanked for his critical revision of an early version of this MS. This study is part of the BIOFUSE project, within the MARBEF network of excellence.

Author Contributions

Conceived and designed the experiments: FR BG. Performed the experiments: FR BG FG VDS. Analyzed the data: FR BG FG VDS. Contributed reagents/materials/analysis tools: JJM. Wrote the paper: FR BG FG VDS JJM.

References

1. Grime JP (1997) Ecology - Biodiversity and ecosystem function: The debate deepens. Science 277: 1260–1261.
2. McCann KS (2000) The diversity-stability debate. Nature 405: 228–233.
3. Worm B, Barbier EB, Beaumont N, Duffy JE, Folke C, et al. (2006) Impacts of biodiversity loss on ocean ecosystem services. Science 314: 787–790.
4. Tilman D, Downing AL (1994) Biodiversity and stability in grasslands. Nature 367: 363–365.
5. Yachi S, Loreau M (1999) Biodiversity and ecosystem productivity in a fluctating environment: The insurance hypothesis. Proceedings of the National Academy of Sciences of the United States of America 96: 1463–1468.
6. Downing A, Leibold M (2010) Species richness facilitates ecosystem resilience in aquatic food webs. Freshwater Biology 55: 2123–2137.
7. Ives A, Carpenter S (2007) Stability and diversity of ecosystems. Science: 58–62.
8. Pfisterer A, Schmid B (2002) Diversity-dependent production can decrease the stability of ecosystem functioning. Nature 416: 84–86.
9. Allison G (2004) The influence of species diversity and stress intensity on community resistance and resilience. Ecological Monographs 74: 117–134.
10. Emmerson MC, Solan M, Emes C, Paterson DM, Raffaelli D (2001) Consistent patterns and the idiosyncratic effects of biodiversity in marine ecosystems. Nature 411: 73–77.
11. Goodsell PJ, Underwood AJ (2008) Complexity and idiosyncrasy in the responses of algae to disturbance in mono- and multi-species assemblages. Oecologia 157: 509–519.
12. Rossi F, Vos M, Middelburg J (2009) Species identity, diversity and microbial carbon flow in reassembling macrobenthic communities. Oikos: 503–512.
13. Solan M, Cardinale BJ, Downing AL, Engelhardt KAM, Ruesink JL, et al. (2004) Extinction and ecosystem function in the marine benthos. Science 306: 1177–1180.
14. Bolam SG, Fernandes TF, Huxham M (2002) Diversity, biomass, and ecosystem processes in the marine benthos. Ecological Monographs 72: 599–615.
15. Romanuk T, Vogt R, Kolasa J (2009) Ecological realism and mechanisms by which diversity begets stability. Oikos 118: 819–828.
16. Jones CG, Lawton JH, Shachak M (1994) Organisms as ecosystem engineers. Oikos 69: 373–386.
17. Berke S (2010) Functional Groups of Ecosystem Engineers: A Proposed Classification with Comments on Current Issues. Integrative and Comparative Biology 50: 147–157.
18. Kristensen E, Penha-Lopes G, Delefosse M, Valdemarsen T, Quintana CO, et al. (2012) What is bioturbation? The need for a precise definition for fauna in aquatic sciences. Marine Ecology Progress Series 446: 285–302.
19. Kristensen E (2001) Impact of polychaetes (Nereis and Arenicola) on sediment biogeochemistry in coastal areas: Past, present, and future developments. Abstracts of Papers of the American Chemical Society 221: U538–U538.
20. Braeckman U, Provoost P, Moens T, Soetaert K, Middelburg J, et al. (2011) Biological vs. Physical Mixing Effects on Benthic Food Web Dynamics. Plos One 6: :-e18078. doi:10.1371/journal.pone.0018078.
21. Lohrer AM, Thrush SF, Gibbs MM (2004) Bioturbators enhance ecosystem function through complex biogeochemical interactions. Nature 431: 1092–1095.
22. Eklof J, van der Heide T, Donadi S, van der Zee E, O'Hara R, et al. (2011) Habitat-Mediated Facilitation and Counteracting Ecosystem Engineering Interactively Influence Ecosystem Responses to Disturbance. Plos One 6: e23229.-doi:10.1371/journal.pone.0023229.
23. Norkko J, Reed DC, Timmermann K, Norkko A, Gustafsson BG, et al. (2012) A welcome can of worms? Hypoxia mitigation by an invasive species. Global Change Biology 18: 422–434. doi:10.1111/j.1365-2486.2011.02513.x.
24. Eisenhauer N, Milcu A, Sabais A, Scheu S (2008) Animal Ecosystem Engineers Modulate the Diversity-Invasibility Relationship. Plos One 3.
25. Cebrian J (2004) Role of first-order consumers in ecosystem carbon flow. Ecology Letters 7: 232–240.
26. Raffaelli DG, Raven JA, Poole LJ (1998) Ecological impact of green macroalgal blooms. Oceanography and Marine Biology, an Annual Review 36: 97–125.
27. Ford RB, Thrush SF, Probert PK (1999) Macrobenthic colonisation of disturbances on an intertidal sandflat: the influence of season and buried algae. Marine Ecology Progress Series 191: 163–174.
28. Rossi F, Underwood AJ (2002) Small-scale disturbance and increased nutrients as influences on intertidal macrobenthic assemblages: experimental burial of wrack in different intertidal environments. Marine Ecology Progress Series 241: 29–39.
29. Kelaher BP, Levinton JS (2003) Variation in detrital enrichment causes spatio-temporal variation in soft-sediment assemblages. Marine Ecology Progress Series 261: 85–97.
30. Rossi F (2006) Small-scale burial of macroalgal detritus in marine sediments: Effects of Ulva spp. on the spatial distribution of macrofauna assemblages. Journal of Experimental Marine Biology and Ecology 332: 84–95.
31. Bolam SG, Fernandez TF, Read P, Raffaelli D (2000) Effects of macroalgal mats on intertidal sandflats: an experimental study. Journal of Experimental Marine Biology and Ecology 249: 123–137.
32. Hull SC (1987) Macroalgal mats and species abundance: a field experiment. Estuarine Coastal and Shelf Science 25: 519–532.
33. Thrush SF (1986) The sublittoral macrobenthic community structure of an irish sea-lough: effect of decomposing accumulatoins of seaweed. Journal of Experimental Marine Biology and Ecology 96: 199–212.
34. Rossi F (2007) Recycle of buried macroalgal detritus in sediments: use of dual-labelling experiments in the field. Marine Biology 150: 1073–1081.
35. Godbold JA, Solan M, Killham K (2009) Consumer and resource diversity effects on marine macroalgal decomposition. Oikos 118: 77–86.
36. Hansen K, Kristensen E (1998) The impact of the polychaete Nereis diversicolor and enrichment with macroalgal (Chaetomorpha linum) detritus on benthic metabolism and nutrient dynamics in organic-poor and organic-rich sediment. Journal of Experimental Marine Biology and Ecology 231: 201–223.
37. O'Brien A, Volkenborn N, van Beusekom J, Morris L, Keough M (2009) Interactive effects of porewater nutrient enrichment, bioturbation and sediment characteristics on benthic assemblages in sandy sediments. Journal of Experimental Marine Biology and Ecology: 51–59.
38. D'Andrea A, DeWitt T (2009) Geochemical ecosystem engineering by the mud shrimp Upogebia pugettensis (Crustacea: Thalassinidae) in Yaquina Bay, Oregon: Density-dependent effects on organic matter remineralization and nutrient cycling. Limnology and Oceanography 54: 1911–1932.
39. Vonk J, Kneer D, Stapel J, Asmus H (2008) Shrimp burrow in tropical seagrass meadows: An important sink for litter. Estuarine Coastal and Shelf Science 79: 79–85.
40. Meysman FJR, Galaktionov ES, Gribsholt B, Middelburg JJ (2006) Bioirrigation in permeable sediments: Advective pore-water transport induced by burrow ventilation. Limnology and Oceanography 51: 142–156.
41. Volkenborn N, Reise K (2006) Lugworm exclusion experiment: Responses by deposit feeding worms to biogenic habitat transformations. Journal Of Experimental Marine Biology And Ecology 330: 169–179.
42. Norkko A, Bonsdorff E (1996) Population responses of coastal zoobenthos to stress induced by drifting algal mats. Marine Ecology Progress Series 140: 141–151.
43. Buchsbaum R, Valiela I, Swain T, Dzierzeski M, Allen S (1991) Available And Refractory Nitrogen In Detritus Of Coastal Vascular Plants And Macroalgae. Marine Ecology-Progress Series 72: 131–143.
44. Nedergaard RI, Risgaard-Peterson N, Finster K (2002) The importance of sulfate reduction associated with Ulva lactuca thalli during decomposition: a mesocosm experiment. Journal of Experimental Marine Biology and Ecology 275: 15–29.

45. Herman PMJ, Middelburg JJ, Widdows J, Lucas CH, Heip CHR (2000) Stable isotopes as trophic tracers: combining field sampling and manipulative labelling of food resources for macrobenthos. Marine Ecology Progress Series 204: 79–92.

46. Mckie B, Schindler M, Gessner M, Malmqvist B (2009) Placing biodiversity and ecosystem functioning in context: environmental perturbations and the effects of species richness in a stream field experiment. Oecologia: 757–770.

47. Papaspyrou S, Kristensen E, Christensen B (2007) Arenicola marina (Polychaeta) and organic matter mineralisation in sandy marine sediments: In situ and microcosm comparison. Estuarine Coastal and Shelf Science 72: 213–222.

48. Needham H, Pilditch C, Lohrer A, Thrush S (2011) Context-Specific Bioturbation Mediates Changes to Ecosystem Functioning. Ecosystems 14: 1096–1109.

49. Huettel M, Roy H, Precht E, Ehrenhauss S (2003) Hydrodynamical impact on biogeochemical processes in aquatic sediments. Hydrobiologia 494: 231–236.

50. Retraubun A, Dawson M, Evans S (1996) The role of the burrow funnel in feeding processes in the lugworm Arenicola marina (L). Journal of Experimental Marine Biology and Ecology 202: 107–118.

51. Fontaine S, Bardoux G, Abbadie L, Mariotti A (2004) Carbon input to soil may decrease soil carbon content. Ecology Letters 7: 314–320.

52. Lennon JT (2004) Experimental evidence that terrestrial carbon subsidies increase CO_2 flux from lake ecosystems. Oecologia 138: 584–591.

53. Hillebrand H, Worm B, Lotze H (2000) Marine microbenthic community structure regulated by nitrogen loading and grazing pressure. Marine Ecology Progress Series 204: 27–38.

54. Riera P (2010) Trophic plasticity of the gastropod Hydrobia ulvae within an intertidal bay (Roscoff, France): A stable isotope evidence. Journal of Sea Research 63: 78–83.

3

Response of the Unicellular Diazotrophic Cyanobacterium *Crocosphaera watsonii* to Iron Limitation

Violaine Jacq[1]*, Céline Ridame[1], Stéphane L'Helguen[2], Fanny Kaczmar[1], Alain Saliot[1]

1 Université Pierre et Marie Curie, UMR LOCEAN -IPSL/CNRS/IRD/MNHN, Paris, France, **2** Université de Brest, CNRS/IRD, UMR 6539, LEMAR, OSU-IUEM, Plouzané, France

Abstract

Iron (Fe) is widely suspected as a key controlling factor of N_2 fixation due to the high Fe content of nitrogenase and photosynthetic enzymes complex, and to its low concentrations in oceanic surface seawaters. The influence of Fe limitation on the recently discovered unicellular diazotrophic cyanobacteria (UCYN) is poorly understood despite their biogeochemical importance in the carbon and nitrogen cycles. To address this knowledge gap, we conducted culture experiments on *Crocosphaera watsonii* WH8501 growing under a range of dissolved Fe concentrations (from 3.3 to 403 nM). Overall, severe Fe limitation led to significant decreases in growth rate (2.6-fold), C, N and chlorophyll *a* contents per cell (up to 4.1-fold), N_2 and CO_2 fixation rates per cell (17- and 7-fold) as well as biovolume (2.2-fold). We highlighted a two phased response depending on the degree of limitation: (i) under a moderate Fe limitation, the biovolume of *C. watsonii* was strongly reduced, allowing the cells to keep sufficient energy to maintain an optimal growth, volume-normalized contents and N_2 and CO_2 fixation rates; (ii) with increasing Fe deprivation, biovolume remained unchanged but the entire cell metabolism was affected, as shown by a strong decrease in the growth rate, volume-normalized contents and N_2 and CO_2 fixation rates. The half-saturation constant for growth of *C. watsonii* with respect to Fe is twice as low as that of the filamentous *Trichodesmium* indicating a better adaptation of *C. watsonii* to poor Fe environments than filamentous diazotrophs. The physiological response of *C. watsonii* to Fe limitation was different from that previously shown on the UCYN *Cyanothece* sp, suggesting potential differences in Fe requirements and/or Fe acquisition within the UCYN community. These results contribute to a better understanding of how Fe bioavailability can control the activity of UCYN and explain the biogeography of diverse N_2 fixers in ocean.

Editor: Douglas Andrew Campbell, Mount Allison University, Canada

Funding: This study was supported by CNRS (http://www.cnrs.fr/) and University Pierre et Marie Curie (http://www.upmc.fr/) funding. The funders had no role in study design, data collection and analysis, decision to publish, or preparation of the manuscript.

Competing Interests: The authors have declared that no competing interests exist.

* E-mail: violaine.jacq@locean-ipsl.upmc.fr

Introduction

In oligotrophic oceanic regions, bioavailable nitrogen (N) concentrations are sufficiently low that they set a constraint on primary productivity [1]. Diazotrophic cyanobacteria are not affected by N limitation due to their ability to use the dinitrogen (N_2) dissolved in oceanic surface waters as an alternative source of N. As N_2 represents an effectively unlimited resource of N, the N_2 fixation ability confers a major ecological advantage to diazotrophic cyanobacteria relative to non-diazotrophic phytoplankton in N depleted tropical and subtropical waters [2]. On a global scale, N_2 fixation represents the largest source of newly-fixed N to the open ocean (120 TgN yr^{-1} [3]), supporting a part of new primary production and influencing the N and carbon (C) cycles [4,5]. In the tropical and subtropical North Atlantic and Pacific oceans, N_2 fixation is estimated to support up to half of the new and export production [5,6], playing a key role in the uptake of atmospheric CO_2 by increasing the strength of the biological pump.

Among environmental factors constraining the distribution of diazotrophic cyanobacteria and the magnitude of N_2 fixation, iron (Fe) is widely believed to be a key controlling factor as the nitrogenase enzyme complex involved in intracellular N_2 reduction in NH_3 is Fe rich [7]. Furthermore, the high energetic cost of N_2 fixation imposes an additional Fe requirement for increase photosynthetic capacity [8,9]. The extremely low Fe solubility in oxic seawater [10] led to dissolved Fe (dFe) concentrations lower than about 1 nM in the open surface ocean [11–13], resulting in the potential limiting role of Fe for marine diazotrophic cyanobacteria. The effects of Fe limitation on the growth and N_2 fixation of the filamentous marine diazotrophic cyanobacteria *Trichodesmium* sp. have been widely evidenced both in artificial [14–16] and natural [17] environments. *Trichodesmium* sp. had been assumed to be the dominant N_2-fixing organism in the open ocean [18,19] until the recent discovery of unicellular diazotrophic cyanobacteria (UCYN, including UCYN-A, -B and -C [20]). Field measurements have highlighted that N_2 fixation rates associated with UCYN probably equal or exceed those associated with *Trichodesmium* sp. at regional scale [4]. N_2 fixation rates associated with UCYN were estimated to be up to 75% of the total N_2 fixation rate in the equatorial Western Pacific under stratified conditions [21]. On a global scale, recent biogeochemical models attribute about 50% of the total oceanic N_2 fixation rate to unicellular analogues [22]. Despite the biogeochemical importance of UCYN, their controlling factors remain poorly known. To date, only one open ocean UCYN species is available in culture: *Crocosphaera watsonii* (UCYN-B). Culture-based and field experiments have shown that light [23,24], temperature [24,25] and phosphorus [26–29] can control the growth of *Crocosphaera*, but the effects of Fe limitation on UCYN have been poorly investigated. The few studies conducted on the impact of Fe limitation on *C. watsonii* highlighted notable change in expression of several proteins under Fe stress [30,31]. Decreases in the N_2 fixation and growth rates of *C. watsonii* have been observed in one Fe-limited culture [32] and recent field enrichment experiments in the tropical Atlantic and Pacific have revealed that abundance of UCYN-B could be Fe limited in their natural habitats [27,28,33]. The response of *C. watsonii* to Fe limitation remains not fully characterized and needs to be quantified. In order to improve our knowledge and understanding of the impact of Fe limitation on UCYN, we conducted trace-metal clean culture experiments of *C. watsonii* WH8501 cultivated under a range of dFe concentrations to quantify for the first time the impact of Fe limitation on the growth, N_2 fixation rate, primary productivity, elemental contents, and cell size of an open ocean UCYN.

Materials and Methods

Culture experiments

All bottles and labware were thoroughly cleaned with suprapur HCl acid and ultra-pure water (>18.2 MΩ). All manipulations were conducted in a clean laboratory within a sterile laminar flow hood (class 100) using sterile and trace metal clean techniques. Batch cultures of *C. watsonii* WH8501 were grown in sterile polycarbonate bottles at 27.5°C, under a 12:12 h light:dark cycle at a light intensity of $\sim$150 µmol photons.m^{-2}.s^{-1}. The cells were cultivated in N free YBCII medium [34], prepared with Suprapur® quality salts and reagents and amended at different dFe concentrations. The medium contained phosphate (20 µM), vitamins (B_{12}, thiamine and biotin) and trace metals (Co, Mo, Cu, Zn and Mn). It was sterilized by autoclaving and 0.2 µm filtration. Cultures were gently mixed using orbital shakers to minimize cell sedimentation. Fe (FeCl$_3$) was added in triplicate cultures to obtain different final dFe concentrations ranging from 0 to 400 nmol.L^{-1} (nM) and was complexed with 2 µM of ethylenediaminetetra-acetic acid (EDTA), a metal ion buffering agent. In order to quantify a potential Fe contamination, dFe was analysed in sterilized YBCII medium before Fe addition by flow injection with online preconcentration and chemiluminescence detection [35] at the LOV laboratory (Villefranche sur mer). A background concentration of 3.3 nM was found in the medium and was systematically included in our results. Consequently, the eight dFe concentrations in the triplicate cultures were 3.3, 5.3, 8.3, 13.3, 23.3, 43.3, 103.3 and 403.3 nM (Table 1). Cells were previously acclimated to these different Fe concentrations for a minimum of 35 generations. Flow cytometry measurements (LOMIC laboratory) showed that our cultures were not axenic and allowed the determination of the abundance and biovolume of bacteria [36]. Using a conversion factor between biovolume and C content of bacteria from [37], we found that the C content associated with bacteria represented on average 0.4% of the total particulate organic carbon (POC) in the cultures. The initial pH in the cultures was 8.15 and variations between the beginning and the end of the growth phase were lower than 0.2 pH units, which avoided CO_2 limitation and pH effects on Fe chelation by EDTA [38].

All the parameters discussed in this study, except cell abundance, were determined during the exponential growth phase. Our results are reported as a function of dFe concentrations as well as dissolved inorganic Fe concentrations, hereafter referred to as Fe' and representing hydrolysed forms of dFe, supposed to be the bioavailable forms of Fe in EDTA buffered artificial seawater [39,40]. Fe' concentrations were computed from the Fe-EDTA complexation data in [38], taking into account influence of pH, light and temperature. The resulting estimated Fe' concentrations in the media ranged from 0.16 to 20.16 nM. The 3 highest Fe' concentrations (2.16, 5.16 and 20.16 nM) are invalid as they exceed the solubility limit for Fe with respect to ferric hydroxide precipitation, which is assumed to be $\sim$1.5 nM based on experimental data from [38].

Cell abundance and growth rate

C. watsonii's abundance was monitored by daily cell counts with an epifluorescence microscope (Nikon Eclipse 50i) using natural fluorescence of chlorophyll a (Chl a). These data were highly similar to those obtained by flow cytometry measurements (data not shown). Specific growth rates in the exponential phase were determined from linear regression of the logarithmic transformed cell abundance versus time.

Cell biovolume

C. watsonii cells were harvested in exponential phase, 2 hours after the beginning of the dark period, onto 0.4 µm polycarbonate membranes, and incubated overnight into a fixative with adjusted osmolarity (3% glutaraldehyde in 0.1M cacodylate pH 7.4, NaCl 1.75%). Membranes were then washed, post-fixed for 1 h with 1% osmium tetroxide in 0.1M cacodylate buffer with 1.75% NaCl, and then dehydrated with graded increasing concentrations of ethanol (50, 70, 96, 100%) and critical point dried (CPD 7501, Quorum Technologies). Finally, membranes were mounted on stubs, gold-sputtered (Scancoat Six, Edwards) and observed with a conventional SEM (Scanning Electron Microscope, Cambridge Stereoscan S260). Pictures were analysed with ImageJ software [41] in order to determine cell diameters and biovolumes. Due to experimental constraints, cell diameters and biovolumes were determined on four cultures (dFe = 3.3, 13.3, 43.3 and 403.3 nM).

Chlorophyll a

Culture samples were gently filtered (pressure<200 mbar) onto 0.7 µm glass microfiber filters (GF/F, Whatman©). Then, the filters were stored at -25°C. After extraction in 90% acetone [42], fluorescence of Chl a was measured at 670 nm on a Hitaschi F-4500 spectrofluorometer. Cellular Chl a content was calculated using the cell abundance at the day of sampling.

CO_2 fixation rate, N_2 fixation rate, C and N content

CO_2 fixation rates were determined using the ^{13}C-tracer addition method [43]. Seven hours after the beginning of the light period, subsamples of cultures (from 25 to 500 ml) were incubated during 3.5 h with a small addition of NaH^{13}CO$_3$ (99%, Eurisotop) in order to obtain a final enrichment of about 10 atom% excess. N_2 fixation rates were determined using the ^{15}N$_2$ gas-tracer addition method [44]. Incubations for CO_2 and N_2 fixation were not performed simultaneously as *C. watsonii* perform a nocturnal N_2 fixation in order to avoid the inhibitory effects of oxygen on nitrogenase due to photosynthesis [45]. Briefly, 2 hours after the onset of the dark period, ^{15}N$_2$ gas (98.3%, EURISOTOP) was added to sub-samples of cultures (from 45 to 630 ml) in polycarbonate bottles equipped with septum caps using a gas-tight

Table 1. Influence of dFe concentrations in the cultures on surface:volume ratio, elemental ratios, cellular and volume-normalized N_2 and CO_2 fixation rates of *C. watsonii* WH8501 (numbers in brackets represent standard deviation).

Media dFe (nM)	3.3	5.3	8.3	13.3	23.3	43.3	103.3	403.3
Surface:volume	3.1	ND	ND	2.9	ND	2.9	ND	2.4
(μm^2:μm^3)	(0.2)			(0.2)		(0.1)		(0.3)
C:N ratios	8.3	8.9	8.8	10.2	9.5	8.3	8.0	9.6
(mol:mol)	(0.5)	(0.3)	(0.6)	(0.4)	(0.9)	(0.3)	(0.8)	(0.5)
Chl *a*:C ratios	51.1	41.0	34.1	36.6	54.7	63.6	66.5	58.1
(μmol:mol)	(8.5)	(3.6)	(7.7)	(4.2)	(6.6)	(6.8)	(8.4)	(5.3)
N_2 fixation	7.2	ND	ND	ND	ND	53.2	ND	65.6
(amol N.μm^{-3}.h^{-1})	(1.2)					(7.3)		(22.2)
CO_2 fixation	1.2	ND	ND	3.1	ND	3.5	ND	3.9
(fmol C.μm^{-3}.h^{-1})	(0.2)			(0.6)		(0.5)		(1.3)

ND : No data.

syringe, and bottles were incubated for 3.5 hours. $^{15}N_2$ tracer was added to obtain a final enrichment of the N_2 pool of about 10 atom% excess. After ^{13}C and $^{15}N_2$-incubations, samples were filtered onto pre-combusted 25 mm GF/F filters and filters were stored at $-25°C$. Prior to analysis, filters were dried at $40°C$ for 48 h. Particulate organic carbon (POC) and nitrogen (PON) concentrations as well as ^{13}C- and ^{15}N-enrichments were quantified with a mass spectrometer (Delta plus, ThermoFisher Scientific, Bremen, Germany) coupled with an elemental analyser (Flash EA, ThermoFisher Scientific) via a type III-interface. Standard deviations were 0.009 µM and 0.004 µM for POC and PON, respectively and 0.0002 atom% and 0.0001 atom% for ^{13}C enrichment and ^{15}N enrichment, respectively. N_2 fixation rates were calculated by isotope mass balanced as described by [44]. Cellular C and N contents as well as molar C:N ratios were estimated using the POC and PON determined during the light period and the cell abundance measured at the day of sampling. Relative N_2 fixation was calculated as the rates of N_2 fixation in the different Fe treatments normalized by the mean rate in the Fe-replete treatment. Relative CO_2 fixation rates were determined using the same calculation.

Statistical analysis

After checking homoscedasticity using a Bartlett test, means were compared using a one-way ANOVA and a pairwise-t-test with the Holm method for *p*-value adjustment ($\alpha = 0.05$). In the case of heterogeneity of the variances, the tests were performed on the log-transformed data. The statistical tests, the Monod non-linear regression and derived growth parameters (maximum growth rate and half saturation constant for growth) were calculated using R software.

Results and Discussion

1. The global influence of Fe limitation

The growth rate of *C. watsonii* was highly dependent on dFe concentrations as shown by the 2.6-fold decrease ($p<0.05$) from 0.52 ± 0.03 d^{-1} under Fe-replete condition to 0.20 ± 0.03 d^{-1} for the lowest dFe concentration (Figure 1). The relationship between specific growth rate and dFe concentrations fits a Monod saturation function ($r^2 = 0.92$) with a maximum specific growth rate (μ_{max}) of 0.54 ± 0.01 d^{-1} and a half-saturation constant for growth with respect to dFe ($K_{\mu dFe}$) of 6.95 ± 0.66 nM dFe

(Figure 1). Pictures of the cells grown under Fe repletion (dFe = 403.3 nM) and severe limitation (dFe = 3.3 nM) (Figure 2A, B) illustrated the dramatic 2.2-fold decrease in the cell size with decreasing dFe concentrations, from 8.4 ± 2.6 μm^3 to 3.8 ± 0.7 μm^3 (Figure 2C). The decrease in biovolume led to a significant increase in the surface to volume (S:V) ratio with Fe stress from 2.4 ± 0.3 μm^{-1} (dFe = 403.3 nM) to 3.1 ± 0.2 μm^{-1} (dFe = 3.3 nM) (Table 1). The mean cellular C and N contents in Fe-replete cultures (dFe = 403.3 nM) were 547 ± 25 fmolC.cell^{-1} and 57 ± 5 fmolN.cell^{-1}, respectively (Figure 3A, B), resulting in a molar C:N ratio of 9.6 ± 0.5 (Table 1). Reducing dFe concentration to 3.3 nM induced a 3.8- and 3.3-fold decreases ($p<0.05$) in the cellular C and N contents, respectively. In all the cultures, C:N was higher than the Redfield ratio (106:16) and there was no correlation between the C:N ratio and dFe concentrations (Table 1). The cellular Chl *a* content strongly declined (4.1-fold, $p<0.05$) from 28 ± 3 fgChla.cell^{-1} to 6.7 ± 1.5 fgChla.cell^{-1} over the whole range of dFe concentrations (Figure 3C) and there was no clear correlation between dFe concentrations and the Chla:C ratio (Table 1). Volume-normalized (V-normalized) C, N and Chl *a* contents decreased significantly between the 2 extreme dFe concentrations by ~1.8-fold ($p<0.05$; Figure 4). Over the range of dFe concentrations, cellular N_2 fixation rates declined by ~17-fold ($p<0.05$; Figure 5A) whereas cellular CO_2 fixation rates decreased by ~7-fold from 29.8 ± 2.1 fmolC.cell^{-1}.h^{-1} to 4.4 ± 0.4 fmolC. cell^{-1}.h^{-1} ($p<0.05$, Figure 5B). The decrease in V-normalized N_2 fixation rates between the two extreme dFe concentrations was much higher (9.1-fold) than that of the CO_2 fixation rates (3.3-fold) (Figure 5, Table 1).

Under Fe-replete conditions, the growth rate, biovolume, cellular N content and C:N ratio of *C. watsonii* reported here were in the range of published data for the WH8501 strain, and Chla:C ratio was lower than the previously published one (Table 2). Cellular C content was higher than those reported in previous studies. CO_2 fixation rates were higher than those obtained by [50] in cultures having a considerably lower growth rate (Table 2). The growth rate obtained under Fe-replete conditions was close to that determined recently in the oligotrophic South Pacific for *C. watsonii* (0.61 d^{-1}, [33]). The N_2 fixation rates we reported are probably underestimated due to the use of the gas bubble enrichment method. Recently, it has been shown that this method may underestimate N_2 fixation rates relative to the enriched $^{15}N_2$ seawater method due to incomplete $^{15}N_2$ gas bubble equilibration

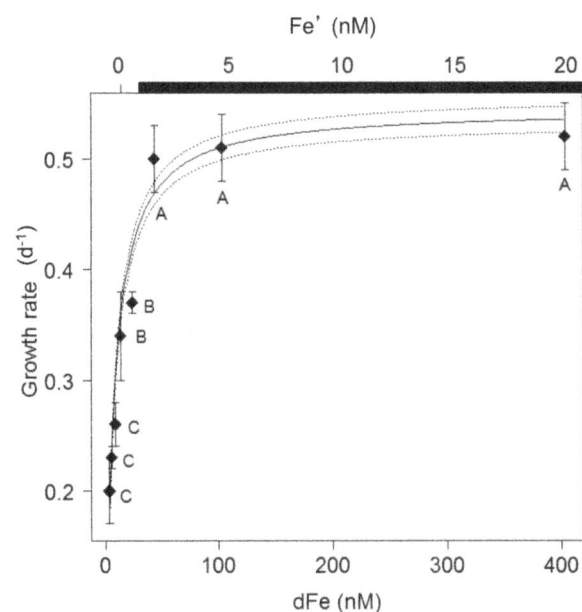

Figure 1. Growth rate of *C. watsonii* related to dFe and Fe′ concentrations. Error bars represent standard deviation; different letters correspond to statistically different means ($p<0.05$) and the black bar indicates the region of expected Fe hydroxide precipitation. The Monod regression, performed with dFe concentrations, is represented by the black line and standard deviation of regression by dotted lines.

[51]. Based on data from these authors, the N_2 fixation rates measured during a short incubation of 3.5 h could be underestimated at least by 70%. However, despite this potential underestimation, the relative N_2 fixation rates should not have been affected.

2. Influence of the degree of Fe-limitation

2.1 Toward a moderate Fe deprivation (from Fe-replete condition to 43.3 nM dFe). Our results showed two distinct responses of *C. watsonii* depending on the degree of Fe limitation (Figure S1). Under a moderate Fe limitation, corresponding to a diminution of dFe concentrations by an order of magnitude, significant decreases in cellular contents (C, N and Chl *a*) and cellular N_2 and CO_2 fixation rates were observed. These decreases were associated with a ~2-fold reduction in biovolume ($p<0.05$), while the growth rate remained unchanged ($p>0.05$, Figure S1). The V-normalized C, N and Chl *a* contents as V-normalized N_2 and CO_2 fixation rates did not significantly change over this range of dFe concentrations ($p>0.05$, Figures 4 and 5). The decrease in the cellular contents and N_2 and CO_2 fixation rates can be attributed to the cell size reduction. A moderate Fe limitation induced a reduction of the cell volume which permitted to *C. watsonii* to maintain maximum C, N and Chl *a* contents as N_2 and CO_2-fixing activities, and hence to keep sufficient energy to sustain optimal growth rates. We suggest that the cell volume reduction of *C. watsonii* represents an adaptive strategy to decreasing Fe availability allowing to a decrease in Fe requirement and to an increase in the S:V ratio (Table 1) which provide an advantage for the Fe uptake by increasing the diffusion-limited uptake rate relative to cell demand [39]. Increase in S:V ratio as an adaptation to Fe limitation was previously evidenced for eukaryotic phytoplankton such as coccolithophores (*Emiliania huxleyi*) [52] and some

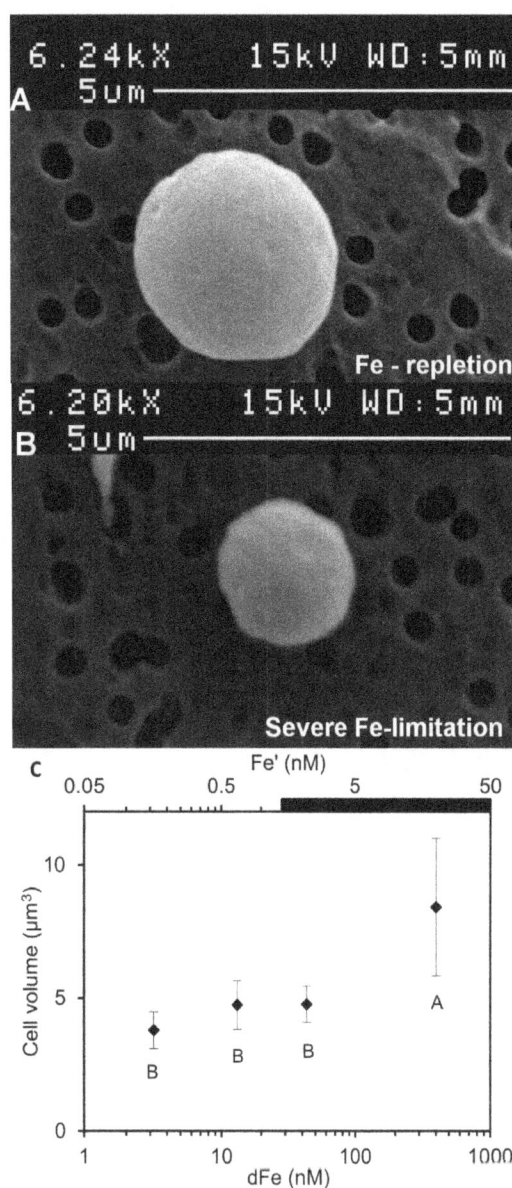

Figure 2. Influence of Fe availability on *C. watsonii* biovolume. Scanning electron microscopy photography of *C. watsonii* growing in (A) Fe-replete condition (dFe = 403.3 nM) and (B) severe Fe-limited condition (dFe = 3.3 nM). (C) Mean biovolume of *C. watsonii* related to dFe and Fe′ concentrations, in log scale. Error bars represent standard deviation; different letters correspond to statistically different means ($p<0.05$) and the black bar indicates the region of expected Fe hydroxide precipitation.

diatoms (*Thalassiosira weissflogii*, *Thalassiosira oceanica* [52] and *Chaetoceros dichaeta* [53]).

Under moderate Fe-limitation conditions, estimated Fe′ concentrations exceeded the solubility limit of Fe with respect to hydroxide precipitation and thus they were not expected to vary despite the reduction of dFe concentrations. Consequently, our observations of significant physiological changes under such conditions suggest that Fe′ was probably not the only available form of Fe for *C. watsonii*. First, we can suspect that Fe from colloidal and/or precipitated amorphous Fe hydroxides is bioavailable. Recently [54] observed that both natural and cultured *Trichodesmium* are able to take up Fe from ferrihydrite

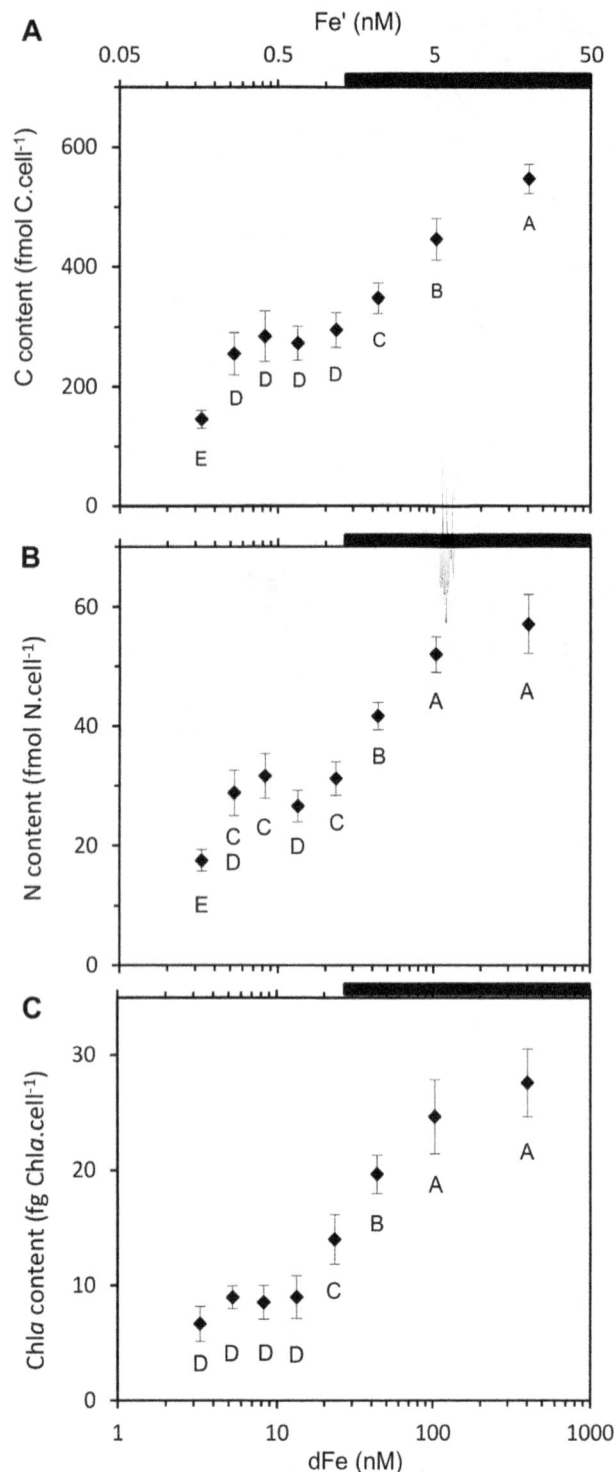

Figure 3. Influence of Fe availability on the elemental composition of *C. watsonii*. Mean cellular content of C (A), N (B) and Chl *a* (C) related to dFe and Fe' concentrations, in log scale. Error bars represent standard deviation; different letters correspond to statistically different means (*p*<0.05) and the black bar indicates the region of expected Fe hydroxide precipitation.

(an amorphous oxidized Fe hydroxide) via cell surface adsorption and biological mediated dissolution. [55] have shown that *Trichodesmium* and the non-diazotrophic unicellular cyanobacteri-

um *Synechoccocus* were able to take up Fe bound to recently formed organic colloids, probably involving biological reduction of colloidal Fe, leading to highly soluble Fe(II) forms [40]. Additionally, *C. watsonii* could acquire Fe from bioreduction of the Fe-EDTA complex into Fe(II) at the cell surface, as the FeEDTA complex could not be transported across the cell membrane [56]. This acquisition strategy has already been shown for the diatom *T. weissflogii*, although reduction rates are widely lower than for Fe' [56]. Based on our results, *C. watsonii* could be able to acquire Fe from other forms than Fe' but its Fe uptake mechanisms to be characterized.

2.2 Toward a more severe Fe deprivation (from 43.3 nM to 3.3 nM dFe). When intensifying Fe deprivation, cellular contents (C, N and Chl *a*) and cellular N_2 and CO_2 fixation rates continued to decrease (Figure 3, 5, S1 A to E). Over this range of Fe concentrations, the growth rate dropped significantly (p<0.05, Figure 1) while the cell volume remained unchanged (Figure 2C, S1F) as well as the S:V ratio (p>0.05, Table 1). This indicates that the cells have reached a minimum volume, and thus a maximum S:V ratio, around 43.3 nM dFe. As a consequence, an ~2-fold decrease in V-normalized contents (C, N and Chl *a*) and a significant drop in V-normalized N_2 and CO_2 fixation rates were observed (Figure 4 and 5). Thus, under severe Fe limitation, the cellular composition and the efficiency of N_2 and CO_2 uptake were strongly affected and the cells were not able to produce sufficient energy to maintain an optimal growth.

In photoautotrophic cells, energy is provided as adenosine triphosphate (ATP) during respiration through the catabolism of carbohydrates produced during photosynthesis. Since a majority of redox metalloenzymes involved in this process are Fe rich proteins [30,57], a severe reduction in bioavailable Fe induced a lower efficiency in the photosynthetic activity, as depicted by the significant decrease of V-normalized CO_2 fixation rates and Chl *a* contents as it is the main light harvesting pigment involved in photosynthesis. Our results showed that V-normalized N_2 fixation rates were more affected by severe Fe limitation than CO_2 fixation rates, as illustrated by ~7.5 and ~3-fold decreases respectively from 43.3 to 3.3 nM dFe. The high Fe content of nitrogenase co-factors associated with the high energetic cost of biological N_2 fixation could explain this pattern. Indeed, recent studies reported that Fe deprivation leads to a down-regulation of nitrogenase expression in both cultured and *in situ Trichodesmium* [58,59]. This could also occur for *C. watsonii*, but it has been not yet evidenced. Additionally, N_2 fixation is the highest energy consuming process in the cell [48,49]. This process is fuelled by catabolism of carbohydrates accumulated during photosynthesis [60]. Thus, decreasing Fe bioavailability also affects the N_2 fixation rates through the photosynthetic deficiency.

3. Comparison of the response of C. watsonii to Fe limitation with other phytoplanktonic species

To date, only two species of UCYN are available in culture: one isolated from the open ocean (*C. watsonii*, UCYN-B [20]) and one from coastal waters (*Cyanothece* WH8904, UCYN-C [61]). A study from [62] has shown no influence of Fe limitation on N_2 fixation rates of *Cyanothece* WH8904 under a wide range of dFe concentrations (from 4 nM to 4 μM complexed with 20 μM EDTA). While the cell diameter of *C. watsonii* (2.5 μm, this study) and *Cyanothece* (~3 μm, determined from [62]) are close under Fe-repletion, differences in Fe requirements and/or Fe acquisition between both species can be strongly suspected. The uncultivated photoheterotrophic UCYN-A do not have photosystem II of the photosynthetic apparatus [63] which contains three Fe atoms [64], and are smaller (diameter<1 μm, [23]) than *C. watsonii*, suggesting

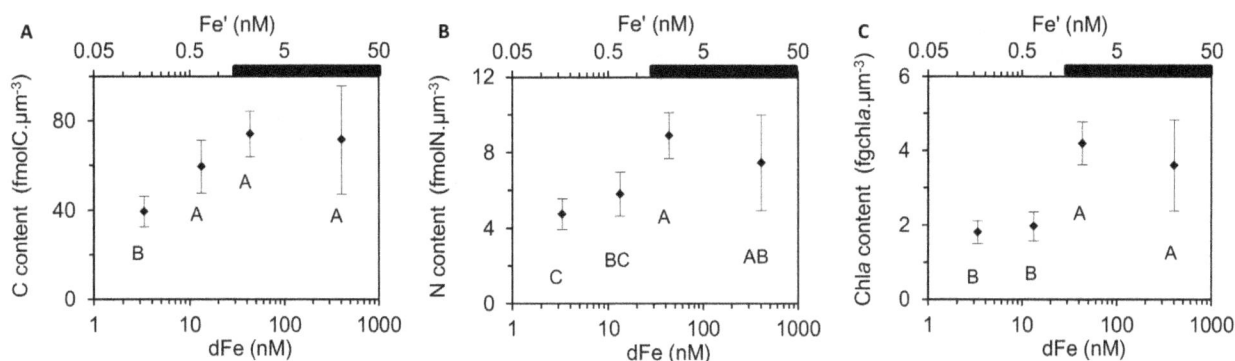

Figure 4. Influence of Fe availability on V-normalized contents. Mean V-normalized contents of C (A), N (B) and Chl a (C) of *C. watsonii* related to dFe and Fe' concentrations, in log scale. Error bars represent standard deviation; different letters correspond to statistically different means ($p<0.05$) and the black bar indicates the region of expected Fe hydroxide precipitation.

that the Fe requirements of UCYN-A are likely lower than those of *C. watsonii*. As a consequence, large differences in Fe requirements and/or Fe acquisition could exist within the UCYN community (UCYN-A, -B, -C). Bioassay experiments in the tropical North Atlantic have shown contrasted responses of UCYN activity to Fe additions. Fe addition stimulated the expression of the *nifH* gene (which encodes for the Fe component of the nitrogenase) from UCYN-B only in the western part, despite detectable dFe concentrations, while *nifH* expression from UCYN-A was not stimulated either in the Western or central part [28]. In the Eastern part, [27] observed a mesoscale variability with either UCYN-A or UCYN-B abundance stimulated by Fe addition for two close sites.

The $K_{\mu dFe}$ of *C. watsonii* was twice as low as that of *Trichodesmium erythraeum* IMS101 growing in the same conditions (13.9±3.3 nM calculated from [16, Ridame and Rochelle-Newall, unpublished data]), indicating that the growth of *C. watsonii* is less impacted by Fe limitation than that of the filamentous *Trichodesmium*. *T. erythraeum* display a much higher biovolume ($\sim$14855 μm^3 under Fe-replete conditions, [16]) than *C. watsonii* (8.4 μm^3), implying a S:V ratio of *T. erythraeum* lower (S:V$\sim$0.55 μm^{-1}, Ridame unpublished data; [62]) than that of *C. watsonii* (S:V = 2.4 μm^{-1}).

The small biovolume and large S:V ratio provide to *C. watsonii* an advantage for Fe and other nutrients uptake. Furthermore, *C. watsonii* perform a nocturnal N_2 fixation [45,46] with a daily synthesis and degradation of Fe-containing proteins involved in photosynthesis and N_2 fixation, in coordination with their utilization [30]. This Fe recycling throughout the diel cycle leads to a reduction in the cellular Fe requirement of *C. watsonii* up to 40% [30]. As *T. erythraeum* perform both photosynthesis and N_2 fixation during the photoperiod, it probably does not employ this Fe conservation strategy to the extent used by *C. watsonii*. The Fe-rich ferredoxin, constitutively used in photosynthetic electron transport, is also an efficient electron donor for nitrogenase of *Trichodesmium* [65]. Under Fe limitation, extra Fe-free flavodoxin could be synthetized to act as an alternative electron donor instead of ferredoxin [66], as previously reported for *T. erythraeum* [67]. In contrast, no increase in flavodoxin in response to Fe stress was observed for *C. watsonii* [30]. The use of flavodoxin at night during N_2 fixation even under Fe replete conditions appears to be an adaptation that allows *C. watsonii* to reduce cellular Fe demand [30]. These physiological characteristics reveal a lower Fe requirements of *C. watsonii* than the filamentous *T. erythraeum*, which is consistent with a higher cellular Fe:C in *T. erythraeum*

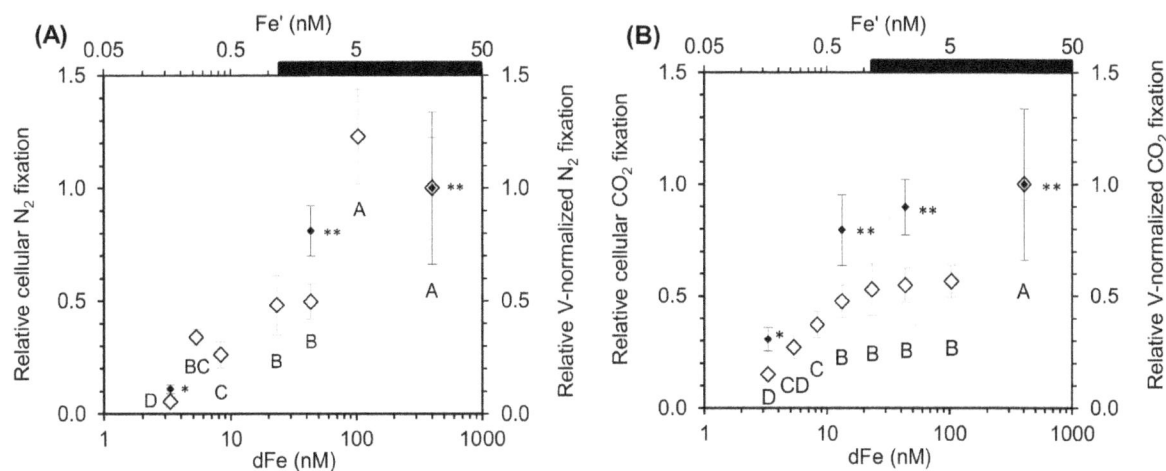

Figure 5. Influence of Fe availability on N_2 and CO_2 fixation rates of *C. watsonii*. (A) Relative N_2 fixation rates and (B) relative CO_2 fixation rates of *C. watsonii* related to dFe and Fe' concentrations, in log scale. Open and filled symbols correspond to the rates per cell and V-normalized rates respectively. Error bars represent standard deviation; different letters and different numbers of star correspond to statistically different means ($p<0.05$) for cellular and V-normalized rates, respectively. The black bar indicates the region of expected Fe hydroxide precipitation.

Table 2. Comparison of growth rate, biovolume, cellular contents, elemental ratio and CO_2 fixation rate of *C. watsonii* WH8501 cultivated under Fe-replete conditions (numbers in brackets represent standard deviation).

Growth rate	Biovolume	C content	N content	C:N	Chl a:C	CO_2 fixation rates	
d^{-1}	μm^3	$fmolC.cell^{-1}$	$fmolN.cell^{-1}$	mol:mol	$\mu mol.mol$	$fmol^1C.cell^{-1}.h^{-1}$	Ref.
0.46			6.9–29.6	8.8 (1.5)			[46]
0.47 (0.01)	4.2–65.4						[23]
				6.9 (0.2)	83 (12)		[31]
0.54	4.2–33.5						[25]
	12–13.6			8.5*			[45]
		500	80	5.2			[47]
0.2	8.2–10.4	140–220	18–40	8.8*			[48]
0.28 (0.02)		120–260	20–35	10.5*			[49]
0.14						~9	[50]
0.52 (0.03)	**8.4 (2.6)***	**547 (25)***	**57 (5)***	**9.6 (0.5)***	**58 (5)***	**29.8 (2.1)**	**This study**

*during light period.

(from 69 to 87 $\mu mol:mol^{-1}$ under Fe replete conditions [14,46]) compared to *C. watsonii* (16 $\mu mol:mol^{-1}$, [46]). Consequently, *C. watsonii* is likely better adapted to the poor Fe natural oceanic waters than *T. erythraeum*.

4. Oceanic relevance and biogeography of N_2 fixers

We quantified for the first time the impact of Fe bioavailability on the growth, cell size, N_2 fixation and photosynthesis of an open ocean UCYN, and demonstrated a physiological response depending on the degree of limitation. As photosynthesis (CO_2 fixation) provides energy for fuelling N_2 fixation, these two processes are tightly related. Thus, the cell response should be globally considered because nutrient limitation, such as Fe deprivation, affects cell metabolism and involves intractable feedbacks.

The quantification of the impact of Fe availability on the N_2 fixation rates of *C. watsonii* contributes to our knowledge about the control of Fe on the N cycle in the tropical and subtropical ocean. As both growth and N_2- and CO_2-fixing activities of *C. watsonii* are highly Fe-dependent, its abundance and activity could be controlled by the atmospheric deposition of aeolian dust which represents the major source of new Fe to the open ocean surface waters [68]. Hence, in the tropical and subtropical oligotrophic ocean, atmospheric Fe input could enhance new production and C export to the deep ocean, through the stimulation of the growth and activity of *C. watsonii*. Based on our findings, the oceanic UCYN like *C. watsonii* could be strongly Fe limited but at a lesser extent that the filamentous diazotrophic cyanobacteria *Trichodesmium*. Thus Fe bioavailability could control the biogeography of these two N_2 fixers. Indeed, in the South-Western Pacific [69] have observed that putative *Crocosphaera* cells were dominant at oceanic stations, while *Trichodesmium* dominated in the more Fe-rich coastal stations. They attributed this relative distribution to Fe availability since Fe is mainly supplied by coastal input in this region. Due to the lack of data about UCYN, parameterizations of N_2 fixation in regional and global ocean models are mostly based on characteristics of *Trichodesmium* sp. (e.g. [70,71]). Hence such biogeochemical models could be improved by the addition of a simulated UCYN, as they are expected to growth in larger niches than *Trichodesmium* regarding Fe availability. Moreover, our study supports the idea that within the UCYN community (UCYN-A,

Crocosphaera, *Cyanothece*) Fe requirements and/or Fe acquisition may strongly vary, meaning that Fe bioavailability could partially explain variabilities in the UCYN community composition. Recent biogeochemical models focused on the biogeography of the N_2 fixers [22,72,73] consider, in addition to a *Trichodesmium* analogue, a UCYN analogue parameterized with growth parameters and cellular contents derived from *Trichodesmium* and non-diazotrophic picophytoplanktonic species. Within such UCYN pool, it could be relevant to consider different types of UCYN (UCYN-A, -B and –C). In this context, the parameters obtained in our study will help to improve the parameterization of N_2 fixation and UCYN distribution in biogeochemical models.

Supporting Information

Figure S1   Two distinct physiological responses of *C. watsonii* to Fe limitation. Cellular contents of C (A), N (B), Chl a (C), cellular N_2 (D) and CO_2 fixation rates (E) and growth rates (F) related to cell volume for 4 dFe concentrations (dFe = 3.3, 13.3, 43.3 and 403.3 nM). Error bars represent standard deviation. Different numbers of stars and different letters correspond to statistically different means for the cell volume and parameters listed above.

Acknowledgments

We would like to thank Manon Tonnard and Benjamin Ledoux for helping with some data acquisition. We also thank Matthieu Bressac and Cécile Guieu (LOV, Villefranche sur Mer, France) for dFe measurements by FIA and Philippe Catala (ARAGO, OOB, Banyuls sur Mer, France) for analyses by flow cytometry. Michaël Trichet (Scanning Electron Microscopy service, FRE3595, UPMC, Paris, France) is acknowledged for the preparation of MEB samples and his help during MEB observation sessions. We are thankful to Hervé Rybarczyk (MNHN, Paris, France) for sharing his helpful knowledge on statistics and Bill Sunda (NOAA, US) for its precious help in the Fe' calculation. Tamara Barriquand and Fauzi Mantoura are acknowledged for their careful proofreading. We are grateful to Sophie Bonnet (MIO, Marseille, France) for providing *Crocosphaera watsonii* WH8501 strain. We would like to thank the two anonymous reviewers for all their relevant comments which allow to improve and precise our works.

Author Contributions

Conceived and designed the experiments: VJ CR. Performed the experiments: VJ CR SL FK. Analyzed the data: VJ CR. Contributed

reagents/materials/analysis tools: VJ CR SL FK. Wrote the paper: VJ CR SL AS.

References

1. Capone DG (2000) Marine nitrogen cycle. In : Kirchman L, editor. Microbial ecology of the ocean. Wiley, New-York.
2. Karl DM, Letelier RM (2008) Nitrogen fixation enhanced carbon sequestration in low nitrate, low chlorophyll seascapes. Mar Ecol Prog Ser 364: 257–268. doi:10.3354/meps07547.
3. Gruber N (2008) The marine nitrogen cycle : overview and challenges. In: Capone DG, Bronk DA, Mulholland MR, Carpenter EJ, editors. Nitrogen in the marine environment (second edition). Elsevier.
4. Montoya JP, Holl CM, Zehr JP (2004) High rates of N2 fixation by unicellular diazotrophs in the oligotrophic Pacific Ocean. Nature 430: 1027–1031. doi:10.1038/nature02744.1.
5. Karl D, Letelier R, Tupas L, Dore J, Christian J, et al. (1997) The role of nitrogen fixation in biogeochemical cycling in the subtropical North Pacific Ocean. Nature 388: 533–538.
6. Gruber N, Sarmiento JL (1997) Global patterns of marine nitrogen fixation and denitrification. Global Biogeochem Cy 11: 235–266.
7. Howard JB, Rees DC (1996) Structural basis of biological nitrogen fixation. Chem rev 96: 2965–2982.
8. Raven JA (1988) The iron and molybdenum use efficiencies of plant growth with different energy, carbon and nitrogen sources. New Phytol 109: 279–287.
9. Kustka A, Sañudo-wilhelmy S, Carpenter EJ, Capone DG, Raven JA (2003) A revised estimate of the iron use efficiency of nitrogen fixation, with special reference to the marine cynobacterium Trichodesmium spp. (cyanophyta). J phycol 39: 12–25.
10. Johnson KS, Gordon RM, Coale KH (1997) What controls dissolved iron concentrations in the world ocean? Mar Chem 57: 137–161.
11. Sarthou G, Baker AR, Blain S, Achterberg EP, Boye M, et al. (2003) Atmospheric iron deposition and sea-surface dissolved iron concentrations in the Eastern Atlantic Ocean. Deep Sea Research I 50: 1339–1352. doi:10.1016/S0967-0637(03)00126-2.
12. Boyd PW, Ellwood MJ (2010) The biogeochemical cycle of iron in the ocean. Nat Geosci 3: 675–682. doi:10.1038/ngeo964.
13. Toulza E, Tagliabue A, Blain S, Piganeau G (2012) Analysis of the global ocean sampling (GOS) project for trends in iron uptake by surface ocean microbes. PLoS One 7: e30931. doi:10.1371/journal.pone.0030931
14. Berman-Frank AI, Cullen JT, Shaked Y, Sherrell RM, Falkowski PG (2001) Iron availability, cellular iron quotas, and nitrogen fixation in Trichodesmium. Limnol Oceanogr 46: 1249–1260.
15. Kustka AB, San SA, Carpenter EJ, Sunda WG (2003b) Iron requirements for dinitrogen- and ammonium-supported growth in cultures of Trichodesmium (IMS 101): Comparison with nitrogen fixation rates and iron : carbon ratios of field populations. Limnol Oceanogr 48: 1869–1884.
16. Bucciarelli E, Ridame C, Sunda WG, Dimier-Hugueney C, Cheize M, et al. (2013) Increased intracellular concentration of DMSP and DMSO in iron-limited oceanic phytoplankton Thalassiosira oceanica and Trichodesmium erythraeum. Limnol and Oceanogr 58: 1667–1679.
17. Moore MC, Mills MM, Achterberg EP, Geider RJ, LaRoche J, et al. (2009) Large-scale distribution of Atlantic nitrogen fixation controlled by iron availability. Nat Geosci 2: 867–871. doi:10.1038/ngeo667.
18. Capone DG, Carpenter E. (1982) Nitrogen fixation in the marine environment. Science 217: 1140–1142.
19. Capone DG, Zehr JP, Paerl HW, Bergman B, Carptenter JE (1997) Trichodesmium, a globally significant marine cyanobacterium. Science 276: 1221–1229. doi/10.1126/science.276.5316.1221.
20. Zehr JP, Waterbury JB, Turner PJ, Montoya JP, Omoregie E, et al. (2001) Unicellular cyanobacteria fix N2 in the subtropical North Pacific Ocean. Nature 412: 635–638.
21. Bonnet S, Biegala IC, Dutrieux P, Slemons LO, Capone DG (2009) Nitrogen fixation in the Western equatorial Pacific: rates, diazotrophic cyanobacterial size class distribution, and biogeochemical significance. Global Biogeochem Cy 23: doi:10.1029/2008GB003439.
22. Monteiro FM, Follows MJ, Dutkiewicz S (2010) Distribution of diverse nitrogen fixers in the global ocean. Global Biogeochem Cy 24: doi:10.1029/2009GB003731.
23. Goebel NL, Edwards CA, Carter BJ, Achilles KM, Zehr JP (2008) Growth and carbon content of three different-sized diazotrophic cyanobacteria observed in the subtropical North Pacific. J Phycol 44: 1212–1220.
24. Moisander PH, Beinart RA, Hewson I, White AE, Johnson KS, et al. (2010) Unicellular cyanobacterial distributions broaden the oceanic N2 fixation domain. Science 327: 1512–1514.
25. Webb EA, Ehrenreich IM, Brown SL, Valois FW, Waterbury JB (2009) Phenotypic and genotypic characterization of multiple strains of the diazotrophic cyanobacterium, Crocosphaera watsonii, isolated from the open ocean. Environ microbio 11: 338–348.
26. Dyhrman ST, Haley ST (2006) Phosphorus scavenging in the unicellular marine diazotroph Crocosphaera watsonii. Appl environ microb 72: 1452–1458. doi:10.1128/AEM.72.2.1452.
27. Langlois RJ, Mills MM, Ridame C, Croot P, LaRoche J (2012) Diazotrophic bacteria respond to Saharan dust additions. Mar Ecol Prog Ser 470: doi: 10.3354/meps10109.
28. Turk-Kubo KA, Achilles KM, Serros TRC, Ochiai M, Montoya JP, et al. (2012) Nitrogenase (nifH) gene expression in diazotrophic cyanobacteria in the Tropical North Atlantic in response to nutrient amendments. Front microbiol 3: 386. doi: 10.3389/fmicb.2012.00386.
29. Garcia NS, Fu F, Hutchins DA (2013) Colimitation of the unicellular photosynthetic diazotroph Crocosphaera watsonii by phosphorus, light and carbon dioxide. Limnol Oceanogr 58: 1501–1512. doi:10.4319/lo.2013.58.4.1501.
30. Saito MA, Bertrand EM, Dutkiewicz S, Bulygin VV, Moran DM, et al. (2011) Iron conservation by reduction of metalloenzyme inventories in the marine diazotroph Crocosphaera watsonii. pnas 108: doi:10.1073/pnas.1006943108.
31. Webb EA, Moffett JW, Waterbury JB (2001) Iron stress in open-ocean cyanobacteria identification of the IdiA protein. Appl environ microb 67: 5444–5452. doi:10.1128/AEM.67.12.5444.
32. Fu F, Mulholland MR, Garcia NS, Beck A, Bernhardt PW, et al. (2008) Interactions between changing pCO2, N2 fixation, and Fe limitation in the marine unicellular cyanobacterium Crocosphaera. Limnol Oceanogr 53: 2472–2484.
33. Moisander PH, Zhang R, Boyle EA, Hewson I, Montoya JP, et al. (2012) Analogous nutrient limitations in unicellular diazotrophs and Prochlorococcus in the South Pacific Ocean. The ISME journal 6: 733–744.
34. Chen YB, Zehr JP, Mellon M (1996) Growth and nitrogen fixation of the diazotrophic filamentous nonheterocystous cyanobacterium Trichodesmium sp. IMS 101 in defined media: evidence for a circadian rhythm. J phycol 32: 916–923.
35. Obata H, Karatani H, Nakayama E (1993) Automated determination of iron in seawater by chelating resin concentration and chemiluminescence detection. Anal chim acta 65: 1524–1528.
36. Troussellier M, Courties C, Lebaron P, Servais P (1999) Flow cytometric discrimination of bacterial populations in seawater based on SYTO 13 staining of nucleic acids. FEMS microbiol Ecol 29: 319–330.
37. Gundersen K, Heldal M, Norland S, Purdie DA, Knap AH (2002) Elemental C, N, and P cell content of individual bacteria collected at the Bermuda Atlantic Time-series Study (BATS) site. Limnol Oceanogr 47: 1525–1530.
38. Sunda WG, Huntsman SA (2003) Effect of pH, light, and temperature on Fe-EDTA chelation and Fe hydrolysis in seawater. Mar Chem 84: 35–47.
39. Hudson RJM, Morel FMM (1990) Iron transport in marine phytoplankton : kinetics of cellular and medium coordination reactions. Limnol Oceanogr 35: 1002–1020.
40. Sunda WG (2001) Bioavailability and Bioaccumulation of Iron in the Sea. In: Turner DR, Hunter KA, editors. The Biogeochemistry of Iron in Seawater. Wiley. pp. 41–84
41. Schneider CA, Rasband WS, Eliceiri KW (2012) NIH Image to ImageJ: 25 years of image analysis. Nat Methods 9: 671–67.
42. Strickland JDH, Parsons TR (1997) A practical handbook of seawater analysis. Bull Fish Res BD Can.
43. Hama T, Miyazaki T, Ogawa Y, Iwakuma T, Takahashi M, et al. (1983) Measurement of photosynthetic production of a marine phytopklankton population using a stable 13C isotope. Mar biol 73: 31–36.
44. Montoya JP, Voss M, Kahler P, Capone DG (1996) A simple, high-precision, high-sensitivity tracer assay for N2 fixation. Appl environ microb 62: 986–993.
45. Mohr W, Intermaggio MP, LaRoche J (2010) Diel rhythm of nitrogen and carbon metabolism in the unicellular, diazotrophic cyanobacterium Crocosphaera watsonii WH8501. Environ microbiol 12: 412–421.
46. Tuit C, Waterbury J, Ravizza G (2004) Diel variation of molybdenum and iron in marine diazotrophic cyanobacteria. Limnol and Oceanogr 49: 978–990.
47. Dekaezemacker J, Bonnet S (2011) Sensitivity of N2 fixation to combined nitrogen forms (NO3− and NH4+) in two strains of the marine diazotroph Crocosphaera watsonii (Cyanobacteria). Mar Ecol Prog Ser 438: 33–46. doi:10.3354/meps09297.
48. Dron A, Rabouille S, Claquin P, Le Roy B, Talec A, et al. (2011) Light-dark (12:12) cycle of carbon and nitrogen metabolism in Crocosphaera watsonii WH8501: relation to the cell cycle. Environ microbiol 14: 967–981.
49. Großkopf T, Laroche J (2012) Direct and indirect costs of dinitrogen fixation in Crocosphaera watsonii WH8501 and possible implications for the nitrogen cycle. Front microbial 3. doi: 10.3389/fmicb.2012.00236.
50. Mohr W, Vagner T, Kuypers MMM, Ackermann M, Laroche J (2013). Resolution of Conflicting Signals at the Single-Cell Level in the Regulation of Cyanobacterial Photosynthesis and Nitrogen Fixation. PLOS one, 8(6), e66060. doi:10.1371/journal.pone.0066060.

51. Mohr W, Grosskopf T, Wallace DWR, La Roche J (2010b) Methodological underestimation of oceanic nitrogen fixation rates. PLOS one 5(9) : e12583. Doi : 10.1371/journal.pone.0012583.

52. Sunda WG, Huntsman SA. (1995) Iron uptake and growth limitation in oceanic and coastal phytoplankton. Mar Chem 50: 189–206. doi:10.1016/0304-4203(95)00035-P.

53. Takeda S (1998) Influence of iron availability on nutrient consumption ratio of diatoms in oceanic waters. Nature 393: 774–777.

54. Rubin M, Berman-frank I, Shaked Y (2011) Dust- and mineral-iron utilization by the marine dinitrogen-fixer Trichodesmium. Nat Geosci 4: 529–534. doi:10.1038/NGEO1181.

55. Wang W, Dei RCH (2003) Bioavailability of iron complexed with organic colloids to the cyanobacteria Synechococcus and Trichodesmium. Aquat Microb Ecol 33: 247–259. DOI: 10.3354/ame033247.

56. Shaked Y, Kustka AB, Morel FMM (2005) A general kinetic model for iron acquisition by eukaryotic phytoplankton. Limnol and Oceanogr 50: 872–882.

57. Behrenfeld MJ, Milligan AJ (2013) Photophysiological expressions of iron stress in phytoplankton. Ann rev mar sci 5: 217–246. 1.1146/annurev-marine-121211-172356.

58. Küpper H, Setlík I, Seibert S, Prásil O, Setlikova E, et al. (2008) Iron limitation in the marine cyanobacterium Trichodesmium reveals new insights into regulation of photosynthesis and nitrogen fixation. The New phytol 179: 784–798.

59. Richier S, Macey AI, Pratt NJ, Honey DJ, Moore CM, et al. (2012) Abundances of iron-binding photosynthetic and nitrogen-fixing proteins of Trichodesmium both in culture and in situ from the North Atlantic. PLOS one 7: e35571. doi:10.1371/journal.pone.0035571.

60. Dron A, Rabouille S, Claquin P, Chang P, Raimbault V, et al. (2012) Light : dark (12:12 h) quantification of carbohydrate fluxes in Crocosphaera watsonii. Aquat Microb Ecol 68: 43–55. doi:10.3354/ame01600.

61. Ehrenreich IM, Waterbury JB, Webb EA (2005) Distribution and diversity of natural product genes in marine and freshwater cyanobacterial cultures and genomes. App environ microb 71: 7401–7413. doi:10.1128/AEM.71.11.7401.

62. Berman-Frank IA, Quigg A, Finkel ZV, Irwin AJ, Haramaty L (2007) Nitrogen-fixation strategies and Fe requirements in cyanobacteria. Limnol and Oceanogr 52: 2260–2269.

63. Zehr JP, Bench SR, Mondragon EA, McCarren J, DeLong EF (2007) Low genomic diversity in tropical oceanic N2-fixing cyanobacteria. Pnas 104: 17807–17812. doi:10.1073/pnas.0701017104.

64. Raven JA (1990) Predictions of Mn and Fe use efficiencies of phototrophic growth as a function of light availability for growth and of C assimilation pathway. New Phytol 116: 1–18. doi.wiley.com/10.1111/j.1469-8137.1990.tb00505.x.

65. LaRoche J, Breitbarth E (2005) Importance of the diazotrophs as a source of new nitrogen in the ocean. J Sea Res 53: 67–91.

66. Ferreira F, Straus NA (1994) Iron deprivation in cyanobacteria. J appl phycol 6: 199–210.

67. Chappell PD, Webb EA (2010) A molecular assessment of the iron stress response in the two phylogenetic clades of Trichodesmium. Environ microbiol 12: doi:10.1111/j.1462-2920.2009.02026.x.

68. Jickells TD, An ZS, Andersen KK, Baker AR, Bergametti G, et al. (2005) Global iron connections between desert dust, ocean biogeochemistry, and climate. Science 308: 67–71.

69. Campbell L, Carpenter E., Montoya JP, Kustka AB, Capone DG (2005) Picoplankton community structure within and outside a Trichodesmium bloom in the Southwestern Pacific Ocean. Vie milieu 55: 185–195.

70. Moore JK (2004) Upper ocean ecosystem dynamics and iron cycling in a global three-dimensional model. Global Biogeochem Cy 18: DOI: 10.1029/2004GB002220.

71. Coles VJ, Hood RR (2007) Modeling the impact of iron and phosphorus limitations on nitrogen fixation in the Atlantic Ocean. Biogeosciences 4: 455–479.

72. Dutkiewicz S, Ward BA, Monteiro F, Follows MJ (2012) Interconnection of nitrogen fixers and iron in the Pacific Ocean: Theory and numerical simulations. Global Biogeochem Cy 26: DOI: 10.1029/2011GB004039.

73. Monteiro FM, Dutkiewicz S, Follows MJ (2011) Biogeographical controls on the marine nitrogen fixers. Global Biogeochemical Cy 25: 1–8. doi:10.1029/2010GB003902.

4

Fluctuations in Species-Level Protein Expression Occur during Element and Nutrient Cycling in the Subsurface

Michael J. Wilkins[1]*, Kelly C. Wrighton[2], Carrie D. Nicora[1], Kenneth H. Williams[3], Lee Ann McCue[1], Kim M. Handley[2], Chris S. Miller[2], Ludovic Giloteaux[4], Alison P. Montgomery[3], Derek R. Lovley[4], Jillian F. Banfield[2], Philip E. Long[3], Mary S. Lipton[1]

1 Biological Sciences Division, Pacific Northwest National Laboratory, Richland, Washington, United States of America, 2 Department of Earth and Planetary Science, University of California, Berkeley, California, United States of America, 3 Earth Sciences Division, Lawrence Berkeley National Laboratory, Berkeley, California, United States of America, 4 Department of Microbiology, University of Massachusetts Amherst, Amherst, Massachusetts, United States of America

Abstract

While microbial activities in environmental systems play a key role in the utilization and cycling of essential elements and compounds, microbial activity and growth frequently fluctuates in response to environmental stimuli and perturbations. To investigate these fluctuations within a saturated aquifer system, we monitored a carbon-stimulated *in situ Geobacter* population while iron reduction was occurring, using 16S rRNA abundances and high-resolution tandem mass spectrometry proteome measurements. Following carbon amendment, 16S rRNA analysis of temporally separated samples revealed the rapid enrichment of *Geobacter*-like environmental strains with strong similarity to *G. bemidjiensis*. Tandem mass spectrometry proteomics measurements suggest high carbon flux through *Geobacter* respiratory pathways, and the synthesis of anapleurotic four carbon compounds from acetyl-CoA via pyruvate ferredoxin oxidoreductase activity. Across a 40-day period where Fe(III) reduction was occurring, fluctuations in protein expression reflected changes in anabolic versus catabolic reactions, with increased levels of biosynthesis occurring soon after acetate arrival in the aquifer. In addition, localized shifts in nutrient limitation were inferred based on expression of nitrogenase enzymes and phosphate uptake proteins. These temporal data offer the first example of differing microbial protein expression associated with changing geochemical conditions in a subsurface environment.

Editor: Karl Rockne, University of Illinois at Chicago, United States of America

Funding: This research was funded by the United States Department of Energy, Office of Science, Environmental Remediation Science Program through the Integrated Field Research Challenge Site at Rifle, Colorado, under contract number DE-AC05-76RL01830 to Pacific Northwest National Laboratory. (http://science.energy.gov/ber/). The funders had no role in study design, data collection and analysis, decision to publish, or preparation of the manuscript.

Competing Interests: The authors have declared that no competing interests exist.

* E-mail: michael.wilkins@pnnl.gov

Introduction

The activities of microbial populations play an important role in the cycling of metals and nutrients in environmental systems, where processes including element uptake, excretion and transformations catalyzed by microorganisms contribute to dynamic fluxes of C, P, N, S, and Fe [1]. Given that these fluxes are closely linked to the physiological state of microbial community members, the interrogation of *in situ* microbial metabolism and activity may offer an opportunity to better understand biogeochemical cycles.

Elucidating microbial metabolic pathways and activity in subsurface environments has traditionally been problematic, with microbial communities located in discrete pore spaces deep underground. Coupled to this, growth rates within complex, low biomass microbial communities are typically slow [2,3], and governed by a range of factors including nutrient availability, limited concentrations of electron donors and acceptors, and other environmental stresses. However, the stimulation of *in situ* microbial activity via carbon amendment allows shifts in global protein and mRNA profiles to be measured under controlled conditions, where the enrichment of specific microbial groups can be predicted and subsequently monitored [4,5].

This approach has been applied at the Rifle Integrated Field Research Challenge (IFRC) site in Western Colorado, where the use of *in situ* carbon amendment experiments over the past decade has allowed both microbiological and geochemical responses to be better predicted [6]. At this site, acetate amendment to the subsurface typically enriches Fe(III)-reducing *Geobacter* spp., both within the planktonic and sediment-associated communities [6]. Understanding the physiology and metabolism of *Geobacter* spp. in environmental systems is important for predicting biogeochemical processes and bioremediation efforts in the subsurface; these species and strains can couple the oxidation of organic carbon to the reduction of a range of metals including Fe, U, V, and Se [7,8,9,10]. By accelerating the rate of these processes via carbon amendment, biogeochemical interactions can be investigated that would be extremely challenging to measure under background rates and conditions. These data then offer the potential to link metabolic and physiological inferences to geochemical measurements, and obtain a greater predictive understanding of subsurface processes.

At the Rifle IFRC site, acetate amendment to the subsurface stimulates Fe(III) reduction for approximately 30 days, during which *Geobacter* populations are thought to catalyze changes in Fe and U biogeochemistry [11]. Following this period, the development of sulfate-reducing conditions in the aquifer is linked to decreasing abundances and activity of planktonic *Geobacter* [11,12]. To date, a number of approaches have been used to interrogate the physiology and ecology of *Geobacter* in the subsurface, including quantification of specific genes associated with N, P and acetate limitation [13,14,15]. During a previous carbon injection field experiment, shotgun proteomic analysis was used to investigate the whole expressed proteome of the stimulated planktonic *Geobacter* population [5]. This technique measures all the expressed proteins within a sample, and can be used to infer activity of specific microbial species. From this study, significant carbon flux through respiratory pathways of *Geobacter* species was inferred, and temporal strain-level shifts within the population were identified. However, due to the small number of proteomic samples recovered, temporal changes in the metabolism and physiology of the *Geobacter* population could not be accurately assessed. Data analysis suggested that the vast majority of *Geobacter* strains in the subsurface at the Rifle IFRC site were most closely related to *G. bemidjiensis*, a member of the "subsurface clade" of *Geobacter* [16].

We have expanded upon this previous work [17], recovering multiple planktonic biomass samples over a period of stimulated Fe(III) reduction in the subsurface during a subsequent carbon amendment experiment. Following the identification of a dominant *Geobacter* sub-population within these samples, shotgun proteomic analyses were used to track protein expression over a 40-day period in the subsurface at the Rifle IFRC. Using these data, we have linked shifting metabolism and physiology with measured geochemical parameters to better understand factors driving subsurface biogeochemical cycles including iron, nitrogen, carbon, and hydrogen transformations.

Materials and Methods

Injection Gallery Design & Operation

The field experiment was carried out during August and September 2010 (23rd August –22nd September) at the Rifle Integrated Field Research Challenge (IFRC) site, located approximately 200 miles west of Denver in Western Colorado (USA) (Coordinates +39° 31′ 45.60″, −107° 46′ 18.50″).

An injection gallery consisting of 10 injection wells, multiple down-gradient monitoring wells arranged in three rows, and three up-gradient monitoring wells was constructed using sonic rotary drilling (Figure S1). Acetate:bromide (50 mM:5 mM) amended groundwater was injected into the subsurface via the injection wells at approximately 16 L per injection well, per day. This rate resulted in a final groundwater concentration of ~5 mM acetate, which served as a carbon source and electron donor over the course of the amendment experiment. Geochemical samples were taken from 5 m depth after purging 12 L groundwater from the sampling well. Ferrous iron and sulfide concentrations were analyzed immediately on site following sampling using the HACH 1,10 Phenanthroline and Methylene Blue colorimetric assays respectively (HACH, CO, USA). Acetate, bromide, and sulfate were also analyzed on site, using a Dionex ICS1000 ion chromatograph equipped with a CD25 conductivity detector and a Dionex IonPac AS22 column (Dionex, CA, USA). U(VI) values were determined using a Kinetic Phosphorescence Analyzer (KPA) (Chemchek, WA, USA). Additional details on geochemical analyses can be found in Williams et al [11].

DNA and Protein Sample Collection

Nine biomass samples for proteomic analyses were recovered from groundwater over the course of the *in situ* biostimulation experiment, on the following days after the start of acetate injection: 5, 8, 10, 13, 15, 17, 29, 36, and 43. Samples for 16S rRNA analysis were recovered after 3, 8, 17, 24, and 29 days. All samples were recovered from well CD01, located 2 m downgradient in the first row of down-gradient monitoring wells (Figure S1). For each sample, between 20–100 L of groundwater pumped at approximately 2 l min^{-1} from well CD01 was filtered through a pre-filter (1.2 μm, 293 mm diameter Supor disc filter, Pall Corporation, NY, USA), followed by 0.2 μm 293 mm diameter Supor disc filter. After filtration, filters were immediately frozen in an ethanol-dry ice mix, and shipped overnight on dry ice to Pacific Northwest National Laboratory for proteomic analysis.

16S rRNA Analysis

DNA was extracted from groundwater filters recovered 3, 8, 17, 24, and 29 days after the start of acetate amendment using the MoBio PowerMax Soil DNA extraction kit (Carlsbad, CA) and triplicate extracts for each sample were combined and concentrated using ethanol precipitation. Extracted DNA was used as a template for amplification of the 16S rRNA gene with the primers 27F (5′-AGAGTTTGATCCTGGCTCAG-3′) and 1492R (5′-GGTTACCTTGTTACGACTT-3′). To minimize PCR bias, amplicons from a gradient PCR reaction (20 cycles) were pooled and used as input for Illumina library preparation and HiSeq 2000 sequencing using standard protocols. EMIRGE analyses were performed as previously reported [18]. Briefly, for each sample (now a bar-coded library), quality-filtered trimmed reads were subsampled (1 million reads) at random without replacement. Each trimmed read subsample was input into an amplicon-optimized version of EMIRGE [18] for assembly into full-length genes. This code is freely available at https://github.com/csmiller/EMIRGE. EMIRGE was run for each subsample for 120 iterations with default parameters designed to merge reconstructed 16S rRNA genes ≥97% identical. Abundance estimates for each assembled 16S rRNA gene were derived by the probabilistic accounting in EMIRGE of how reads map to each assembled rRNA sequence [18]. The starting rRNA database was derived from version 102 of the SILVA SSU database, while taxonomy was assigned to each OTU using SILVA 108.

To examine the overall diversity and relative abundance in 16S rRNA sequences, results for organisms with relative abundance greater than 0.01% were included, with relative abundance data calculated by EMIRGE for the entire dataset. To demonstrate the relative abundance and phylogenetic affiliation of the most abundant *Geobacter spp.*, 16S rRNA sequences above 0.5% relative abundance were aligned using MUSCLE [19], and then incorporated into a neighbor-joining tree using MEGA [20]. Bootstrap analysis was carried out using 100 iterations. Relative abundance data was added to the phylogenetic tree using ITOL [21]. The 16S rRNA sequences used in this study are included in Table S1.

Proteomic Sample Preparation

Rifle groundwater sample filters (0.2 μm) were removed from −80°C freezer individually for proteomics sample preparation as follows: frozen filters were crushed into small pieces and 6 g of the pieces were placed into a 50 mL Falcon tube. Lysis buffer (2% (w/v) SDS, 100 mM DTT in 100 mM ammonium bicarbonate pH 7.6, (Sigma-Aldrich, St. Louis, MO)) was added to the sample and vortexed and incubated at 95°C for 5 minutes. The tube was vortexed and spun to reduce the bubbles and the supernatant was added to 3 barocycle Pulse tubes (with rinsing of the filter pieces) and barocycled for 10 cycles (20 seconds at 35,000 psi back down to ambient pressure for 10 seconds) (Pressure Biosciences Inc.,

South Easton, MA). The sample was removed from the Pulse tubes and spun at 15,000×g for 5 minutes to pellet debris. The Filter-Aided Sample Preparation (FASP) [22] technique was used to remove the SDS from the sample. Two 15 mL 30 K MWCO spin filters (Millipore, Billerica, MA) were filled with 13 mL of 8 M urea in 100 mM ammonium bicarbonate, pH 8.5 (Sigma-Aldrich, St. Louis, MO) and 2 mL of sample in each. The spin filters were spun at 4000×g for 40 minutes to the dead volume of ~200 µl and 10 mL of 8 M urea, pH 8.0 was added and spun again at 4000×g for 40 minutes. This step was repeated 3 times. Finally, 100 mM ammonium bicarbonate, pH 8.0 was added and spun at 4000×g for 40 minutes, with the step repeated once. A Coomassie Plus (Thermo Scientific, Rockford, IL) assay was used to determine protein concentration and 100 mM ammonium bicarbonate pH 8.0 was added to the sample to cover the filter. Tryptic digestion (Promega, Madison, WI) was performed at a 1:50 (w/w) trypsin to protein ratio with the addition of 1 mM $CaCl2$ to stabilize the trypsin and reduce autolysis. The collection tube was cleaned and the sample was incubated overnight at 37°C. The following day the spin filter was spun at 4000×g for 30 minutes to collect the peptides. The filter was rinsed once with 100 mM ammonium bicarbonate, pH 8.0 and spun again at 4000×g for 30 minutes. The sample was cleaned via strong cation exchange (SCX) solid phase extraction (SPE) (Supelco, Bellefonte, PA) and dried in a speed-vac to 100µl and assayed with Bicinchoninic acid (BCA) (Thermo Scientific, Rockford, IL) to determine the final peptide concentration and vialed for 2D-LC-MS/MS analysis.

2D-LC-MS/MS Analysis

The 2D-LC system was custom built using two Agilent 1200 nanoflow pumps and one 1200 capillary pump (Agilent Technologies, Santa Clara, CA), various Valco valves (Valco Instruments Co., Houston, TX), and a PAL autosampler (Leap Technologies, Carrboro, NC). Full automation was made possible by custom software that allows for parallel event coordination providing near 100% MS duty cycle through use of two trapping and analytical columns. All columns were manufactured in-house by slurry packing media into fused silica (Polymicro Technologies Inc., Phoenix, AZ) using a 1 cm sol-gel frit for media retention. Column dimensions are as follows: first dimension SCX column; 5-µm PolySULFOETHYL A (PolyLC Inc., Columbia, MD), 15-cm ×360 µm outer diameter (o.d.) ×150 µm inner diameter (i.d.). Trapping columns; 5-µm Jupiter C_{18} (Phenomenex, Torrence, CA), 4-cm ×360 µm o.d. ×150 µm i.d. Second dimension reversed-phase columns; 3-µm Jupiter C_{18} (Phenomenex, Torrence, CA), 35-cm ×360 µm o.d. ×75 µm i.d. Mobile phases consisted of 0.1 mM NaH_2PO_4 (A) and 0.3 M NaH_2PO_4 (B) for the first dimension and 0.1% formic acid in water (A) and 0.1% formic acid in acetonitrile (B) for the second dimension.

MS analysis was performed using a LTQ Orbitrap Velos ETD mass spectrometer (Thermo Scientific, San Jose, CA) outfitted with a custom electrospray ionization (ESI) interface. Electrospray emitters were custom made using 150 µm o.d. ×20 µm i.d. chemically etched fused silica [23]. The heated capillary temperature and spray voltage were 275°C and 2.2 kV, respectively. Data were acquired for 100 min, beginning 65 min after sample injection and 15 min into gradient. Orbitrap spectra (AGC 1×10^6) were collected from 400–2000 m/z at a resolution of 60 k followed by data dependent ion trap CID MS/MS (collision energy 35%, AGC 3×10^4) of the ten most abundant ions. A dynamic exclusion time of 60 sec was used to discriminate against previously analyzed ions.

Data Analysis

MS/MS data was searched using SEQUEST against a peptide database constructed from the *Geobacter bemidjiensis* genome, using relatively conservative filters [Xcorr values of 1.9 (+1), 2.2 (+2), and 3.5 (+3)]. Resulting peptide identifications were filtered using an MSGF cutoff value to $1\,e^{-10}$ [24]. Peptides identified by only one spectral count were discarded. Spectral count data for each identified protein were normalized for protein length by dividing spectral counts by the amino acid protein length. These values were subsequently log transformed. These abundance values were converted to z-score values (also called the Standard Row Function). Z-scores were calculated by taking the mean protein abundance across all conditions, subtracting from this the individual protein abundance, and dividing this value by the standard deviation of the values. Where data was missing from one condition, the absent value was assigned a score equal to the lowest Z-score in the data matrix divided by 1.5. Likewise, the Z-scores for that protein in the other conditions were assigned a value equal to the highest Z-score in the data matrix multiplied by 1.5. This resulted in a presence/absence appearance in subsequent heat maps. Z-score values can be used to determine proteins showing significant changes from their average values. In this study, z-score values were considered significantly different if the difference was at least 2 or greater. Heat maps were generated using the TIGR software MeV (http://www.tm4.org/mev/). Probable orthologous proteins were identified using the protocol identified in Callister et al. [25], with version 4.1 of INPARANOID used in this study.

Results and Discussion

Nine proteomic samples were collected from well CD-01 (located 2.5 m downgradient from the region of injection) and binned into three different phases of carbon amendment; Early (samples collected after 5, 8, and 10 days), Middle (samples collected after 13, 15, and 17 days), and Late (samples collected after 29, 36, and 43 days). (Figure 1A). Bins were assigned via hierarchical clustering of samples based on a suite of geochemical measurements (acetate, Fe(II), U(VI), sulfate, sulfide) taken at each time point (Figure S2). Acetate and bromide concentrations in groundwater were monitored in the sampling well to determine the start of biostimulation in that region of the aquifer. Increases in aqueous Fe(II) concentrations following the arrival of acetate in the first row of downgradient monitoring wells (Figure S1) (approximately 5 days after the start of injection) likely indicated the start of stimulated enzymatic Fe(III) reduction. Fe(II) values increased from background concentrations of ~50 µM to between 100–150 µM during the early stages of the experiment. The middle stage of biostimulation was characterized by elevated Fe(II) concentrations (between 150–200 µM), before fluctuating and decreasing concentrations were monitored during the later stages of the experiment (Figure 1A). Concurrent to this, acetate concentrations followed similar trends (Figure 1B), while aqueous U(VI) concentrations increased slightly following the stimulation of microbial activity, and then decreased rapidly over a 10-day period to levels below the U.S. Environmental Protection Agency's Maximum Contaminant Level (MCL) for uranium (Figure 1A). Finally, sulfide (S^{2-}) concentrations were below detection for the first 30 days of the experiment, and only after 37 days were concentrations of aqueous S^{2-} measured (Figure 1B).

To confirm *Geobacter* dominance in the samples, as well as identify the temporal distribution of *Geobacter* strains, 16S rRNA gene sequences from 5 biomass samples (collected on days 3, 8, 17,

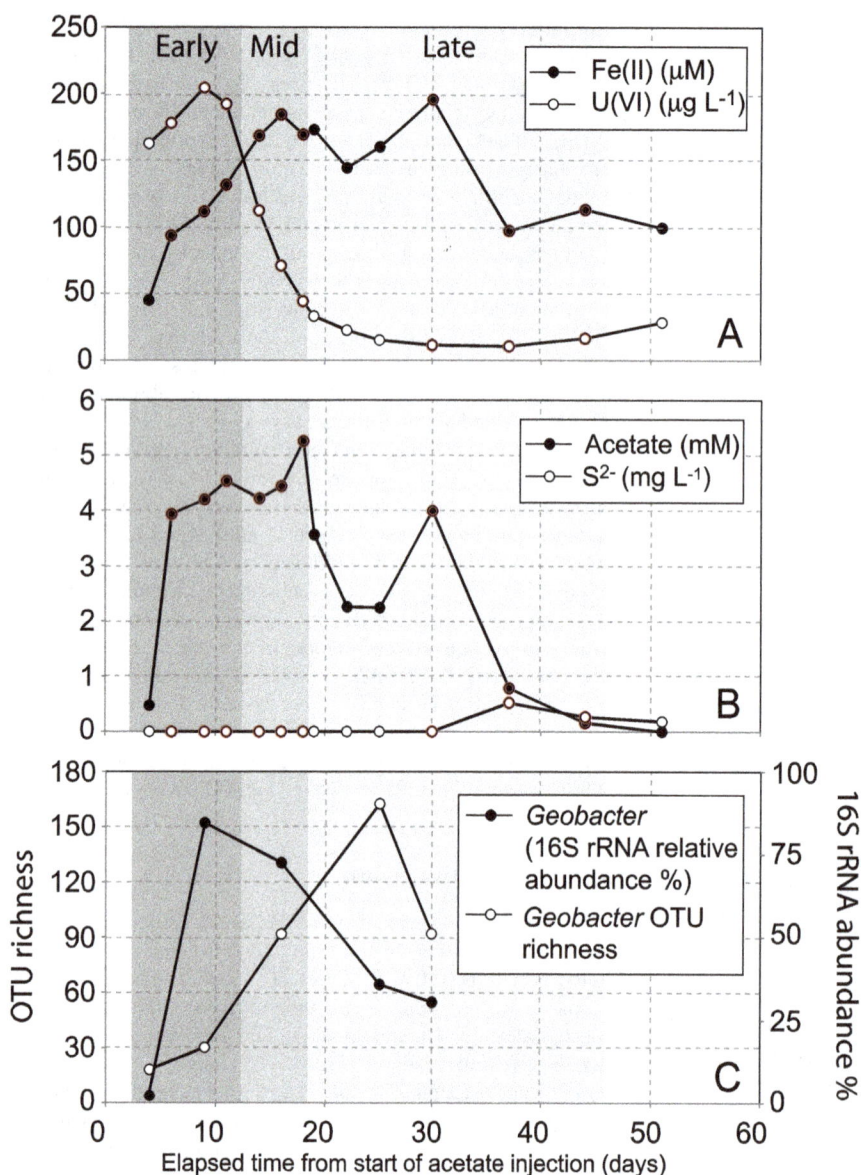

Figure 1. Geochemical and microbiological data obtained from downgradient well CD01 at the Rifle site, showing (A/B) the concurrent increase in Fe(II) and decrease in aqueous U(VI) associated with acetate arrival in the downstream monitoring well, and (C) the relative abundance of members of the *Geobacteraceae*, and *Geobacter* strain richness over time. Red circles around a data point indicate that a proteomic sample was collected.

24, and 29) were analyzed. Results revealed that *Geobacter* strains were rapidly enriched within the microbial community following the arrival of acetate in the subsurface; the relative abundance of *Geobacter* 16S rRNA sequences increased from 2% to 85% over a 5 day period, before gradually decreasing over the remaining time points (Figure 1C). Complementary groundwater cell count data from an adjacent well confirmed that *Geobacter* cell numbers rapidly increased during the first ~8 days of carbon amendment before leveling off [26]. Although *Geobacter* strain richness also increased over time, only a few strains were responsible for the majority of *Geobacter* dominance over the course of the experiment. This observation suggests that a small number of fast-growing *Geobacter* strains responded to the presence of acetate, and were subsequently complemented by strains that either exhibited slower growth rates, or were able to occupy specific biogeochemical niches during the later period of carbon amendment (Figures 1C

and 2). This finding is supported by previous studies demonstrating that *Geobacter* strains were significantly enriched following acetate amendment at the Rifle IFRC site [6,11,27]. Within the *Geobacteraceae*, phylogenetic placement of the recovered 16S rRNA sequences revealed that indigenous *Geobacter* strains closely related to *G. bemidjiensis* were the dominant members of this population (Figure 2, Table S2). Other more distantly related strains emerged in later time points, but contributed a much smaller relative fraction of 16S rRNA sequences (<5%) (Figure 2).

To investigate how dominant *Geobacter* strains responded to excess carbon flux into the local environment, planktonic biomass was sampled at nine time points and analyzed using high-resolution proteomic 2D-LC-MS/MS measurements. In instances such as this, where metagenomic sequence data is unavailable, genomic information from sequenced strains closely-related to environmental species can be used to search

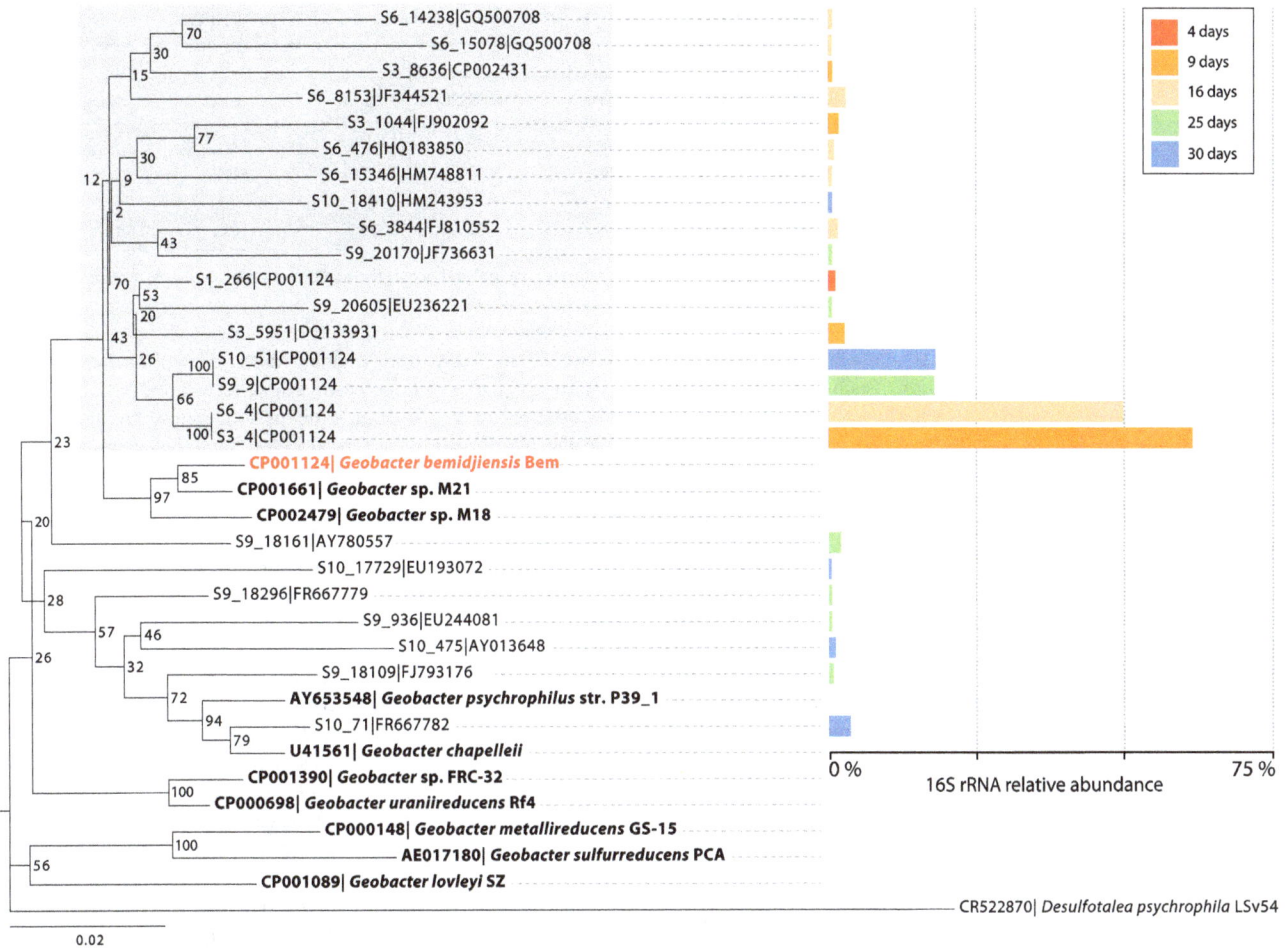

Figure 2. Neighbor-joining phylogenetic tree showing the placement of *Geobacter***-like environmental 16S rRNA sequences recovered from planktonic biomass at five time-points during carbon amendment.** Bolded sequences show the placement of isolate *Geobacter* 16S rRNA sequences, in the context of the environmental sequences. Sequences within the grey box fall within the *G. bemidjiensis*/M21/M18 clade, and account for the majority of environmental *Geobacter* sequences recovered during this study. Accession numbers associated with the environmental sequences correspond to the best match when aligned to SILVA, Greengenes, and the RDP databases.

mass spectrometry data; conserved protein sequences between closely-related strains allow predicted peptides (from a sequenced isolate) to be matched to measured mass spectra (from an environmental sample) [17,28]. This study was aided by the availability of genomic information from multiple sequenced *Geobacter* isolates, including *Geobacter* species strains M18 and M21, and *G. uraniireducens* that were all isolated from the Rifle site. Within the *Geobacter*, a fraction of proteins encoded by these isolate genomes are conserved; patterns of orthologous proteins were assessed, and used to identify 1116 orthologous proteins across all eight genomes that represented a *Geobacter* "core" proteome that would likely be present in environmental strains (Table S3). While this number represented a significant fraction of protein coding genes within each organism (Table 1), the number of orthologs was even higher between a few closely related *Geobacter* strains; *G. bemidjiensis* and the Rifle site isolate strain M21 share 2561 orthologous proteins, with 94% amino acid similarity across these orthologs [17]. Given (1) the phylogenetic similarity of the majority of dominant environmental strains to the *G. bemidjiensis*/M18/M21 clade (grey box in Figure 2), (2) the high number of orthologs shared between *G. bemidjiensis*, M18 and M21, (3) the desire to limit redundancy with the search database, and (4) the well annotated and

curated nature of the *G. bemidjiensis* genome, predicted peptides from the *G. bemidjiensis* genome were used to search the proteomic MS/MS data.

Across nine planktonic biomass samples, over 900 proteins from environmental *Geobacter* strains that match the *G. bemidjiensis* proteome were subsequently detected (Table S4). Despite the challenges associated with measuring peptides from environmental samples, 530 of the 1116 proteins (~47%) comprising a "core" *Geobacter* proteome were detected in all nine samples. These results confirm (1) the ability to identify and detect a significant number of conserved proteins from environmental strains using closely-related isolate genomic data in search databases, and (2), that the activity of the identified core enzymes extends to maintaining growth and survival of environmental strains in subsurface environments (Table S4). In total, 718 of ~900 *G. bemidjiensis* proteins detected within these samples have orthologs in at least one other *Geobacter* strain (Figure 3), with the highest number shared between *G. bemidjiensis* and strain M21 (712 detected orthologs).

Temporal Geochemical-Proteomic Analyses

Significant protein abundance shifts were investigated across the three different geochemical stages (early, middle, late) of

Table 1. Number of orthologous proteins across eight *Geobacter* genomes that comprise a "core" proteome.

	Gbem	GM21	GM18	Gura	GFRC32	Gsul	Gmet	Glov
Protein-coding genes	4034	4152	4523	4430	3839	3465	3576	3725
Fraction of orthologous genes common to all *Geobacter* strains	28%	27%	25%	25%	29%	32%	31%	30%
Expressed fraction	13%	13%	12%	12%	14%	15%	15%	14%

The expressed fraction refers to *Geobacter bemidjiensis* proteins detected within this data set, extrapolated across the additional seven *Geobacter* genomes.

biostimulation (Figure S2). Overall trends were indicative of a population responding to stimulation (such as the sudden availability of carbon), and were similar to growth patterns measured for *Geobacter* strains within laboratory settings in batch cultures [9]; initial rapid growth of the Geobacter population was inferred by statistically significant ($P<0.05$) abundance increases (relative to later stages of biostimulation) for proteins associated with biogenesis (Figure 4, Table S5). As an example, 34% of the detected ribosomal proteins (**C**luster of **O**rthologous **G**ene (COG) category J) were at greater abundances in samples recovered during the early stage of biostimulation, compared to 12% in the middle stage. This carbon usage results in a *Geobacter* biomass "bloom" within biostimulated regions of the aquifer, as inferred by 16S rRNA relative abundances (Figure 2), and cell count data [26]. In addition, similar observations have been reported in earlier carbon amendment experiments in the Rifle subsurface [6,11]. During the middle stage of biostimulation, slowing of *Geobacter* growth was inferred from decreasing abundances of proteins associated with biogenesis (as described above) (Figure 4). Conversely however, abundance increases ($P<0.05$) were observed in proteins associated with energy generation (COG category C) and amino acid metabolism and transport (COG category E) (Figure 4) over the same time period, consistent with some level of increasing respiration and cell maintenance. Finally, measured abundances decreased for large numbers of proteins between the subsequent middle and late stages, indicating that significant losses in growth and activity occur in the planktonic *Geobacter* population over this time period.

It is worth noting that the shifts in protein abundances reported here do not simply correspond to changes in organism abundances, as displayed in Figure 2. The fraction of *Geobacter* 16S rRNA sequences as a total of the whole microbial population decreases

between the early and middle stages of carbon amendment (Figure 1C), and yet increases are observed in certain protein abundances over this same time period. These changes within specific pathways are presented below, and reveal physiological shifts occurring over the period of carbon amendment within the Rifle aquifer.

Acetate Activation and Utilization

A key characteristic of *Geobacter* strains is their efficient uptake and use of acetate. This carbon compound is utilized via two different pathways, both of which activate acetate to acetyl-CoA. The first pathway involves the enzyme acetyl-CoA transferase (ATO) (Gbem_0468, Gbem_0795), which has two functions in *Geobacter* strains: the activation of acetate to acetyl-CoA, and the conversion of succinyl-CoA to succinate as part of the tricarboxylic acid (TCA) cycle [29]. Because of this coupling (Figure 5), acetyl-CoA produced via this mechanism can be completely consumed via condensation with citrate to form oxaloacetate (Figure 5). Additional acetyl-CoA must therefore be synthesized for biosynthetic reactions via a two-step reaction involving acetate kinase (ACK) and phosphotransacetylase (PTA). This acetyl-CoA is then converted to pyruvate via a pyruvate ferredoxin oxidoreductase (PFOR) operating in reverse (Figure 5) [30].

There is proteomic support for both acetate-activation pathways throughout the datasets, with ATO pathway components (Gbem_0468, Gbem_0795) present at greater abundances than ACK (Gbem_2277) and PTA (Gbem_2276) across all phases of carbon amendment (Figure 6A). Significant shifts in protein abundances ($P<0.05$) between the sample stages were inferred using Z score calculations (Figure 6B) [31], and revealed changing trends in carbon utilization. Both ATO enzymes (Gbem_0795, Gbem_0468) increased in abundance between the early and

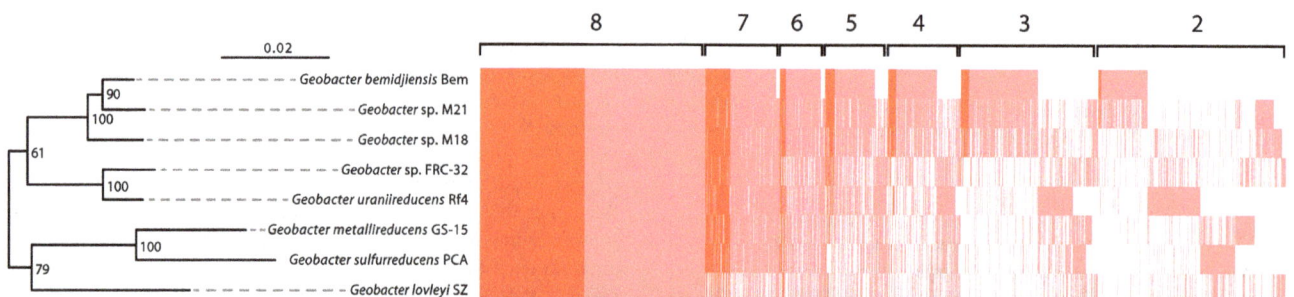

Figure 3. Distribution and expression of orthologous proteins across eight *Geobacter* genomes. Pink shading illustrates the presence of orthologous proteins within genomic data, while red shading indicates expression of an orthologous protein by an environmental *Geobacter* strain, identified using predicted peptides from *G. bemidjiensis*. Values along the top of the chart indicate the number of *Geobacter* strains the orthologs are distributed over. For the core proteome (as identified by orthologs present in all eight *Geobacter* genomes), ~47% expression is detected within biomass recovered from the Rifle subsurface. The data is coupled to a neighbor-joining tree constructed using 16S rRNA sequences from eight sequenced *Geobacter* genomes, and illustrates the correlation between inferred evolutionary distance and the distribution of orthologous proteins.

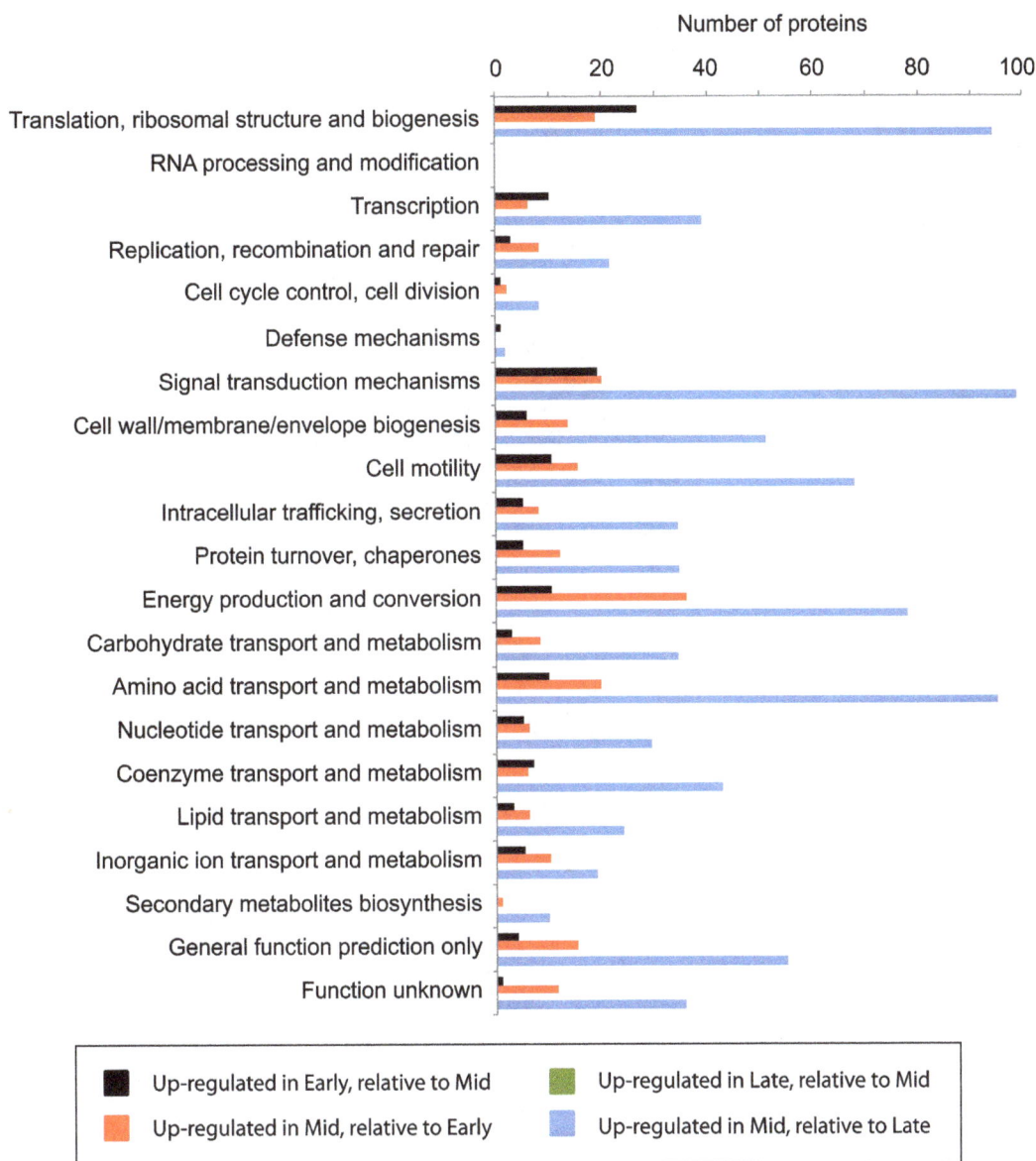

Figure 4. Significant shifts in protein abundances between the three stages of carbon amendment, binned into COG categories. Significant protein abundance increases and decreases between the stages were inferred using Z-score calculations.

middle stages of biostimulation, indicative of increasing flux through respiratory pathways. Further emphasizing the importance of energy generation, many TCA cycle enzymes were highly abundant across all three sample stages, with citrate synthase (Gbem_3905, Gbem_1652), isocitrate dehydrogenase (Gbem_2901), aconitate hydratase (Gbem_1294), and succinate dehydrogenase (Gbem_3332) all increasing in abundance from the early to middle period of biostimulation (Figure 6B). Indeed, these three enzymes contribute to the ~17% of proteins showing abundance increases across this period that are associated with energy generation and conversion (COG category C) (Figure 4). Inferred high fluxes of carbon through respiratory pathways are supported by *in silico* predictions for the closely related species *Geobacter sulfurreducens*. Data from the in silico study suggests that >90% of consumed acetate is directed to the TCA cycle for respiration when growing on Fe(III) [32].

Mirroring trends identified within the COG classification data (Figure 4), proteomic data suggests that while energy generation was presumably increasing over this time period, carbon flux to biosynthesis was not concurrently up regulated. Neither ACK nor PTA enzymes showed significant abundance increases between the early and middle stages of the experiment. A similar trend was observed for the PFOR enzyme (Gbem_0209), that converts acetyl-CoA to pyruvate (Figure 6B). The activity of this enzyme is the primary mechanism for generating 4-carbon compounds that are necessary for growth when acetate is the primary carbon source [30]. From these and other protein abundances associated with central metabolism, we can infer that (1) the flux passing through respiratory pathways may increase during the middle state of biostimulation, and (2) consequently, a larger fraction of carbon flux occurs towards biosynthesis during the early period of the experiment relative to later stages, indicative of *Geobacter* cell growth and proliferation following the initial arrival of carbon in

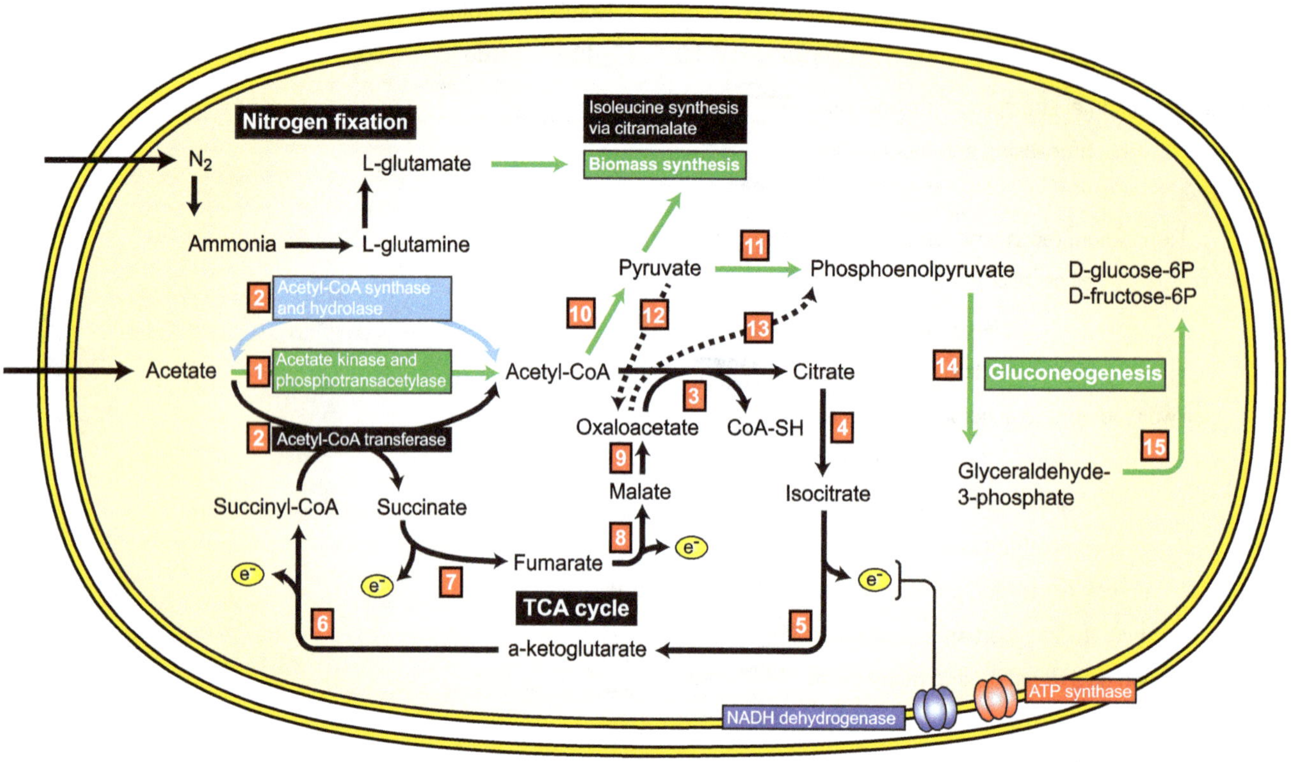

Figure 5. Central metabolism in indigenous *Geobacter* strains as inferred from proteomic data. Red number-containing boxes refer to specific enzymes in figure 6. Adapted from Mahadevan et al. [30].

the subsurface. However, it is worth noting that pyruvate carboxylase (Gbem_0273) can channel pyruvate synthesized via PFOR into respiratory pathways (via conversion to oxaloacetate). Given that this enzyme was detected within the proteomic results, the flux of carbon towards respiratory pathways may be even greater than is reflected within these data.

The identification of potential shifts in carbon flux through central metabolism has implications for metal biogeochemical cycles and bioremediation. Our results hint at complex linkages between cellular metabolism and the extracellular environment. Here proteomic inferences suggest a larger fraction of carbon was shunted to respiratory reactions rather than anapleurotic reactions during the middle stage of carbon amendment, when U(VI) was effectively removed from solution (Figure 1A). While these results may indicate that these metabolic shifts play a direct role in the efficiency of enzymatic U(VI) reduction, we note the lag in U(VI) reduction may also be indirectly impacted by biostimulation activities. Specifically, higher amounts of reactive Fe(III) present in early biostimulation could abiotically re-oxidize U(IV) phases in the aquifer, thereby masking active U(VI) reduction [33]. As carbon amendment progresses, the disappearance of more reactive Fe(III) phases (due to biological enzymatic dissolution) and increased concentrations of U(IV) (potentially due to shift in central metabolism from biosynthesis to respiration) may dilute these U(IV) reoxidation effects.

Alternative Electron Donors

While these data suggest that a significant fraction of carbon flux is directed towards respiration in subsurface *Geobacter* strains, uptake hydrogenases may also play a role in driving respiratory processes in the Rifle aquifer. *Geobacter bemidjiensis* contains multiple genes encoding uptake hydrogenases [34], potentially expanding

the range of electron donors that can be utilized for respiration. Both small and large subunits of NiFe hydrogenases (Gbem_3139, Gbem_3136, Gbem_3884) were detected within the proteomic samples, and as with other enzymes associated with energy generation, increases in hydrogenase abundance were observed between the early and middle sample stages (supplementary information). These hydrogenases therefore presumably contribute to increased rates of respiratory processes that were already inferred from protein abundances. The potential for hydrogenase activity within this population is perhaps unexpected; given the relatively high concentrations of aqueous carbon that can be utilized as an electron donor, the additional utilization of hydrogen for respiration may not be essential for survival. However, hydrogenase expression in this instance may be an example of this population maximizing energy generation during exposure to relatively carbon rich environmental conditions. In addition, recent studies have suggested a role for hydrogenases as part of the oxidative stress response in *Geobacter* species [35]. A function in oxidative stress would correlate to the central metabolism carbon flux profiles we infer here, when there is a shift from anabolic to respiratory processes, thus increasing oxidative stress. The increased abundance of a manganese and iron superoxide dismutase (Gbem_2204) over this time period may reflect another response to this stress.

Nutrient Limitation

During the middle and late stages of Fe(III) reduction, any increase in the ratio of respiration/biosynthesis may reflect a slowing growth rate and could be associated with limiting nutrient concentrations that limit biomass production. As biomass is synthesized within the aquifer, essential nutrients and elements may become growth limiting. *Geobacter* utilize a number of

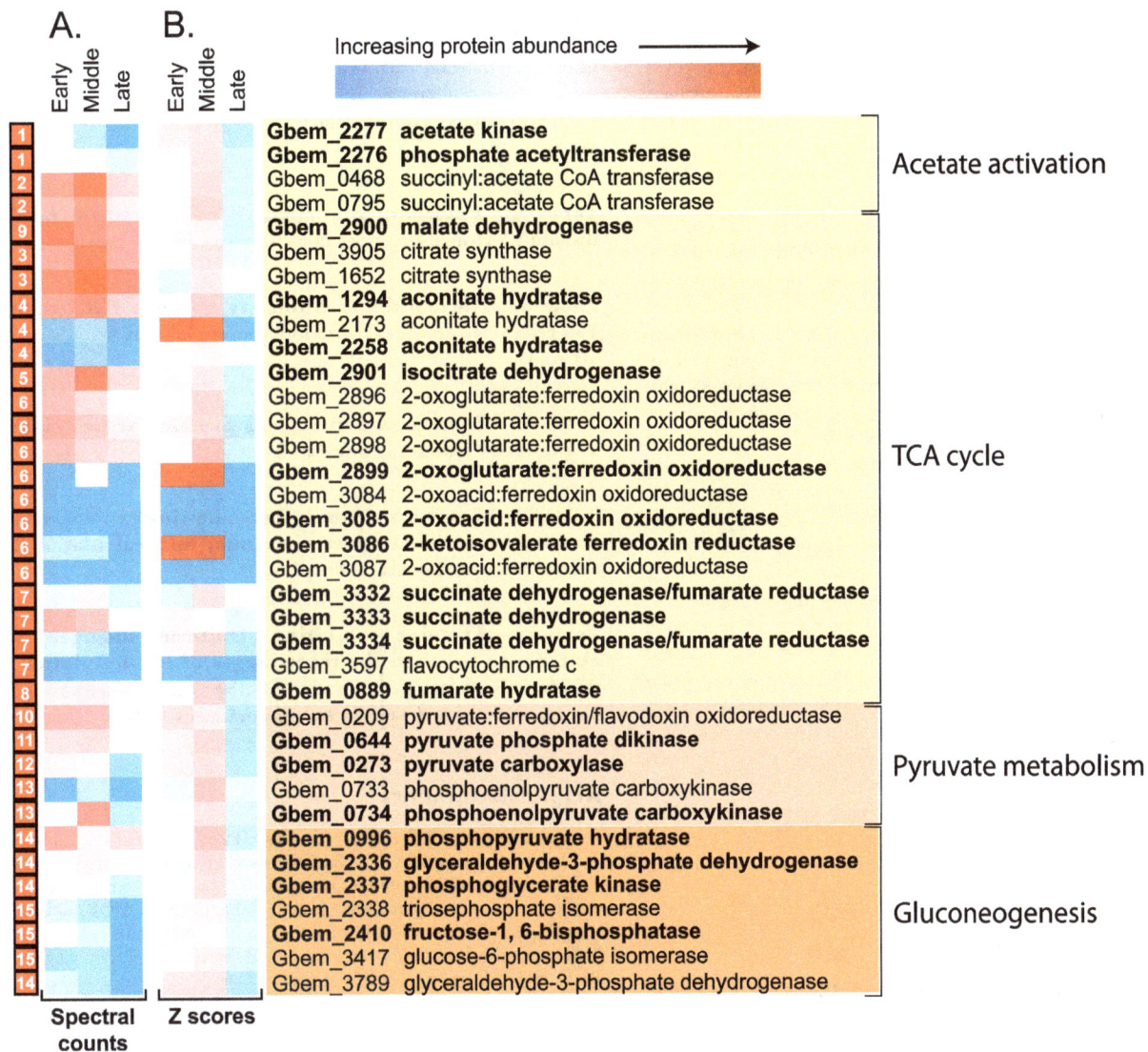

Figure 6. Relative abundance data for central metabolic pathways outlined in figure 4, using both log transformed spectral count information (A), and Z-scores (B) to better identify relative abundance shifts across the three stages of carbon amendment. Proteins that are orthologous across all eight sequenced *Geobacter* species are highlighted bold.

common strategies for coping under these conditions, including fixing atmospheric N_2 via nitrogenase activity [36], and expressing P uptake mechanisms [14]. However, no clear patterns of nitrogenase expression were identified in this data set. Although nitrogenase enzymes (NifK, NifD, NifH) were detected across all three stages of biostimulation (Table S4), previous measurements of bulk aqueous ammonium concentrations from nearby wells in the Rifle aquifer have suggested that non-limiting N concentrations are present during carbon amendment [13]. However, given the heterogeneous nature of the subsurface at Rifle [37], nitrogenase expression may reflect the development of local N-limiting regions within the aquifer around high biological activity. If limiting N concentrations are present in the subsurface during carbon amendment, the ability to fix nitrogen potentially offers *Geobacter* species a competitive advantage over other subsurface bacterial strains that are unable to carry out this process.

Given the low P concentrations within Rifle groundwater [14], indigenous *Geobacter* strains express high-affinity phosphate ABC transporters for P uptake. Both ATPase subunits and periplasmic binding proteins encoded by the *pst-pho* operon were observed within the dataset, while a phosphate selective porin (Gbem_4031) increased in abundance from the early to middle phase of carbon amendment. Interestingly, one phosphate ABC transporter increased in abundance over this same time period (Gbem_1847), while another decreased in abundance (Gbem_1710). Given that both transporters are associated with the high-affinity *pst* system, these differing expression patterns suggest that they may occupy different physiological roles in the subsurface. Ultimately however, the expression of components of the *pst-pho* operon across all stages of carbon amendment indicates that phosphate limitation is likely a key process affecting biostimulated microorganisms.

Conclusions

Proteomic investigations had previously focused on acetate-stimulated planktonic biomass at the Rifle IFRC [5]. While these results had identified potential strain level shifts within the microbial community, and allowed central metabolism to be studied, the lack of a temporal series of samples had precluded statistical analyses of shifts in protein expression over the duration of biostimulation. In this study, we have utilized a greater number of samples to investigate the *in situ* temporal response of a microbial population to increased carbon availability during a biostimulation experiment. *Geobacter* strains were rapidly enriched within the planktonic microbial community upon acetate amendment and likely contributed to rapidly increasing aqueous Fe(II) concentrations over the first 15 days of the experiment. Physiological inferences suggest that the "bloom" of *Geobacter* biomass within the aquifer was associated with the efficient utilization of acetate for both respiration and biosynthesis, with potential shifts in carbon flux through anabolic and catabolic reactions over time. These temporal physiological changes have direct impacts on the aquifer biogeochemistry; the potential for increasing flux through respiratory pathways at certain time points has significant implications for elemental cycling in subsurface environments; electrons are thought to be primarily transferred to oxidized iron minerals, liberating soluble Fe(II) and any other adsorbed compounds into groundwater. However, these strains have the potential to dump electrons onto a wide range of redox-active metals and compounds, including organic matter (humic compounds), vanadium, and uranium, and therefore alter their physical and chemical behavior. Concurrently, decreasing biosynthesis in *Geobacter* strains may be linked indirectly to increasing activity of sulfate-reducing bacteria (SRB), as has been reported previously [12,38]. Greater activity of SRB results in rising aqueous sulfide concentrations which can subsequently react with other metal cations to form precipitates and clog pore networks, catalyze the dissolution of Fe(III) phases, and release adsorbed metal cations from Fe(III) mineral surfaces. These data emphasize the tight biogeochemical linkages that exist between microbial assemblages and the surrounding local environment, and the metabolic shifts that occur within a population in response to these environmental stimuli.

Supporting Information

Figure S1 Plot layout at the Rifle IFRC. Acetate injection wells are labeled CG-01 thru CG-10. Downgradient monitoring well CD-01 is highlighted with a red box.

Figure S2 Proteomic sample clustering for quantitative analysis. Samples collected after 5, 8, and 10 days were grouped into the "early" phase of biostimulation, 13, 15, and 17 days into

the "middle" stage, and 29, 36, and 43 days into the "late" stage. Clustering was performed using geochemical data (square root transformed) from each sampling time point (Fe(II), S^{2-}, U, Acetate, and Sulfate) in R using the dist and hclust functions. Euclidean distances were calculated with average linkages between samples.

Table S1 Environmental 16S rRNA sequences used during phylogenetic tree construction in this study.

Table S2 Nucleotide % similarity between full length 16S rRNA sequences from environmental and sequenced strains.

Table S3 Predicated orthologous proteins across eight sequenced *Geobacter* genomes.

Table S4 Shotgun proteomic data, showing raw spectral counts, normalized spectral counts, and calculated Z scores across the nine samples.

Table S5 Proteins exhibiting significant changes in abundance between the three stages of biostimulation, displayed as a percentage of the total number of proteins detected across the experiment (925).

Acknowledgments

We thank the city of Rifle, CO, the Colorado Department of Public Health and Environment, and the U.S. Environmental Protection Agency, Region 8, for their cooperation in this study. Portions of this work were performed at the Environmental Molecular Sciences Laboratory, a DOE national scientific user facility located at the Pacific Northwest National Laboratory. This material is based upon work supported through the Integrated Field Research Challenge Site (IFRC) at Rifle, Colorado.

Author Contributions

Operated the Rifle IFRC field site where the experiments were carried out: PL KHW. Conceived and designed the experiments: MJW KCW KHW MSL LM. Performed the experiments: MJW CDN KCW KHW AM LG. Analyzed the data: MJW KCW CM KMH. Contributed reagents/ materials/analysis tools: DL PL JFB. Wrote the paper: MJW KCW.

References

1. Konopka A Ecology, Microbial. In: Schaechter M, editor. Encyclopedia of Microbiology. Oxford: Elsevier. 91–106.

2. Lin B, Westerhoff HV, Röling WFM (2009) How *Geobacteraceae* may dominate subsurface biodegradation: physiology of *Geobacter metallireducens* in slow-growth habitat-simulating retentostats. Enviro Microbiol 11: 2425–2433.

3. Mailloux BJ, Fuller ME (2003) Determination of In Situ Bacterial Growth Rates in Aquifers and Aquifer Sediments. Appl Environ Microbiol 69: 3798–3808.

4. Holmes DE, O'Neil RA, Chavan MA, N'Guessan LA, Vrionis HA, et al. (2008) Transcriptome of *Geobacter uraniireducens* growing in uranium-contaminated subsurface sediments. ISME J 3: 216–230.

5. Wilkins MJ, Verberkmoes NC, Williams KH, Callister SJ, Mouser PJ, et al. (2009) Proteogenomic monitoring of *Geobacter* physiology during stimulated uranium bioremediation. Appl Environ Microbiol 75: 6591–6599.

6. Anderson RT, Vrionis HA, Ortiz-Bernad I, Resch CT, Long PE, et al. (2003) Stimulating the In Situ Activity of *Geobacter* Species To Remove Uranium from the Groundwater of a Uranium-Contaminated Aquifer. Appl Environ Microbiol 69: 5884–5891.

7. Lovley DR, Phillips ER (1988) Novel mode of microbial energy metabolism: organic carbon oxidation coupled to dissimilatory reduction of iron or manganese. Appl Environ Microbiol 54: 1472–1480.

8. Lovley DR, Phillips EJP, Gorby YA, Landa E (1991) Microbial reduction of uranium. Nature 350: 413–416.

9. Ortiz-Bernad I, Anderson RT, Vrionis HA, Lovley DR (2004) Vanadium Respiration by *Geobacter metallireducens*: Novel Strategy for In Situ Removal of Vanadium from Groundwater. Appl Environ Microbiol 70: 3091–3095.

10. Pearce CI, Pattrick RAD, Law N, Charnock JC, Coker VS, et al. (2009) Investigating different mechanisms for biogenic selenite transformations: *Geobacter sulfurreducens, Shewanella oneidensis and Veillonella atypica*. Environ Technol 30: 1313–1326.

11. Williams KH, Long PE, Davis JA, Wilkins MJ, N'Guessan AL, et al. (2011) Acetate availability and its influence on sustainable bioremediation of uranium-contaminated groundwater. Geomicro J 28: 519–539.

12. Barlett M, Zhuang K, Mahadevan R, Lovley DR (2011) Integrative analysis of the interactions between *Geobacter* spp. and sulfate-reducing bacteria during uranium bioremediation. Biosciences Discuss 8: 11337–11357.

13. Mouser PJ, N'Guessan AL, Elifantz H, Holmes DE, Williams KH, et al. (2009) Influence of heterogeneous ammonium availability on bacterial community structure and the expression of nitrogen fixation and ammonium transporter genes during in situ bioremediation of uranium-contaminated groundwater. Environ Sci Technol 43: 4386–4392.

14. N'Guessan AL, Elifantz H, Nevin KP, Mouser PJ, Methe B, et al. (2010) Molecular analysis of phosphate limitation in *Geobacteraceae* during the bioremediation of a uranium-contaminated aquifer. ISME J 4: 253–266.

15. Elifantz H, N'Guessan LA, Mouser PJ, Williams KH, Wilkins MJ, et al. (2010) Expression of acetate permease-like (apl) genes in subsurface communities of

Geobacter species under fluctuating acetate concentrations. FEMS Microbiol Ecol 73: 441–449.

16. Holmes DE, O'Neil RA, Vrionis HA, N'Guessan LA, Ortiz-Bernad I, et al. (2007) Subsurface clade of *Geobacteraceae* that predominates in a diversity of Fe(III)-reducing subsurface environments. ISME J 1: 663–677.

17. Wilkins MJ, Callister SJ, Miletto M, Williams KH, Nicora CD, et al. (2010) Development of a biomarker for *Geobacter* activity and strain composition; Proteogenomic analysis of the citrate synthase protein during bioremediation of U(VI). Microb Biotech: 4: 55–63.

18. Miller CS, Handley KM, Wrighton KC, Frischkorn KR, Thomas BC, et al. (2013) Short-read assembly of full-length 16S amplicons reveals bacterial diversity in subsurface sediments. PLoS ONE In Press.

19. Edgar RC (2004) MUSCLE: multiple sequence alignment with high accuracy and high throughput. Nucleic Acids Res 32: 1792–1797.

20. Tamura K, Peterson D, Peterson N, Stecher G, Nei M, et al. (2011) MEGA5: Molecular Evolutionary Genetics Analysis Using Maximum Likelihood, Evolutionary Distance, and Maximum Parsimony Methods. Mol Biol Evol 28: 2731–2739.

21. Letunic I, Bork P (2011) Interactive Tree Of Life v2: online annotation and display of phylogenetic trees made easy. Nucleic Acids Res 39: W475–W478.

22. Wisniewski JR, Zougman A, Nagaraj N, Mann M (2009) Universal sample preparation method for proteome analysis. Nat Meth 6: 359–362.

23. Kelly RT, Page JS, Luo Q, Moore RJ, Orton DJ, et al. (2006) Chemically Etched Open Tubular and Monolithic Emitters for Nanoelectrospray Ionization Mass Spectrometry. Anal Chem 78: 7796–7801.

24. Kim S, Gupta N, Pevzner PA (2008) Spectral Probabilities and Generating Functions of Tandem Mass Spectra: A Strike against Decoy Databases. J Prot Res 7: 3354–3363.

25. Callister SJ, McCue LA, Turse JE, Monroe ME, Auberry KJ, et al. (2008) Comparative Bacterial Proteomics: Analysis of the Core Genome Concept. PLoS ONE 3: e1542.

26. Holmes DE, Giloteaux L, Barlett M, Chavan MA, Smith JA, et al. (2013) Molecular Analysis of the In Situ Growth Rate of Subsurface *Geobacter* Species. Appl Environ Microbiol In Press.

27. Vrionis HA, Anderson RT, Ortiz-Bernad I, O'Neill KR, Resch CT, et al. (2005) Microbiological and geochemical heterogeneity in an in situ uranium bioremediation field site. Appl Environ Microbiol 71: 6308–6318.

28. Denef VJ, Shah MB, VerBerkmoes NC, Hettich RL, Banfield JF (2007) Implications of strain- and species-level sequence divergence for community and isolate shotgun proteomic analysis. J Prot Res 6: 3152–3161.

29. Segura D, Mahadevan R, Juarez K, Lovley DR (2008) Computational and experimental analysis of redundancy in the central metabolism of *Geobacter sulfurreducens*. PloS Comput Biol 4.

30. Mahadevan R, Palsson Bò, Lovley DR (2011) In situ to in silico and back: elucidating the physiology and ecology of *Geobacter* spp. using genome-scale modelling. Nat Rev Micro 9: 39–50.

31. Ding YHR, Hixson KK, Aklujkar MA, Lipton MS, Smith RD, et al. (2008) Proteome of *Geobacter sulfurreducens* grown with Fe(III) oxide or Fe(III) citrate as the electron donor. Biochim Biophys Acta 1784: 1935–1941.

32. Tang YJJ, Chakraborty R, Martin HG, Chu J, Hazen TC, et al. (2007) Flux analysis of central metabolic pathways in *Geobacter metallireducens* during reduction of soluble Fe(III)-nitrilotriacetic acid. Appl Environ Microbiol 73: 3859–3864.

33. Finneran KT, Anderson RT, Nevin KP, Lovley DR (2002) Potential for bioremediation of uranium-contaminated aquifers with microbial U(VI) reduction. Soil Sed Contam 11: 339–357.

34. Coppi MV (2005) The hydrogenases of *Geobacter sulfurreducens*: a comparative genomic perspective. Microbiol 151: 1239–1254.

35. Tremblay P-L, Lovley DR (2012) Role of the NiFe hydrogenase Hya in oxidative stress defense in *Geobacter sulfurreducens*. J Bact 194: 2248–2253.

36. Bazylinski DA, Dean AJ, Schuler D, Phillips EJP, Lovley DR (2000) N-2-dependent growth and nitrogenase activity in the metal-metabolizing bacteria, *Geobacter* and *Magnetospirillum* species. Environ Microbiol 2: 266–273.

37. Campbell KM, Kukkadapu RK, Qafoku NP, Peacock AD, Lesher E, et al. (2012) Geochemical, mineralogical and microbiological characteristics of sediment from a naturally reduced zone in a uranium-contaminated aquifer. Appl Geochem 27: 1499–1511.

38. Druhan JL, Steefel CI, Molins S, Williams KH, Conrad ME, et al. (2012) Timing the Onset of Sulfate Reduction over Multiple Subsurface Acetate Amendments by Measurement and Modeling of Sulfur Isotope Fractionation. Environ Sci Technol 46: 8895–8902.

5

Phylogenetic Patterns in the Microbial Response to Resource Availability: Amino Acid Incorporation in San Francisco Bay

Xavier Mayali*, Peter K. Weber, Shalini Mabery, Jennifer Pett-Ridge

Physical and Life Science Directorate, Lawrence Livermore National Laboratory, Livermore California, United States of America

Abstract

Aquatic microorganisms are typically identified as either oligotrophic or copiotrophic, representing trophic strategies adapted to low or high nutrient concentrations, respectively. Here, we sought to take steps towards identifying these and additional adaptations to nutrient availability with a quantitative analysis of microbial resource use in mixed communities. We incubated an estuarine microbial community with stable isotope labeled amino acids (AAs) at concentrations spanning three orders of magnitude, followed by taxon-specific quantitation of isotopic incorporation using NanoSIMS analysis of high-density microarrays. The resulting data revealed that trophic response to AA availability falls along a continuum between copiotrophy and oligotrophy, and high and low activity. To illustrate strategies along this continuum more simply, we statistically categorized microbial taxa among three trophic types, based on their incorporation responses to increasing resource concentration. The data indicated that taxa with copiotrophic-like resource use were not necessarily the most active, and taxa with oligotrophic-like resource use were not always the least active. Two of the trophic strategies were not randomly distributed throughout a 16S rDNA phylogeny, suggesting they are under selective pressure in this ecosystem and that a link exists between evolutionary relatedness and substrate affinity. The diversity of strategies to adapt to differences in resource availability highlights the need to expand our understanding of microbial interactions with organic matter in order to better predict microbial responses to a changing environment.

Editor: Stefan Bertilsson, Uppsala University, Sweden

Funding: This work was supported by Lawrence Livermore National Laboratory (LLNL) Laboratory Directed Research and Development (LDRD) funding (grant # 11-ERD-066) and the Department of Energy (DOE) Office of Biological and Environmental Research (OBER) Genome Sciences Program (grant # SCW1039). The funders had no role in study design, data collection and analysis, decision to publish, or preparation of the manuscript.

Competing Interests: The authors have declared that no competing interests exist.

* E-mail: mayali1@llnl.gov

Introduction

Microbes dominate the biomass and biogeochemical activity of aquatic environments on a global scale [1]. Niche differentiation for resource acquisition, among other factors, allows the co-existence of a high diversity of aquatic microbes in the same volume of water, as different organisms are adapted to utilize the same resources but at different concentrations [2]. For example, ammonium and phosphate uptake at micromolar concentrations is dominated by larger organisms (mostly eukaryotes), whereas at nanomolar concentrations, organisms smaller than 1 micron (bacteria and archaea) dominate uptake [3]. A similar phenomenon has been described in marine ammonia oxidizing communities, where ammonia-oxidizing bacteria dominate activity at higher ammonium concentrations and ammonia-oxidizing archaea dominate activity at lower concentrations [4]. Niche differentiation likely also occurs at lower taxonomic levels; for example some microbial phyla appear to respond differently to varying concentrations of leucine, in particular during different stages of an algal bloom [5].

Niche differentiation may occur because most aquatic microbes acquire nutrients from their environment one molecule at a time through substrate-specific transporters often coupled with cell-bound extracellular enzymes that interact directly with dissolved

substrates on a molecular level [6]. Different transporters have substrate-binding efficiencies adapted to different concentrations [7], allowing microorganisms to respond to the resource patchiness of seawater [8]. In the context of both inorganic and organic nutrient acquisition, microbes have been classified into two main guilds with respect to their adaptation to substrate concentrations: oligotrophs and copiotrophs [9], the latter also referred to as opportunitrophs [2]. Oligotrophs are generally small cells with small genomes, are characterized by slow, steady growth, and numerically dominate in low nutrient environments such as sub-tropical oceanic gyres [10]. Copiotrophs, with larger cells and more extensive genomes [11], subsist on a feast-or-famine approach (fast growth interspersed with inactivity) and numerically dominate in high nutrient coasts and estuaries. While the use of these two terms has been quite beneficial for our conceptual understanding of microbial biogeochemical cycling, a more nuanced theoretical framework is needed to reflect the complexity of resource heterogeneity in nature [6]. For example, we might consider that trophic strategy is distributed on a continuum between oligotrophy and copiotrophy, with several intermediate states. An improved understanding of trophic strategies is important to reliably predict microbial responses to changes in resource availability likely to be caused by anthropogenic forces such as eutrophication and pollution.

Here, we expand on the classification scheme of oligotroph vs. copiotroph to more fully categorize the range of microbial substrate utilization strategies that enable microbes to co-exist through trophic niche differentiation. To develop these categories, we tested how microbial taxa responded in their incorporation of amino acids (AAs), organic substrates commonly utilized by aquatic microorganisms [12,13]. A number of previous studies (using bulk methods) have shown that AAs provide a large fraction of C and N requirements to aquatic bacterial communities [14,15]. Subsequent studies, with more modern techniques, have shown evidence that dominant marine taxa (such as SAR11) have the genetic capability to uptake AAs [16,17] and express them [13,18], and incubation experiments have validated these hypotheses with cell-specific methods [19,20]. To quantify AA incorporation at three different concentrations, we used Chip-SIP, a high-throughput method of quantifying taxon-specific incorporation of stable isotope labeled substrates [21]. Our results revealed a relatively high degree of complexity in the microbial community response to varying AA concentrations, and suggest that substrate affinity may be an evolutionarily conserved trait.

Materials and Methods

Incubation of Field Samples

Surface water was collected at the public pier in Berkeley, CA USA (37°51'46.67"N, 122°19'3.23"W) on 03/17/2011 and brought back to the laboratory within one hour in a cooler. No specific permissions were required for collection of seawater at this location and our studies did not involve endangered or protected species. Glass bottles (500 ml) were filled without air space and dark incubated at 14°C. Samples were incubated in triplicate bottles with 5 μM (High), 500 nM (Medium), and 50 nM (Low) mixed amino acids (99 atm % ^{15}N labeled; Omicron Biochemicals Inc., South Bend, IN, USA), collected by filtration after 12 hrs and frozen at −80°C. RNA extracts from triplicate incubations were combined and hybridized to a system-specific high-density 16S microarray (see description below) and subsequently analyzed by isotopic imaging with a nano secondary ion mass spectrometer (NanoSIMS 50, Cameca, France). RNA extracts from all three treatments (H = High, M = Medium, and L = Low) were combined for fluorescent labeling (see below) in order to compare the three concentration treatments to one another.

RNA Extraction and Labeling

The RNA samples were separately analyzed for fluorescence (with a microarray scanner) and isotopic enrichment (with the NanoSIMS), because fluorescence labeling causes dilution of the isotopic signal. RNA from frozen filters was extracted with the RNEasy kit according to manufacturer's instructions (Qiagen, Hilden, Germany). Alexafluor 532 labeling was done with the Ulysis kit (Life Technologies, Carlsbad, CA, USA) on the samples for fluorescent labeling for 10 min at 90°C (2 μL RNA, 10 μL labeling buffer, 2 μL Alexafluor reagent), followed by fragmentation. All RNA (fluorescently labeled or not) was fragmented using 1X fragmentation buffer (Affymetrix, Santa Clara, CA, USA) for 10 min at 90°C and concentrated by isopropanol precipitation to a final concentration of 500 ng μL^{-1}.

Microarray Hybridization and NanoSIMS

A phylogenetic microarray designed to target San Francisco Bay microbial communities (Figure S3, Table S1) included probes specific to ribosomal RNA operational taxonomic units (OTUs) as well as more general probes targeting the three domains of life (Bacteria, Archaea, Eukarya), two abundant marine bacterial orders (*Alteromonadales* and *Rhodobacterales*) and the genus *Polaribacter* [21]. Due to the variability in 16S diversity in different parts of the 16S phylogeny, there was no standard % similarity or taxonomic classification (genus, species, strain, etc.) that we could use to describe the lowest phylogenetic level targeted by the array. In general, taxa were targeted at the lowest possible phylogenetic level, subordinate to the genus level. To synthesize the microarrays, glass slides coated with indium-tin oxide (ITO; Sigma-Aldrich, St. Louis, MO, USA) were coated with silane Super Epoxy 2 (Arrayit Corporation, Sunnyvale, CA, USA) to provide a starting matrix for DNA synthesis. Custom-designed microarrays (spot size = 17 μm) were synthesized using a photolabile deprotection strategy [22] on the LLNL Maskless Array Synthesizer (Roche Nimblegen, Madison, WI, USA). Reagents for synthesis were delivered through an Expedite system (PerSeptive Biosystems, Framingham, MA, USA). For array hybridization, RNA samples (1 μg) in 1X Hybridization buffer (Roche Nimblegen, Madison, WI, USA) were placed in Nimblegen X4 mixer slides and incubated inside a Maui hybridization system (BioMicro Systems, Salt Lake City, UT, USA) for 18 hrs at 42°C and subsequently washed according to the manufacturer's instructions (Roche Nimblegen, Madison, WI, USA). Arrays with fluorescently labeled RNA were imaged with a Genepix 4000B fluorescence scanner (Molecular Devices, Sunnyvale, CA, USA) at pmt = 650 units. Secondary ion mass spectrometry (SIMS) analysis of microarrays hybridized with ^{15}N rRNA was performed at LLNL with a Cameca NanoSIMS 50 (Cameca, Gennevilliers, France). A Cs+ primary ion beam was used to enhance the generation of negative secondary ions. Nitrogen isotopic ratios were determined by electrostatic peak switching on electron multipliers in pulse counting mode, measuring ^{12}C^{14}N^{-} and ^{12}C^{15}N^{-} simultaneously. More details of the instrument parameters are provided elsewhere [21]. Ion images were stitched together and processed to generate isotopic ratios with custom software (LIMAGE, L. Nittler, Carnegie Institution of Washington). Isotopic ratios were converted to delta (permil) values using $\delta = [(R_{meas}/R_{standard}) - 1] \times 1000$, where R_{meas} is the measured ratio and $R_{standard}$ is the ratio measured in unhybridized locations of the sample. All fluorescence and NanoSIMS data have been deposited to NCBI's Gene Expression Omnibus archive under record number GSE56119.

Data Analyses

For each phylotype, isotopic enrichment of individual probe spots was plotted versus probe fluorescence and a linear regression slope was calculated. This slope (permil/fluorescence), which we refer to as the hybridization-corrected enrichment (HCE), is a metric that can be used to compare the relative incorporation of a given substrate by different taxa, or the relative incorporation by one taxon across different treatments [21]. Two procedures were carried out to assign phylotypes to guilds according to AA incorporation patterns. First, taxa were assigned to groups based on activity level, defined by HCE values for a single concentration. We note that this assignment does not take into consideration the response to different substrate concentrations. The HCE values at a single concentration followed a lognormal distribution with a small number of high values and mostly low to intermediate values (Fig. S1a). While the distribution of HCEs represents a continuum of activity, for simplicity we aimed to separate taxa into two groups: the first with high activity and the second with low activity. We set the cutoff response to distinguish highly active and less active taxa at a value of 50% of the maximum, which corresponded to a local high in the percent change along the distribution (Fig. S1b). The second procedure to assign phylotypes to trophic guilds involved the examination of individual taxa and their incorporation response (i.e., HCE values) under the three different concentrations (H, M, L) tested. For each taxon, we used an analysis of covariance (ANCOVA with a standard least squares model) to determine if concentration (or concentration and fluorescence together in a nested test) had a significant effect on isotopic enrichment (p<0.05), with the null hypothesis being no concentration effect (H≈M≈L). This procedure tested whether the slope of enrichment/fluorescence (HCE) or the y-intercept (isotopic enrichment) were significantly different for the different substrate concentrations. For each taxon, if the ANCOVA was not significant, it meant that the Null hypothesis (H≈M≈L) was not rejected (treatment H/M/L had no effect on the relationship between fluorescence and enrichment). If the ANCOVA was significant, then it means that at least 1 of the 3 treatments had a significant effect on isotopic enrichment. In theory, all the possibilities were: H>M, H>L, M>L, L>H, L>M, and/or M>H. However, since lower substrate concentrations should not lead to higher isotope incorporation, the actual possibilities were H>M~L, H~M>L, or H>M>L. With a post-hoc test, we examined each ANCOVA result (for each taxon) individually to find out which treatments were significantly different. We did not find any taxa that showed the pattern H>M~L, thus the only trophic strategies identified were the Null hypothesis (H≈M≈L), H>M>L and H≈M>L.

HCE values for each treatment were normalized to the sum of the HCE values for the three concentration treatments and were plotted on a barycentric ternary diagram (taxa not isotopically enriched were not plotted). Trophic strategies were mapped onto a maximum parsimony phylogenetic tree of representative 16S sequences targeted by the array to examine the phylogenetic distribution of these traits. To statistically test the phylogenetic distribution of trophic strategy on the tree, individual characters states were randomly reshuffled 1000 times, and for each reshuffling, a parsimony score (number of character state changes) was calculated. This was carried out with the software package Mesquite [23]. The parsimony score of the real character distribution was compared to this null distribution with an F-Test.

Results and Discussion

Establishment of Discrete Categories of Functional Responses to Resource Availability

In our initial analysis, we restricted our examination of the data to pairwise comparisons of two concentrations to start developing a classification scheme for microbial response to varying substrate concentrations (Fig. 1). Generally, incorporation at high substrate concentrations was weakly but significantly correlated to incorporation at low concentration, meaning that rRNA phylotypes with high activity at the high substrate concentration also had high activity at the low concentration (and vice-versa). This is exemplified by statistically significant linear regression analyses (e.g. $R^2 = 0.62$, p<0.0001 for H vs. L; black line in Fig. 1). This initial analysis delineated two groups of organisms: the first exhibited high isotopic incorporation (i.e. high activity), and the other group lower isotopic incorporation (i.e. lower activity). We considered taxa to exhibit high activity if their measure of isotopic incorporation was more than 50% of the highest incorporation recorded for a given concentration (Fig. S1). In this set of experiments, all highly active taxa (4) were members of the *Rhodobacteriaceae*, represented by probe sets targeting 3 taxa and the family more broadly. These marine bacteria belong to a subclade of the Alpha Proteobacteria, play key biogeochemical roles in the ocean [24] and are believed to be substrate generalists [25]. We note that relatively few taxa were classified as highly active (4 out

of 107) based on our 50% cutoff criterion, in agreement with the theory that only a few members of a microbial community are very active at any one time [26]. It is unlikely that this result was caused by bottle effects [27] or selection of taxa during batch culture incubations [28], as previous work has shown that marine communities incubated in bottles for extended periods become dominated by *Alteromonadales* [29], not *Rhodobacteriaceae*.

The response of the domain-specific probes indicated that Bacteria and Archaea increased their AA incorporation with increasing concentrations (Fig. 2a, 2b). In other words, as the concentration of added labeled substrate increased, isotopic incorporation increased. Eukaryotic AA incorporation was near the detection limit at the low concentration (and not considered significantly enriched by our conservative criteria) and was positive at the medium and high concentrations (Fig. 2c). The data for the three probe sets for bacterial orders/genus (*Alteromonadales*, *Rhodobacterales*, *Polaribacters*) allowed us to begin to split the bacterial response detected at the domain level into more specific components. These data revealed that the probes for these lower phylogenetic groups did not necessarily respond similarly to the Bacteria-specific probe set. *Alteromonadales* increased their isotopic incorporation with increasing substrate concentration (Fig. 2d), as the general Bacterial domain probes did. However, *Rhodobacterales* and *Polaribacters* showed no significant increase in isotopic incorporation from the medium to the high concentration treatment (Fig. 2e, 2f).

In general, the taxon-specific isotope incorporation responses to increasing AA availability spanned a continuous range, from organisms that exhibited no increased incorporation with increasing concentrations, to those with increased incorporation as AA availability increased. For the sake of conceptual understanding, we split this continuum into discrete categories, which were identified based on statistical tests of significantly different isotopic incorporation for the three AA concentrations. The maximum number of potential categories was inherently limited by the number of substrate concentrations tested. With a greater number of AA concentrations, we likely would have identified more response categories. The statistical tests that we carried out led to the classification of three main responses to increasing substrate concentrations. First, about one third of the taxa (32/107) exhibited no difference in isotopic incorporation among the three concentrations (Fig. 3a). We classified this strategy as "H≈M≈L" (high≈medium≈low). In some of these cases, isotopic incorporation was very low (near our detection limit) due to low activity by those taxa, so our method likely could not have detected differential incorporation even if it were occurring. However, in other cases, activity was substantially above background and also did not increase with increased substrate availability. Our interpretation of this phenomenon is that the microbial populations were saturated at the lower concentration, and adding more substrate did not increase incorporation. Another non-mutually exclusive possibility is that another nutrient was limiting to their growth, although this is less likely in a nutrient-rich eutrophic ecosystem such as San Francisco Bay. We interpret this strategy (H≈M≈L) to be analogous to oligotrophy, keeping in mind that trophic strategy is a relative measure. In other words, an organism showing this response is only an oligotroph when compared to another with a different response, and this would change depending on the ecosystem, the substrate, and the concentrations tested. For example, a copiotroph in our sampled eutrophic estuary would be considered an oligotroph relative to *E. coli* growing in the laboratory, and an oligotroph in SF Bay might be considered a copiotroph compared to a very slow growing microbe in the open ocean.

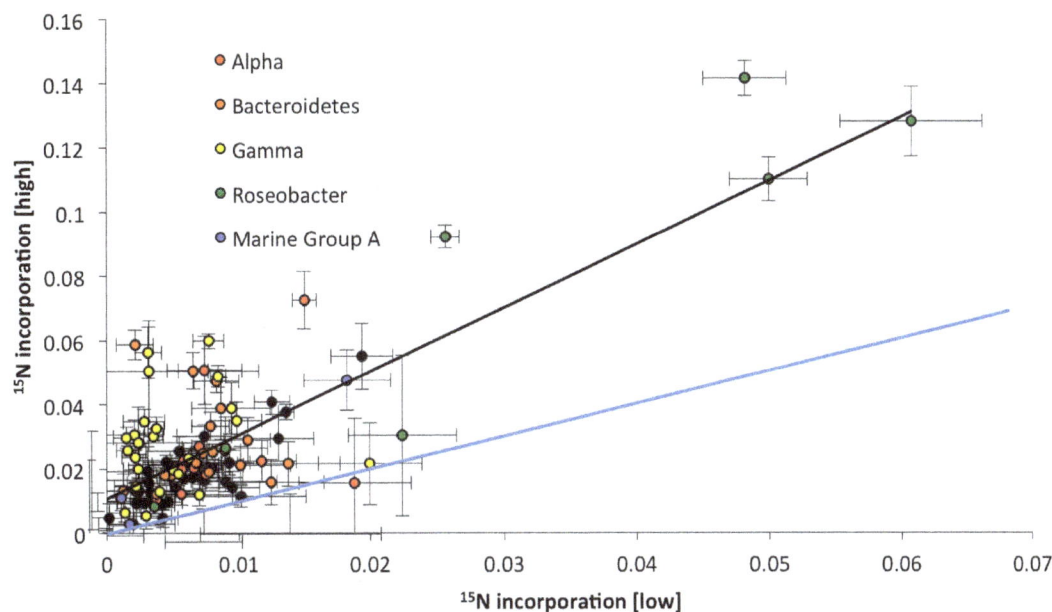

Figure 1. Pairwise comparisons of isotopic incorporation of ^{15}N labeled AAs by 107 16S rRNA phylotypes from SF Bay at two concentrations (high, 5 micromolar and low, 50 nanomolar). Each data point represents the HCE (hybridization corrected enrichment) for a probe set (the slope of delta permil divided by fluorescence). Error bars indicate two standard errors of the slope calculation. The black line represents the linear regression and the blue the 1 to l line.

The second identified trophic strategy included 40 phylotypes that exhibited increased incorporation at the medium (M) concentration compared to the low (L) but then were saturated or limited by another nutrient (H≈M>L, Figs. 3b). Physiologically, this indicates that AA incorporation was not saturated at the two lower concentrations but only at the highest concentration. We consider these taxa to be of intermediate trophy, able to respond to additions of 500 nM AAs, as compared to the relatively oligotrophic H≈M≈L strategy. The third identified group included 25 phylotypes with increased incorporation as more substrates were added (H>M>L, Figs. 3c). These organisms were not saturated at any of the concentrations tested, and we consider them to be relatively copiotrophic, able to respond to additions of 500 nM and 5 μM AAs. Since in general, additions of AAs to seawater lead to increased bacterial growth rates [30], one of our hypotheses was that highly active taxa (i.e., those with high AA incorporation) would be copiotrophs (H>M>L). We found this to be the case for two out of the three *Rhodobacteriaceae* taxa, while the other taxon and the *Rhodobacteriaceae* family exhibited intermediate trophy (H≈M>L). We originally assumed that metabolic activity and the ability to incorporate increased amounts of resources might be related, because fast-growing organisms might be expected to outgrow their competitors if they are able to incorporate resources when they are in excess. Thus we hypothesized that taxa with high AA incorporation would also be able to increase their incorporation as AA concentrations increased. This hypothesis was not fully supported by the data since not all high activity taxa were categorized as H>M>L (though many of them were).

After assigning taxa to a trophic strategy as outlined above, we plotted the isotopic incorporation for the three tested AA concentrations on a ternary diagram (Fig. 4). This plot graphically depicts the ratios of three variables on an equilateral triangle. In this case, the variables are the HCE values (a measure of isotopic incorporation) under 5 μM, 500 nM, and 50 nM added AAs, and for each phylotype, the denominator is the sum of the three

responses. We note that activity level (i.e., active vs. less active) is not taken into account here, but only the response to substrate concentrations (H≈M>L, etc.). Data on ternary plots were color-coded according to the defined trophic strategies as described above. The ternary plots allowed the assigned categories to be visualized, and demonstrated a clear graphical partition of H≈M>L and H>M>L, while the null hypothesis H≈M≈L was less clearly distinct.

Is Trophic Strategy Related to Phylogeny?

A common assumption in microbial ecology, although controversial [31], is that evolutionarily-related organisms share some physiological attributes with one another. To examine if such a pattern existed here, we color-coded the ternary plots according to bacterial taxonomy (not shown). No obvious taxonomic pattern was evident based on a visual examination of this plot. We also statistically tested the hypothesis that trophic strategy was related to phylogeny by mapping the identified trophic guilds onto a phylogenetic tree of the 16S rRNA gene (Fig. 5). We determined whether the three trophic strategies were statistically restricted to certain parts of the phylogeny using a reshuffling analysis. Two of the trophic strategies (H≈M>L and H>M>L) showed a statistically significant difference from the random distribution (Fig. S2), meaning that they were phylogenetically clustered. This implies that these strategies may be under positive selective pressure in this ecosystem, an environment rich in nutrients and particles [32,33]. In terms of taxonomic information, the intermediate strategy (H≈M>L) was more frequent in the Gamma Proteobacteria while the more copiotrophic strategy (H>M>L) was more frequent within the Bacteroidetes and chloroplasts (the latter is a marker for phototrophic eukaryotes). This is consistent with previous findings that Bacteroidetes are numerically enriched during algal blooms, when organic matter is in high supply [34,35]. The finding that eukaryotes were copiotrophic for AA incorporation was more unexpected. It is

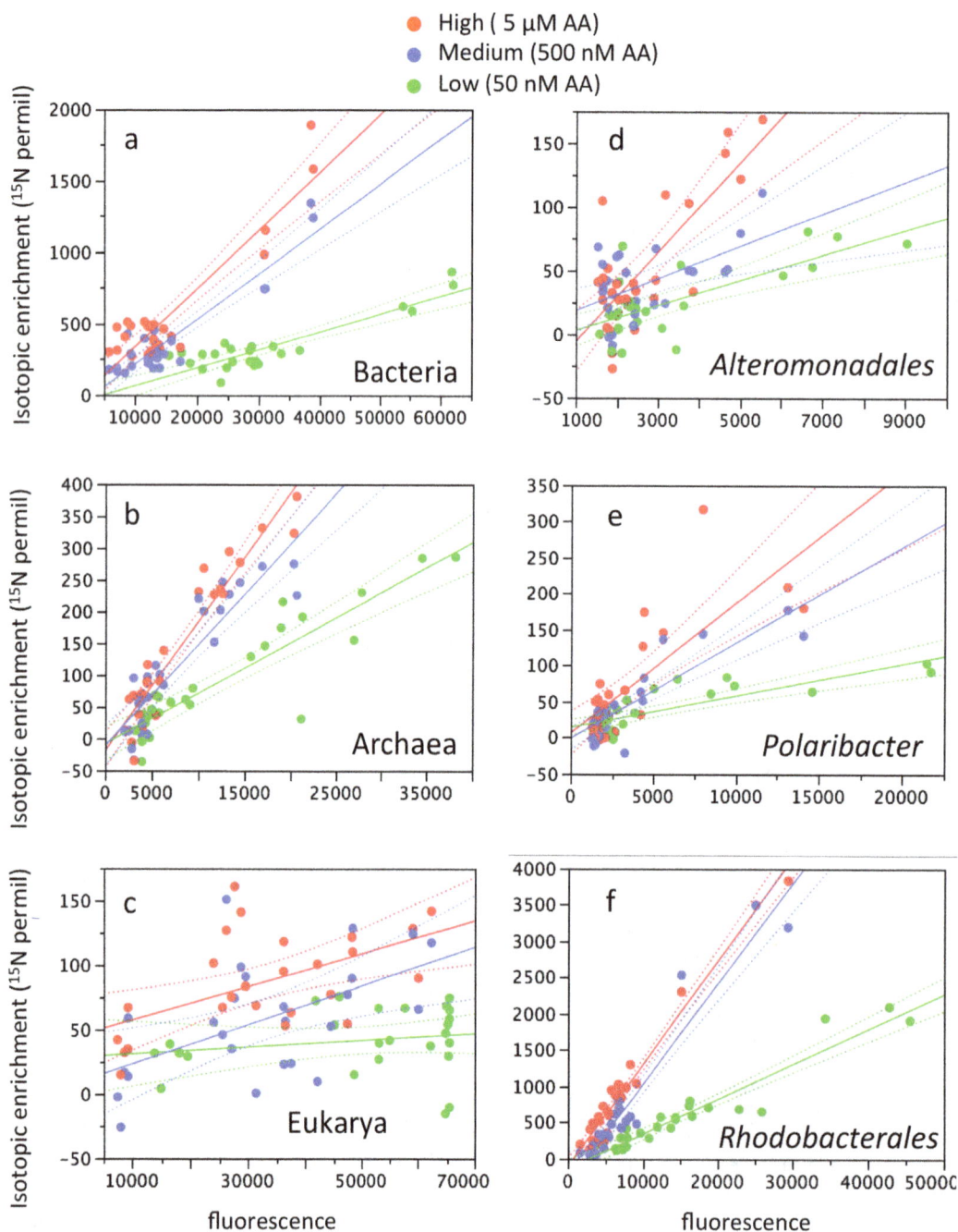

Figure 2. Response of taxa targeted by domain-specific (a–c) and genus or order specific (d–f) probes to increasing amino acid concentrations (red = high, blue = medium, green = low). Data points are for individual probes. Solid lines represent the linear regression and dotted lines are 95% confidence intervals.

known that eukaryotes incorporate dissolved organic nitrogen [36,37], and previous work has shown copiotrophy by larger size fractions (mostly eukaryotes) for ammonium and phosphate [3]. Taken together, these previous observations are consistent with our data. More generally, the finding of some correlation between trophic strategy and 16S phylogeny is noteworthy, as it suggests that the substrate affinities of microbial transporters (at least those for AAs) are evolutionarily conserved and results in evolutionarily related organisms having similar responses to increasing substrate concentrations.

Increased Categorization Enhances Conceptual Interpretation of Microbial Resource Utilization

One of the outcomes of our study is an increased number of categories of microbial resource acquisition strategies. Categorizing organisms with similar functional traits into more resource acquisition strategies allows us to strengthen our conceptual understanding of the processes they mediate, such as C and N cycling. However, the number of boxes needs to be relatively low or we lose our ability to use them conceptually. For example, we can easily conceptualize microbes based on their temperature

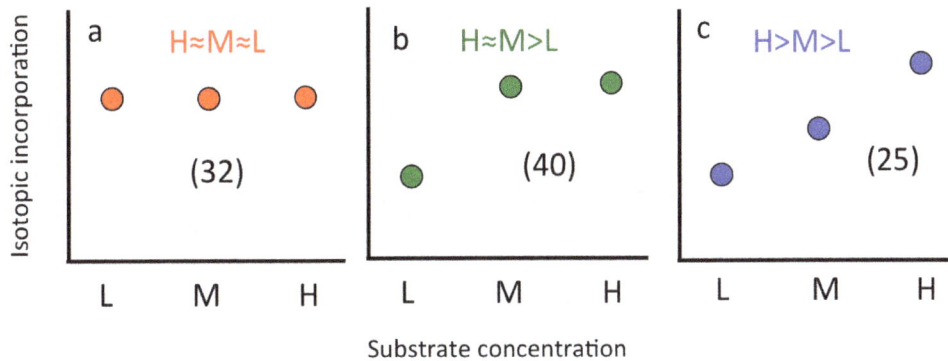

Figure 3. Three different types of microbial responses to resource availability identified by measuring isotopic incorporation of amino acids at High, Medium, and Low concentrations in SF Bay, with the numbers of taxa identified in parentheses.

tolerance, such as thermophiles, mesophiles, and psychrophiles [38], being adapted to high, medium, and low temperature, respectively. However, it is difficult to conceptualize ten or twenty such groups (e.g. those that prefer 70–80°C, 60–70°C, 50–60°C, etc.). Our analysis resulted in three trophic strategy guilds and two activity guilds, which allows us to retain a conceptual understanding of these functional categories without the high number of categories being too detailed to be useful.

The results presented here show that in addition to spatial and resource partitioning [39], microbial niche differentiation based on substrate availability is a factor that contributes to the maintenance of diversity in aquatic environments. In a eutrophic ecosystem, with three concentrations of one type of organic substrate, we defined more complex substrate acquisition strategies than the previously identified dichotomy of oligotrophy vs. copiotrophy. We note that although AAs are incorporated by most microbial groups, previous work has shown that assimilation of individual AAs can be quite variable and some taxa do not incorporate amino acids at detectable levels [40]. Since our experiments used a mixture of AAs, it is conceivable that our guild classifications might have been different if individual AAs were tested. The classifications would also potentially be different with other substrates, at different concentrations, during different times of the year, or in other ecosystems. Hence, the conclusions presented here are not meant to represent an absolute classification of these organisms into trophic guilds but instead to demonstrate that the well-documented concepts of oligotrophy and copiotrophy are simplifications of a continuum of responses, and perhaps more importantly, implies a relationship between activity and trophic strategy that is not universal.

The guilds identified here, which were based on both activity and response to increased substrate concentration, are critical microbial adaptations to an environment with constantly changing resource availability such as an estuary. Our analysis demonstrates that the complex substrate incorporation patterns of natural mixed microbial communities can be quantified and categorized in moderately simple classification schemes, leading to an improvement in how microbial populations are assigned to functional guilds. The categorization of microbes into an increasing number of "boxes" could be particularly valuable for biogeochemical models, where the vast diversity of bacterial heterotrophic processes [41] are typically represented in only one or two boxes. Expanding the bacterial "black box" (i.e., increasing the number of boxes to more accurately reflect the diversity of microbial responses to changing resources) should increase the accuracy and usefulness of such models. Furthermore, our quantitative approach would eventually allow these boxes to be dropped for a quantitative modeling approach, no longer needing discrete categorization. In particular, knowledge about microbial response to varying nutrient concentrations would be useful, as these are expected to change under a variety of climate change scenarios [42]. Together with new approaches to chemical characterization of organic matter complexity [43], the types of data shown here offer the possibility of testing hypotheses about critical microbial biogeochemical function. Our results quantifying the phylotype-specific response of microbial AA incorporation with increasing concentrations represents only the beginning of what we hope are many future experiments. For example, do microbial taxa exhibit the same resource use strategies for different substrates? Do taxa change their resource use strategies over time, space, or under different environmental conditions? It is clear that many of these questions must be answered before we have enough functional understanding of microbial biogeochemistry to be able to predict ecosystem-level responses to changing environmental conditions driven by both natural and anthropogenic forces.

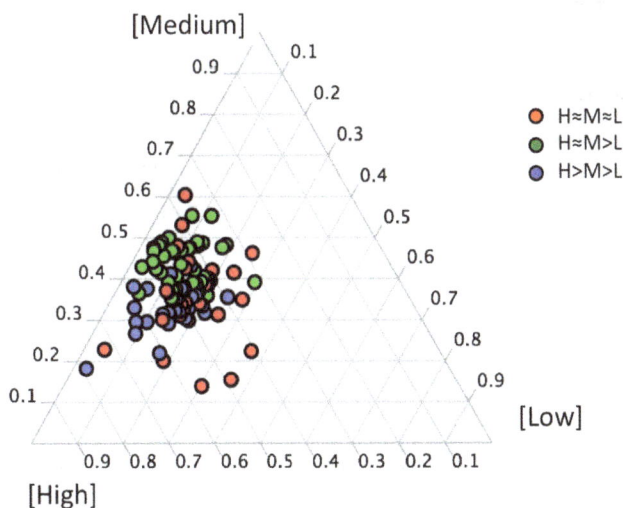

Figure 4. Ternary plot graphically depicting the ratios of the rRNA phylotype-specific incorporation to varying AA concentrations added to SF Bay water. Data are color-coded according to the trophic strategies identified in Fig. 3. The position of each data point in relation to the three corners represents the relative contribution of each concentration response.

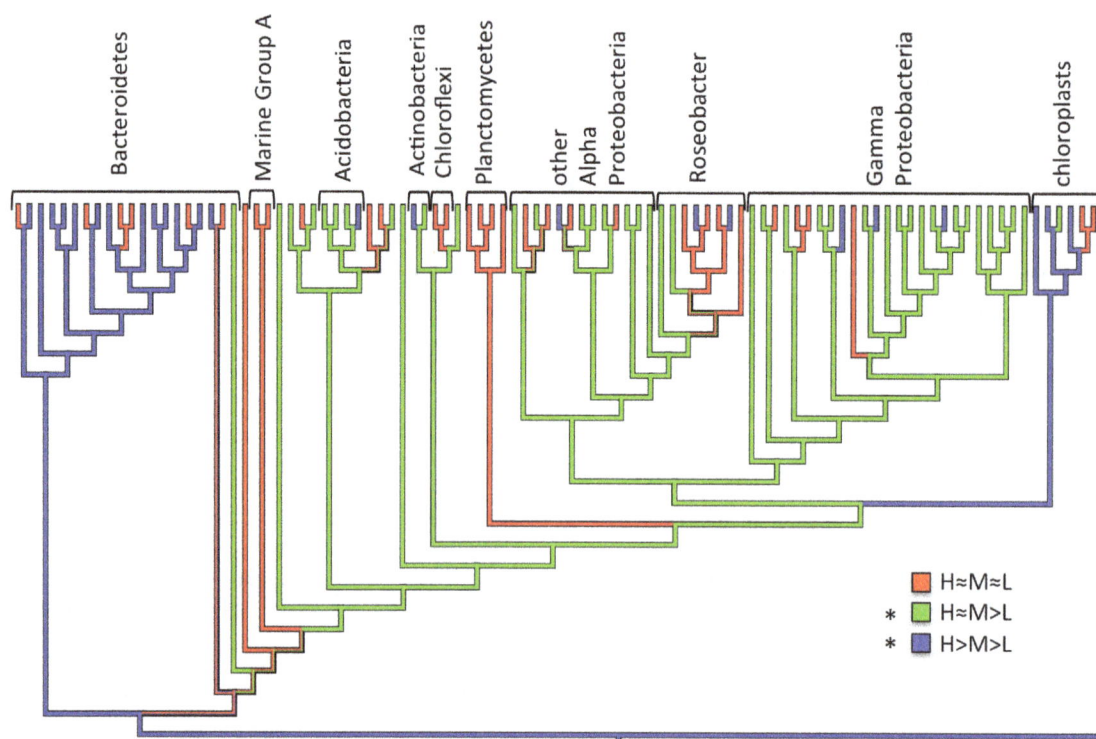

Figure 5. Amino acid incorporation trophic strategies mapped onto a maximum parsimony unrooted 16S rRNA gene phylogeny of taxa from a SF Bay seawater sample. Ancestral states were identified by parsimony. Asterisks indicate strategies with a statistically clustered distribution indicating a phylogenetic signal.

Supporting Information

Figure S1 (a) Hybridization Corrected Enrichment (HCE) values and (b) percent change from one value to the next lowest, for 107 taxa at the high amino acid concentration, ranked from high to low. Arrow indicates the cutoff used to delineate highly active versus less active tax, which corresponds to 50% of the maximum.

Figure S2 Result of phylogenetic distribution of taxa assigned to the "H≈M>L" strategy (a) and the "H>M>L" strategy (b) on the 16S phylogeny, showing the frequency distribution of parsimony scores from 1000 randomly shuffled character states. The parsimony score for the actual dataset is indicated by the arrow and was significantly different from the null distribution based on an F-test, indicating a phylogenetic signal.

Figure S3 Ribosomal Operational Taxonomic Units (OTUs) targeted by Chip-SIP phylogenetic microarray, including Genbank accession numbers of representative sequences, taxonomy,

trophic strategy identified by amino acid incorporation at 3 concentrations, and heat map measures of relative isotopic incorporation (blue = low, black = medium, and yellow = high). *denotes highly active taxa as defined in Figure 1.

Table S1 List of probes specific for San Francisco Bay natural community used for Chip-SIP analyses.

Acknowledgments

We thank L. Nittler for software development and R. Paerl, E. Ramon, and L. Gomez-Consarnau for insightful discussions. Work was performed under the auspices of the U.S. Department of Energy at Lawrence Livermore National Laboratory under Contract DE-AC52-07NA27344.

Author Contributions

Conceived and designed the experiments: XM. Performed the experiments: XM. Analyzed the data: XM PKW. Contributed reagents/materials/analysis tools: SM. Wrote the paper: XM PKW JPR.

References

1. Whitman WB, Coleman DC, Wiebe WJ (1998) Prokaryotes: the unseen majority. Proc Natl Acad Sci USA 95: 6578–6583.
2. Polz MF, Hunt DE, Preheim SP, Weinreich DM (2006) Patterns and mechanisms of genetic and phenotypic differentiation in marine microbes. Phil Trans R Soc B 361: 2009–2021.
3. Suttle C, Fuhrman JA, Capone DG (1990) Rapid ammonium cycling and concentration-dependent partitioning of ammonium and phosphate: implications for carbon transfer in planktonic communities. Limnol Oceanogr 35: 424–433.

4. Martens-Habbena W, Berube PM, Urakawa H, de la Torre JR, Stahl DA (2009) Ammonia oxidation kinetics determine niche separation of nitrifying Archaea and Bacteria. Nature 461: 976–979.
5. Alonso C, Pernthaler J (2006) Concentration-dependent patterns of leucine incorporation by coastal picoplankton. Appl Environ Microbiol 72: 2141–2147.
6. Azam F (1998) Microbial control of oceanic carbon flux: the plot thickens. Science 280: 694–696.
7. Nissen H, Nissen P, Azam F (1984) Multiphasic uptake of D-glucose by an oligotrophic marine bacterium. Mar Ecol Prog Ser 16: 155–160.

8. Stocker R, Seymour JR, Samadani A, Hunt DE, Polz MF (2008) Rapid chemotactic response enables marine bacteria to exploit ephemeral microscale nutrient patches. Proc Natl Acad Sci USA 105: 4209–4214.

9. Koch AL (2001) Oligotrophs versus copiotrophs. BioEssays 23: 657–661.

10. Morris RM, Rappe MS, Connon SA, Vergin KL, Siebold WA, et al. (2002) SAR11 clade dominates ocean surface bacterioplankton communities. Nature 420: 806–810.

11. Lauro FM, McDougald D, Thomas T, Williams TJ, Egan S, et al. (2009) The genomic basis of trophic strategy in marine bacteria. Proc Natl Acad Sci USA 106: 15527–15533.

12. Fuhrman JA, Azam F (1980) Bacterioplankton secondary production estimates for coastal waters of British Columbia, Canada, Antarctica, and California, USA. Appl Environ Microbiol 39: 1085–1095.

13. Poretsky RS, Sun S, Mou X, Moran MA (2010) Transporter genes expressed by coastal bacterioplankton in response to dissolved organic carbon. Envir Microbiol 12: 616–627.

14. Jørgensen NOG, Kroer N, Coffin RB, Yang X-H, Lee C (1993) Dissolved free amino acids, combined amino acids, and DNA as sources of carbon and nitrogen to marine bacteria. Mar Ecol Prog Ser 98: 135–148.

15. Coffin RB (1989) Bacterial uptake of dissolved free and combined amino acids in estuarine water. Limnol Oceanogr 34: 531–542.

16. Giovannoni SJ, Tripp HJ, Givan S, Podar M, Vergin KL, et al. (2005) Genome streamlining in a cosmopolitan oceanic bacterium. Science 309: 1242–1245.

17. Yooseph S, Nealson KH, Rusch DB, McCrow JP, Dupont CL, et al. (2010) Genomic and functional adaptation in surface ocean planktonic prokaryotes. Nature 468: 60–66.

18. Sowell SM, Abraham PE, Shah M, Verberkmoes NC, Smith DP, et al. (2011) Environmental proteomics of microbial plankton in a highly productive coastal upwelling system. ISME J 5: 856–865.

19. Malmstrom RR, Kiene RP, Cottrell MT, Kirchman DL (2004) Contribution of SAR11 Bacteria to Dissolved Dimethylsulfoniopropionate and Amino Acid Uptake in the North Atlantic Ocean. Appl Environ Microbiol 70: 4129–4135.

20. Alonso-Sáez L, Gasol JM (2007) Seasonal Variations in the Contributions of Different Bacterial Groups to the Uptake of Low-Molecular-Weight Compounds in Northwestern Mediterranean Coastal Waters. Appl Environ Microbiol 73: 3528–3535.

21. Mayali X, Weber PK, Brodie EL, Mabery S, Hoeprich P, et al. (2012) High-throughput isotopic analysis of RNA microarrays to quantify microbial resource use ISME J 6: 1210–1221.

22. Singh-Gasson S, Green RD, Yue Y, Nelson C, Blattner F, et al. (1999) Maskless fabrication of light-directed oligonucleotide microarrays using a digital micromirror array. Nat Biotech 17: 974–978.

23. Maddison WP, Madison DR (2011) Mesquite: a modular system for evolutionary analysis. 2.75 ed.

24. Buchan A, Gonzàlez JM, Moran MA (2005) Overview of the marine Roseobacter lineage. Appl Environ Microbiol 71: 5665–5677.

25. Newton RJ, Griffin LE, Bowles KM, Meile C, Gifford S, et al. (2010) Genome characteristics of a generalist marine bacterial lineage. ISME J 4: 784–798.

26. Smith EM, del Giorgio PA (2003) Low fractions of active bacteria in natural aquatic communities? Aquat Microb Ecol 31: 203–208.

27. Hammes F, Vital M, Egli T (2010) Critical evaluation of the volumetric "bottle effect" on microbial batch growth. Appl Environ Microbiol 76: 1278–1281.

28. Eilers H, Pernthaler J, Amann R (2000) Succession of Pelagic Marine Bacteria during Enrichment: a Close Look at Cultivation-Induced Shifts. Appl Environ Microbiol 66: 4634–4640.

29. Schafer H, Servais P, Muyzer G (2000) Successional changes in the genetic diversity of a marine bacterial assemblage during confinement. Arch Microbiol 173: 138–145.

30. Kirchman DL, Rich JH (1997) Regulation of Bacterial Growth Rates by Dissolved Organic Carbon and Temperature in the Equatorial Pacific Ocean. Microb Ecol 33: 11–20.

31. Doolittle WF, Zhaxybayeva O (2009) On the origin of prokaryotic species. Genome Res 19: 744–756.

32. Kranck K, Milligan TG (1992) Characteristics of suspended particles at an 11-hour anchor station in San Francisco Bay, California. Journal of Geophysical Research: Oceans 97: 11373–11382.

33. Hollibaugh JT, Wong PS, Murrell MC (2000) Similarity of particle-associated and free-living bacterial communities in northern San Francisco Bay, California. Aquat Microb Ecol 21: 103–114.

34. Kirchman DL (2002) The ecology of Cytophaga-Flavobacteria in aquatic environments. FEMS Microbiol Ecol 39: 91–100.

35. Teeling H, Fuchs BM, Becher D, Klockow C, Gardebrecht A, et al. (2012) Substrate-controlled succession of marine bacterioplankton populations induced by a phytoplankton bloom. Science 336: 608–611.

36. Bronk DA, See JH, Bradley P, Killberg L (2007) DON as a source of bioavailable nitrogen for phytoplankton. Biogeosciences 4: 283–296.

37. Mulholland MR, Lee C, Glibert PM (2003) Extracellular enzyme activity and uptake of carbon and nitrogen along an estuarine salinity and nutrient gradient. Mar Ecol Prog Ser 258: 3–17.

38. Ratkowsky DA, Olley J, McMeekin TA, Ball A (1982) Relationship between temperature and growth rate of bacterial cultures. J Bacteriol 149: 1–5.

39. Hunt DE, David LA, Gevers D, Preheim SP, Alm EJ, et al. (2008) Resource partitioning and sympatric differentiation among closely related bacterioplankton. Science 320: 1081–1085.

40. Salcher MM, Posch T, Pernthaler J (2013) In situ substrate preferences of abundant bacterioplankton populations in a prealpine freshwater lake. ISME J 7: 896–907.

41. Eichinger M, Poggiale JC, Sempéré R (2011) Toward a mechanistic approach to modeling bacterial DOC pathways: a review. In: Jiao N, Azam F, editors. Microbial carbon pump in the ocean. Washington D.C.: Science/AAAS.

42. Harley CDG, Randall Hughes A, Hultgren KM, Miner BG, Sorte CJB, et al. (2006) The impacts of climate change in coastal marine systems. Ecol Lett 9: 228–241.

43. Kujawinski EB (2011) The impact of microbial metabolism on marine dissolved organic matter. Ann Rev Mar Sci 3: 567–599.

6

Bioavailable Soil Phosphorus Decreases with Increasing Elevation in a Subarctic Tundra Landscape

Andrea G. Vincent[1,2]*, **Maja K. Sundqvist**[2], **David A. Wardle**[2], **Reiner Giesler**[1,3]

1 Department of Ecology and Environmental Sciences, Umeå University, Umeå, Sweden, **2** Department of Forest Ecology and Management, Swedish University of Agricultural Sciences, Umeå, Sweden, **3** Climate Impacts Research Centre, Department of Ecology and Environmental Sciences, Umeå University, Abisko, Sweden

Abstract

Phosphorus (P) is an important macronutrient in arctic and subarctic tundra and its bioavailability is regulated by the mineralization of organic P. Temperature is likely to be an important control on P bioavailability, although effects may differ across contrasting plant communities with different soil properties. We used an elevational gradient in northern Sweden that included both heath and meadow vegetation types at all elevations to study the effects of temperature, soil P sorption capacity and oxalate-extractable aluminium (Al_{ox}) and iron (Fe_{ox}) on the concentration of different soil P fractions. We hypothesized that the concentration of labile P fractions would decrease with increasing elevation (and thus declining temperature), but would be lower in meadow than in heath, given that N to P ratios in meadow foliage are higher. As expected, labile P in the form of Resin-P declined sharply with elevation for both vegetation types. Meadow soils did not have lower concentrations of Resin-P than heath soils, but they did have 2–fold and 1.5–fold higher concentrations of NaOH-extractable organic P and Residual P, respectively. Further, meadow soils had 3-fold higher concentrations of Al_{ox} + Fe_{ox} and a 20% higher P sorption index than did heath soils. Additionally, Resin-P expressed as a proportion of total soil P for the meadow was on average half that in the heath. Declining Resin-P concentrations with elevation were best explained by an associated 2.5–3.0°C decline in temperature. In contrast, the lower P availability in meadow relative to heath soils may be associated with impaired organic P mineralization, as indicated by a higher accumulation of organic P and P sorption capacity. Our results indicate that predicted temperature increases in the arctic over the next century may influence P availability and biogeochemistry, with consequences for key ecosystem processes limited by P, such as primary productivity.

Editor: Xiujun Wang, University of Maryland, United States of America

Funding: This project was funded by a Wallenberg Scholars award to DAW (http://www.wallenberg.com/kaw/) and a grant from the Centre for Environmental Research in Umeå (CMF) to DAW and RG (grant number 0632292; http://www8.umu.se/cmf/). The funders had no role in study design, data collection and analysis, decision to publish, or preparation of the manuscript.

Competing Interests: The authors have declared that no competing interests exist.

* E-mail: vincent.andrea@gmail.com

Introduction

Phosphorus (P) is an important macronutrient in subarctic tundra, where it is often co-limiting with nitrogen (N), and is sometimes the main limiting nutrient for plant growth [1–3]. The main source of plant-available, inorganic P (hereafter 'available P') in subarctic tundra is the biological mineralization of organic P [1,4,5]. Most soil P in the surface of tundra soils is organic [1,4,6] and is dominated by highly labile compounds [7]. Temperature is one of the main controls of organic matter decomposition in the arctic [8–10], meaning it is likely to constrain organic P mineralization and the supply of available P for plants. As such, warming experiments often show increases in P mineralization and/or plant P availability [9,11,12]. Likewise, an increase in foliar and litter P concentrations, together with a decrease in foliar and litter N to P ratios (indicative of greater relative P availability), have been observed with decreasing elevation (thus increasing temperature) in subarctic tundra [13]. The predicted annual average air temperature increases of 3–5°C in the subarctic during this century [14–16] could therefore influence the availability of P. Current knowledge on the distribution of different P forms in tundra landscapes and how they may be affected by temperature is limited, despite this information being crucial for understanding future temperature effects on bioavailable P.

Elevational gradients are powerful tools for studying how temperature and associated climatic factors, which shift with elevation, influence ecosystems properties and processes [17–21]. As such, an increasing number of studies in a wide range of ecosystems have used elevational gradients to study how temperature affects ecological processes [22–24] including in the subarctic [13,25–28]. Elevational gradients also provide excellent opportunities for exploring the impacts of temperature on the availability of soil P over larger spatial scales and timeframes than what is possible through conventional experiments [19]. There are, to our knowledge, no studies available on the responses of different P pools to elevation in subarctic or arctic tundra. These landscapes are biogeochemically heterogeneous as a consequence of spatial variation in topography and plant community structure [29,30]. This results in high spatial variation in P availability [6,25] and in the concentrations of Al and Fe [6,31], which influence P availability [32,33]. The Fennoscandian tundra therefore consists of a mosaic of highly contrasting vegetation types. Specifically, heath vegetation occurs on soils with low pH and N availability and is dominated by slow-growing dwarf-shrubs, while meadow vegetation grows on soils that are more N-rich and is dominated by faster-growing herbaceous species [25,29,30]. The biogeochemistry of P differs between the two vegetation types [6] and soil phosphate concentrations and plant foliar N to P ratios suggest that the relative importance of P versus N limitation is greater in meadow than in heath [6,13,25]. As such, obtaining a representative picture of how P biogeochemistry varies in the subarctic requires explicit recognition of both vegetation types.

In this study, we used a well-established elevational gradient [13,25,26,34] in which both heath and meadow vegetation types occur at all elevations, to study the effects of elevation-associated variation in temperature on P availability and biogeochemistry in a subarctic ecosystem. We also used this gradient to examine whether previously reported changes in foliar and litter P contents and N to P ratios with elevation are matched by shifts in the concentration of P fractions of different lability. To determine P fractions we used the Hedley fractionation method [35–38], an approach widely used to determine landscape-level variation in P availability and dynamics [6,39,40–42]. Additionally, we examined if other known drivers of P availability, such as soil P sorption capacity and Al and Fe concentration, influence the distribution of P fractions across the gradient for both the heath and meadow vegetation. Specifically, we tested the hypotheses that (1) The concentrations of labile P fractions decline with elevation (and therefore temperature) regardless of vegetation type, (2) Across all elevations, meadow soils have consistently lower concentrations of labile P than heath soils, together with higher concentrations of Al and Fe and higher soil P sorption capacity. By addressing these hypotheses we aim to better understand how temperature changes, such as those that are expected through climate warming, may affect P availability across two dominant vegetation types in subarctic tundra ecosystems.

Materials and Methods

Ethics statement

This study was carried out across an elevational gradient ranging from 500 to 1000 m above sea level (a.s.l.) along the north-east facing slope of Mount Suorooaivi (1193 m), located approximately 20 km south-east of Abisko, northern Sweden (68°21′ N, 18°49′ E), as described in [25]. Figure 1 shows a map with the location of the study site and the elevational gradient. No part of this gradient is located within national reserves and the land is public and not government protected. We confirm that all national and international rules were observed during the field work. This investigation did not involve measurements on animals or humans. The soils collected for this research were sampled at very small spatial scales and thus had negligible effects on ecosystem functioning. We have no commercial interests or conflicts of interest in performing this work.

Study site

The mean annual precipitation in the area, measured at the Abisko Scientific Research Station, was 310 mm for the period 1913–2000, with the highest mean monthly precipitation in July (51 mm) and the lowest in April (12 mm) [43]. The treeline in this area is formed by *Betula pubescens* spp. *czerepanovii* (mountain birch) and is located at 500–650 m a.s.l. at the study site. Two types of vegetation, heath and meadow, grow in a mosaic across the study area and co-occur on all elevations, with the meadow generally found in shallow depressions. The heath is dominated by ericaceous dwarf-shrubs such as *Vaccinium vitis-idaea*, *V. uliginosum* spp. *uliginosum*, *Empetrum hermaphroditum* and *Betula nana*. The meadow vegetation is dominated by graminoids such as *Deschampsia flexuosa* and *Anthoxanthum alpinum*, herbs such as

Saussurea alpina, *Viola biflora* and *Solidago virgaurea*, and sedges, notably *Carex bigelowii* [25]. The bedrock is comprised of salic igneous rocks and quartic and phyllitic hard schists. The soils are podzols at lower elevations and cryosols at higher elevations. For more details on the study system see Table S1 and [13,25].

In the summer of 2007, four replicate plots (2×2 m) were established in each of the two vegetation types in each of six elevations (every 100 m ranging from 500 to 1000 m) rendering a total of 48 plots as described by [13,25]. To minimize pseudoreplication within each elevation, the average distance between each plot and the next nearest plot was c. 15 m (with the mean distance between the two most distant plots in each vegetation type being c. 100 m). Because the microtopography, hydrology, and soil fertility of these communities is highly spatially heterogeneous over short distances (i.e. in the order of a few metres) [29], it is expected that the 15 m distance among plots is enough to ensure adequate independence among them [13,25]. Plots at the 500 m elevation site were located in open birch forest, plots at 600 m were situated immediately above the forest line, and plots from 700 m to 1000 m were devoid of trees [25]. Monthly mean air temperatures during August 2009 at 500 m, 700 m and 1000 m at the study site were 12.1°C, 11.8°C, and 9.9°C, respectively. The daily mean temperature across the elevational gradient during the growing season of 2009 is given in Figure S1; similar data for the previous year (2008) is also given in Figure S2 [25]. Elevational gradients of this type serve as useful natural experiments to inform on the effects of temperature on ecological properties when other potentially co-varying factors can be kept constant [17,18]; as such, all plots in our study have the same aspect (north-east facing slope), parent material, and slopes of 4–18°.

Soil sampling

Humus soils were sampled on August 4, 2009, in a 1×1 m subplot inside each 2×2 m plot. For each plot, a minimum of four 4.5 cm diameter cores were sampled to the full humus depth to ensure a total sample volume of approximately 0.3 L humus, and humus depth was recorded. The humus depth (mean ± standard error) across all heath plots was 5.4±0.2 cm (for each elevation mean depths are 6.3, 6.1, 5.0, 4.0, 5.8, and 5.4 starting at 500 m.a.s.l. in ascending order), and for meadow plots it was 3.0±0.6 cm (starting at the 500 m.a.s.l. site and in ascending order, 7.2, 2.2, 1.7, 1.1, 3.2, and 2.8 cm). Within each plot, the cores were sieved (2 mm mesh) in the field to homogenize the samples, and combined to yield a single bulked sample per plot. Samples were sealed in polyethylene bags and transported to the laboratory on the same day as sampling. From each sample, a subsample was immediately stored at 2°C (<48 h) and the remaining portion was frozen at −20°C.

Hedley fractionation

In order to characterize soil P composition, we performed a five-step sequential extraction [38] with some modifications [44,45], outlined in Figure 2. This method was chosen because it provides a direct estimate of the lability of different operationally-defined P pools, and is the most commonly used method to investigate differences in P availability and dynamics in natural soils [35–37,46]. In Step 1, 2 g (dry weight) of humus soil were combined with 180 mL deionized water and one 9×62 mm anion exchange membrane (hereafter called 'resin') (55164 2S, BDH Laboratory Supplies, Poole, England) [47] in a 250 mL centrifuge bottle and shaken overnight (16 h, 150 rpm). The following day, the resins were removed and the sample centrifuged (14 000 g, 15 min, 10°C), after which the supernatant was discarded and the

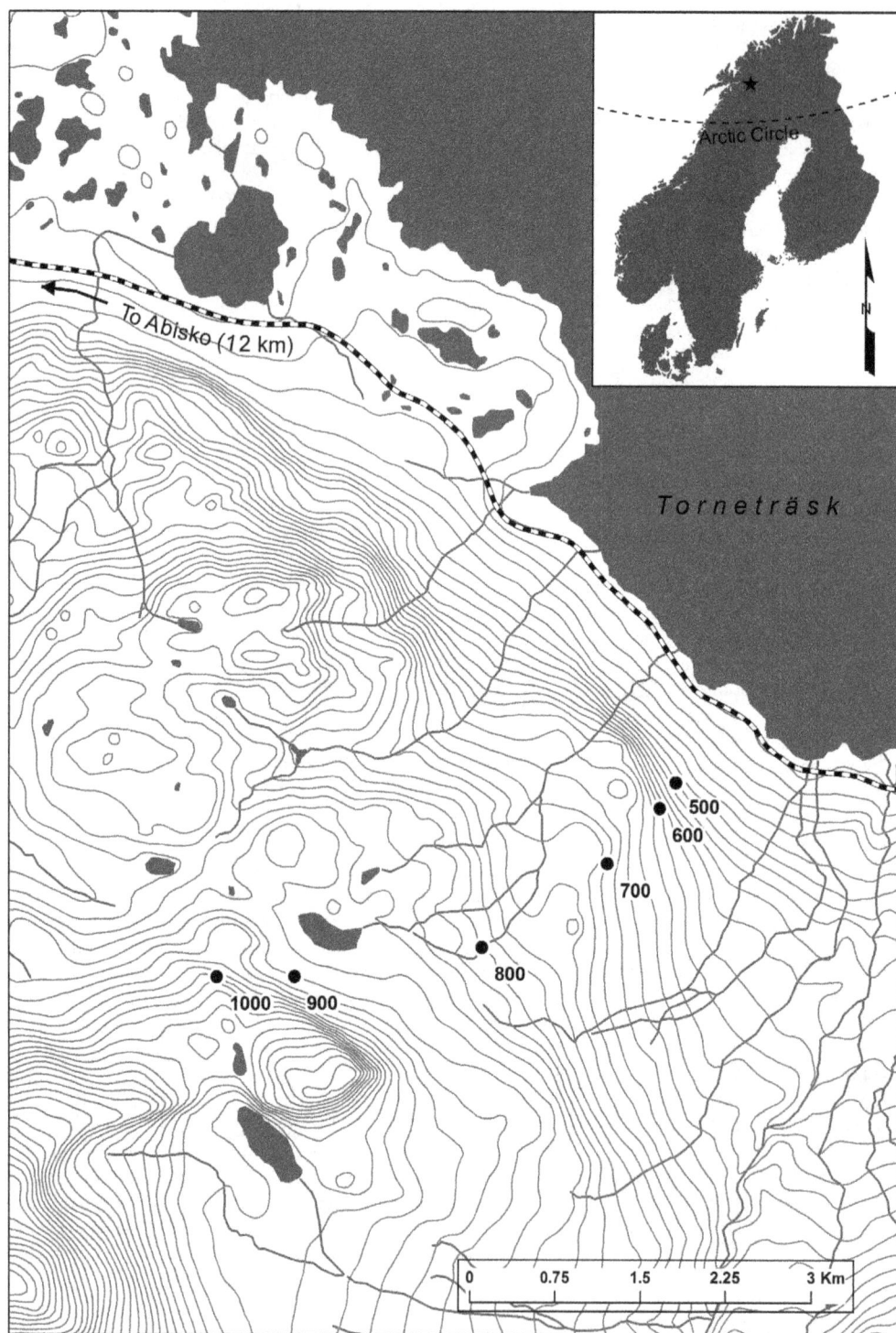

Figure 1. Location of the study elevational gradient. Filled black circles indicate each of the six study sites, ranging in elevation from 500 to 1000 m above sea level. The nearest town is Abisko, and its position in Sweden is indicated by a star in the map inset.

remaining soil used in Step 2. The resin was transferred to a bottle and eluated on a shaker (1 h, 150 rpm) with 40 mL NaCl. The eluate was immediately stored at −20°C until further analysis of this resin-extractable P fraction, hereafter referred to as 'Resin-P'. In Step 2, the soil remaining from Step 1 was combined with 180 mL 0.5 M NaHCO₃, set on a shaker (16 h, 150 rpm), centrifuged (14 000 g, 15 min, 10°C), and 40 mL of the

supernatant removed and stored at −20°C for determination of organic and inorganic NaHCO₃-extractable P (hereafter referred to as 'Bic-P$_o$' and 'Bic-P$_i$', respectively). In Step 3, the soil remaining from Step 2 was extracted with 0.2 M NaOH following the same procedure as in Step 2 and stored at −20°C for further analysis of organic and inorganic NaOH-extractable P (hereafter referred to as 'NaOH-P$_o$' and 'NaOH-P$_i$', respectively). In Step 4, the soil remaining from Step 3 was combined with 1.0 M HCl, following the exact same extraction procedure as in Steps 2 and 3 and the supernatant stored at $-20°C$ for further analysis of HCl-extractable P (hereafter, 'HCl-P'). In the final and fifth step, the soil remaining from Step 4 was washed with 180 mL of deionised water by shaking for 1 h, centrifuging, and discarding the supernatant, after which the soil was set to air dry at room temperature. This residual air-dried soil was ground in a ball mill and a 200 mg subsample was combined with 4 mL of concentrated nitric acid (HNO_3) and 1 mL of hydrogen peroxide (H_2O_2) (soil to solution ratio 1:40) and digested in a microwave (Mars XPress, CEM, Germany). This constitutes the 'Residual-P' fraction.

Total labile P consists of resin-extractable and bicarbonate-extractable P [48]. Resin-P is well correlated with P uptake by plants [49,50], has a rapid turnover and high bioavailability, and consists of P in a form that can exchange freely between the solid phase of the soil and the soil solution [35,51]. Bicarbonate-P_i (Bic-P_i) has similar sources to resin-P, turns over fast and is also bioavailable in the short-term [35], while bicarbonate-P_o (Bic-P_o) is easily mineralizable and supplies plant-available P. Hydroxide-extractable P_i (NaOH-P_i) and P_o (NaOH-P_o) are associated with Al and Fe phosphates, have lower plant availability, and longer turnover times. HCl-extractable P represents calcium-bound P_i and is often taken to represent P associated with primary minerals; it is also considered to be more stable [48]. By using extractants of different strength, this method is considered to quantify pools according to their 'lability' to plants [48].

Sorption index

In order to estimate the relative P sorption capacity of each soil sample from each plot, we used a single point P sorption method [52]. For each sample, 2 g (dry weight) of soil were weighed into each of two 60 mL bottles respectively, and 40 mL of 100 mM KCl was added to each bottle. For one of the two bottles, 1.6 mg P

g soil^{-1} (dry weight basis, equivalent to 50 mmol kg^{-1} soil) of phosphate (KH_2PO_4) was added, which yielded two suspensions per soil sample: one with added P (spiked) and one without added P (unspiked). The suspensions were shaken for 24 h, filtered (Munktell 00H filter paper, pore size approx. 1 μm; Grycksbo, Sweden) and the amount of phosphate in each of the two filtrates was determined colorimetrically. The amount of sorbed P was estimated as the difference between the phosphate concentration in spiked and unspiked samples. We calculated the P sorption capacity using the equation

$$Sorption = \frac{x}{\log c}$$

where x is the quantity of P sorbed onto soil constituents and c is the equilibrium P concentration in the soil +P solution [52]. Thus, a high index indicates a high P sorption capacity.

Determination of P in extracts

All Hedley fractionation and sorption index extracts were analysed for molybdate-reactive P using a flow injection analyzer (FIA) (FIAstarTM 5000 Analyzer, FOSS Analytical AB, Höganäs, Sweden). The NaHCO$_3$ and NaOH solutions were diluted by a factor of 5–10 and amended with sulphuric acid (20 μl concentrated H_2SO_4 to 5 mL diluted extract) in order to precipitate organic material, after which they were centrifuged and the supernatant analyzed by FIA. Since the NaHCO$_3$ and NaOH extracts were colored, the measured P concentration was corrected by subtracting the effect of the color in the analysis. Total P in the NaHCO$_3$ and NaOH extracts was determined after digestion with acidified potassium persulfate ($K_2S_2O_8$), and inorganic P was analysed as above. The concentration of organic P was calculated as the difference between total and inorganic P. Phosphorus in the digests was analysed as above. Total soil P was calculated as the sum of all P fractions measured in the Hedley fractionation [6,39,44,53] and we refer to it hereafter as 'Total P'. Since all of our samples are from organic soils, we express the concentrations of all P fractions as mg kg^{-1} soil dry weight, which would be equivalent to expressing them on a per unit organic matter basis.

Aluminium and iron

Aluminium (Al) and iron (Fe) concentrations in the soil samples were determined following extraction by 0.2 M acid oxalate ($C_2H_8N_2O_4$) adjusted to a pH of 3 [54]. A subsample of 0.5 g dry weight of soil from each sample was combined with acid oxalate solution in a 1:30 soil to solution ratio and shaken on an orbital shaker for 4 h in the dark. Extracts were subsequently filtered (00H, Munktell Filter AB, Grycksbo, Sweden) and stored at 5°C (~2 months) following analysis of oxalate extractable Al and Fe (hereafter referred to as Al$_{ox}$ and Fe$_{ox}$, respectively) determined by inductively-coupled plasma optical-emission spectroscopy (ICP-OES) (Perkin Elmer). The oxalate extractant is assumed to release exchangeable Al and Fe and dissolve non-crystalline and poorly crystalline oxides of Al and Fe (i.e. organic and amorphous Al and Fe forms), which are major P sorbents in non-calcareous soils [55]. Concentrations of Al$_{ox}$, Fe$_{ox}$, and all P fractions are expressed as mg kg^{-1} soil dry weight on the basis of oven-dried soils (105°C, 24 h).

Statistical analysis

We used multivariate ANOVA (MANOVA) for the entire dataset given the large number of response variables measured per soil sample. We followed a significant MANOVA with a two-way ANOVA to test for the effect of elevation and vegetation type and

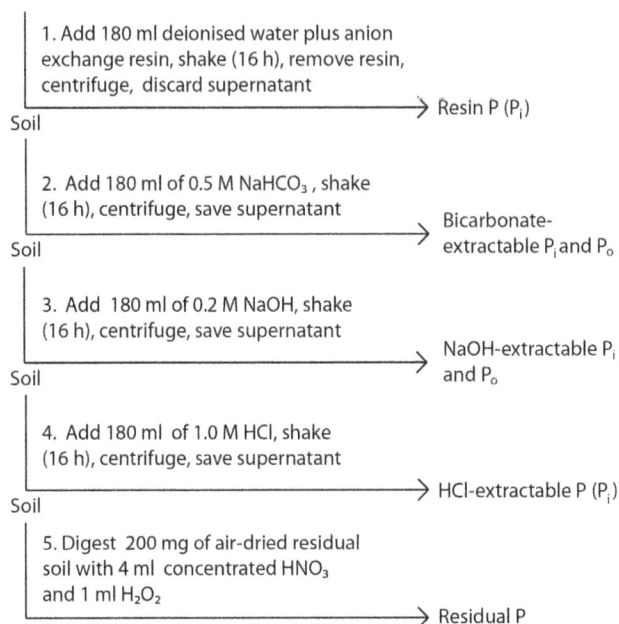

Add 2 g (dry weight) soil to a 250 ml centrifuge bottle

1. Add 180 ml deionised water plus anion exchange resin, shake (16 h), remove resin, centrifuge, discard supernatant
→ Resin P (P$_i$)

2. Add 180 ml of 0.5 M NaHCO$_3$, shake (16 h), centrifuge, save supernatant
→ Bicarbonate-extractable P$_i$ and P$_o$

3. Add 180 ml of 0.2 M NaOH, shake (16 h), centrifuge, save supernatant
→ NaOH-extractable P$_i$ and P$_o$

4. Add 180 ml of 1.0 M HCl, shake (16 h), centrifuge, save supernatant
→ HCl-extractable P (P$_i$)

5. Digest 200 mg of air-dried residual soil with 4 ml concentrated HNO$_3$ and 1 ml H$_2$O$_2$
→ Residual P

Figure 2. Flow chart showing the five steps involved in the sequential phosphorus fractionation used in this study.

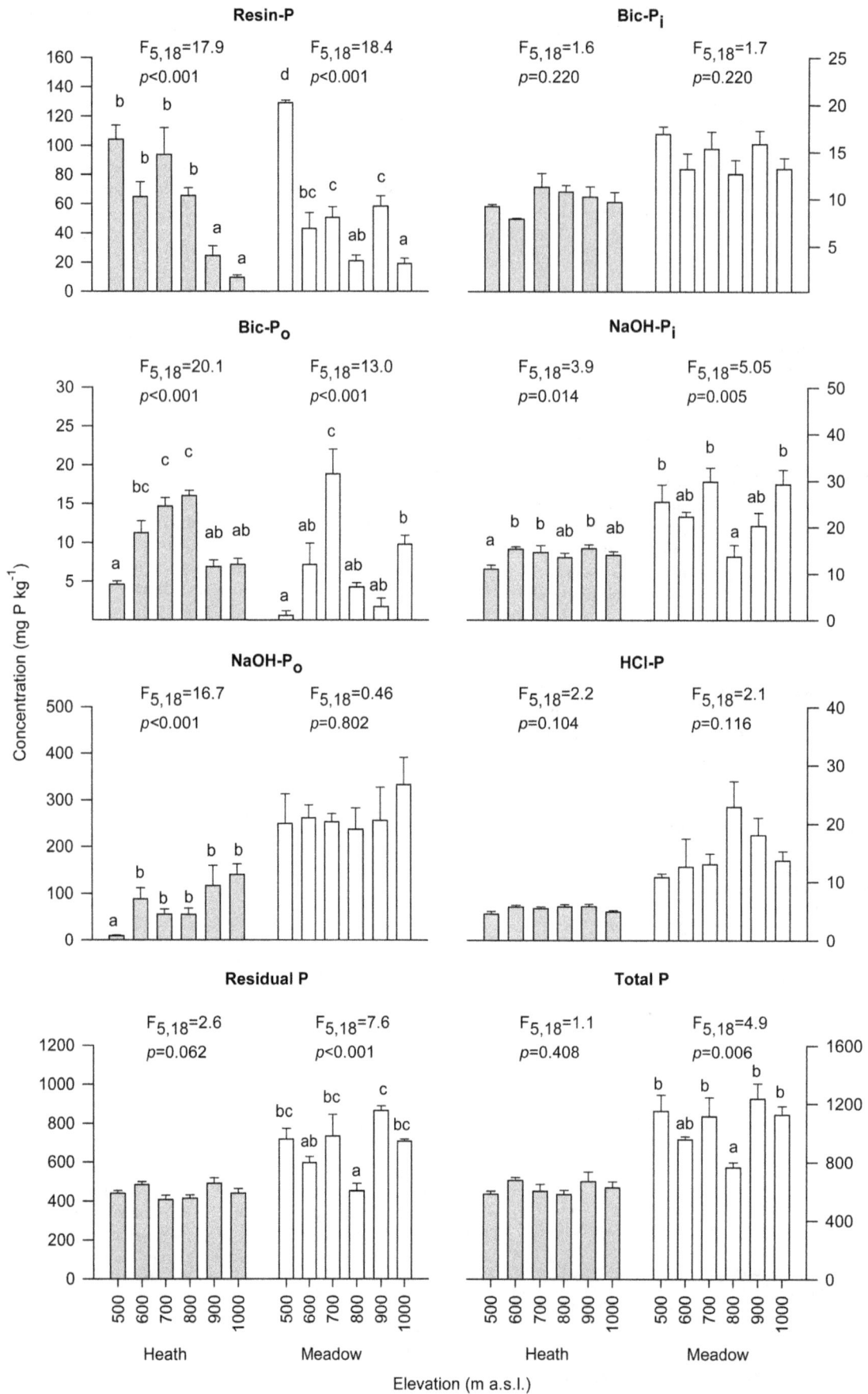

Figure 3. Concentration of phosphorus fractions in humus soils in contrasting vegetation types across an elevational gradient. Soils were collected in subarctic heath and meadow vegetation along an elevational gradient (500–1000 m) in Abisko, Sweden. Panels represent phosphorus fractions extractable with: anion-exchange resins (Resin-P); $NaHCO_3$ (inorganic fraction - Bic-P_i; and organic fraction - Bic-P_o); NaOH (inorganic fraction - NaOH-P_i; and organic fraction - NaOH-P_o); HCl (HCl-P fraction); non-extractable P (Residual P fraction); and Total P (arithmetic sum of all P fraction). Bars represent mean concentration (+1 SE) for four plots; for each P fraction and within each vegetation type, F and p values (with d.f.) are from a one-way ANOVA testing for the effect of elevation within each vegetation type, and bars topped with the same letter do not differ at $p = 0.05$ (Tukey's h.s.d.). Note the difference in y-axis scales.
doi:10.1371/journal.pone.0092942.g003

their interaction on each response variable [13,25,26]. We chose ANOVA because it is the most powerful way of detecting significant responses to the underlying gradient even when these responses are not unidirectional or simple [56]. To further explore the effects of elevation within vegetation types, one-way ANOVA testing for the effect of elevation was performed separately for both heath and meadow. Where significant effects of elevation were found, data were further analyzed for differences among means using Tukey's honestly significant difference (h.s.d.) at $p = 0.05$. Tukey's h.s.d. was chosen because it reduces Type I error when multiple comparisons are being performed [57,58].

In order to account for potential effects of co-variation of soil P sorption capacity with elevation on labile P concentrations, we divided the concentration of each of the three labile components (i.e. Resin-P, Bic-P_i, and Bic-P_o) for each soil sample by the sorption index for that sample. We then performed separate one-way ANOVA testing for the effect of elevation on the transformed data, followed by Tukey's h.s.d. at $p = 0.05$ after a significant ANOVA result. We used Pearson's correlation to test for the relationship between Al_{ox} and Fe_{ox}; linear regression was used to test for the relationship between soil P fractions with $Al_{ox} + Fe_{ox}$, sorption index and temperature. When required, data were transformed to conform to the assumptions of parametric tests. For statistical analyses we used SPSS PASW Statistics 18.0 and R version 3.3.0 (www.r-project.org).

Results

Effects of elevation

Overall, Residual-P was the most abundant P fraction in humus across all elevations and vegetation types (59–76% of total soil P), followed by NaOH-P_o (1.6–31%), and Resin-P (1.5–18%) (Figure 3, Table S2). Total labile P (i.e. the sum of Resin-P, Bic-P_i and Bic-P_o) represented 4–20% of total soil P (Table S2). The concentrations of the remaining fractions ranged from 0.8–3.0% of total soil P (Table S2). The most unidirectional effect of elevation was found for Resin-P, for which the highest concentrations were recorded at the lowest elevation and the lowest concentrations at the highest elevation, for both vegetation types (Figure 3, Table 1). Resin-P concentrations at the highest elevation were 7–fold and 11–fold lower than at the lowest elevation for heath and meadow, respectively. Total labile P trends mirrored those of Resin P, concentrations at the highest elevation (1000 m) in heath were less than one fifth of those recorded at the 500 m (lowest elevation) and 700 m sites. In meadow, Total labile P concentrations at the highest elevation were less than one third of those recorded at the lowest elevation. While elevation also had a significant effect on all other soil properties except Bic-P_i and HCl-P (which was almost significant) (Table 1, Table 2), there were no simple unidirectional trends with elevation for any of the other P fractions measured (Figure 3) or for $Al_{ox} + Fe_{ox}$ or the sorption index (Figure 4). After concentrations were divided by the sorption index, the unidirectional elevational trend remained for Resin-P and the non-unidirectional effect remained for Bic-P_o; further, a significant effect of elevation emerged for Bic-P_i; through it being highest at the lowest elevation for the meadow (Figure S3). When

all elevations and vegetation types were combined, Resin-P was significantly and negatively related with both $Al_{ox} + Fe_{ox}$ and the Sorption Index, although the relationship was weak (Table 3). Significant positive relationships were observed between most other P fractions and $Al_{ox} + Fe_{ox}$ and Sorption Index, the strongest being for NaOH-P_o and HCl-P (Table 3). Finally, Resin-P and Total labile P were significantly positively related with temperature in both the heath and meadow vegetation (Table S3), as was the case for pH in the heath. NaOH-P_o was negatively correlated with temperature in the heath (Table S3).

Effects of vegetation

The concentrations of all P fractions except Resin and Total labile P were significantly different between meadow and heath,

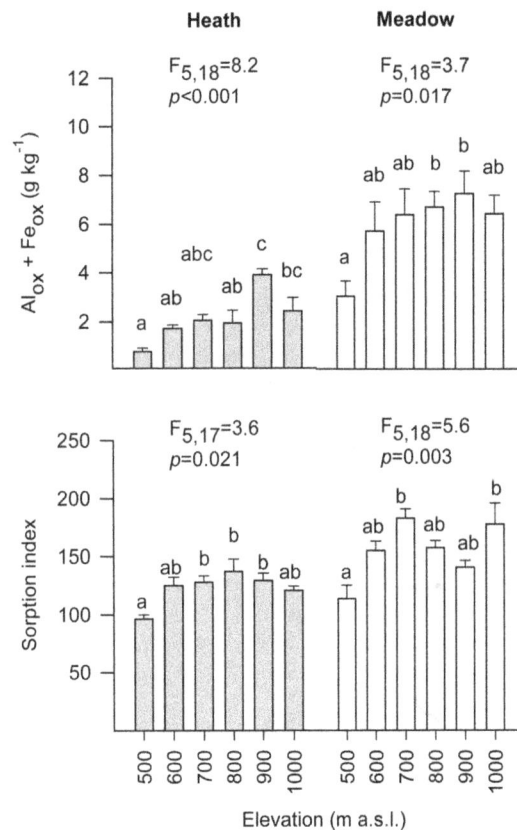

Figure 4. Concentration of aluminum and iron, and phosphorus sorption index for humus soils across an elevational gradient. Soils were collected in subarctic heath and meadow vegetation types along an elevational gradient (500–1000 m) in Abisko, Sweden. The top panel represents the concentration of the sum of oxalate-extractable Al and Fe ($Al_{ox} + Fe_{ox}$), the bottom panel represents the soil phosphorus sorption index. Bars represent mean values (+1 SE) for four plots; within each vegetation type, bars toped with the same letter do not differ at $p = 0.05$ (Tukey's h.s.d.) after a significant ANOVA (ANOVA results in Table 2).

Table 1. Effect of vegetation type and elevation as determined by multivariate analysis of variance (MANOVA) (F- values, with p in parenthesis) and two-way ANOVA on the concentration (mg kg^{-1}) of different phosphorus (P) fractions in humus soils along an elevational gradient in Abisko, Sweden.

	ANOVA results		
Variables	**Vegetation type (V)**	**Elevation (E)**	**V x E interaction**
Multivariate analyses			
MANOVA	5.602 (<0.001)	23.34 (<0.001)	2.89 (<0.001)
Univariate analysis			
Resin P[a]	0.140 (0.710)	27.4 (<0.001)	8.8 (<0.001)
Bic-extractable P_i	47.5 (<0.001)	1.8 (0.147)	1.4 (0.255)
Bic-extractable P_o	12.7 (0.001)	21.5 (<0.001)	7.6 (<0.001)
Total labile P[b]	0.94 (0.339)	27.6 (<0.001)	7.6 (<0.001)
NaOH-extractable P_i[a]	89.3 (<0.001)	4.6 (0.002)	5.1 (0.001)
NaOH-extractable P_o[a]	147.8 (<0.001)	11.5 (<0.001)	8.2 (<0.001)
HCl-extractable P	59.0 (<0.001)	2.4 (0.058)	1.8 (0.142)
Residual P[a]	104.7 (<0.001)	7.9 (<0.001)	4.7 (0.002)
Total P[a,c]	160.5 (<0.001)	4.0 (0.005)	2.7 (0.038)

For all variables, degrees of freedom for V = 1,36, E = 5,36, V*E = 5, 36.
Bic = Bicarbonate; P_i = inorganic P; P_o = organic P.
[a]Data were log transformed prior to analysis.
[b]Sum of Resin P, Bic-extractable P_i and Bic-extractable P_o.
[c]Sum of all sequentially extracted P fractions.

and of these all were highest in the meadow except for Bic-P_o (Table 1, Figure 3). The NaOH-P_o fraction had the largest difference in concentration between the two vegetation types (Figure 3); it was on average three–fold higher in meadow than in heath and represented ~25% of total P in the meadow but only ~13% in the heath. The concentrations of Al_{ox} and Fe_{ox} were highly correlated with each other ($R^2 = 0.951$, $p<0.001$), and their sum ($Al_{ox} + Fe_{ox}$) was on average three times higher in meadow than in heath (Figure 4). The sorption index was also significantly higher in meadow than in heath, on average by 20% (Figure 4, Table 2). Additionally, sorption index was significantly positively related to $Al_{ox} + Fe_{ox}$ (Figure 5). After concentrations were divided by the sorption index, Bic-P_i and Bic-P_o remained highest in the meadow and heath, respectively, and Resin-P was also highest in the heath (Figure S3). There was an interactive effect between vegetation type and elevation for all P fractions except Bic-P_i and HCl-P, meaning that soil P composition responds differently to changes in elevation depending on vegetation type (Table 1, Figure 3).

Discussion

Effects of elevation on the concentration of labile P fractions

The concentrations of labile P fractions were hypothesized to decline with increasing elevation and associated declines in temperature, regardless of vegetation type. We consider Resin-P, Bic-P_i and Bic-P_o to represent the most labile P fractions [35]; Resin-P is well correlated with plant P uptake [49,59], and is considered to be the most highly bioavailable fraction [35,51]. Further, Bic-P_i is considered to derive from similar sources to Resin-P, while Bic-P_o to be easily mineralized [35,48]. These three fractions are discussed separately given the differences in their concentration and dynamics observed in this study. Partially in line with our predictions, Resin-P concentration declined with elevation in both heath and meadow. However, for this P form there was also a strong interactive effect of vegetation × elevation, meaning that this pattern of decline differed between the two vegetation types and that elevation (and thus temperature) effects on P availability therefore depend on vegetation type. Concen-

Table 2. Effects of vegetation type and elevation as determined by ANOVA (F- values, with p values in parenthesis) on phosphorus sorption index and the concentration of oxalate-extractable Al and Fe ($Al_{ox} + Fe_{ox}$) in humus soils along an elevational gradient in Abisko, Sweden.

	ANOVA results		
Variables	**Vegetation type (V)**	**Elevation (E)**	**V x E interaction**
$Al_{ox} + Fe_{ox}$ (mg kg^{-1})[a]	107.9 (<0.001)	8.8 (<0.001)	0.6 (0.666)
Sorption index	36.3 (<0.001)	7.2 (<0.001)	2.4 (0.060)

Degrees of freedom (d.f.) for V = 1,36, E = 5,36, V*E = 5, 36 for $Al_{ox} + Fe_{ox}$; and d.f. for V = 1,35, E = 5,35, V*E = 5, 35 for Sorption index.
[a]Data were square-root transformed prior to analysis.

Table 3. Linear regressions between the sum of oxalate-extractable aluminium and iron (Al_{ox}+ Fe_{ox}) and the Sorption Index with phosphorus (P) fractions in humus soils along an elevational gradient in Abisko, Sweden.

P fraction	Al_{ox}+Fe_{ox}		Sorption Index	
	R^2 (*p*-value)	Direction	R^2 (*p*-value)	Direction
Resin P	0.165 (0.002)	Negative	0.136 (0.006)	Negative
Bic-P_i	0.264 (<0.001)	Positive	0.124 (0.009)	Positive
Bic-Po	0.002 (0.305)	N/A	0.078 (0.032)	N/A
Total labile P[a]	0.138 (0.005)	Negative	0.070 (0.039)	N/A
NaOH-P_i	0.241 (<0.001)	Positive	0.385 (<0.001)	Positive
NaOH-P_o	0.467 (<0.001)	Positive	0.356 (<0.001)	Positive
HCl P	0.654 (<0.001)	Positive	0.223 (<0.001)	Positive
Residual P	0.352 (<0.001)	Positive	0.118 (0.010)	Positive
Total P[b]	0.419 (<0.001)	Positive	0.218 (<0.001)	Positive

Values are for heath and meadow vegetation data combined. Degrees of freedom for all P pools are 1,46 for Al_{ox}+Fe_{ox}; and 1,45 for the Sorption Index due to one missing value.
[a]Sum of Resin P, Bic-extractable P_i and Bic-extractable P_o.
[b]Sum of all sequentially extracted P fractions.

trations of Bic-P_i and P_o did not show any unidirectional trends with elevation, but given that they occurred in much lower concentrations, the overall trend is still one of declining bioavailable P with increasing elevation. In some organic soils, the concentration of bioavailable inorganic P has been shown to be influenced by Al and/or Fe concentration and soil P sorption capacity [60–62]. While Al_{ox} + Fe_{ox} concentrations and the sorption index in our study sites differed across elevations (Figure 4) and were both weakly negatively correlated with Total labile P (Table 3), the elevational trends in Resin-P remained even after correcting for the sorption index (Figure S3), suggesting that they are largely explained by factors other than sorption.

A number of factors could explain the observed decrease in Resin-P with elevation. Organic P (as NaOH-extractable and Residual P) is the dominant form of P in these humus soils and enzymatic hydrolysis of organic P is a likely driver for the release of bioavailable inorganic P, as has been shown for Alaskan tundra

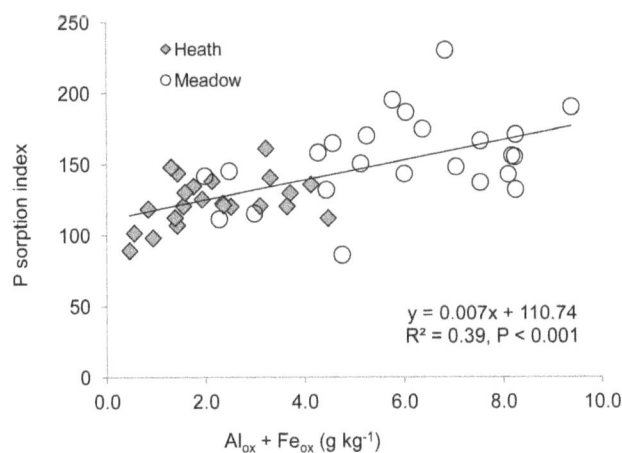

Figure 5. Relationship between phosphorus sorption capacity and metal concentration in humus soils across an elevational gradient. Sorption index versus the sum of oxalate-extractable Al and Fe (Al_{ox} + Fe_{ox}) in subarctic heath and meadow vegetation types across an elevational gradient in Abisko, Sweden.
doi:10.1371/journal.pone.0092942.g005

[1]. Temperature is the strongest driver of soil enzyme activity in the subarctic [63] and warming experiments show that even a relatively small increase in temperature (1.2–1.7°C) can cause a large increase in organic P mineralization in alpine ecosystems [64]. The average air temperature difference across our elevational gradient during the vegetation growing season is about 2.5–3.0°C (Figures S1 and S2), which would be sufficient to explain the differences in labile P we observed. This suggests that the decline in temperature associated with increasing elevation is an important driver of the elevational decline in Resin-P observed and is consistent with our hypothesis. However, further information on organic P mineralization processes, for example using mineralization and soil enzyme studies together with ^{31}P nuclear magnetic resonance (NMR) spectroscopy to characterize soil organic P [65,66], would be useful to increase our mechanistic understanding of the trends observed here. Our results are in line with the few studies that have measured labile inorganic P concentrations at different elevations in subarctic tundra, as they report lower concentrations of 1.0 M KCl-extractable P at high elevations (1150 and 1000 m) than at low elevations (450 and 500 m) [25,67]. Our results are also consistent with [13] who showed a decrease in foliar P concentrations and an increase in foliar N to P ratios with elevation independent of vegetation type. Decreasing organic P mineralization with increasing elevation should lead to the accumulation of soil organic P, which was the case for NaOH-P_o in heath (Figure 3). However, a greater accumulation of soil organic P with elevation may have been negated to some extent by declining primary productivity, as is observed with increasing elevation both in the vicinity of the study area [28] and globally [19].

Temperature can affect P availability directly via effects on microbial mineralization [68] and soil process rates [69] but also indirectly by influencing factors that affect soil processes such as plant [70] and microbial [71] community composition. Temperature variation across the elevational gradient studied here has previously been shown to be related to a range of variables including soil pH, total N and ammonium concentration, C to N ratio, vegetation density, plant and microbial community composition, and fungal to bacterial ratios [25]. As such, many of the soil and vegetation properties that vary along the elevational gradient are likely to represent indirect temperature controls on P availability. Our interpretation is supported by many other studies that have used elevational gradients to understand how temperature affects ecological properties and processes [17,22–25,72–75].

To our knowledge, this is the first time that a Hedley P fractionation analysis has been carried out along a subarctic elevational gradient [6] and the two most comprehensive reviews on P fractionation lack data for these ecosystems [35,36]. These data are necessary to constrain soil P pools in terrestrial biogeochemical models, which are invaluable to understand the processes controlling P cycling and the role of P in driving terrestrial plant productivity [36]. The dominance of organic and residual P fractions (~87% of total soil P on average) and the concentrations of Resin-P measured in these soils are comparable with previous findings for other high latitude tundra ecosystems [1,4,6]. Additionally, the highest Resin-P concentrations measured (i.e. at the lowest elevations) were around two thirds of those reported for a Swedish boreal forest humus with low P sorption capacity [39]. While the concentration of Total labile P (sum of Resin P and Bicarbonate-extractable P) is considered to be low in tundra relative to other ecosystems [1,4–6], our results show that this may not necessarily hold when a wide range of elevations is considered. The 3–5 fold variation in Total labile P concentrations that we found along this elevational gradient encompasses the whole range of Total labile P concentrations recently reported in a world synthesis of Hedley P fractionation studies in natural ecosystems that spans 11 soil orders [36]. Our results further suggest that because small changes in elevation (and thus temperature) were associated with large changes in available P, increases in temperature according to current climate change predictions [14–16] may have a significant impact on future P availability in arctic tundra.

Vegetation differences and distribution of different P fractions

We hypothesized that the concentration of labile P fractions (Resin-P, Bic-P_i and Bic-P_o) would be lower in soils under meadow than heath vegetation, concomitant with higher Al and Fe concentrations and soil P sorption capacity. Our results partially support this, as overall Bic-P_o concentration was significantly lower, and sorption index and $Al_{ox} + Fe_{ox}$ concentration were significantly higher, in meadow than in heath soils. Nevertheless, the concentration of Bic-P_i showed the opposite pattern and the concentration of Resin-P, the largest labile P pool, was not significantly different between vegetation types. Because higher Al and/or Fe has been shown to be positively correlated with P retention in organic soils [60,76–78], we expected that it would also lead to lower concentrations of Resin-P, which has a high propensity for sorption. However, Al_{ox} and Fe_{ox} concentrations in our soils were much lower than those of other organic soils for which negative correlations between labile inorganic P and Al and/or Fe have been reported [60,61,76]. No correlations have been reported for other organic soils that have Al_{ox} and Fe_{ox} concentrations more within the range of what we measured here [6,40]. Taken together, our results suggest that the Al and Fe concentrations in meadow soils are insufficient to exert a strong control on the concentration of labile inorganic P.

The higher $Al_{ox} + Fe_{ox}$ concentrations and sorption index values in the meadow soils had little apparent effect on the concentration of Resin-P, but they may have resulted in higher sorption of organic P. This is suggested by the fact that concentrations of NaOH-P_o and Residual P (which in our soils is mostly organic) were 2.0- and 1.5- fold higher in meadow than in heath. Some organic P compounds have a high affinity for Al and Fe oxides

[79–81], and soil organic P is often strongly positively related with Al and Fe concentration in organic soils, as seen in both this (Table 3) and other studies [40,45,61,77,82,83]. Soil organic P is generally considered to be less prone to sorption than are labile inorganic P forms such as Resin-P [84], but in some humus soils organic P is correlated with Al_{ox} and Fe_{ox} concentration while labile inorganic P is not [40], which is consistent with our findings. A higher organic P sorption capacity in meadow soils could lower P availability indirectly by protecting organic P from microbial mineralization. This is supported by our results showing that Resin-P concentration expressed as a percentage of total soil P was approximately half in meadow than in heath soils (Table S2), and is consistent with findings of relatively higher P limitation in meadow than in heath vegetation [6,13,25].

Conclusions

A 500 m decrease in elevation accompanied by a 2.5–3.0°C increase in temperature resulted in approximately 10–fold higher Resin-P concentration in subarctic soils under contrasting tundra vegetation types. However, the specific way in which Resin-P concentration changed with elevation (and thus temperature) differed with vegetation type. In meadow soils, the higher concentrations of oxalate-extractable Al and Fe, higher P sorption capacity, higher accumulation of organic P and proportionally lower concentrations of Resin-P than in heath soils is consistent with previous reports of higher relative P limitation in meadow vegetation. Our results suggest that predicted temperature increases of 3–5°C for the arctic in the next century could increase the concentrations of labile P in soils, but that the specific pattern of this increase is likely to depend on vegetation type. This is supported by previous work showing significant relationships between temperature and soil and vegetation properties known to influence P availability. Our results also emphasize the need for a better mechanistic understanding of P dynamics in arctic environments. Specifically, knowledge of organic P forms and of the processes that affect their mineralization is crucial for furthering our understanding of how temperature and vegetation affect P availability. New methods which apply 2-D ^1H-^{31}P NMR spectroscopy on soils [65] used in combination with enzyme assays and tracer experiments with ^{32}P and ^{33}P radioisotopes would be a natural follow-up of this study. This knowledge is important because future increases in soil P availability may affect key ecosystem processes such as primary production in these highly nutrient-limited tundra ecosystems.

Supporting Information

Figure S1   Temperature along the elevational gradient in 2009. Daily mean temperature (°C) in July and August 2009 at 500 m, 700 m and 1000 m, along the elevational study gradient.

Figure S2   Temperature along the elevational gradient in 2008. Daily mean temperature (°C) from 28 June to 31 August 2008 measured at 400, 700 and 1000 m along the elevational study gradient.

Figure S3   Concentration of soil phosphorus fractions divided by phosphorus sorption index along the elevational gradient. Panels represent phosphorus (P) fractions extractable with: anion-exchange resins (Resin-P) and $NaHCO_3$ (inorganic fraction – Bic-P_i; and organic fraction – Bic-P_o). Bars represent mean values (+1 SE) for four plots; for each P fraction and within each vegetation type, F and p values (with d.f.) are from a one-way ANOVA testing for the effect of elevation within each vegetation type, and bars topped with the same letter do not differ at $p = 0.05$ (Tukey's h.s.d.). Note the difference in y-axis scales.

Table S1 Selected properties of humus soils in contrasting vegetation types along the elevational gradient. Values represent the mean (± 1 SE) of four plots, data from [13,25].

Table S2 Concentrations of different phosphorus fractions across an elevational gradient. Concentrations are expressed as proportions (%) of total soil phosphorus (P) in humus soils in contrasting vegetation types (heath and meadow). $P_o =$ organic P, $P_i =$ inorganic P. Values represent the means (± 1 SE) of four replicate plots.

Table S3 Linear regressions between soil phosphorus fractions and air temperature. Temperature is the August

2009 mean, and phosphorus (P) is measured on humus soils collected under each of two vegetation types along an elevational gradient.

Acknowledgments

We are grateful to the staff at the Climate Impacts Research Centre (CIRC) for help in the field, to Keith Larson for producing Figure 1, and to Leo Condron, Jonathan De Long, and three anonymous reviewers for helpful comments on previous versions of this manuscript.

Author Contributions

Conceived and designed the experiments: AGV MKS DAW RG. Performed the experiments: AGV MKS DAW RG. Analyzed the data: AGV MKS RG. Contributed reagents/materials/analysis tools: RG. Wrote the paper: AGV MKS DAW RG. Wrote the grant application that funded the work: DAW RG.

References

1. Chapin FS, Barsdate RJ, Barel D (1978) Phosphorus cycling in Alaskan coastal tundra - hypothesis for the regulation of nutrient cycling. Oikos 31: 189–199.
2. Nadelhoffer KJ, Johnson L, Laundre J, Giblin AE, Shaver GR (2002) Fine root production and nutrient content in wet and moist arctic tundras as influenced by chronic fertilization. Plant and Soil 242: 107–113.
3. Shaver GR, Chapin FS (1995) Long-term responses to factorial NPK fertilizer treatment by Alaskan wet and moist tundra wet sedge species. Ecography 18: 259–275.
4. Giblin AE, Nadelhoffer KJ, Shaver GR, Laundre JA, McKerrow AJ (1991) Biogeochemical diversity along a riverside toposequence in Arctic Alaska. Ecological Monographs 61: 415–435.
5. Weintraub MN (2011) Biological phosphorus cycling in arctic and alpine soils. In: Bünemann E, Oberson A, Frossard E, editors. Phosphorus in action. Berlin Heidelberg: Springer-Verlag. pp. 295–316.
6. Giesler R, Esberg C, Lagerström A, Graae B (2012) Phosphorus availability and microbial respiration across different tundra vegetation types. Biogeochemistry 108: 429–445.
7. Turner BL, Baxter R, Mahieu N, Sjögersten S, Whitton BA (2004) Phosphorus compounds in subarctic Fennoscandian soils at the mountain birch (*Betula pubescens*) - tundra ecotone. Soil Biology and Biochemistry 36: 815–823.
8. Rustad LE, Campbell JL, Marion GM, Norby RJ, Mitchell MJ, et al. (2001) A meta-analysis of the response of soil respiration, net nitrogen mineralization, and aboveground plant growth to experimental ecosystem warming. Oecologia 126: 543–562.
9. Schmidt IK, Jonasson S, Shaver GR, Michelsen A, Nordin A (2002) Mineralization and distribution of nutrients in plants and microbes in four arctic ecosystems: responses to warming. Plant and Soil 242: 93–106.
10. Stark S (2007) Nutrient cycling in the tundra. In: Marschner P, Rengel Z, editors. Nutrient cycling in terrestrial ecosystems. Heidelberg: Springer-Verlag. pp. 309–330.
11. Chapin FS, Shaver GR, Giblin AE, Nadelhoffer KJ, Laundre JA (1995) Responses of arctic tundra to experimental and observed changes in climate. Ecology 76: 694–711.
12. Jonasson S, Michelsen A, Schmidt IK, Nielsen EV (1999) Responses in microbes and plants to changed temperature, nutrient, and light regimes in the arctic. Ecology 80: 1828–1843.
13. Sundqvist MK, Giesler R, Wardle DA (2011b) Within- and across-species responses of plant traits and litter decomposition to elevation across contrasting vegetation types in subarctic tundra. PLoS ONE 6(10): e27056. doi:10.1371/journal.pone.0027056.
14. ACIA (2005) Arctic Climate Impact Assessment. Cambridge, U.K: Cambridge University Press.
15. IPCC (2007a) Climate Change 2007: Impacts, Adaptation and Vulnerability. Contribution of Working Group II to the Fourth Assessment Report of the Intergovernmental Panel on Climate Change. Cambridge, UK: Cambridge University Press.
16. IPCC (2007b) Climate Change 2007: The Physical Science Basis. Contribution of Working Group I to the Fourth Assessment Report of the Intergovernmental Panel on Climate Change. Cambridge, UK and New York, USA: Cambridge University Press.
17. Fukami T, Wardle DA (2005) Long-term ecological dynamics: reciprocal insights from natural and anthropogenic gradients. Proceedings of the Royal Society B-Biological Sciences 272: 2105–2115.
18. Körner C (2007) The use of 'altitude' in ecological research. Trends in Ecology and Evolution 22: 569–574.
19. Sundqvist MK, Sanders NJ, Wardle DA (2013) Community and ecosystem responses to elevational gradients: processes, mechanisms and insights for global change. Annual Review of Ecology, Evolution and Systematics (in press).
20. Vitousek PM, Matson PA, Turner DR (1988) Elevational and age gradients in Hawaiian montane rainforest: foliar and soil nutrients. Oecologia 77: 565–570.
21. Vitousek PM, Turner DR, Parton WJ, Sanford RL (1994) Litter decomposition on the Mauna Loa environmental matrix, Hawai'i: Patterns, mechanisms and models. Ecology 75: 418–429.
22. Bragazza L, Parisod J, Buttler A, Bardgett RD (2013) Biogeochemical plant-soil microbe feedback in response to climate warming in peatlands. Nature Climate Change 3: 273–277.
23. Salinas N, Malhi Y, Meir P, Silman M, Cuesta RR, et al. (2011) The sensitivity of tropical leaf litter decomposition to temperature: results from a large-scale leaf translocation experiment along an elevation gradient in Peruvian forests. New Phytologist 189: 967–977.
24. Girardin CAJ, Aragao LEOC, Malhi Y, Huaraca Huasco W, Metcalfe DB, et al. (2013) Fine root dynamics along an elevational gradient in tropical Amazonian and Andean forests. Global Biogeochemical Cycles 27: 252–264.
25. Sundqvist MK, Giesler R, Graae BJ, Wallander H, Fogelberg E, et al. (2011a) Interactive effects of vegetation type and elevation on aboveground and belowground properties in a subarctic tundra. Oikos 120: 128–142.
26. Sundqvist MK, Wardle DA, Olofsson E, Giesler R, Gundale MJ (2012) Chemical properties of plant litter in response to elevation: subarctic vegetation challenges phenolic allocation theories. Functional Ecology 26: 1090–1099.
27. Karlsson J, Jonsson A, Jansson M (2005) Productivity of high-latitude lakes: climate effect inferred from altitude gradient. Global Change Biology 11: 710–715.
28. Jansson M, Hickler T, Jonsson J, Karlsson J (2008) Links between terrestrial primary production and bacterial production and respiration in lakes in a climate gradient in subarctic Sweden. Ecosystems 11: 367–376.
29. Björk RG, Klemedtsson L, Molau U, Harndorf J, Odman A, et al. (2007) Linkages between N turnover and plant community structure in a tundra landscape. Plant and Soil 294: 247–261.
30. Eskelinen A, Stark S, Männistö M (2009) Links between plant community composition, soil organic matter quality and microbial communities in contrasting tundra habitats. Oecologia 161: 113–123.
31. Litaor MI (1992) Aluminium mobility along a geochemical catena in an alpine watershed, Front Range, Colorado. Catena 19: 1–16.
32. Guzman G, Alcantara E, Barron V, Torrent J (1994) Phytoavailability of phosphate adsorbed on ferrihydrite, hematite, and goethite. Plant and Soil 159: 219–225.
33. Brady NC, Weil RR (1999) The nature and properties of soils. New Jersey: Prentice-Hall.
34. Milbau A, Shevtsova A, Osler N, Mooshammer M, Graae BJ (2013) Plant community type and small-scale disturbances, but not altitude, influence the invasability in subarctic ecosystems. New Phytologist 197: 1002–1011.
35. Cross AF, Schlesinger WH (1995) A literature review and evaluation of the Hedley fractionation: Applications to the biogeochemical cycle of phosphorus in natural ecosystems. Geoderma 64: 197–214.
36. Yang X, Post WM (2011) Phosphorus transformations as a function of pedogenesis: A synthesis of soil phosphorus data using Hedley fractionation method. Biogeosciences 8: 2907–2916.
37. Johnson AH, Frizano J, Vann DR (2003) Biogeochemical implications of labile phosphorus in forest soils determined by the Hedley fractionation procedure. Oecologia 135: 487–499.

38. Hedley MJ, Stewart JWB, Chauhan BS (1982) Changes in inorganic and organic phosphorus fractions induced by cultivation practices and by laboratory incubations. Soil Science Society of America Journal 46: 970–976.

39. Lagerström A, Esberg C, Wardle DA, Giesler R (2009) Soil phosphorus and microbial response to a long-term wildfire chronosequence in northern Sweden. Biogeochemistry 95: 199–213.

40. Achat DL, Bakker MR, Augusto L, Derrien D, Gallegos N, et al. (2013) Phosphorus status of soils from contrasting forested ecosystems in southwestern Siberia: effects of microbiological and physicochemical properties. Biogeosciences 10: 733–752.

41. Dieter D, Elsenbeer H, Turner BL (2010) Phosphorus fractionation in lowland tropical rainforest soils in central Panama. Catena 82: 118–125.

42. Kitayama K, Majalap-Lee N, Aiba S (2000) Soil phosphorus fractionation and phosphorus-use efficiencies of tropical rainforests along altitudinal gradients of Mount Kinabalu, Borneo. Oecologia 123: 342–349.

43. Kohler J, Brandt O, Johansson M, Callaghan T (2006) A long-term Arctic snow depth record from Abisko, northern Sweden, 1913-2004. Polar Research 25: 91–113.

44. Binkley D, Giardina C, Bashkin MA (2000) Soil phosphorus pools and supply under the influence of *Eucalyptus saligna* and nitrogen-fixing *Albizia facaltaria*. Forest Ecology and Management 128: 241–247.

45. Giesler R, Satoh F, Ilstedt U, Nordgren A (2004) Microbially available phosphorus in boreal forests: Effects of aluminum and iron accumulation in the humus layer. Ecosystems 7: 208–217.

46. Condron LM, Newman S (2011) Revisiting the fundamentals of phosphorus fractionation of sediments and soils. Journal of Soils and Sediments 11: 830–840.

47. Saggar S, Hedley MJ, White RE (1990) A simplified resin membrane technique for extracting phosphorus from soils. Fertilizer Research 24: 173–180.

48. Tiessen H, Moir JO (1993) Characterization of available P by sequential extraction. In: Carter MR, editor. Soil Sampling and Methods of Analysis. Ann Arbor, Michigan: Lewis Publishers. pp. 75–86.

49. Sibbesen E (1978) An investigation of anion exchange resin method for soil phosphorus extraction. Plant and Soil 50: 305–321.

50. Tran TS (1992) A comparison of four resin extractions and 32P isotopic exchange for teh assessment of plant available P. Canadian Journal of Soil Science 72: 281–294.

51. Frossard E, Condron LM, Oberson A, Sinaj S, Fardeau JC (2000) Processes governing phosphorus availability in temperate soils. Journal of Environmental Quality 29: 15–23.

52. Bache BW, Williams EG (1971) A phosphate sorption index for soils Journal of Soil Science 22: 289–301.

53. Hedley MJ, Kirk GJD, Santos MB (1994) Phosphorus efficiency and the forms of soil phosphorus utilized by upland rice cultivars. Plant and Soil 158: 53–62.

54. Buurman P, van Lagen B, Velthorst EJ (1996) Manual for soil and water analyses. Leiden: Blackhuys.

55. McKeague JA, Day JH (1966) Dithionite- and oxalate-extractable Fe and Al as aids in differentiating various classes of soils. Canadian Journal of Soil Science 46: 13–22.

56. Wardle DA, Bardgett RD, Walker LR, Peltzer DA, Lagerstrom A (2008) The response of plant diversity to ecosystem retrogression: evidence from contrasting long-term chronosequences. Oikos 117: 93–103.

57. Zar JH (2010) Biostatistical Analysis: Prentice-Hall/Pearson. 944 p.

58. Quinn GP, Keough MJ (2002) Experimental Design and Data Analysis for Biologists. New York: Cambridge University Press.

59. Tran TS (1992) A comparison of four resin extractions and 32P isotopic exchange for the assessment of plant available P. Canadian Journal of Soil Science 72: 281–294.

60. Dell'Olio LA, Maguire RO, Osmond DL (2008) Influence of Mehlich-3 extractable aluminum on phosphorus retention in organic soils. Soil Science 173: 119–129.

61. Giesler R, Petersson T, Hogberg P (2002) Phosphorus limitation in boreal forests: Effects of aluminum and iron accumulation in the humus layer. Ecosystems 5: 300–314.

62. Nieminen M, Jarva M (1996) Phosphorus adsorption by peat from drained mires in southern Finland. Scandinavian Journal of Forest Research 11: 321–326.

63. Wallenstein MD, McMahon SK, Schimel JP (2009) Seasonal variation in enzyme activities and temperature sensitivities in Arctic tundra soils. Global Change Biology 15: 1631–1639.

64. Rui YC, Wang YF, Chen CR, Zhou XQ, Wang SP, et al. (2012) Warming and grazing increase mineralization of organic P in an alpine meadow ecosystem of Qinghai-Tibet Plateau, China. Plant and Soil 357: 73–87.

65. Vestergren J, Vincent AG, Jansson M, Persson P, Istedt U, et al. (2012) High-Resolution Characterization of Organic Phosphorus in Soil Extracts Using 2D H-1-P-31 NMR Correlation Spectroscopy. Environmental Science & Technology 46: 3950–3956.

66. Vincent AG, Vestergren J, Gröbner G, Persson P, Schleucher J, et al. (2013) Soil organic phosphorus transformations in a boreal chronosequence. Plant and Soil 367: 149–162.

67. Jonasson S, Havstrom M, Jensen M, Callaghan TV (1993) In-situ mineralization of nitrogen and phosphorus of arctic soils after perturbations simulating climate change. Oecologia 95: 179–186.

68. Schimel JP, Bilbrough C, Welker JM (2004) Increased snow depth affects microbial activity and nitrogen mineralization in two Arctic tundra communities. Soil Biology & Biochemistry 36: 217–227.

69. Jenny H (1994) Climate as a soil-forming factor. In: Jenny H, editor. Factors of soil formation: A system of quantitative pedology. New York: Dover Publications. pp. 104–196.

70. Kardol P, Cregger MA, Campany CE, Classen AT (2010) Soil ecosystem functioning under climate change: plant species and community effects. Ecology 91: 767–781.

71. Strickland MS, Lauber C, Fierer N, Bradford MA (2009) Testing the functional significance of microbial community composition. Ecology 90: 441–451.

72. Sveinbjörnsson B, Davis J, Abadie W, Butler A (1995) Soil carbon and nitrogen mineralization in the Chugach Mountains of South-Central Alaska, USA. Arctic Antarctic and Alpine Research 27: 29–37.

73. Kitayama K, Aiba S-I, Majalap-Lee N, Ohsawa M (1998) Soil nitrogen mineralization rates of rainforests in a matrix of elevations and geological substrates on Mount Kinabalu, Borneo. Ecological Research 13: 301–312.

74. Wang S, Ruan H, Wang B (2009) Effects of soil microarthropods on plant litter decomposition across an elevation gradient in the Wuyi Mountains. Soil Biology & Biochemistry 41: 891–897.

75. Hoch G, Körner C (2012) Global patterns of mobile carbon stores in trees at the high-elevation tree line. Global Ecology and Biogeography 21: 861–871.

76. Giesler R, Andersson T, Lovgren L, Persson P (2005) Phosphate sorption in aluminum- and iron-rich humus soils. Soil Science Society of America Journal 69: 77–86.

77. Kang J, Hesterberg D, Osmond DL (2009) Soil organic matter effects on phosphorus sorption: a path analysis. Soil Science Society of America Journal 73: 360–366.

78. Richardson CJ (1985) Mechanisms controlling phosphorus retention capacity in freshwater wetlands. Science 228: 1424–1427.

79. Celi L, Lamacchia S, Marsan FA, Barberis E (1999) Interaction of inositol hexaphosphate on clays: adsorption and charging phenomena. Soil Science 164: 574–585.

80. McBridge M, Kung K (1989) Complexation of glyphosate and related ligands with iron (III). Soil Science Society of America Journal 53: 1668–1673.

81. Ognalaga M, Frossard E, Thomas F (1994) Glucose-1-phosphate and myo-inositol hexaphosphate adsorption mechanisms on goethite. Soil Science Society of America Journal 58: 332–337.

82. Gerke J (2010) Humic (organic matter)-Al(Fe)-phosphate complexes: An underestimated phosphate form in foils and source of plant-available phosphate. Soil Science 175: 417–425.

83. Vincent AG, Schleucher J, Grobner G, Vestergren J, Persson P, et al. (2012) Changes in organic phosphorus composition in boreal forest humus soils: the role of iron and aluminium. Biogeochemistry 108: 485–499.

84. Celi L, Barberis E (2005) Abiotic stabilization of organic phosphorus in the environment. In: Turner BL, Frossard E, Baldwin DS, editors. Organic Phosphorus in the Environment. Wallingford, UK.: CAB International, pp. 113–132.

7

Sea Ice Biogeochemistry: A Guide for Modellers

Letizia Tedesco[1]*, Marcello Vichi[2,3]

1 Marine Research Centre, Finnish Environment Institute, Helsinki, Finland, 2 Istituto Nazionale di Geofisica e Vulcanologia, Bologna, Italy, 3 Centro Euro-Mediterraneo sui Cambiamenti Climatici, Bologna, Italy

Abstract

Sea ice is a fundamental component of the climate system and plays a key role in polar trophic food webs. Nonetheless sea ice biogeochemical dynamics at large temporal and spatial scales are still rarely described. Numerical models may potentially contribute integrating among sparse observations, but available models of sea ice biogeochemistry are still scarce, whether their relevance for properly describing the current and future state of the polar oceans has been recently addressed. A general methodology to develop a sea ice biogeochemical model is presented, deriving it from an existing validated model application by extension of generic pelagic biogeochemistry model parameterizations. The described methodology is flexible and considers different levels of ecosystem complexity and vertical representation, while adopting a strategy of coupling that ensures mass conservation. We show how to apply this methodology step by step by building an intermediate complexity model from a published realistic application and applying it to analyze theoretically a typical season of first-year sea ice in the Arctic, the one currently needing the most urgent understanding. The aim is to (1) introduce sea ice biogeochemistry and address its relevance to ocean modelers of polar regions, supporting them in adding a new sea ice component to their modelling framework for a more adequate representation of the sea ice-covered ocean ecosystem as a whole, and (2) extend our knowledge on the relevant controlling factors of sea ice algal production, showing that beyond the light and nutrient availability, the duration of the sea ice season may play a key-role shaping the algal production during the on going and upcoming projected changes.

Editor: João Miguel Dias, University of Aveiro, Portugal

Funding: LT acknowledges the support from the Maj and Tor Nessling Foundation (http://www.nessling.fi), project 2010357, and the Estonian Science Foundation (http://www.etf.ee), project MJD62. MV acknowledges the support of the Italian Ministry of Education, University and Research and Ministry for Environment, Land and Sea through the project GEMINA. The funders had no role in study design, data collection and analysis, decision to publish, or preparation of the manuscript.

Competing Interests: The authors have declared that no competing interests exist.

* E-mail: letizia.tedesco@environment.fi

Introduction

Sea ice plays a key role in the climate system [1], mainly due to the albedo positive feedback [2] and to the amplified climate changes undergoing in sea ice-covered regions [3]. In parallel, the ice-associated (sympagic) biology has a key role in winter ecology of ice-covered waters [4]. During winter months sea ice algae are essential to overwintering zooplankton, being the only food source available [5]. Most of the regions seasonally covered by sea ice are the most productive of the oceans, with a shorter production season but more intense algal blooms (e.g. [6]). The highest algal cell and chlorophyll (Chl) concentrations of any aquatic environment has been found in sea ice [7]. Besides the qualitative and quantitative relevance of sea ice algae, recent works have highlighted the importance of sea ice for e.g. dimethyl sulfide production [8], source/sink of CO_2 [9], bioaccumulation of iron [10], and enhanced $CaCO_3$ precipitation [11]. Indirectly, the presence of sea ice also affects the pelagic dynamics: under-ice phytoplankton blooms can be massive when compared to adjacent open water areas [12].

While biogeochemical models of the pelagic ecosystem are commonly developed in the Arctic Ocen [13] and in the Southern Ocean [14], and more recently used to assess potential changes in the ecosystem dynamics under future climate change scenarios [15], the same cannot be said of sea ice biogeochemical models, mostly excluded in large-scale studies except in rare cases [16–19].

Ignoring to include the sea ice biogeochemical component in modelling studies of polar oceans implies neglecting the quantitative and qualitative importance that we currently know sea ice biogeochemistry holds.

Sea ice has been long time considered an impermeable layer between the ocean and the atmosphere, and a rather thin layer when compared with the depths of the oceans. More recently the biogeochemical importance of sea ice in global biogeochemical cycles has been reviewed [20] and large scale Chl data collection has been organized in the Southern Ocean [21], showing the large spatial and temporal patchiness of the observations. Sea ice sampling presents several difficulties: weather conditions often limit data collection, while sampling methods are either time consuming and/or expensive. Comprehensive modelling studies may thus be the most suitable method to integrate among sparse observations, contributing to the understanding of the role that sea ice biogeochemistry plays in the past, present and future state of the polar oceans ([20–24]).

A major aim of this paper is to formulate a theoretical background for the construction of applicative models of sea ice ecosystems. The presented conceptual study stems from a previous application that was thoroughly tested against observations [25]. The model used in [25] demonstrated to satisfactorily capture the specific environmental features of a typical Arctic site in the Greenland Sea as well as the more variable conditions in a Baltic Sea location. By distilling from this previously validated model the theoretical relationships of the dependence on external forcing functions, we aim at making more evident the major factors controlling sea ice algae dynamics. We develop a methodology to build sea ice biogeochemical models starting from a pelagic biogeochemical model and including the key functional types found in sea ice. This methodology is highly flexible and can be applied to any existing ocean model, despite its resolution and complexity. Our current knowledge and application of one-dimensional coupled sea ice physical-biogeochemical models span from prescribed sea ice physical properties to mushy layer theory [23], from simple nutrient-phytoplankton-detritus (NPD) models to stoichiometrically-flexible multiple Plankton Functional Type models [25] (see Table 1 for a list of sea ice biogeochemical models). Since the aim of this work is to encourage modelers of the marine ecosystem to include a new component to their modelling framework wherever and whenever sea ice is part of it, we chose to apply our general methodology to a model of intermediate complexity, yet computationally feasible for large-scale coupled configurations, but with realistic biological and physical descriptors.

Among the ice-covered oceans, the Arctic is the one facing the most dramatic changes: the sea ice pack has decreased by more than 40% in the last three decades [26], though modulated by large variability. Between the 2012 new summer minimum and the 2013 winter maximum, the Arctic Ocean has registered the largest increase in ice extent in the satellite records (NOAA press release, April 2, 2013). First-year ice is rapidly replacing multi-year ice and projections show that the Seasonal Ice Zone might cover the entire Arctic as early as the 2020s [27]. It is therefore needed to foster the development of models on the study of the controlling factors of nutrients and algae in a typical season of Arctic first-year ice, as the first-year ice is the most common type of ice that we are expected to encounter in the near future. We show how even a simple conceptual and numerical model exercise can help to further extend our knowledge on the limiting factors controlling sea ice algae growth, and we will do it by proposing specific ecological indicators. We will analyze in particular the different

effect of some selected physical (i.e. snow cover) and biological (i.e. nutrient availability) controlling factors on the growth of sea ice algae in the current state and considering a shortening of the ice seasons. We will finally show that the benefits of modelling coupled sea ice-pelagic biogeochemistry are manifold: it allows to simulate all-year-around biogeochemistry without the need of seasonal initial conditions, it will ensure a more adequate representation of the ecosystem as a whole with a more realistic representation of the oceanic spring bloom [28] and always ensuring a strict mass conservation, the only but extremely important law that marine ecosystem models have.

Methods: Setting the Stage

A conceptual sea ice ecosystem

The limited number of biogeochemical models implemented for the sea ice ecosystem with respect to those implemented for the adjacent pelagic ecosystem is generally attributed to large uncertainties in sea ice biogeochemical processes. While this may be true for specific processes that are still difficult to quantify in a general framework (e.g. calcium carbonate precipitation, [11]), most main physiological and ecological processes that occur in sea ice are the same as in seawater: e.g. photosynthesis, respiration, exudation, remineralization.

Sea ice is made of a pure solid matrix, liquid saline brines, gas bubbles and impurities. The biological community that is found as sea ice grows is composed of bacteria, microalgae and heterotrophic protists that live in brine pockets and channels, the liquid saline fraction of sea ice. Sea ice is characterized by steep gradients in temperature, salinity, light and space (brines) availability, and those are the major physical constrains to microalgae's growth, making the sea ice bottom a more similar habitat to that of seawater, thus more suitable for the biological community. On the other hand, light is strongly attenuated from surface to bottom sea ice, especially if sea ice is snow-covered. Thus photosynthetic organisms require adaptation/acclimation to low temperature, high salinity and/or low light intensities in order to survive and/or

Table 1. List of sea ice biogeochemical models and their components, revised and extended after [20].

Reference	N. of groups	Constit.	Layers	Ice-ocean fluxes	Ocean
Arrigo et al, 1993 [40]	3N-1P	n p s	ML	Diffusion	n.a.
Arrigo et al., 1997 [58]	3N-1P	n p s	1L static	Diffusion	n.a.
Lavoie et al., 2005 [37]	1N-1P	s	1L static	Diffusion	n.a.
Nishi and Tabeta, 2005 [38]	2N-1P-1Z-2D	n s c l	1L static	Diffusion, Convection	1D
Jin et al., 2006 [36]	3N-1P	n s	1L static	Diffusion	1D
Lavoie et al., 2009 [59]	1N-1P	s	1L static	Diffusion	1D
Tedesco et al., 2010 [25]; Tedesco et al., 2012 [28]	4N-2P-2D	n p s l c	1L dynamic	Growth/melt	1D
Vancoppenolle et al., 2010 [39]	1N	s	ML	Growth/melt, brine transport	n.a.
Pogson et al., 2011 [60]	1N-1P	s	ML	Diffusion	n.a.
Deal et al., 2011 [16]; Jin et al., 2012 [17]	3N-1P	n s	1L static	Diffusion	3D
Sibert et al, 2011 [19]	1N-1P-1Z	n	1L static	Turbolent	3D
Elliott et al., 2012 [61]	3N-1P-1Z-1D	n s l c f	1L static	Diffusion	3D
Saenz and Arrigo, 2012 [62]	3N-1P	n p s	ML	Desalination	n.a.
Rubao et al., 2013 [15]	3N-1P	n s f	1L static	Diffusion	3D

N = nutrient; P = algae; D = detritus; Z = fauna; n = nitrogen; p = phosphorous; s = silicon; l = chlorophyll; c = carbon; f = sulfur; ML = multi-layer; 1L = 1 layer.

bloom in sea ice. In general, rates of physiological processes such as photosynthesis increase with temperature up to some point and this is true also for polar species [29]. The Q_{10}, a measure of the rate of change of algal growth as a consequence of increasing the temperature by 10 degree Celsius, originally proposed by [30], is often used to quantify these metabolic rates in models. Ice algae have been found to have a Q_{10} ranging between 1.0 and 6.0 [31], indicating high potential of acclimation [32]. In addition to temperature, as salinity diverges from sea water values, growth rates, photosynthetic efficiency and capacity of sea ice algae are reduced [33]. Sea ice algae living under several meters of ice and snow have shown to have some of the most extreme low-light adaptation [34], while low-light acclimation is accomplished by e.g. an increase in photosynthetic efficiency [35] and in photosynthetic units per cell [14].

Dissolved nutrients are found in the brines, thus their concentration is usually higher at the bottom of sea ice than at the surface. Occasionally, heavy snow loads lower the freeboard bringing seawater at the interface between snow and ice where surface communities might flourish. Similarly, during rafting events, some seawater can be placed in the interior of sea ice where the internal communities can develop. Thus, depending on their ability to adapt/acclimate, photosynthetic organisms are differently distributed in the bottom, interior and surface of sea ice (Fig. 1). Those differences can be addressed by the choice of different functional types of organisms or by different parameter values. However, whatever diversity is found in sea ice, the same diversity characterizes the ocean underneath, simply by considering mass conservation. Thus, independently of the ecosystem complexity chosen for modelling the pelagic ecosystem, a similar level of complexity should be used to model the sea ice community, and this is true for simple and for more comprehensive models: the numbers of functional groups in sea ice must be the same or smaller than in seawater and the biogeochemical processes that characterizes the chosen functional groups will be the same in both environments. This is easily done when coupling the two components and letting each chemical functional group (or a reduced number) found in seawater to be able to be transferred to sea ice and viceversa, as we will see in detail in the following sections.

Vertical representation

Biogeochemical reactions occur in brines, which are the liquid fraction of sea ice. Brines are not uniformly distributed in sea ice. In an ideal sea ice season, during which first-year ice does not deform and does not depress, sea ice brines are constantly larger at the bottom - and are likely to exceed the permeability threshold - and become smaller towards the surface. In this ideal ice season the development of the biological community is only at the sea ice bottom and a 2-layers model may be considered sufficient to realistically represent the biological and non-biological fraction of sea ice. The simplest approach is to consider a single sea ice layer of constant thickness where biology is active (Fig. 2a), as for example the lowermost 0.02 m of multiyear Arctic sea ice [36–38].

This simple method tends to underestimate primary production in first-year ice, as for example shown in [25]. A possible solution to overcome this problem yet maintaining a single model layer is to compute the biologically-active fraction of sea ice as function of sea ice permeability controlled by the vertical distribution of sea ice temperature and salinity. The Biologically-Active Layer (BAL, [25], Fig. 2b) is that part of sea ice where the liquid fraction (brines) are interconnected (relative volume larger than 5 %) and where the biological community may develop, in contrast to the

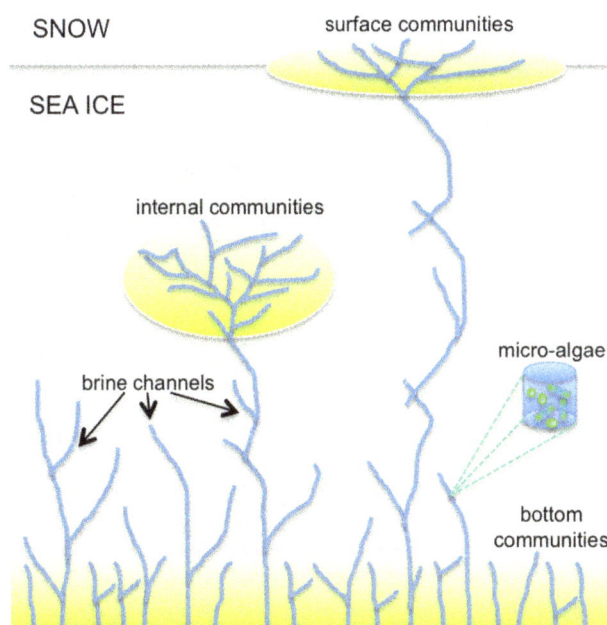

Figure 1. Schematic distribution of ice algae in sea ice. Sea ice algae are generally found at the bottom of sea ice, where temperature and salinity are more similar to those of seawater, and where light is the most likely limiting factor to growth. If snow ice is formed, seawater might reach the sea ice/snow interface and some communities develop at the surface. During rafting and ridging, occasionally seawater can be placed at intermediate depths and the internal communities grow.

rest of the ice where brines are not connected and the biological community is less likely to survive.

In other cases, sea ice can depress and seawater can flood due to snow-ice formation, and deform due to converging or diverging of ice floes. Those are typical features of Antarctic sea ice. In such conditions brines may be temporarily large also at the surface or at intermediate depths, allowing the development of the so-called surface and internal communities, respectively, as described in Fig. 1. In these conditions a detailed vertical resolution is relevant and a multiple layer model (Fig. 2c) such as that of [39] has been shown to reproduce the process. It is important in this last case to consider that the possibility to have interior and surface communities may require a further addition of physiological traits, which may lead to a necessary increase of complexity within the sea ice ecosystem and hence the pelagic ecosystem.

Coupling the sea ice and the pelagic systems

While sea ice and seawater are certainly characterized by similar biogeochemical processes, there are relevant physical differences between the two systems. In the ocean, particles move in all directions in space, while in sea ice particles are trapped in a semi-solid matrix and organisms are distributed in the available volume. Because the volume of brines changes with time, sea ice models are better described with one or more layers of variable thickness, where only the last one is connected with the underlying ocean model. Fig. 3 presents a simplified representation of the physical interface between a sea ice model and a typical ocean model discretized in terms of vertical levels. For completeness, we note that also the ocean model may be expressed in terms of layers of variable depth as in the case of an isopycnal model.

One dimensional (vertical) sea ice algae models have represented the coupling and the boundary fluxes at the ice-ocean interface

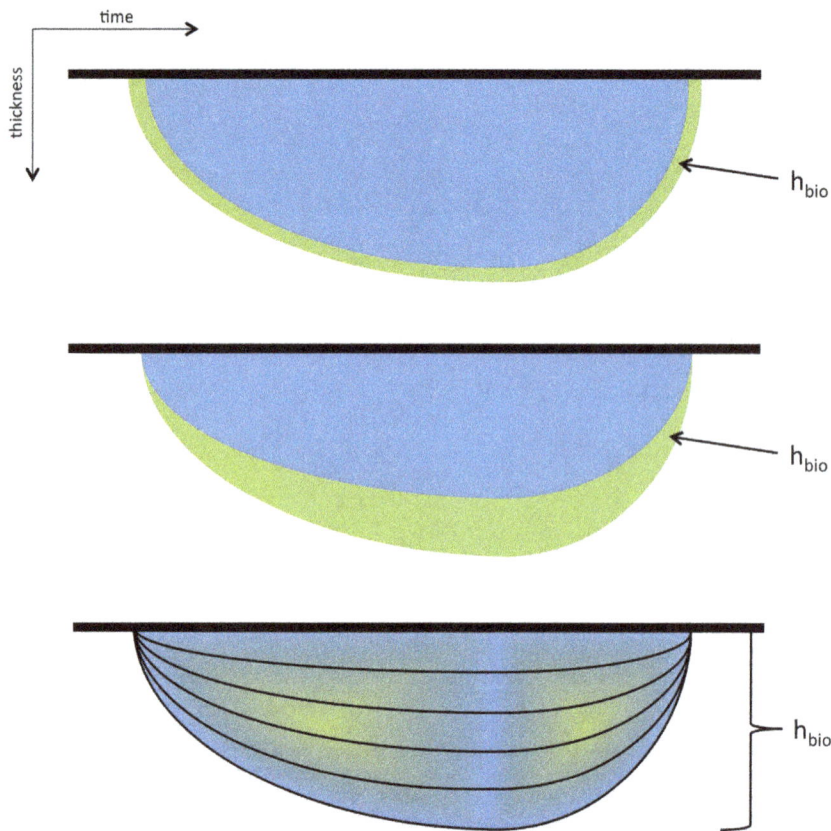

Figure 2. The choice of the vertical representation of the sea ice biogeochemical model. A comparison between three different layer models: a) skeletal layer: a bottom layer of prescribed thickness; b) Biologically-Active Layer (BAL): the bottom sea ice layer that is permeable (relative brine volumes larger than 5%) during the entire ice season; 3) multi-layer: a prescribed number of ice layers of the same thickness.

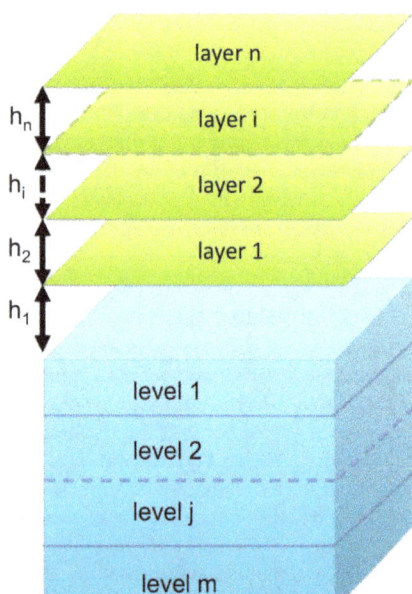

Figure 3. Coupling structure between a level model of the ocean with a layer model of sea ice. The coupling is done between the bottom sea ice layer n1 and the first ocean level m1, independently of the number of layers n and levels m of the model.

in several ways. The more intuitive approaches are to prescribe a constant diffusivity value at the sea ice-ocean interface [37,38] or to impose empirical functions of brine volume flux [36,40]. A more dynamical representation may involve the computation of prognostic fluxes as a function of sea ice growth/melt velocities [25,41] or the inclusion of mushy layer theory concepts [39,42]. In any of these cases, the most important constraint is mass conservation. Due to the seasonal nature of first-year ice, initial concentrations of sea ice variables will only be due to the ocean boundary flux with the seawater counterparts and sea ice must be totally emptied before complete melting.

Simulation of a Typical Sea Ice Season

Forcing functions

Based on a published time series [43] and on previous model simulations [25,28,44], we designed an idealized typical season of first-year ice in the low-latitude Arctic. The proposed time evolution of the sea ice boundary conditions were linearized as much as possible to minimize the noise due to forcing and therefore highlight the major controlling factors for algal growth.

Model simulations are ideally located at 65°N, and surface irradiance values range sinusoidally between 60 and 600 μE m^{-2} s^{-1} as in [43] (Fig. 4A). Seawater salinity is fixed at 32 and seawater freezing temperature is -1.728°C. Sea ice is prescribed to grow from day 1 (i.e. December, 1st) to day 120 to a maximum thickness of 0.6 m. At day 121 sea ice starts melting and the ocean is ice-free by day 170 (i.e. May, 20th). A cubic function is used to simulate the reduced sea ice growth rate as sea ice thickens and ages with similar dynamics as in [43] (Fig. 4B). Snow thickness is generally highly variable in coastal locations as found in [43] (Fig. 4B). The high frequency variability and small spatial scales are likely to affect the local behaviour of the sea ice ecosystem and therefore the choice of an idealized forcing function representing this latitude is more difficult. We chose to use a simple linear function for snow accumulation and melt and that resembles the mean snow thickness that was measured in [43]: snow accumulates on sea ice from day 1 until day 120 and reaches a maximum thickness of 0.10 m, then it melts completely by day 160 (Fig. 4B). Seawater flooding does not occur since the ratio between snow and ice thicknesses never exceeds 1:3 (e.g.[43]), assuming an average density of 300 kg m^{-3} for snow, 900 kg m^{-3} for sea ice and 1000 kg m^{-3} for seawater. The surface temperature linearly decreases from freezing temperature on day 1 to $-20°C$ on day 60, then linearly increases until day 120 when it reaches again the freezing point. The surface temperature is then fixed at the freezing point until the ocean is ice-free (Fig. 4C). For sake of simplicity, sea ice is isosaline (3.0) during the whole ice season.

Model construction

The steps to be taken in the development of the sea ice biological model are:

(i)   the choice of the vertical representation

(ii) the choice of the ecosystem complexity

(iii) the coupling with the ocean.

The choice of the model vertical representation must be made first, as this will determine the other steps. As our simulation considers undeformed first-year ice with no flooding event, sea ice communities will be growing only at the bottom. As a compromise between single and multiple layer models we used the approach of [25], which defines a dynamically-varying bottom layer (the Biologically-Active Layer BAL, h_{bio} of Fig. 2b). This two-layer model allows to represent the abiotic fraction of sea ice and the permeable part characterized by a relative brine volume larger than 5%.

The time-evolution of the physical properties of the BAL were computed by the sea ice thermodynamic model of [46], a refined Semtner 0-layer ice model [47] with detailed snow physics, with the addition of a halodynamic component that describes salinity

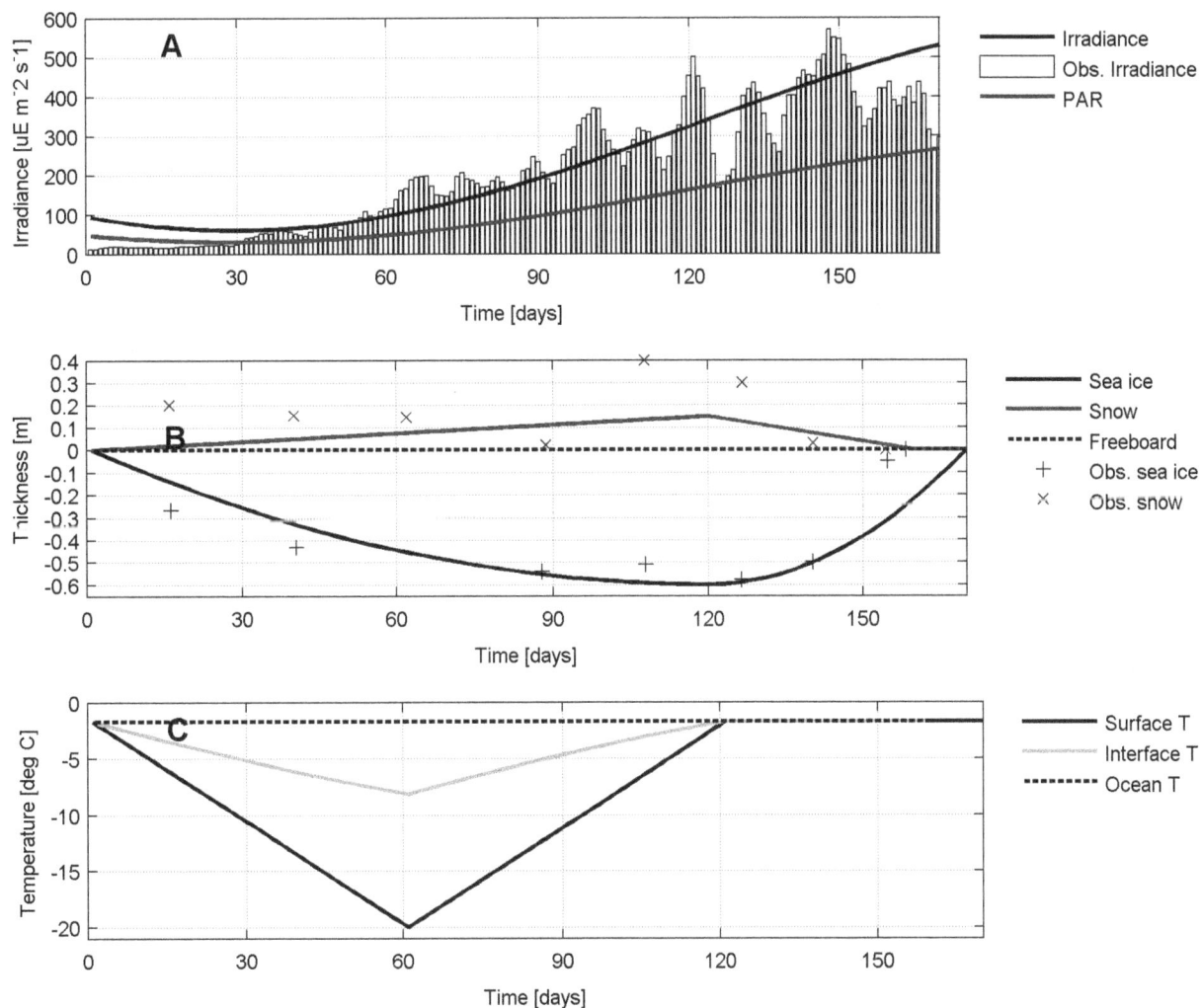

Figure 4. Model set up. Prescribed properties (lines) of a typical season of first-year ice in the low Arctic based on the observations (markers) reported by [43]: A) irradiance and Photosynthetic Available Radiation (PAR), B) snow and ice thicknesses, C) surface, snow/sea ice interface and ocean temperatures.

variations in sea ice (salt entrapment, gravity drainage and flushing) and even more detailed snow physics [25]. A brief mathematical description of the model is provided in section S1 in File S1 (on-line supporting information). The thermodynamic component of the model describes the heat conduction through sea ice and several snow and intermediate snow/ice layers. Given the constant sea ice bulk salinity and the surface and ocean temperatures, brine salinity and volumes can be computed at different depths according to [48], and in particular within the BAL. A simple parameterization based on the Bouguer-Lambert law [49] was used for the computation of the Photosynthetically Active Radiation (PAR) penetrating and reaching the BAL (see File S1). The BAL thickness increases as long as sea ice grows to a maximum of about 0.11 m, then it decreases similarly to sea ice until the end of the ice season (Fig. 5A). The relative brine volume of the BAL decreases until a minimum of about 6.5% (Fig. 5B), larger than 5% by definition, while brines and sea ice temperatures decrease (Fig. 5C) and brine salinity increases (Fig. 5D). The brine temperature is close to the freezing point of seawater and does not go below $-2.5°C$, suggesting no temperature limitation to algal growth. The brines salinity range is also small, between 32 and 46, indicating no regulating role of salinity. Finally, the amount of PAR that reaches the BAL continuously decreases to very low values until snow melting, despite the increase in surface irradiance, pointing to a potential limiting role of light. As snow begins melting, the amount of PAR increases exponentially (Fig. 5E).

We also choose an intermediate approach for the ecosystem complexity, simplifying the comprehensive model used in [25] that requires a large number of initialization and validation data that are yet far to be available in all sea ice observational sites. In a stoichiometrically-flexible network the model of [25] describes inorganic nutrients (NO_3+NO_2, NH_4, PO_4, SiO_4), 2 functional groups of algae (adapted and survivors), particulate and dissolved organic and inorganic matter, and gases such as carbon dioxide and oxygen, for a total number of 22 state variables. The simplified version of the model presented here features one single limiting macronutrient (SiO_4) and one single group of sea ice algae, i.e. diatoms, generally dominant in the sea ice habitat [50], detritus and gases for totally 9 state variables. A schematic diagram of the model is presented in Fig. 6, model's variables and parameters are reported in Table S2 and Table S3 in File S1, while a mathematical description of the model is given in section S2 in File S1. The limiting nutrient is silicate, but any other nutrient can be chosen as model's currency. Silicon was chosen because the functional group of algae is made of diatoms that require silicate uptake. If the model must have one single chemical component as currency, then silicon is likely to be the most appropriate for the sea ice system. However, many oceanic models use nitrogen as model's currency since it often the most limiting in the oceans. In this latter case, modellers can choose if either increasing the number of state of variables of their model including both silicon and nitrogen components, either if using a N:Si conversion factor. Silicate dynamics differentiate from nitrate and phosphate dynamics as silicate does not accumulate in the cell and it is more likely to be parameterized with a simple Michaelis-Menten function (e.g [51]) and thus directly controls carbon photosynthesis. If nitrate or phosphate are instead chosen as most limiting nutrient, those are decoupled from carbon uptake because of the existence of cellular storage capabilities. The co-limitation from all nutrients can be done with a threshold method, as in [25], and it is considered in the parameterization of some processes such as chlorophyll synthesis and sinking. Multiple nutrient limitation is different for nutrients that can be stored in the cell (nitrate and

phosphate) and nutrients that cannot (silicate). [25] allows three alternative ways to combine N and P limitation: the minimum among the two nutrients, a threshold combination (Liebig-like) and a multiplicative approach [51].

[28] showed that sea ice diatoms need to be both photoadapted and photoacclimated to the sea ice light environment. The same optimum Chl:C ratio for sea ice diatoms (0.03 mg Chl/mg C) is thus kept here, able to change according to the organisms' requirements. The simplified model is thus the same as that of [28], but it is characterized by only three basic constituents (C, Chl and Si) and by the physiological rates of photosynthesis, respiration, mortality/excretion and nutrient uptake as presented in Fig. 6. A comparison between this simplified version of the model and the standard and more comprehensive model of [25] is highlighted in the next section.

We coupled the sea ice model with a simple slab ocean, which is meant to represent the mixed layer depth under sea ice (15 m). We defined in seawater the same constituents and processes that we find in sea ice with the addition of bacteria and microzooplankton, as in [28]. The ocean model was initialized according to typical winter mixing conditions: 8.0 mmol Si m^{-3} of dissolved silicate and 1 mg C m^{-3} of diatoms in seawater. The sea ice model did not need to be initialized because it is controlled by the exchanges of dissolved and particulate matter between sea ice and seawater. As a coupling method between the ocean and the sea ice habitat we choose again a method of intermediate complexity [25,41], which defines the fluxes at the interface as a function of sea ice growth/melt velocities and thickness of the BAL. A complete mathematical description of the coupling fluxes are given in section S2 in File S1 (on-line supporting information).

Reference simulation

The reference simulation (S0) reproduces an enrichment of dissolved silica during the sea ice growth season, followed by a sharp decrease due to the combined action of silica uptake by algae and brine loss due to melting (Fig. 7). The algae bloom reaches its peak at day 134 when nutrients are about to be exhausted (Fig. 7) and the eventual depletion is the combined results of nutrient utilization and volume loss in the biological layer. To show the coexistence of growth and habitat loss processes we also show the silicate curve obtained with a simulation that does not include biological uptake (abiotic simulation, Fig. 7). The difference between the curves represents the amount of nutrient used for sea ice algal growth.

The potential error that is made by using this simplified model rather than the more comprehensive model of [25] can be estimated by comparing results given for S0 by both models (Fig. 8). As the model of [25] requires a larger set of variables to be initialized (nitrate, phosphate and survivor algae), several values were considered. The same initial concentration in seawater as for sea ice diatoms was given to survivor algae (1 mg C m^{-3}), while for nutrients initialization we compared: (i) a Redfield-like initialization 15 Si: 16 C: 1 P given 8 mmol m^{-3} of initial silicate as in S0; (ii) non-Redfield typical concentrations of the open Barents Sea [52], with silicate ranging between 6 and 8 mmol m^{-3}, 12 mmol m^{-3} of nitrate and 0.85 mmol m^{-3} of phosphate (iii) average concentrations reported for the whole Arctic in the Hydrochemical Atlas of the Arctic Ocean [53] (13.20 mmol m^{-3} of silicate, 3.28 mmol m^{-3} of nitrate and 0.83 mmol m^{-3} of phosphate). All simulations reproduce a similar bloom timing but different bloom magnitude (Fig. 8): while the "Redfield-like" and "Barents 2" runs have a smaller peak, the "Barents 1" and "Arctic" are more similar to the S0 peak. Changes in the initial pelagic nutrient conditions have more a direct effect on the

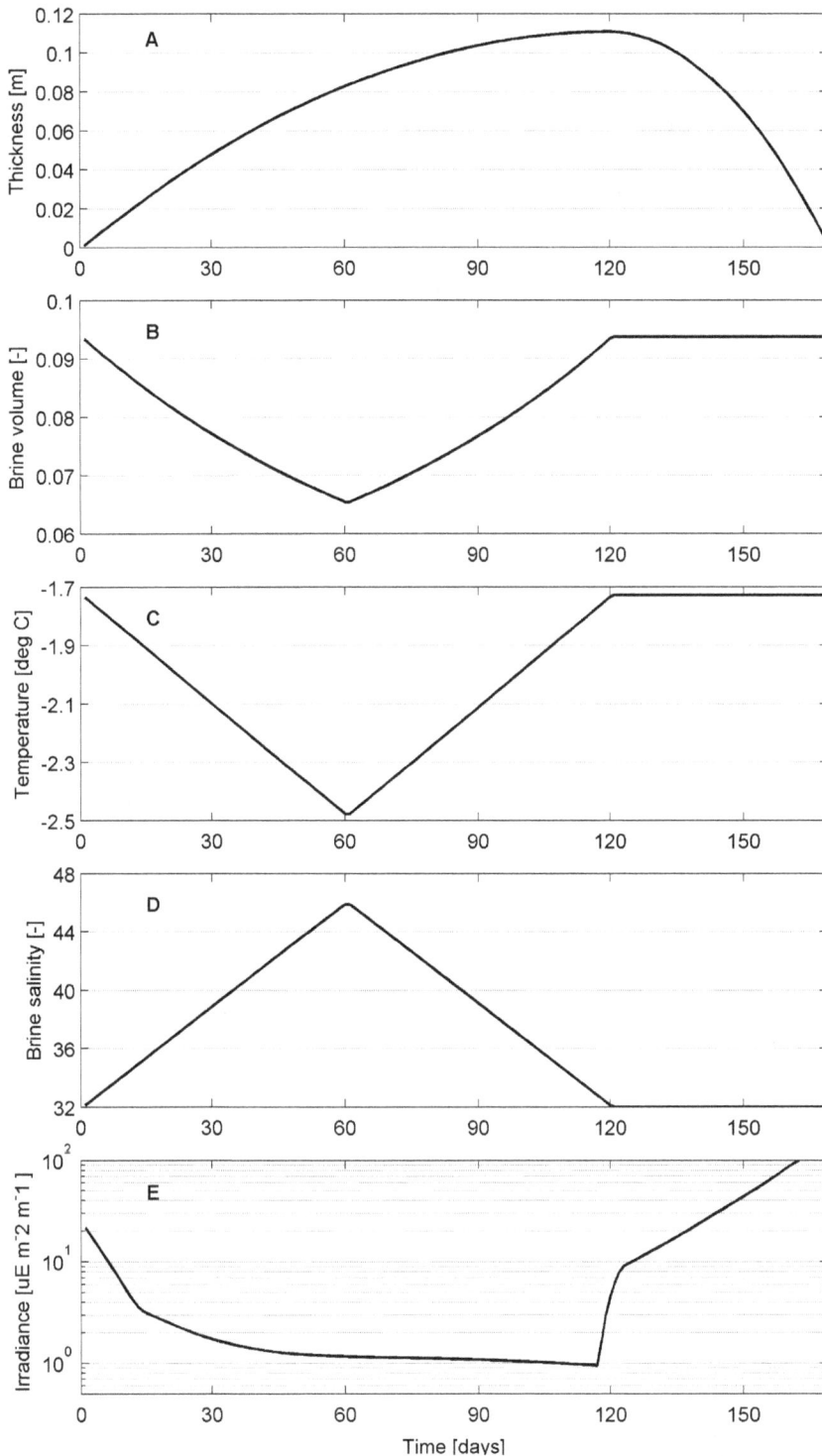

Figure 5. BAL properties along a typical season of first-year ice in the low Arctic. A) thickness, B) relative brine volume, C) sea ice and brines temperature, D) brines salinity, E) average PAR.

magnitude of the bloom rather than on its timing as also shown in [28].

Sea ice bloom indicators

Although the maximum amplitude (a_m) and the time of the peak (t_m) of an algal bloom may be identified with frequent observations,

the exact extent is a matter of definition. Time of initiation and termination may be individuated according to a threshold criterion that can be set in absolute terms (concentration) or in relative terms (some fraction of the maximum amplitude, [54]). As for phytoplankton, the relevant phases of ice algae phenology are the time of initial growth t_i, the time of maximum amplitude t_m

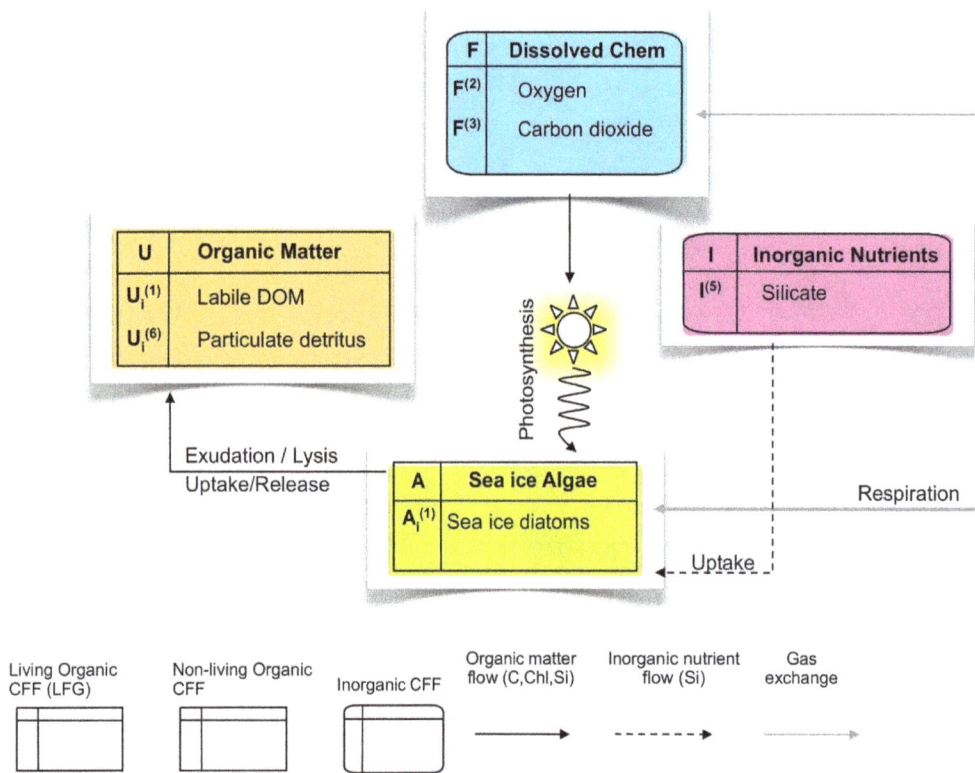

Figure 6. Scheme of the intermediate complexity sea ice biogeochemical model presented in this work. The model includes: 3 inorganic Chemical Functional Families (CFF) i.e. 2 gases (F, oxygen and carbon dioxide) in dissolved forms and 1 macronutrient (I, silicate); 1 non-living CFF encompassing dissolved and particulate organic matter (U); and 1 living CFF of sea ice algae (A), i.e. diatoms. Organic matter flows are due to photosynthesis of sea ice algae and to exudation, lysis, uptake and release of DOM and POM. Inorganic nutrient flows is the silica uptake. Gas exchange is due to oxygen production and carbon dioxide consumption by sea ice algae.

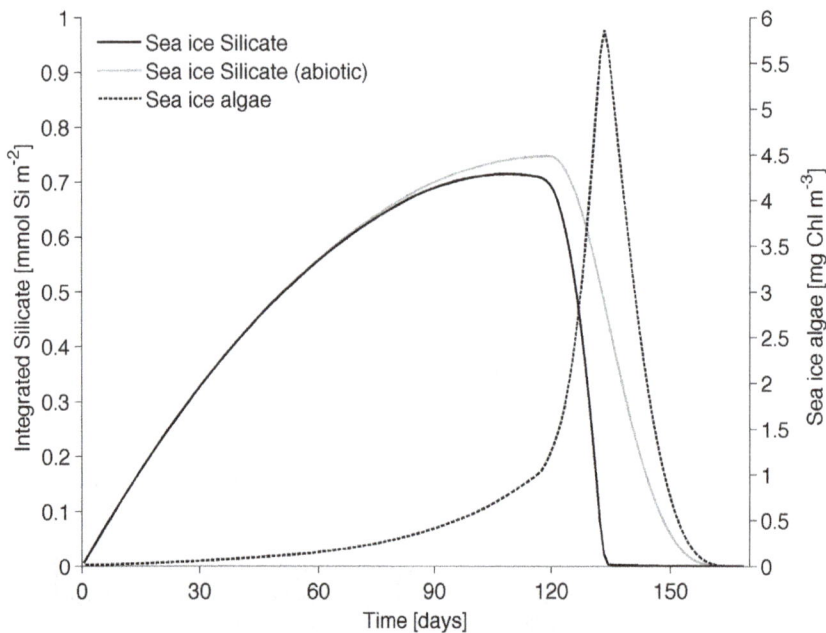

Figure 7. Reference simulation for the typical season of first-year ice in the low Arctic. Vertically integrated dissolved silica concentration in sea ice for the reference and abiotic simulations and Chl volume concentration of sea ice algae in the BAL. Silicate concentration is presented with the integrated value to show the progressive increase of nutrients in the BAL.

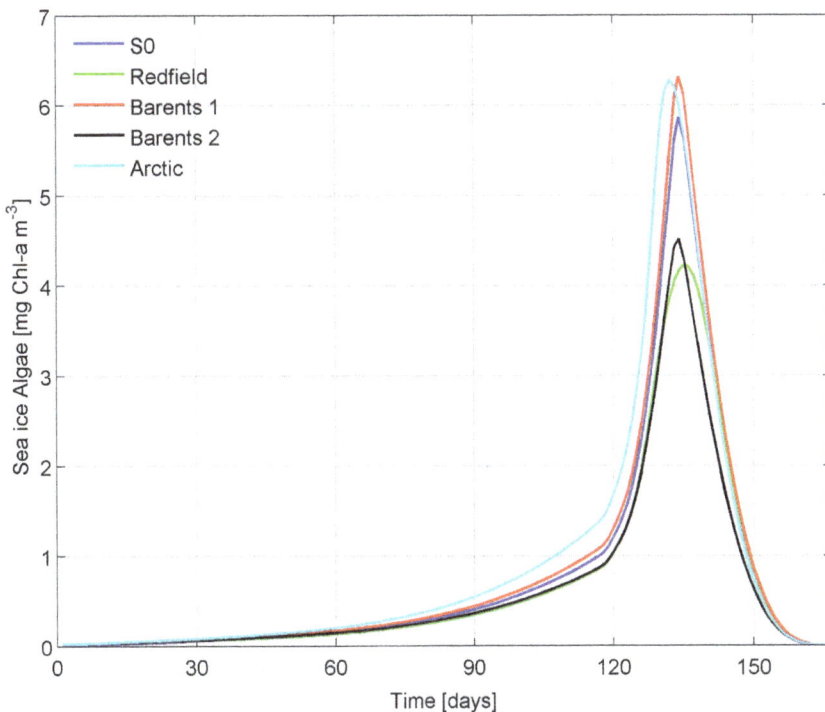

Figure 8. Chl comparison between the standard model of [25] and the simplified model described in this work. Chl comparison between the normal idealized case (S0) produced by the simplified model presented in this work, and by the more comprehensive model of [25] when nutrients are initialized: i) with Redfield ratio with respect to silicate (15 Si: 16 N: 1 P), ii) with non-Redfield values reported from the Barents Sea (Barents 1 = 8 mmol Si m^{-3}; Barents 2 = 6 mmol Si m^{-3}), and iii) using average values reported for the entire Arctic Ocean.

and the duration t_d. Such kind of ecological indicators are objective-oriented metrics and different methods have been described to investigate phytoplankton phenology. [15] recently studied the Arctic sea ice algal and phytoplankton phenology in terms of maximum amplitude and timing of the peak. However, the usage of indicators of phenological phases, such as time of initiation and duration, in ice algae modelling is yet to start. Considering the large variability in maximum chlorophyll amplitude that can be found in sea ice (e.g. Table 8.1 in [7]), an index that considers a relative threshold rather than an absolute value for indicating the bloom initiation and termination seems more appropriate. Among the potential candidates, we propose to use the anomalies given by the amplitude of the concentration minus its standard deviation. Positive values will thus represent the period of bloom activity in sea ice. This index is independent of the length of the time series (sea ice seasons may present a large variability in the world ocean) and it is less biased by the long period of quiescence that may be observed when light penetration is the limiting factor and the only variability is controlled by the boundary flux with the ocean.

The indices computed from the reference simulation S0 – and from all the other experiments that are following presented – are given in Table 2. The bloom initiation is well at the end of the sea ice season and starts some days after the idealized snow function begins to decrease (day 120). Over a period of about 170 days of sea ice, the bloom period derived from the proposed indicator is a small portion of 26 days. We notice that the bloom indices can be computed as anomalies respect to Chl or carbon content and the results are slightly different due to the parameterization of the natural process of light acclimation. The Chl bloom is longer and starts earlier than the C bloom, implying that there may be a

mismatch between indicators computed in terms of biomass or abundances and the ones derived from bulk chlorophyll values.

Results and Discussion: Response to Sea Ice Scenarios

The chosen evolution of sea ice environmental conditions in the Arctic is an idealized case that cannot encompass all the variability that may be observed in reality. We have therefore prepared a set of "scenarios" considering the possible events that may be typically found in this kind of ecosystem to show how a rather simple model can give hints on the underlying real-world processes. The set of sensitivity experiments investigates the dependence on the following parameters:

- S1: change in the pelagic nutrient availability
- S2: variation of snow thickness
- S3: shift in the day of snow melt
- S4: reduction of the length of the ice season

The resulting indices of algal phenology for each scenario are reported in Table 2 and a detailed explanation of the model response is given in the next sections. Fig. 9 presents an overview of the changes in the BAL thickness, silicate concentrations and sea ice algae Chl, the latter shown as anomalies from the respective standard deviation. Scenarios S1 and S3 assume no change in the length of the idealized ice season and the BAL remains unchanged from the reference simulation (Fig. 9A). The thickness of the biotic layer is mainly controlled by snow thickness as further analyzed later on. Nutrient concentrations vary in each scenario, except in S3 (Fig. 9B). In S1, silicate is prescribed to be smaller (S1a) and larger (S1b), while in S2 and S4 the dynamics of the nutrient is

Table 2. Phenology indicators from the reference and scenario simulations.

Simulation	t_i	t_m	t_d	a_m
S0: Reference	121(124)	134	26(23)	5.9(335.4)
S1a: Nutrient reduction	121(125)	135	27(23)	2.6(150.9)
S1b: Nutrient increase	124(126)	135	24(22)	11.2(656.8)
S2a: Snow reduction	94(95)	110	40(41)	17.1(596.1)
S2b: Snow increase	134(137)	149(150)	25(21)	1.1 (85.7)
S3a: Early snow melt	121(124)	134	26(23)	5.6(342.3)
S3b: Late snow melt	121(124)	134	26(23)	6.0(330.7)
S4a: 25% shorter ice season	96(98)	106	22(20)	4.3(291.2)
S4b: 50% shorter ice season	70(72)	79	15(14)	2.4(180.2)

The indices are: day of initial growth t_i, day of maximum amplitude t_m, duration t_d and maximum amplitude a_m over the bloom duration. Values are referred to the sea ice algae Chl content and in brackets to carbon content. If no value is given in brackets the numbers coincide.
doi:10.1371/journal.pone.0089217.t002

driven by the fluxes at the ice-ocean boundary in combination with the respective biological processes of uptake and remineralization. The timing and duration of the bloom in each scenario are identified in Fig. 9C using the proposed indicator, where we observe a substantial effect on the bloom features, with the largest impact driven by variations of snow thickness and the length of the ice season.

Sensitivity to nutrient concentrations

The most simple sensitivity experiment is related to the initial nutrient concentration, which in our case represent the winter background of dissolved silica that may be found in different Arctic regions. We perform such experiment by halving the seawater concentration to a typical value of an open ocean (4 mmol m^{-3}, scenario S1a, Fig. 9B) compared to the reference S0 of 8 mmol m^{-3} (Fig. 9B, S0) and by doubling it as it may be found in coastal waters with land-fast ice (16.0 mmol m^{-3}, S1b, Fig. 9B). The response of sea ice algae to the variation in silicate concentration is almost linear (Fig. 9C): larger concentrations are associated with larger blooms and viceversa. No effect is seen on the timing of the bloom but only on the amplitude (Table 2). Among all scenarios, the smallest nutrient concentration (Fig. 9B, S1a) results in one of the lowest biomasses (Fig. 9C, S1a) and the highest silicate concentration (Fig. 9B, S1b) among the largest blooms (Fig. 9C, S1b).

Sensitivity to snow thickness

Light extinction through snow is extremely high (see the supplementary Table 1 in File S1) and a different snow thickness is expected to highly affect the amount of light reaching the bottom sea ice and the response of primary producers. In scenario S2, the initial day of solid deposition, the day of maximum snow cover and the day of complete snow melt are not changed. We only consider a different maximum value of snow thickness, from a minimum of 0.05 m (S2a), to a maximum of 0.15 m (S2b, Fig. 10A). There are large differences both in the timing and the magnitude of the bloom (Fig. 9C), with thinner snow associated to 3-times larger Chl concentration and 24 days earlier bloom (Table 2). Since the bloom starts at day 94 and peaks at day 110 (Table 2), well before snow starts melting (day 120), we conclude that a snow cover of 0.05 m is not light-limiting. This is also confirmed by the temporal

dynamics of the algae Chl:C ratio, which is stable during the whole season (Fig. 10C), indicating no specific acclimation needs. This is not true instead in S0 and S2b, where we observe a gradually increasing acclimation to dark conditions (larger Chl:C ratio) until snow melts and later an opposite type of acclimation, i.e. to high light conditions and thus a smaller Chl:C ratio, also in this case more pronounced in S2b than in S0.

Sensitivity to the day of snow melt

Scenario S3 maintains the maximum snow thickness of 0.10 m as in the reference simulation S0, but varies the day of complete snow melt, shifting it backward to day 150 (S3a) and postponing it to day 170 (S3b). As described in the previous scenario, light attenuation by snow is proportional to the amount of light that reaches the surface and at the same latitude this depends only on the day of the year. The sooner snow completely melts, the sooner larger amount of light penetrates sea ice and reaches its bottom. And the longer the ice season the larger will be the amount of incoming short-wave radiation until the maximum reached at the summer solstice. By changing this parameter we do not observe any significant difference in the peak timing (Table 2, day 134 as in S0) and amplitude (not more than about 5% difference with the maximum amplitude in terms of Chl in S3b, Fig. 9C). We attribute this to the fact that snow cover is light-limiting for most of the ice season, as found in analyzing scenario S2. Even though an earlier snow melt provides the bottom communities with a larger amount of light, there is not enough time for the community to develop further as the sea ice habitat is already shrinking by melting underneath and the scenario presenting a later day of snow melt is still not sufficient to significantly change the bloom characteristics.

Sensitivity to the length of the ice season

It was pointed out in the Introduction that ice seasons in the Arctic are generally projected to be shorter. We look in scenario S4 at the response of sea ice algae to shorter ice seasons, while keeping the same amount of maximum snow cover (0.10 m) as in the reference simulation S0. We do that by reducing the ice season length of an equal amount of days at the beginning and at the end of the season. In scenario 4a the ice season is shorter of about 25% (129 days) and in scenario 4b it is 50 % shorter (87 days, Fig. 10D). Accordingly, the ice thickness is also reduced. An important feature of this experiment is the fact that the maximum ice and snow thicknesses are reached earlier in the year, allowing us to analyze how the biota respond to different sunlight availability, depending only on the day of the year rather than on the snow thickness itself, which is 0.10 m in all scenarios. A 25% shorter ice season (S4a) produces a 28-days earlier and about 30 % smaller bloom (Table 2 and Fig. 9C). The 50% shorter ice season (S4b) shows a similar response, producing an even more accentuated smaller bloom (less than half of the one in S0 (Fig. 9C). The Chl:C ratio (Fig. 10E) clearly shows that in both S4a and S4b the process of light acclimation to darkness is interrupted by the onset of melting snow never reaching the values found in S0, and the shorter is the season the earlier is the interruption. Despite the larger PAR (Fig. 10E), space (Fig. 9A), and nutrient availability (Fig. 9B), the shorter the ice season the smaller the bloom. The restricted time window during which the bloom occurs appears to be the main regulating factor for biomass to have sufficient time to be built.

Conclusions

The qualitative and quantitative importance of the sea ice biota was shortly reviewed and a general framework to develop a sea ice

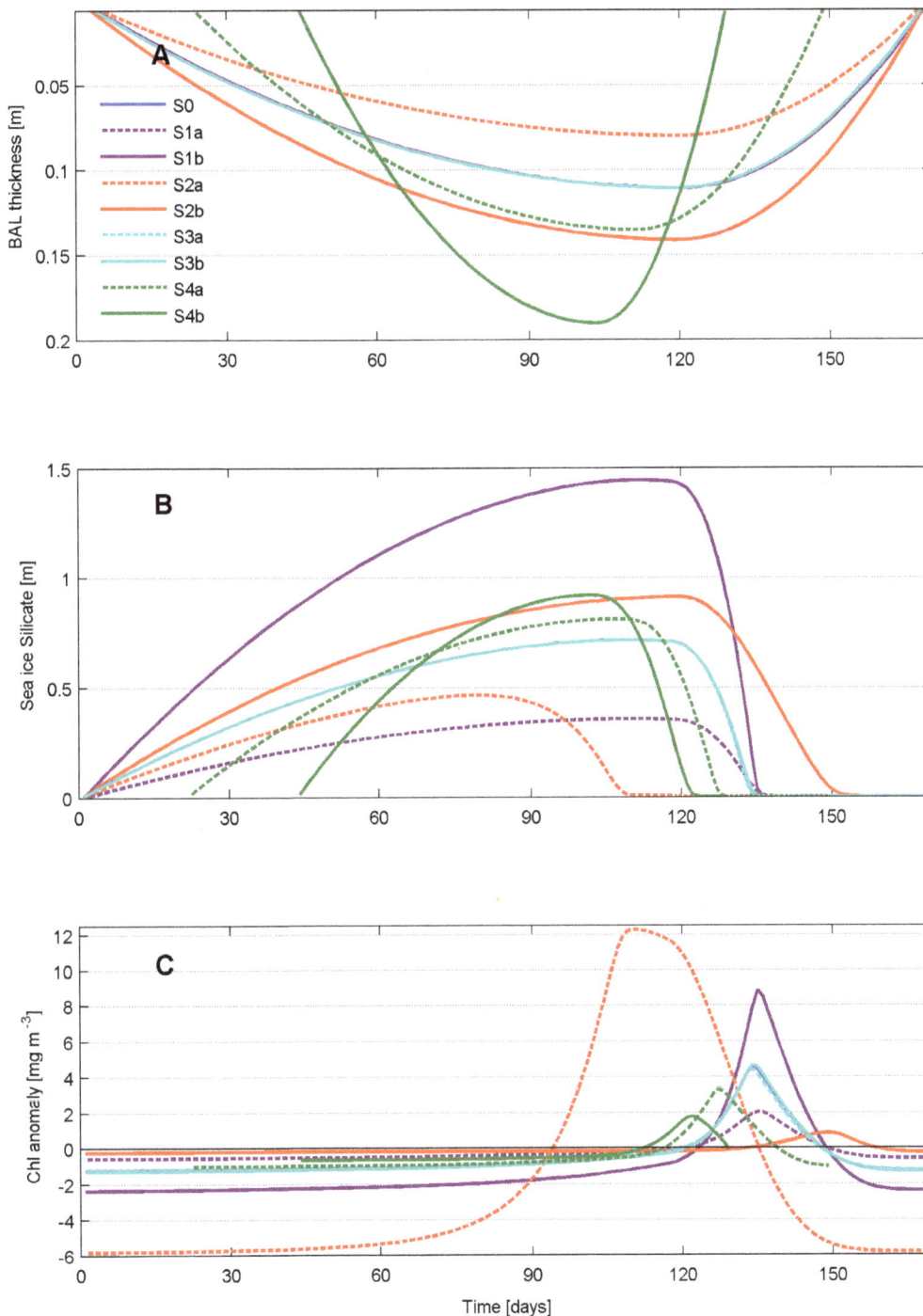

Figure 9. Model response to different scenarios. Temporal evolution of: A) the BAL thickness (note the curves of experiments S0, S1 and S3 coincide); B) silicate concentration (note the curves of experiments S3 coincide); and C) sea ice algae Chl anomalies computed by subtracting the respective standard deviation. Positive values identify the bloom period in each scenario.

biogeochemical model from the knowledge on pelagic modelling was presented. We considered different levels of ecosystem model complexity (from simple NPZD models to multiple PFT stoichiometric models) and vertical representation (from single to multilayer models). This list of options is intended to help the modeller to choose the most appropriate set up under different conditions of applicability. In large-scale simulations and coupled configurations, compromises have to be made and model complexity may

be reduced for computational reasons. The applied model will be a compromise between resolving the vertical variability of the sea ice biota, the complexity of the food web, the extent of the spatial and temporal scales and the overall computational constraints. In this work we applied this methodology to build an intermediate complex sea ice biogeochemical model from a previously validated realistic application and using a dynamic single-layer vertical representation together with a simplified plankton functional type

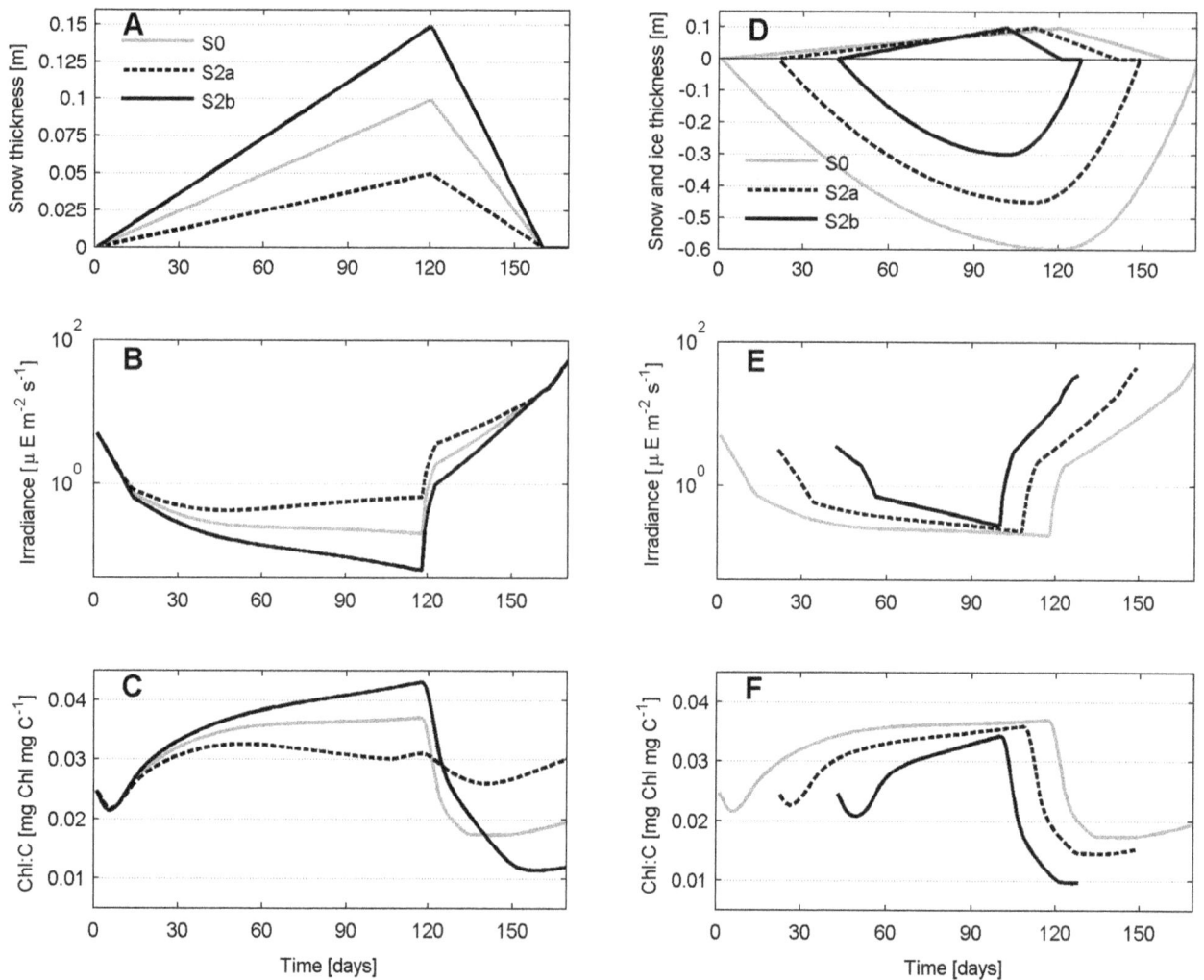

Figure 10. Sensitivity to snow thickness (S2) and season length (S4). Thickness (top), irradiance (middle) and Chl:C ratio (bottom) for S2 (A, B and C) and S4 (D, E and F). Note the log scale in panels B and E.

model with variable stoichiometry. We coupled the sea ice model to the ocean by computing the boundary fluxes as function of growth/melt velocities and sea ice thickness. We presented a coupling strategy between the bottom sea ice layer and the first ocean level that relies on simple concepts of boundary fluxes and ensures strict mass conservation. We emphasize the importance of developing models that clearly obey to such law.

We used in this work some ecological indicators for sea ice algae commonly used to describe phytoplankton phenology, such as the time of initial growth, the time of maximum amplitude, the maximum amplitude value, and the bloom duration. We proposed a modified criteria to detect the timing of the bloom in terms of initiation and duration, based on arguments that refer to the specific seasonal nature of first-year ice. Ecological indicators are objective-oriented tools that can help quantifying temporal and spatial variations, which are very much needed at this time of rapid changes. More studies to test different indices suitable for sea ice biogeochemical modelling are thus envisaged.

Polar regions are rapidly changing and the Arctic is currently facing unpredicted changes. First-year ice is rapidly replacing multi-year ice and our understanding of the changes in primary production patterns in sea ice and in the oceans is mostly a speculative topic [55]. We have used our idealized Arctic ice season not only to show how to develop a sea ice biogeochemical model but also to investigate the relevance of several factors affecting sea ice algae production. Beyond the light limitation due to snow cover and the control exerted by initial nutrients, we showed in this study that the duration of the ice season appears to be an important co-regulating factor. Given the same conditions, excessively short ice seasons would reduce the entire biological community because of lower light availability, as earlier melting dates will occur with fewer hours of sunlight, and because of reduced time to build up biomass. We summarize the studied scenarios and their combination in Fig. 11 (see also figures S1 and S2 in File S1). Favorable conditions such as fewer solid precipitation might counteract the predicted decrease to a certain extent, for instance in the case of a 25% reduction of the sea ice season duration. While the general trend is towards shorter ice seasons, there might be some local variability and some hot spots of productivity might still be found. However, climate projections generally indicate not only shorter ice seasons but also more abundant precipitations in the Arctic [56], thus we can expect that

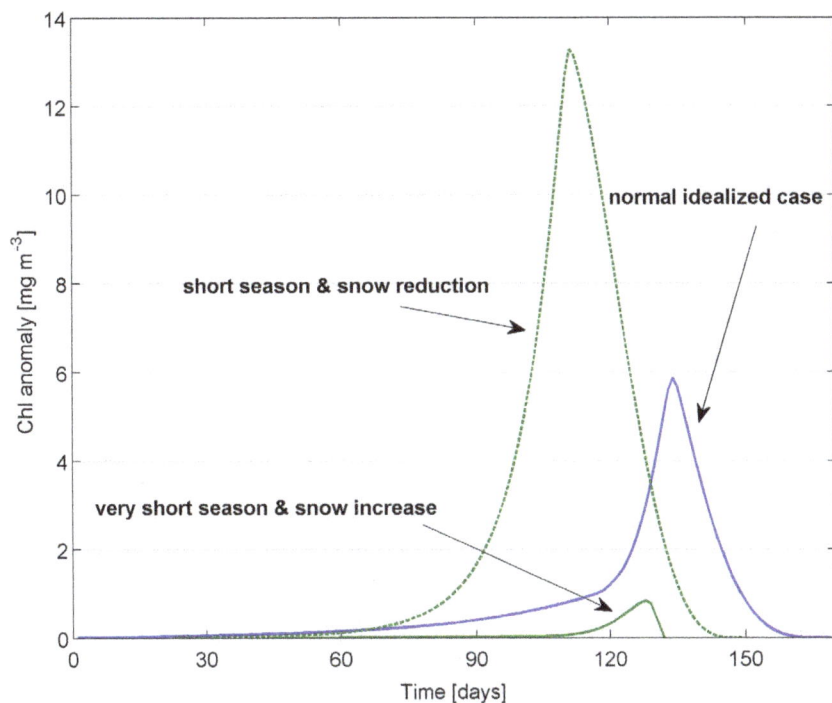

Figure 11. Combined scenarios. Comparison between Chl biomass in the normal idealized case (S0), in a shorter ice season (25% reduction as in S4a) and reduced snow cover (max 5 cm as in S2a), and in a very short ice season (50% reduction as in S4b) and increase snow cover (max 15 cm as in S2b).

the impoverishment of sea ice would occur at an even faster rate (Fig. 11). We hypothesize that, despite any other positive conditions, excessively short ice seasons would likely reduce the sea ice biological community of producers. Given the same latitude and approximately the same amount of sunlight, this critical condition may be expected to arrive earlier in regions that experience shorter ice seasons. Instead, given the same length of the ice season, this condition is expected to arrive earlier in regions that experience less hours of sunlights.

These considerations are based on the biological production of sea ice diatoms, which we earlier pointed out to be the most abundant and productive group of algae found in sea ice. However, sea ice is inhabited also by other algal groups that have been demonstrated to respond differently to varying light regimes (e.g.[57]). Given enough time, we may either expect the development of new adaptation patterns by diatoms or the dominance of new groups of algae, which should both be accomplished with an increase of model complexity to an extent not considered in the present study.

References

Acknowledgments

This work was written under the auspices of BEPSII - Biogeochemical Exchange Processes at the Sea-Ice Interfaces - SCOR Working Group 140. The authors are grateful to Ditte Marie Mikkelsen for providing the data from Kobbefjord ice sampling station. The authors thank all the sea ice biologists, ecologists, geophysicists and modellers they met and discussed with in the last few years for having inspired large part of this work. The authors are grateful to the editor and to two anonymous reviewers, whose comments helped to better address and clarify some fundamental issues discussed in this work.

Author Contributions

Conceived and designed the experiments: LT MV. Performed the experiments: LT MV. Analyzed the data: LT MV. Contributed reagents/materials/analysis tools: LT MV. Wrote the paper: LT MV.

1. Maykut G (1978) Energy exchange over young sea ice in the central Arctic. Journal of Geophysical Research 83(C7): 3646–3658.
2. Curry JA, Schramm JL, Ebert E (1995) Sea ice-albedo climate feedback mechanism. Journal of Climate 8: 240–247.
3. Comiso JC (2010) Variability and Trends of the Global Sea Ice Cover. In: Thomas DN, Dieckmann GS, editors, Sea ice, Oxford, UK: Wiley-Blackwell. 2nd edition, pp. 205–245. doi: 10.1002/9781444317145.ch6.
4. Thomas DN, Papadimitriou S, Michel C (2010) Biogeochemistry of Sea Ice. In: Thomas DN, Dieckmann GS, editors, Sea ice, Oxford, UK: Wiley-Blackwell. 2nd edition, pp. 425–467. doi: 10.1002/9781444317145.ch12.
5. Bluhm BA, Gradinger RR, Schnack-Schiel SB (2010) Sea Ice Meio- and Macrofauna. In: Thomas DN, Dieckmann GS, editors, Sea ice, Oxford, UK: Wiley-Blackwell. 2nd edition, pp. 357–393. doi:10.1002/9781444317145.ch10.

6. Thomas DN (2012) Chapter 4 - Sea ice. In: Bell EM, editors, Life at Extremes: Environments, Organisms and Strategies for Survival, Wallingford, UK: CABI Publishing. pp.62–80.
7. Arrigo KR, Mock T, Lizotte MP (2010) Primary Producers and Sea Ice. In: Thomas DN, Dieckmann GS, editors, Sea ice, Oxford, UK: Wiley-Blackwell. 2nd edition, pp. 283–325. doi: 10.1002/9781444317145.ch8.
8. Stefels J, Steinke M, Turner S, Malin G, Belviso S (2007) Environmental constraints on the production and removal of the climatically active gas dimethylsulphide (DMS) and implications for ecosystem modeling. Biogeochemistry 112: 245–275.
9. Tison JL, Worby A, Delille B, Brabant F, Papadimitriou S, et al. (2008) Temporal evolution of decaying summer first-year sea ice in the Western Weddell Sea, Antarctica. Deep Sea Research Part II: Topical Studies in Oceanography 55: 975–987.

10. Lannuzel D, Schoemann V, de Jong J, Pasquer B, van der Merwe P, et al. (2010) Distribution of dissolved iron in Antarctic sea ice: Spatial, seasonal, and inter-annual variability. Journal of Geophysical Research: Biogeosciences 115 (G3).

11. Rysgaard S, Glud RN, Sejr MK, Bendtsen J, Christensen PB (2007) Inorganic carbon transport during sea ice growth and decay: A carbon pump in polar seas. Journal of Geophysical Research 83(C03016).

12. Arrigo KR, Perovich DK, Pickart RS, Brown ZW, van Dijken GL, et al. (2012) Massive Phytoplankton Blooms Under Arctic Sea Ice. Science 336: 1408.

13. Wassmann P, Slagstad D, Riser CW, Reigstad M (2006) Modelling the ecosystem dynamics of the Barents Sea including the marginal ice zone: II. Carbon ux and interannual variability. Journal of Marine Systems 59(1-2): 1–24.

14. Arrigo KR, Worthen DL, Robinson DH (2003) A coupled ocean-ecosystem model of the Ross Sea: 2. Iron regulation of phytoplankton taxonomic variability and primary production. Journal of Geophysical Research: Oceans 108(C7).

15. Rubao J, Meibing J, Varpe O (2013) Sea ice phenology and timing of primary production pulses in the Arctic Ocean. Global Change Biology 19(3): 734–741.

16. Deal C, Jin M, Elliott S, Hunke E, Maltrud M, et al. (2011) Large-scale modeling of primary production and ice algal biomass within arctic sea ice in 1992. Journal of Geophysical Research: Oceans 116(C7).

17. Jin M, Deal C, Lee SH, Elliott S, Hunke E, et al. (2012) Investigation of Arctic sea ice and ocean primary production for the period 1992–2007 using a 3-D global ice-ocean ecosystem model. Deep Sea Research Part II: Topical Studies in Oceanography 81–84: 28–35.

18. Lancelot C, de Montety A, Goosse H, Becquevort S, Schoemann V, et al. (2009) Spatial distribution of the iron supply to phytoplankton in the Southern Ocean: a model study. Biogeosciences 6(12): 2861–2878.

19. Sibert V, Zakardjian B, Goosse H, Gosselin M, Starr M, Senneville S, et al. (2011) 3D bio-physical model of the sympagic and planktonic productions in the Hudson Bay system. Journal of Marine Systems 88(3): 401–422.

20. Vancoppenolle M, Meiners KM, Michel C, Bopp L, Brabant F, et al. (2013) Role of sea ice in global biogeochemical cycles: emerging views and challenges. Quaternary Science Reviews In press.

21. Meiners KM, Vancoppenolle M, Thanassekos S, Dieckmann GS, Thomas DN, et al. (2012) Chlorophyll a in Antarctic sea ice from historical ice core data. Geophysical Research Letters 39(21).

22. Eilola K, Martensson S, Meier HEM (2013) Modeling the impact of reduced sea ice cover in future climate on the Baltic Sea biogeochemistry. Geophysical Research Letters 40(1): 149–154.

23. Hunke EC, Notz D, Turner AK, Vancoppenolle M (2011) The multiphase physics of sea ice: a review for model developers. The Cryosphere 5(4): 989–1009.

24. Popova EE, Yool A, Coward AC, Dupont F, Deal C, et al. (2012) What controls primary production in the Arctic Ocean? Results from an intercomparison of five general circulation models with biogeochemistry. Journal of Geophysical Research: Oceans 117(C8).

25. Tedesco L, Vichi M, Haapala J, Stipa T (2010) A dynamic Biologically Active Layer for numerical studies of the sea ice ecosystem. Ocean Modelling 35: 89–104.

26. Stroeve JC, Kattsov V, Barrett A, Serreze M, Pavlova T, et al. (2012) Trends in Arctic sea ice extent from CMIP5, CMIP3 and observations. Geophysical Research Letters 39(16).

27. Overland JE, Wang M (2013) When will the summer Arctic be nearly sea ice free? Geophysical Research Letters 40(10): 2097–2101.

28. Tedesco L, Vichi M, Thomas DN (2012) Process studies on the ecological coupling between sea ice algae and phytoplankton. Ecological Modelling 226: 120–138.

29. Ralph PJ, McMinn A, Ryan KG, Ashworth C (2005) Short-term effect of temperature on the photokinetics of microalgae from the surface layers of Antarctic pack ice. Journal of Phycology 41(4): 763–769.

30. Eppley RW (1972) Temperature and phytoplankton growth in the sea. Fis Bull 70: 1063–1085.

31. Arrigo KR, Sullivan CW (1992) The inuence of salinity and temperature covariation on the photophysiological characteristics of Antarctic sea ice microalgae. Journal of Phycology 28(6): 746–756.

32. Mock T, Hoch N (2005) Long-Term Temperature Acclimation of Photosynthesis in Steady-State Cultures of the Polar Diatom Fragilariopsis Cylindrus. Photosynthesis Research 85(3): 307–317.

33. Krell A, Funck D, Plettner I, John U, Dieckmann G (2007) Regulation of proline metabolism under salt stress in the psychrophilic diatom Fragilariopsis Cylindrus (Bacillariophyceae). Journal of Phycology 43(4): 753–762.

34. Arrigo KR, Dieckmann G, Gosselin M, Robinson D, Fritsen C, et al. (1995) High resolution study of the platelet ice ecosystem in McMurdo Sound, Antarctica: biomass, nutrient, and production profiles within a dense microalgal bloom. Marine Ecology Progress Series 127: 255–268.

35. Palmisano AC, SooHoo JB, Sullivan CW (1985) Photosynthesis-irradiance relationships in sea ice microalgae from McMurdo Sound, Antarctica. Journal of Phycology 21(3): 341–346.

36. Jin M, Deal CJ, Wang J, Shin KH, Tanaka N, et al. (2006) Controls of the landfast ice-ocean ecosystem offshore Barrow, Alaska. Annals of Glaciology 44: 63–72.

37. Lavoie D, Denman K, Michel C (2005) Modeling ice algae growth and decline in a seasonally ice- covered region of the Arctic (Resolute Passage, Canadian Archipelago). Journal of Geophysical Research 110(C11009).

38. Nishi Y, Tabeta S (2005) Analysis of the contribution of ice algae to the ice-covered ecosystem in Lake Saroma by a coupled ice-ocean ecosystem model. Journal of Marine Systems 55: 249–270.

39. Vancoppenolle M, Goosse H, de Montety A, Fichefet T, Tremblay B, et al. (2010) Modelling brine and nutrient dynamics in Antarctic sea ice: the case of dissolved silica. Journal of Geophysical Research 115(C02005).

40. Arrigo KR, Kremer JN, Sullivan CW (1993) A simulated Antarctic fast ice ecosystem. Journal of Geophysical Research: Oceans 98(C4): 6929–6946.

41. Tedesco L, Vichi M (2010) BFM-SI: a new implementation of the Biogeochemical Flux Model in sea ice. CMCC Research Papers: pp. 17.

42. Jeffery N, Hunke EC, Elliott SM (2011) Modeling the transport of passive tracers in sea ice. Journal of Geophysical Research: Oceans 116(C7).

43. Mikkelsen D, Rysgaard S, Glud R (2008) Microalgal composition and primary production in Arctic sea ice a seasonal study from Kobberfjord/Kangerluar-sunnguaq, West Greenland. Mar Ecol Prog Ser 368(65–74).

44. Thomas DN, Kaartokallio H, Tedesco L, Majaneva M, Piiparinen J, et al. (2014) Biological Oceanography of the Baltic Sea. In: Snoeijs P, Schubert H, Radziejewska T, editors, Sea ice, Springer. p. 400. In press.

45. Sturm M, Massom RA (2010) Snow and Sea Ice. In: Thomas DN, Dieckmann GS, editors, Sea ice, Oxford, UK: Wiley-Blackwell. 2nd edition, pp. 153–204. doi:10.1002/9781444317145.ch12.

46. Tedesco L, Vichi M, Haapala J, Stipa T (2009) An enhanced sea ice thermodynamic model applied to the Baltic sea. Boreal Environ Res 14: 68–80.

47. Semtner AJ (1976) A Model for the Thermodynamic Growth of Sea Ice in Numerical Investigations of Climate. J Phys Oceanogr 6: 379–389.

48. Assur A (1958) Composition of sea ice and its tensile strength. In: Arctic Sea Ice, Washington, DC: National Acad. Sci.- Nat. Res. Council, Publication 598. pp.106–138.

49. Untersteiner N (1964) Calculations of temperature regime and heat budget of sea ice in the central Arctic. Journal of Geophysical Research 69: 4755–4766.

50. Poulin M, Daugbjerg N, Gradinger R, Ilyash L, Ratkova T, et al. (2011) The pan-Arctic biodiversity of marine pelagic and sea-ice unicellular eukaryotes: a first-attempt assessment. Marine Biodiversity 41(1): 13–28.

51. Flynn KJ (2003) Modelling multi-nutrient interactions in phytoplankton; balancing simplicity and realism. Progress in Oceanography 56: 249–279.

52. Sakshaug E (2004) Primary and secondary production in Arctic seas. In: Stein R, MacDonald RW, editors, The Organic Carbon Cycle in the Arctic Ocean, New York, USA: Springer. pp.57–81.

53. Nikiforov SL, Colony R, Timokhov L (2001) Hydrochemical Atlas of the Arctic Ocean. St. Petersburg and Fairbanks: State Research Center of the Russian Federation the Arctic and Antarctic Research Institute of the Russian Federal Service for Hydrometeorology and Environmental Monitoring, and International Arctic Research Center, University of Alaska, Fairbanks, 50 pp.

54. Platt T, Sathyendranath S (2008) Ecological indicators for the pelagic zone of the ocean from remote sensing. Remote Sensing of Environment 112(8): 3426–3436.

55. Post E, Bhatt US, Bitz CM, Brodie JF, Fulton TL, et al. (2013) Ecological Consequences of Sea-Ice Decline. Science 341(6145): 519–524.

56. Christensen J, Hewitson B, Busuioc A, Chen A, Gao X, et al. (2007) Regional Climate Projections. In: Climate Change 2007: The Physical Science Basis. Contribution of Working Group I to the Fourth Assessment Report of the Intergovernmental Panel on Climate Change, Cambridge. Cambridge, United Kingdom and New York, NY, USA: University Press pp. 847–940.

57. Alou-Font E, Mundy C, Roy S, Gosselin M, Agust S (2013) Snow cover affects ice algal pigment composition in the coastal Arctic Ocean during spring. Mar Ecol Prog Ser 474: 89–104.

58. Arrigo K,Worthen DL, Lizotte MP, Dixon P, Dieckmann G (1997) Primary production in Antarctic sea ice. Science 276: 394–397.

59. Lavoie D, Macdonald R, Denman K (2009) Primary productivity and export uxes on the Canadian shelf of the Beaufort Sea: a modelling study. Journal of Marine Systems 75: 17–32.

60. Pogson L, Tremblay B, Lavoie D, Michel C, Vancoppenolle M (2011) Development and validation of a one-dimensional snow-ice algae model against observations in Resolute Passage, Canadian Arctic Archipelago. Journal of Geophysical Research: Oceans 116(C04010).

61. Elliott S, Deal C, Humphries G, Hunke E, Jeffery N, et al. (2012) Pan-Arctic simulation of coupled nutrient-sulfur cycling due to sea ice biology: Preliminary results. Journal of Geophysical Research 117(G01016).

62. Saenz BT, Arrigo KR (2012) Simulation of a sea ice ecosystem using a hybrid model for slush layer desalination. Journal of Geophysical Research 117(C05007).

Effects of Soil Data and Simulation Unit Resolution on Quantifying Changes of Soil Organic Carbon at Regional Scale with a Biogeochemical Process Model

Liming Zhang[1,2], Dongsheng Yu[2]*, Xuezheng Shi[2], Shengxiang Xu[2], Shihe Xing[1]*, Yongcong Zhao[2]

1 College of Resource and Environment, Fujian Agriculture and Forestry University, Fuzhou, China, **2** State Key Laboratory of Soil and Sustainable Agriculture, Institute of Soil Science, Chinese Academy of Sciences, Nanjing, China

Abstract

Soil organic carbon (SOC) models were often applied to regions with high heterogeneity, but limited spatially differentiated soil information and simulation unit resolution. This study, carried out in the Tai-Lake region of China, defined the uncertainty derived from application of the DeNitrification-DeComposition (DNDC) biogeochemical model in an area with heterogeneous soil properties and different simulation units. Three different resolution soil attribute databases, a polygonal capture of mapping units at 1:50,000 (P5), a county-based database of 1:50,000 (C5) and county-based database of 1:14,000,000 (C14), were used as inputs for regional DNDC simulation. The P5 and C5 databases were combined with the 1:50,000 digital soil map, which is the most detailed soil database for the Tai-Lake region. The C14 database was combined with 1:14,000,000 digital soil map, which is a coarse database and is often used for modeling at a national or regional scale in China. The soil polygons of P5 database and county boundaries of C5 and C14 databases were used as basic simulation units. Results project that from 1982 to 2000, total SOC change in the top layer (0–30 cm) of the 2.3 M ha of paddy soil in the Tai-Lake region was +1.48 Tg C, −3.99 Tg C and −15.38 Tg C based on P5, C5 and C14 databases, respectively. With the total SOC change as modeled with P5 inputs as the baseline, which is the advantages of using detailed, polygon-based soil dataset, the relative deviation of C5 and C14 were 368% and 1126%, respectively. The comparison illustrates that DNDC simulation is strongly influenced by choice of fundamental geographic resolution as well as input soil attribute detail. The results also indicate that improving the framework of DNDC is essential in creating accurate models of the soil carbon cycle.

Editor: Jose Luis Balcazar, Catalan Institute for Water Research (ICRA), Spain

Funding: The work was funded by National Natural Science Foundation of China (No. 41001126), and the National Basic Research Program of China (973 Program) (2010CB950702). The funders had no role in study design, data collection and analysis, decision to publish, or preparation of the manuscript.

Competing Interests: The authors have declared that no competing interests exist.

* E-mail: dshyu@issas.ac.cn (DY); fafuxsh@126.com (SX)

Introduction

An estimated 1500 Pg of C is held in the form of soil organic carbon (SOC), representing 2/3 of the global terrestrial organic carbon pool [1–3]. SOC plays a vital role in the global carbon cycle, where a slight alteration of the soil carbon pool can cause profound changes in atmospheric CO_2 concentrations. Agro-ecosystems, accounting for 10% of the total terrestrial area, are one of the most sensitive terrestrial ecosystems subject to heavy human activity [3]. Increasing agricultural soil C sequestration is recognized as one strategy for achieving food security and improving soil quality.

Paddy soil is a major cultivated soil in China, and a unique type of anthropogenic soil recognized by Chinese Soil Taxonomy [3–5]. The total area of paddy soils is 45.7 M ha, which accounts for 34% of the total cultivated land in China [6]. This area also accounts for 22% of the total waterlogged farming area worldwide and produces about 44% of all grain in China [4]. Therefore, accurate estimation of paddy soil SOC change in China is vitally important for a comprehensive understanding of SOC dynamics and agro-ecosystem sustainability.

Recently, scientists have applied modeling to estimate SOC change in cropping systems [7–14]. The DeNitrification-DeComposition

(DNDC) model, developed by Li et al. [15,16], is a process-based model focused on agrosystem carbon and nitrogen cycling and has been widely used for regional studies in the USA [17], China [11], India [18] and Europe [19]. Recently the DNDC model was determined to be one of the well performing models based on seven long-term experiments selected by the Global Change and Terrestrial Ecosystems Soil Organic Matter Network (GCTE SOMNET), which evaluated model performance using three different land uses, a range of climatic conditions within the temperate region, and different treatments [11,14].

In China, scientists have studied SOC change using the DNDC model for many years. At the regional scale, Tang et al. [11] simulated SOC changes for cropland in China for 1998 using the DNDC model, and they found that SOC would be lost at a rate of 78.89 Tg C year^{-1}. Zhang et al. [20] linked the DNDC model and 1: 14,000,000 soil database to estimate SOC stock changes for the year 2000 in Northwest China, revealing a decline in SOC stock. At the field scale, Wang et al. [21] tested DNDC uncertainty based on six long-term (10–20 year) SOC datasets from the Northeast, North, Northwest, Central South, East, and Southwest China. Results from the six validation tests supported the previous

Figure 1. Geographical location of the study area in China.

conclusions that the DNDC model was capable of quantifying SOC change in the agroecosystems across the entire area of China.

To date, the county boundary was used as the basic simulation unit in most DNDC simulations conducted at regional scale [11,20]. As a result, these simulations are often subject to great uncertainties since the soil property data were averaged for the area, which greatly ignore the impacts of soil heterogeneity therein [18,22]. Moreover, many researchers used coarse soil attribute data obtained from the books such as Soil in China (Vol. 1–6) and 1: 14,000,000 soil maps at national or a regional scale in China [11,20]. However, studies have already pointed out that the effect of soil heterogeneity on SOC change estimation is a major source of uncertainty when using the DNDC model at the regional scale [18,22,23].

This study, which was carried out in the rice-dominated Tai-Lake Region of China, provides a chance to test the uncertainty of the DNDC model caused by different precisions of soil data and basic simulation unit. The goals of this study were to: (1) compare SOC changes modeled with different resolutions of soil databases and varied basic simulation units, (2) assess the uncertainty derived from these soil databases with different resolutions and basic simulation units, and (3) give some suggestions for improving the performance of the biogeochemical DNDC model applied at the regional scale.

Materials and Methods

Study area

The Tai-Lake region (118°50'-121°54'E, 29°56'-32°16'N), an area of intensive rice cultivation, is located in the middle and lower reaches of the Yangtze River paddy soil region of China. The region includes the entire Shanghai City administrative area and a part of Jiangsu and Zhejiang provinces, and covers a total area of 36,500 km^2 (**Fig. 1**) [4]. The Tai-Lake region mainly consists of plains formed on deltas with numerous rivers and lakes. The climate is warm and moist with abundant sunshine and a long growing season. Annual rainfall is 1,100–1,400 mm, with a mean temperature of 16°C, and average annual sunshine of 1,870–2,225 hours. The frost-free period is over 230 days. The study area is one of the oldest agricultural regions in China, with a long history of rice cultivation spanning several centuries. Most

cropland in the region is managed as a rice and winter wheat rotation. Rice is planted in June and harvested in October and wheat is planted in November and harvested in May [24].

Approximately 66% of the total land area is covered with paddy soils [24]. Paddy soils in the Tai Lake area are derived mostly from loess, alluvium, and lacustrine deposits, and are classified into 6 soil subgroups according to the Genetic Soil Classification of China (GSCC) system which are represented in the 1:50,000 digital soil map (**Table 1**). As map scale decreased, the soil subgroups of submergenic, bleached, percogenic and degleyed on the 1:50,000 soil map was eliminated and emerged into the soil subgroups of degleyed and hydromorphic in the 1:14,000,000 soil map. Therefore, those were only two paddy soil subgroups of degleyed and hydromorphic in the 1:14,000,000 soil map. The GSCC nomenclature as well as the subgroup's reference name in US Soil Taxonomy (ST) include; Hydromorphic (Typic Epia-quepts), Submergenic (Typic Endoaquepts), Bleached (Typic Epiaquepts), Gleyed (Typic Endoaquepts), Percogenic (Typic Epiaquepts), and Degleyed (Typic Endoaquepts) [25,26].

Description of the DNDC model

The DNDC model (Version 9.1) is a process-based soil biogeochemical research tool that was developed to estimate the impact of management strategies on the fate of nitrogen (N) and carbon (C) in agroecosystems. It integrates crop growth and soil biogeochemical processes on a daily time step and simulates N and C cycles in plant-soil systems.

The model contains six interacting sub-models which describe the generation, decomposition, and transformation of organic matter, and outputs the dynamic components of SOC and greenhouse gas fluxes. The six sub-models include: 1) a soil climate component which use soil physical properties, air temperature, and precipitation data to calculate soil temperature, moisture, and redox potential (Eh) profiles and soil water fluxes through time. The results of the calculation are then fed to the other sub-models; 2) a nitrification component; 3) a denitrification module, which calculates hourly denitrification rates and N_2O, NO, and N_2 production during periods when the soil Eh decreases due to rainfall, irrigation, flooding, or soil freezing; 4) simulation of SOC decomposition and CO_2 production through soil microbial respiration; 5) a plant growth component, which calculates daily

Table 1. The subgroups of paddy soil in the Tai-Lake region, China.

Subgroups	Horizonation*	Descriptions
Bleached	A-P-E-C	Mainly distributed in foothills, usually no underground water, impervious layer at 60 cm depth, soil reaction close to neutral or slightly acid.
Gleyed	Aa-Ap-G-C	Mainly distributed in depressional areas, high underground water level, poorly drained, distinct gleyization, soil reaction was slightly acid.
Percogenic	Aa-Ap-C	Mainly distributed on gentle hill slopes, no underground water, associated with rain-fed paddy fields, soil reaction was neutral to slightly acid.
Degleyed	Aa-Ap-Gw-G	Same distribution area as Gleyed paddy soils, after man-made drainage the underground water level decreases leading to degley processes, soil reaction was slightly acid.
Submergenic	A-Ap-P-C	Mainly distributed in alluvial plain or low flat ground, moderate drainage, underground water level was below 60 cm, soil reaction was neutral.
Hydromophic	Aa-Ap-P-W-G-C	Mainly distributed in floodplain, long cultivation history, well-drained, underground water level was below 90 cm, soil reaction was neutral.

*According to GSCC (Genetic Soil Classification of China), Aa means arable layer, Ap plow pan, C undeveloped parent material, Ds fragmental deposit horizon, E bleached horizon, G gley horizon, Gw degley horizon, P percogenic horizon, W waterlogogenic horizon.

root respiration, water, and N uptake by plants, and plant growth; and 6) a fermentation module, which calculates daily methane (CH_4) production and oxidation. The DNDC model can simulate C and N biogeochemical cycles in paddy rice ecosystems, as the model has been modified by adding a series of anaerobic processes [15,16,22,23,27,28,29,30].

At present, the DNDC model has been utilized by scientists in many countries, for example, the model is applied to simulate the carbon cycle in paddy field in Italy, China and Germany, in wheat fields in Canada, and it has been used to simulate the dynamics of soil organic matter in a 100 year experimental field in Rothamsted Experimental Station in England [14,31]. At the international conference on global change in Asia-Pacific areas in 2000, the DNDC model was recommended as the primary method for SOC studies in the in the Asia-Pacific region [31].

Database development

A major challenge for using an ecosystem model at regional scale is to assemble adequate datasets required to initialize and run the model. We examined the influence of database choices by executing simulation runs with different input sets using individual or combinations of databases. The geographic resolution or fundamental simulation unit could be represented by any of three assessment unit format datasets, polygon-based database of 1:50,000 (P5), county-based database of 1:50,000 (C5), and county-based database of 1:14,000,000 (C14). The three soil datasets covered 37 counties in Tai-Lake region.

The polygon-based database of 1:50,000 (P5) was linked a digital soil map (1:50,000), the most detailed of the three databases, in the Tai-Lake region contains 52,034 paddy soil polygons (**Table 2**). The polygons were derived from 1,107 soil profiles extracted from the latest national soil map (1:50,000), the Second National Soil Survey of China in the 1980s-1990s, with attribute assignment using the Pedological Knowledge Based (PKB) method based on GSCC [32]. The 1:50,000 digital soil database consists of many soil attributes, such as soil name, horizon thickness, bulk density, organic carbon content, clay content, pH, etc.

Soil parameters in C5 were derived from the 1:50,000 digital soil map (**Fig. 2 and Table 2**). However the attributes for C14 were derived from different sources than C5, primarily the 1:14,000,000 national soil map [33,34] (**Fig. 2**). C14 was widely used when the DNDC model was applied to national or regional scale in China [11,20]. The C14 in the Tai-Lake region contained 8 polygons of paddy soils representing 49 paddy soil profiles, and was also compiled via the Pedological Knowledge Based (PKB) method based on GSCC [32].

The C5 and C14 were built from the default method developed for DNDC, in which the maximum and minimum values of soil texture, pH, bulk density, and organic carbon content were recorded for each county (**Fig. 2**). So, the DNDC modeling of C5 and C14 methods conducted have used counties as the basic simulation unit in the Tai-Lake region (**Fig. 2**). After regional runs with C5 and C14 database, the DNDC model produced two SOC

Table 2. Characteristics of different resolution soil attribute databases of paddy soils in GSCC in the Tai-Lake region, China.

Soil database	Map scale	Source of soil maps	Source of soil data	Basic map units	Number of soil profiles	Number of polygons	Simulation unit
P5	1:50,000	Soil Survey Office of County in Jiangsu Province, Zhejiang Province and Shanghai City	Soil Series of County in Jiangsu Province, Zhejiang Province and Shanghai City	Soil Species	1,107	52,034	polygon
C5	1:50,000	Soil Survey Office of County in Jiangsu Province, Zhejiang Province and Shanghai City	Soil Series of County in Jiangsu Province, Zhejiang Province and Shanghai City	Soil Species	1,107	52,034	county
C14	1:14,000,000	Institute of Soil Science, Chinese Academy of Sciences	Soil Series of China	Subgroups	49	8	county

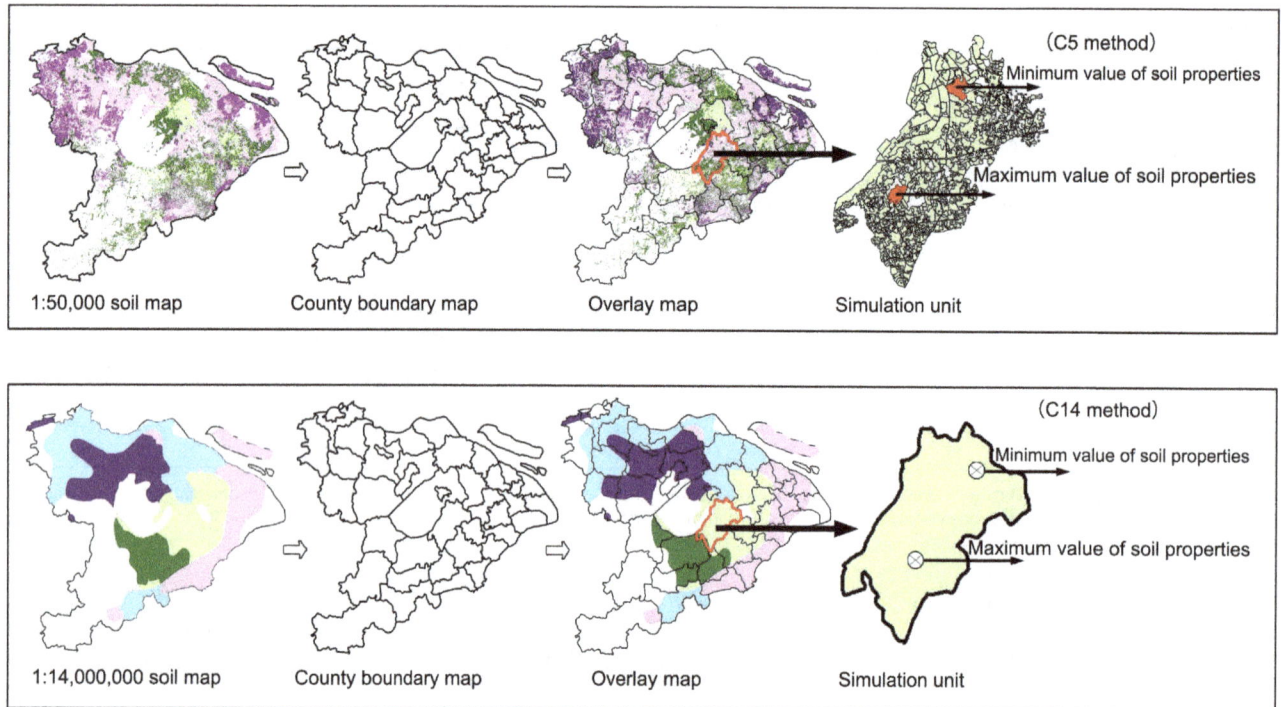

Figure 2. Description of C5 and C14 methods in the Tai-Lake region of China.

change (0–30 cm) resulting from two runs with the maximum and minimum soil values in each county. In this paper we present the mean results (average of maximum and minimum estimates) [11]. The DNDC modeling of P5 method conducted has used polygon as the basic simulation unit in the Tai-Lake region (**Table 2**). Therefore, the DNDC model runs with P5 database produced a single annual SOC change (0–30 cm) for each polygon. The total

Figure 3. Geographical location of weather stations across or near the Tai-Lake region, China.

SOC change of each county in the P5 was calculated by summing the SOC change of all polygons in a county. For a more complete description of P5 method see Zhang et al [35,36] and Xu et al [37].

For comparison in this study, both the polygon-based (P5) and county-based (C5 and C14) soil databases in the Tai-Lake region were run concurrently so the DNDC model could generalize regional SOC change from 1982 to 2000. The results simulated by DNDC with the two types of databases were compared to assess the advantages of using detailed, polygon-based 1:50,000 soil dataset (P5) [35,36,38,39].

In this study, the crop dataset included physiological data for summer rice and winter wheat in the Tai-Lake region. The crop parameters were obtained from thorough testing with that reflected the typical conditions of Tai-Lake region, which were founded on a wide range of information form Chinese literature published during the past decade and a publication of Gou et al [40,41].

Daily meteorological data (precipitation, maximum and minimum air temperature) for 1982–2000 from 13 weather stations across and near the Tai-Lake region were acquired from the National Meteorological Information Center, China Meteorological Administration (**Fig. 3**) [42]. Each county in the simulation was assigned to the nearest weather station [11,20,31].

The agricultural management dataset included sowing acreage, nitrogen fertilizer application rates, livestock, planting and harvest dates, and agricultural population at the county level from 1982 to 2000 in three resolution databases. The crop management practices of different counties were almost the same because the Tai-Lake region was a plain in topography. The main measures of farming management in the study area included: (1) fertilizer application: nitrogen synthetic fertilizer was applied for 6 times in the basal, tillering and heading stage for rice, and in the basal, jointing and heading stage for wheat; and organic manure (20% of livestock wastes and 10% of human wastes) was applied twice as base fertilizer for rice and wheat at the rates calculated based on the local livestock numbers (866, 44, 95, and 23 kg C head^{-1} yr^{-1} for cattle, sheep, swine and human, respectively); and N concentration in rainfall was 2.07 ppm; (2) crop residue management: 15% of aboveground crop residue was returned to the soil; (3) water management: one time of midseason and 5 time of shallow flooding (from June 17 to July 23, from July 28 to August 12, from August 24 to September 11, from September 18 to September 25, and from September 27 to October 2, respectively) were applied at summer rice; (4) tillage: twice at the 20 cm tilling depth for rice and 10 cm for wheat on the planting dates before 1990; and no-till applied for wheat after 1990; (5) growing period: rice is planted in June and harvested in October and wheat is planted in November and harvested in May; (6) optimum yield: rice is 7500 kg dry matter ha^{-1} and wheat is 3750 kg dry matter ha^{-1} [11,14,24,35,41,43]. All simulation methods within a certain county have the same feature input value such as crops, agricultural management, and climate, except soil feature [38,39].

Evaluation of simulation accuracy in three resolution databases

In order to evaluate the accuracy in three resolution databases, the simulated results of DNDC model were tested against measured data from paddy soils of the Tai-Lake region, which is the same area examined here.

From the perspective of previous studies, most dynamic models were only tested or validated with static long-term field-scale observations due to a lack of available soil data with temporal and spatial variation. Since these models have not yet been validated by regional scale data, uncertainty concerning their accuracy exists when they were applied to larger area dynamic SOC simulation [3].This study compared simulation results with the spatial distribution of SOC measurements from 1033 paddy soil sampling sites acquired in 2000, to validate and assess model performance in different simulation methods (P5, C5, and C14). The bias in the total difference between simulation and measurement were determined by calculating the correlation coefficient (r), the

relative error (E), the mean absolute error (MAE) and the root mean square error (RMSE), as follows: [8,44].

$$r = \frac{\sum_{i=1}^{n} \left(V_{oi} - \overline{V_{oi}}\right)\left(V_{Si} - \overline{V_{Si}}\right)}{\left[\sum_{i=1}^{n} \left(V_{oi} - \overline{V_{oi}}\right)^2\right]^{1/2}\left[\sum_{i=1}^{n} \left(V_{Si} - \overline{V_{Si}}\right)^2\right]^{1/2}} \quad (1)$$

$$E = \frac{100}{n} \times \sum_{i=1}^{n} \frac{V_{oi} - V_{Si}}{V_{oi}} \quad (2)$$

$$\text{RMSE} = \sqrt{\frac{1}{n}\sum_{i=1}^{n}(V_{oi} - V_{Si})^2} \quad (3)$$

$$MAE = \frac{1}{n}\sum_{i=1}^{n} ABS(V_{oi} - V_{Si}) \quad (4)$$

Where V_{oi} are the observed values, $\overline{V_{oi}}$ is the mean of the observed data, V_{Si} is the simulated value, $\overline{V_{Si}}$ is the mean of the simulated value, $Si \in (P5,C5,C14)$, and n is the number in the sequence of the data pairs. If E is less than 5% or between 5% and 10%, the simulation is satisfactory or acceptable, respectively; otherwise, it is unacceptable [44]. The greater r value is and the smaller RMSE or MAE value is, the greater prediction accuracy is. Conversely, a lower r value and more elevated RMSE or MAE value, the lower prediction accuracy is.

Data comparison and analysis

SOC change as quantified by DNDC modeling with the P5 assessment unit data set are recognized as a benchmark for comparison with the results of the DNDC model runs with the other two assessment unit data sets as input. The P5 are thought

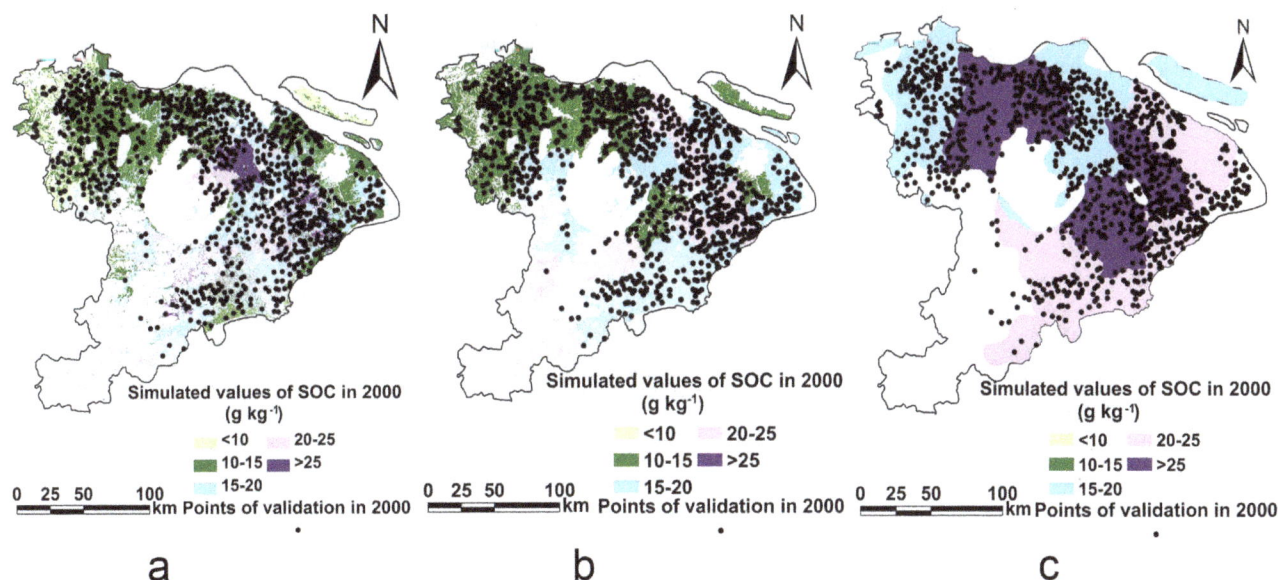

Figure 4. Spatial distribution of validation points and simulated SOC values from different simulation methods for the Tai-Lake region for 2000 (a: P5, b: C5, and c: C14).

theoretically to be more accurate than the C5 and C14 because of their relative greater detail and accuracy [35,36,38,39]. Relative variation of an index value (VIV, %) of C5 and C14 methods is calculated as the formula (1). The index values (IV) were quantified from the P5 data set (IV_{P5}) and other data sets (IV_{Ci}) to support data set comparison [23,38,39].

$$VIV(\%) = ABS\left(100 \times (IV_{P5} - - IV_{Ci})/IV_{P5}\right) \qquad (5)$$

Where ABS is the absolute value function, IV_{P5} is the total SOC change with P5, and IV_{Ci} is the total SOC change produced by C5 (or C14).

Previous results from the sensitivity tests of the DNDC model indicated that the spatial heterogeneity of soil properties (e.g. texture, SOC content, bulk density, and pH) are the major sources of uncertainty for simulating SOC changes under specific management conditions at regional scale [18,20,22,23]. In order to test the most sensitive soil properties factor, the correlation of soil properties and average annual SOC changes were determined by step-wise regression analysis by using SPSS statistical software [37,45]. The step-wise regression is useful in checking how entering each variable affects the overall regression model, which begins by entering the variable with the largest partial statistic and checking the importance of the coefficient of the variable [45,46]. This method keeps adding more variables, each time recalculating the coefficients. During the incorporation of a variable into the model, the partial statistic of the already entered variable changes and might cause it to be unimportant. The operation stops when the model has incorporated the variables with the most significant contribution and discarded the least significant ones [47].

Results and Discussion

Difference of simulation accuracy in three resolution databases

Three maps of average SOC content for paddy soils at surface layers (0–15 cm) in the study area in 2000 were constructed on the basis of simulated data in different simulation methods (P5, C5, and C14) (**Fig. 4**). Also, corresponding SOC validation points were constructed from measurements of the surface layer (0–15 cm) of 1033 paddy soil samples taken in the study area in 2000. Fig. 4 demonstrates that the observed SOC in 2000 varied from 1.9 g kg^{-1} to 36 g kg^{-1}. By comparison, Fig. 4 also illustrates that

simulated SOC in 2000 varied from 5.1 g kg^{-1} to 34 g kg^{-1} in P5, from 11 g kg^{-1} to 24 g kg^{-1} in C5, and from 17 g kg^{-1} to 28 g kg^{-1} in C14; where 99.6%, 84.1% and 57.1% of simulated paddy soil samples in P5, C5 and C14 were within the ranges produced by the observed SOC data. Furthermore, the relative errors (E) of P5 and C5 were 6.4% and 5.0%, respectively; and within the range of 5%–10%, demonstrating that the DNDC model in P5 and C5 were acceptable for modeling SOC of paddy soils in the Tai-Lake region according to the evaluation criteria described earlier (**Fig. 5a and b**) [8,44]. Moreover, the small values of MAE (4.0 g kg^{-1}) and RMSE (5.0 g kg^{-1}) in P5 and C5 also indicated that the modeled results were encouragingly consistent with observations in the Tai-Lake region (**Fig. 5a and b**). However, the E, MAE and RMSE of C14 reached -33%, 6.0 g kg^{-1} and 7.0 g kg^{-1}, respectively, suggesting that the simulated results of C14 were not suitable for simulating paddy soils in the Tai-Lake region (**Fig. 5c**).

Overall, though the values of E, MAE and RMSE between P5 and C5 had no significant differences, P5 was recognized better due to high correlation coefficient (0.5) and accurate simulation range (99.6%) (**Fig. 5a and b**). Furthermore, the simulation of P5 can differentiate the difference of paddy soil type within a county. Some studies showed that SOC content spatial variability was correlated with soil type spatial variability(**Fig. 5a**) [32,48,49]. Compared to the SOC validation of DNDC model in cropland by other scientists, accurate simulation of P5 (r = 0.50**; E = 6.4%; MAE = 4.0 g kg^{-1}; RMSE = 5.0 g kg^{-1} and n = 1033) and C5 (r = 0.40**; E = 5.0%; MAE = 4.0 g kg^{-1}; RMSE = 5.0 g kg^{-1} and n = 1033) are higher than those of Liu et al [50] (r = 0.25–0.66 and n = 68), Liu et al [51] (E = 27.6% and n = 49), and Xu et al [52] (r = 0.22**; RMSE = 4.4 g kg^{-1} and n = 243); and are almost similar to that of the Xu et al [52] (r = 0.52**; RMSE = 4.1 g kg^{-1} and n = 1385). However, the SOC accurate simulation of DNDC model in P5, C5 and C14 are lower than those of Studdert et al [53] (r = 0.73** and n = 286) by using the RothC model and Yu et al [54] (r = 0.98** and n = 349) by using the Agro-C model. Therefore, the results mentioned above suggest that modification of the DNDC model is necessary to better simulate SOC change from cropping systems. With continued modification, DNDC model could become a powerful tool for estimating SOC change at regional and national scales.

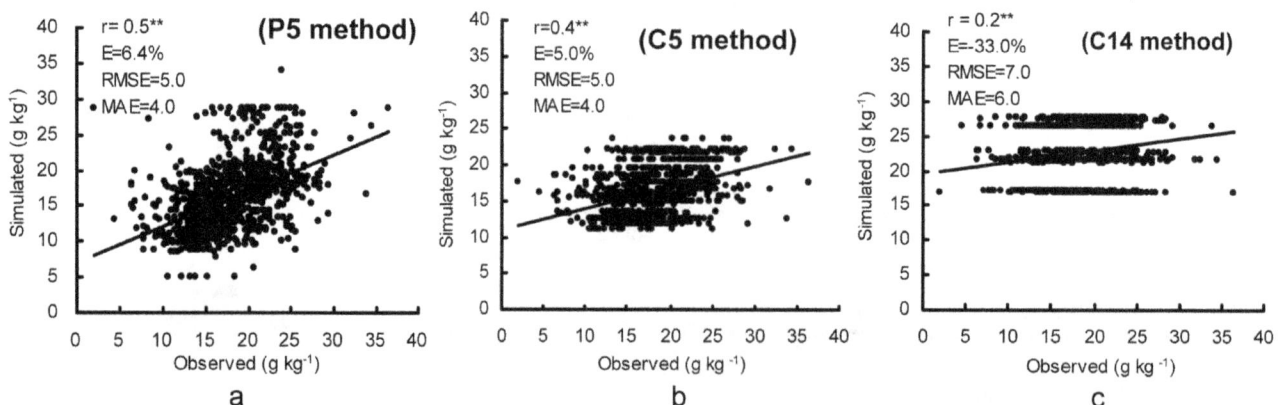

Figure 5. Comparison between simulated and observed SOC values from different simulation methods of the Tai-Lake region for 2000 (a: P5, b: C5, and c: C14).

Table 3. Soil properties at three resolution soil attribute databases contributing to the variability of average annual SOC change in Tai-Lake region paddy soils from 1982 to 2000.

Soil database	Number of simulation units	$\triangle R^{2a}$				Adjusted R^2
		Initial SOC (g kg^{-1})	Clay(%)	pH	Bulk density (g cm^{-3})	
P5	52,034	0.778***	0.066***	0.009***	0.025***	0.878***
C5	37	0.881***			0.062***	0.939***
C14	37	0.185***		0.757***		0.938***

***significant at 0.001 probability levels, respectively.
[a]The change in the R^2 statistic is produced by adding a soil property into stepwise multiple regressions.

Variation of soil properties derived as input for DNDC modeling in three resolution databases in Tai-Lake region

Results of the contribution of soil properties to the variability of average annual SOC change are given in Table 3. All variables (i.e., initial SOC content, pH, bulk density, and clay content) were included in the step-wise regression analysis. For the P5 and C5 resolution databases, initial SOC content accounted for 77.8%–88.1% of the difference of average annual SOC change for paddy soils from 1982 to 2000, while other soil parameters only accounted for less than 6.6% of the difference. For the C14 resolution database, initial SOC content accounted for 18.5% of the difference of average annual SOC change for paddy soils from 1982 to 2000, and soil pH accounted for 75.7% of the difference. Therefore, it could be inferred that the differences in SOC change modeled with the three resolution databases were primarily due to the differences in initial SOC content and pH.

Table 4 shows the initial SOC content (0–5 cm), clay content (0–10 cm), pH (0–10 cm), and bulk density (0–10 cm) derived as input for DNDC modeling, from P5, C5 and C14 for the Tai-Lake region. As for the entire Tai-Lake region, the average initial SOC values sourced from P5 was lower than that from C5 and C14. Another difference is that the average values of clay content and pH sourced from C14 were also higher than those from P5 and C5. The average bulk density sourced from C5 was higher than that from P5 and C14.

The differentiation of soil properties was also shown at the county scale in the Tai-Lake region (**Table 4**). The average values of initial SOC content and bulk density sourced from C5 for 24 counties were higher than those from P5; the other was that the average values of clay content for 24 counties and pH for 20 counties in C5 were lower than those from P5. Although the average clay content sourced from P5 for 25 counties was slightly lower than that from C14, but the average initial SOC content sourced from C14 for 34 counties was obviously higher than that of P5. According to statistics describing the 1:50,000 digital soil database of the Tai-Lake region, initial SOC content of six paddy soil subgroups, namely submergenic, bleached, percogenic, hydromorphic, degleyed and gleyed, were 10 g kg^{-1}, 10 g kg^{-1}, 11 g kg^{-1}, 15 g kg^{-1}, 19 g kg^{-1}, and 25 g kg^{-1}, respectively. As map scale decreased from 1:50,000 to 1:14,000,000, the submergenic, bleached, percogenic and degleyed subgroups on the 1:50,000 digital soil map were eliminated and merged into the hydromorphic and degleyed subgroups in the 1:14,000,000 digital soil map [32,39]. The initial SOC content of the hydromorphic and gleyed subgroups in the 1:14,000,000 digital soil database were 17 g kg^{-1} and 28 g kg^{-1}, respectively, which were higher than most paddy soil subgroups in the 1:50,000 digital soil

database. Therefore, the average initial SOC content of most counties in C14 was significantly higher than that from P5, while the average values of bulk density for 20 counties and pH for 24 counties in C14 was lower than those from P5. The results demonstrated that the soil properties (i.e., texture, SOC content, bulk density, and pH) in three resolution databases methods had large differences in the Tai-Lake region. Many studies have showed that SOC spatial variability is expressed by map delineations and map unit composition which varied with scales, resulting in the assignment of different soil properties at each scale of aggregation [32,48,49]. As such, an improper of soil map scales and simulation unit may lead to SOC estimation inaccuracy.

Variation of the average annual-, total SOC change modeled with the three resolution databases in Tai-Lake region

Similar trends can be observed in estimates of average annual-, total SOC change over the 19 year study period for three resolution databases decreased from P5 to C14 (**Fig. 6**). Simulation results demonstrate that total SOC change of P5 in the top layer (0–30 cm) of the 2.3 M ha of paddy rice fields in the Tai-Lake region was +1.48 Tg C from 1982 to 2000, with the annual SOC change ranging from -45 kg C ha^{-1} yr^{-1} to 92 kg C ha^{-1} yr^{-1} (**Fig. 6**). From 1982 to 1988, the SOC change modeled with P5 inputs was almost negative with annual changes ranging from -3.2 kg C ha^{-1} yr^{-1} to -45 kg C ha^{-1} yr^{-1}. According to agricultural statistical data, chemical fertilizer application rate ranged from 180 kg N ha^{-1} yr^{-1} to 350 kg N ha^{-1} yr^{-1}, which is a relatively low value. Low fertilizer application rates often result in reduced SOC sequestration [31,55]. From 1989 to 2000, rural economic development led to increased fertilizer application from 350 kg N ha^{-1} yr^{-1} to 400 kg N ha^{-1} yr^{-1}. Increasing fertilizer application results in enhanced crop production and residue accumulation, and the latter leads to an increase of SOC. Further, much of the region has been utilizing no-tillage practices in planting wheat since 1991, which contribute to reduced SOC decomposition [35].

Although three resolution databases within a certain county have the same feature input value such as crops, agricultural management, and climate; SOC balance of C5 (or C14) in the Tai-Lake region was almost negative with annual changes ranging from 86 kg C ha^{-1} yr^{-1} to -205 kg C ha^{-1} yr^{-1} (or -185 kg C ha^{-1} yr^{-1} to -693 kg C ha^{-1} yr^{-1}) from 1982 to 2000 (**Fig. 6**). The total SOC changes of C5 and C14 in the Tai-Lake region were −3.99 Tg C and −15.38 Tg C, respectively, from 1982 to 2000. With the total SOC change as modeled with P5 inputs as the baseline, the relative deviation of C5 and C14 were 368% and 1126%, respectively.

Table 4. Statistics for soil properties derived as input for DNDC modeling in different counties, from P5, C5 and C14 for the Tai-Lake region.

County	P5 SOC (WA)	P5 Clay	P5 BD	P5 pH	C5 SOC Range	C5 SOC Ave	C5 Clay Range	C5 Clay Ave	C5 BD Range	C5 BD Ave	C5 pH Range	C5 pH Ave	C14 SOC Range	C14 SOC Ave	C14 Clay Range	C14 Clay Ave	C14 BD Range	C14 BD Ave	C14 pH Range	C14 pH Ave
Zhangjiagang	14	28	1.22	7.7	10–17	14	5–32	19	1.16–1.33	1.25	7.4–8.0	7.7	12–21	17	24–31	28	1.19–1.23	1.21	6.0–7.4	6.7
Changshu	17	25	1.23	7.0	9–38	24	9–34	22	1.05–1.46	1.26	5.5–8.1	6.8	12–21	17	24–31	28	1.19–1.23	1.21	6.0–7.4	6.7
Taicnang	14	33	1.20	7.7	9–20	15	23–42	33	1.11–1.38	1.25	7.4–8.6	8.0	11–31	21	18–48	33	1.12–1.27	1.20	5.5–7.4	6.5
Kunshan	19	35	1.15	7.1	11–34	18	22–44	33	0.94–1.40	1.17	6.4–7.6	7.0	23–33	28	32–56	44	1.14–1.14	1.14	6.2–6.9	6.6
Wuxian	24	41	1.08	6.6	6–30	23	26–47	37	0.97–1.47	1.22	3.4–7.4	5.6	12–21	17	24–31	28	1.19–1.23	1.21	6.0–7.4	6.7
Wujiang	17	36	1.06	5.9	3–26	15	17–58	38	0.89–1.67	1.28	4.9–6.9	5.9	23–33	28	32–56	44	1.14–1.14	1.14	6.2–6.9	6.6
Wuxi	14	28	1.16	6.7	4–17	11	14–34	24	1.09–1.39	1.24	5.3–7.2	6.3	21–33	27	25–31	28	1.13–1.23	1.18	6.0–6.3	6.1
Jiangyin	13	13	1.28	6.2	6–17	12	8–25	17	0.99–1.54	1.27	5.4–8.0	6.7	21–33	27	25–31	28	1.13–1.23	1.18	6.0–6.3	6.1
Wujin	12	9	1.22	6.8	7–18	13	4–13	9	1.08–1.51	1.30	6.2–7.9	7.1	21–33	27	25–31	28	1.13–1.23	1.18	6.0–6.3	6.1
Jintan	10	9	1.32	6.8	7–14	11	4–13	9	1.17–1.58	1.38	5.5–7.6	6.6	12–21	17	24–31	28	1.19–1.23	1.21	6.0–7.4	6.7
Liyang	10	10	1.23	6.2	6–17	12	7–12	10	1.05–1.37	1.21	6.0–7.2	6.6	12–21	17	24–31	28	1.19–1.23	1.21	6.0–7.4	6.7
Yixing	13	27	1.17	6.0	3–30	17	10–53	32	1.11–1.58	1.35	4.4–8.5	6.5	21–33	27	25–31	28	1.13–1.23	1.18	6.0–6.3	6.1
Dantu	7	36	1.25	6.6	2–19	11	12–49	31	1.07–1.39	1.23	5.8–8.0	6.9	12–21	17	24–31	28	1.19–1.23	1.21	6.0–7.4	6.7
Jurong	10	30	1.23	5.4	6–13	10	15–38	27	1.10–1.29	1.20	5.1–7.4	6.3	12–21	17	24–31	28	1.19–1.23	1.21	6.0–7.4	6.7
Danyang	12	29	1.24	6.7	8–16	12	16–53	35	1.07–1.36	1.22	5.8–7.8	6.8	12–21	17	24–31	28	1.19–1.23	1.21	6.0–7.4	6.7
Jiaxing	19	34	1.19	6.5	10–26	18	20–56	38	0.98–1.34	1.16	5.8–7.6	6.7	23–33	28	32–56	44	1.14–1.14	1.14	6.2–6.9	6.6
Jiashan	21	39	1.23	6.2	15–27	21	22–44	33	1.03–1.34	1.19	5.7–7.0	6.4	23–33	28	32–56	44	1.14–1.14	1.14	6.2–6.9	6.6
Pinghu	15	35	1.10	6.6	9–24	17	22–43	33	0.92–1.48	1.20	6.3–7.2	6.8	11–31	21	18–48	33	1.12–1.27	1.20	5.5–7.4	6.5
Haiyan	17	40	1.17	6.7	7–25	16	22–52	37	0.92–1.51	1.22	5.7–7.3	6.5	11–31	21	18–48	33	1.12–1.27	1.20	5.5–7.4	6.5
Haining	14	36	1.19	6.6	7–24	16	19–52	36	0.92–1.51	1.22	6.0–7.5	6.8	11–31	21	18–48	33	1.12–1.27	1.20	5.5–7.4	6.5
Tongxiang	14	30	1.05	6.5	7–29	18	20–52	36	0.91–1.34	1.13	6.0–7.4	6.7	14–33	24	25–34	30	1.12–1.13	1.13	6.3–6.6	6.5
Huzhou	23	30	1.10	6.2	13–37	25	7–42	25	0.99–1.37	1.18	5.6–6.7	6.2	14–33	24	25–34	30	1.12–1.13	1.13	6.3–6.6	6.5
Changxing	17	31	1.14	5.8	6–31	19	9–47	28	0.84–1.53	1.19	3.6–7.1	5.4	14–33	24	25–34	30	1.12–1.13	1.13	6.3–6.6	6.5
Anji	18	22	1.16	6.0	12–34	23	12–42	27	0.84–1.44	1.14	5.4–6.7	6.1	14–33	24	24–35	30	1.12–1.13	1.13	6.3–6.6	6.5
Deqing	19	32	1.12	6.3	7–26	17	18–38	28	0.87–1.53	1.20	5.2–7.2	6.2	14–33	24	24–35	30	1.12–1.13	1.13	6.3–6.6	6.5
Yuhang	15	5	1.16	6.6	9–21	15	16–48	32	0.95–1.34	1.15	5.9–7.3	6.6	11–31	21	18–48	33	1.12–1.27	1.20	5.5–7.4	6.5
Linan	22	22	1.09	6.2	18–27	23	8–29	19	0.91–1.14	1.03	5.5–7.8	6.7	11–31	21	18–48	33	1.12–1.27	1.20	5.5–7.4	6.5
Minhang	13	26	1.18	7.6	10–18	14	17–46	32	1.11–1.30	1.21	6.4–8.0	7.2	11–31	21	18–48	33	1.12–1.27	1.20	5.5–7.4	6.5
Jiading	13	28	1.10	7.6	9–20	15	13–44	29	0.94–1.24	1.09	6.5–8.1	7.3	11–31	21	18–48	33	1.12–1.27	1.20	5.5–7.4	6.5
Chuangsha	12	29	1.15	7.6	9–20	15	17–36	27	1.06–1.33	1.20	7.3–8.0	7.7	11–31	21	18–48	33	1.12–1.27	1.20	5.5–7.4	6.5
Nanhui	16	31	1.18	7.4	13–22	18	8–35	22	1.11–1.21	1.16	6.5–8.1	7.3	11–31	21	18–48	33	1.12–1.27	1.20	5.5–7.4	6.5
Qingpu	21	27	1.15	7.1	7–33	20	11–36	24	0.94–1.53	1.24	5.6–8.3	7.0	23–33	28	32–56	44	1.14–1.14	1.14	6.2–6.9	6.6
Songjiang	23	26	1.20	6.8	10–33	22	8–37	23	1.03–1.47	1.25	5.6–8.1	6.9	23–33	28	32–56	44	1.14–1.14	1.14	6.2–6.9	6.6
Jinshan	20	29	1.22	7.0	11–37	24	18–36	27	1.11–1.47	1.29	4.6–8.3	6.5	11–31	21	18–48	33	1.12–1.27	1.20	5.5–7.4	6.5
Fengxian	15	25	1.20	7.4	12–18	15	19–39	29	1.11–1.49	1.30	6.9–8.1	7.5	11–31	21	18–48	33	1.12–1.27	1.20	5.5–7.4	6.5
Baoshan	11	23	1.21	7.9	9–19	14	8–44	26	1.11–1.28	1.20	7.2–8.2	7.7	11–31	21	18–48	33	1.12–1.27	1.20	5.5–7.4	6.5
Chongming	10	17	1.11	8.1	9–13	11	15–29	22	1.11–1.21	1.12	7.8–8.1	8.0	12–16	14	24–39	31	1.17–1.27	1.22	7.3–7.4	7.5
Tai-Lake region	15	26	1.18	6.7	2–38	16	4–58	27	0.84–1.54	1.23	3.4–8.3	6.7	11–33	22	18–56	32	1.12–1.27	1.18	5.5–7.4	6.5

WA = Weighted average of soil properties by the area of each polygon; SOC = Initial SOC content (g kg^{-1}); Clay = Clay content (%); BD = Bulk Density (g cm^{-3}); Range = Range of maximum and minimum soil properties; Ave = Average of maximum and minimum soil properties.

As Table 3 illustrated, initial SOC content was the most sensitive parameter controlling SOC change among all soil factors in P5 and C5 [20,22]. The average initial SOC value of P5 and C5 were 15 g kg^{-1} and 16 g kg^{-1} for the entire Tai-Lake region, respectively. Furthermore, the average initial SOC content sourced from P5 for 24 counties was lower than that from C5, while the average clay content sourced from P5 for 24 counties was also higher than that from C5. Many previous studies showed that soils with lower initial organic carbon and higher clay content tended to sequester C [20,22,35]. The high SOC sequestration

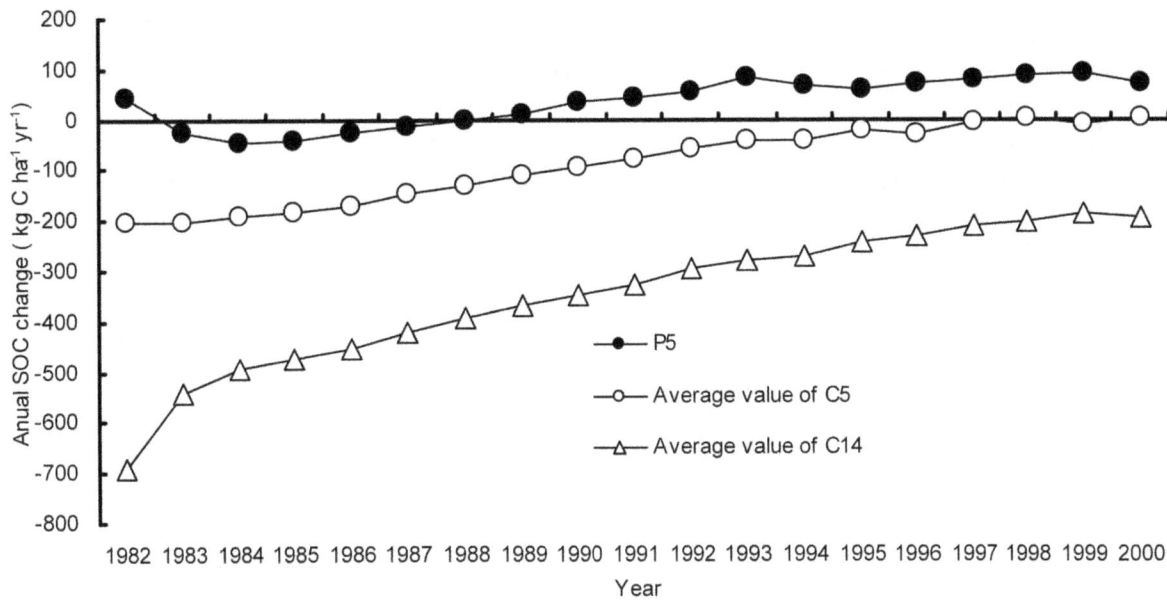

Figure 6. Temporal distribution of average annual SOC change modeled with P5, C5 and C14 from 1982 to 2000 in the Tai-Lake region, China.

rate (34 kg C ha^{-1} yr^{-1}) was thus associated with P5 (**Fig. 7a**). Conversely, the high SOC losses rate (-91 kg C ha^{-1} yr^{-1}) was associated with C5 (**Fig. 7b**). The SOC losses rate (-349 kg C ha^{-1} yr^{-1}) in C14 was the highest in the three resolution databases (**Fig. 7c**). Table 3 demonstrates that pH and initial SOC content are the most sensitive parameters controlling SOC change among all soil factors in C14. The average initial SOC value (22 g kg^{-1}) of C14 was significantly higher than that of P5 (15 g kg^{-1}) and C5 (16 g kg^{-1}) for the entire Tai-Lake region. In addition, the average pH value of C14 for 34 counties was close to neutral (6.5–7.5), and the average initial SOC contents of C14 for 28 counties were higher than 20 g kg^{-1}. Some studies showed that soils with neutral pH value and higher organic carbon content were favorable for CO_2 production by providing more substrates and better living environment for microbes [22,56].

The comparison illustrates that using different basic simulation units and soil data sources will produce different conclusions as to C sequestration or C liberation in the same study area. The implication is that more precise soil data and high resolution simulation units were necessary for better simulating regional scale SOC dynamics. The simulation outcome can be attributed to how the databases represent soil types and spatial heterogeneity, which is more precisely done with larger scale soil data and high resolution simulation units (e.g., 1:50,000 soil database).

Distribution of the average annual-, total SOC change modeled with the three resolution databases in different counties

The differentiation of the average annual-, total SOC change in P5, C5 and C14 was also shown at the county scale in the Tai-Lake region (**Table 5 and Fig. 7**). In the modeled domain, there were 26 counties that gained SOC and 11 counties that lost SOC from 1982 to 2000 in P5. The highest SOC sequestration rate of P5 were in Dantu, Jurong, Jiading and Baoshan counties which was higher than 200 kg C ha^{-1} yr^{-1}, due to the low initial SOC content (7.1 g kg^{-1}, 9.5 g kg^{-1}, 13 g kg^{-1} and 11 g kg^{-1}, respectively). In addition, the clay content of P5 in Dantu and

Jurong counties were 36% and 30%, respectively. High clay content is associated with high SOC sequestration [22,57,58]. By contrast, the greatest SOC loss rate of P5 in the Huzhou, Songjiang, Linan and Wuxian county was more than 170 kg C ha^{-1} yr^{-1}, due to the high initial SOC content (23 g kg^{-1}, 23 g kg^{-1}, 22 g kg^{-1} and 24 g kg^{-1}, respectively). Moreover, the clay content of P5 in Linan and Songjiang counties were only 22% and 26%, respectively. Low clay content is linked to high CO_2 emissions [22].

However, under the same agricultural practice, there were only 14 counties that gained SOC and 23 counties that lost SOC from 1982 to 2000 in C5. The highest SOC sequestration rate of C5 were in Dantu, Jurong, Jintan, Chongming, and Baoshan counties which was higher than 150 kg C ha^{-1} yr^{-1}. The main reason was that the initial SOC content of C5 in Dantu, Jurong, Jintan, Chongming, and Baoshan counties were 11 g kg^{-1}, 10 g kg^{-1}, 11 g kg^{-1}, 11 g kg^{-1} and 14 g kg^{-1}, respectively; the other was that the average clay content of C5 in Dantu, Jurong, and Baoshan counties ranged from 26% to 31%. Some studies showed that low initial SOC value and high clay content were linked to low CO_2 emissions [22,57,58]. In contrast, the greatest SOC loss rate of C5 in Jinshan, Changshu, Huzhou, Anji, and Kunshan county were more than 400 kg C ha^{-1} yr^{-1}, which possessed high initial SOC and low bulk density [22,35]. Compared with the P5 resolution database, the average annual-, total SOC change modeled with C5 for 28 counties was lower than that from P5. With the total SOC change as modeled with P5 inputs as the baseline, the relative deviations of counties in Jiangyin, Zhangjiagang and Kunshan were relatively high (>1000%). The relative deviations ranged from 50% to 250% in most counties. Only fifteen counties (Wuxian, Wujin, Jintan, Liyang, Dantu, Jurong, Huzhou, Yuhang, Linan, Minhang, Jiading, Chuangsha, Songjiang, Baoshan, and Chongming) had relatively low value of relative deviation (<100%). The SOC changes for the two resolution databases are almost in agreement with the soil feature across the 37 simulated counties (**Table 4 and Table 5**). The average initial SOC content sourced from C5 for 24 counties was higher than

Table 5. Distribution of the average annual SOC change (kg C ha^{-1} yr^{-1}) and the total SOC change (Gg C) in different counties of the Tai-Lake region, China modeled with P5, C5 and C14 from 1982 to 2000.

County	Area 10⁴ha	P5 ASC WA	P5 TSC	C5 ASC Max	C5 ASC Min	C5 ASC AVE	C5 TSC Max	C5 TSC Min	C5 TSC AVE	C14 ASC Max	C14 ASC Min	C14 ASC AVE	C14 TSC Max	C14 TSC Min	C14 TSC AVE
Zhangjiagang	2.54	2	1	138	−50	44	66	−24	21	57	−418	−181	27	−202	−87
Changshu	7.55	−64	−92	246	−1438	−596	353	−2063	−855	92	−404	−156	132	−580	−224
Taicnang	6.14	43	50	241	−360	−60	281	−420	−70	192	−993	−400	224	−1158	−467
Kunshan	7.57	38	55	252	−1057	−402	362	−1520	−579	−75	−838	−456	−108	−1205	−656
Wuxian	14.78	−172	−483	356	−876	−260	998	−2458	−730	72	−403	−166	201	−1132	−466
Wujiang	9.79	53	99	586	−717	−65	1091	−1333	−121	−59	−792	−425	−110	−1472	−791
Wuxi	9.77	49	91	478	−196	141	888	−364	262	−256	−958	−607	−476	−1778	−1127
Jiangyin	8.69	2	4	313	−101	106	517	−167	175	−262	−951	−606	−432	−1569	−1000
Wujin	14.85	91	256	248	−57	96	701	−160	270	−274	−969	−621	−774	−2734	−1754
Jintan	7.10	146	197	245	81	163	331	110	220	136	−364	−114	184	−492	−154
Liyang	10.84	177	365	302	−115	94	623	−237	193	127	−386	−129	263	−796	−267
Yixing	10.34	147	289	514	−1011	−248	1011	−1987	−488	−262	−950	−606	−514	−1867	−1191
Dantu	5.07	371	357	615	−276	170	592	−266	163	141	−367	−113	136	−353	−109
Jurong	8.03	238	363	403	−8	198	614	−12	301	125	−374	−124	191	−570	−190
Danyang	9.58	47	85	416	−152	132	758	−276	241	130	−364	−117	237	−663	−213
Jiaxing	6.57	−124	−155	412	−523	−55	514	−652	−69	−11	−752	−381	−13	−938	−476
Jiashan	4.13	−103	−81	110	−631	−260	87	−494	−204	−48	−786	−417	−38	−616	−327
Pinghu	4.81	127	116	349	−466	−58	319	−425	−53	274	−923	−324	250	−843	−296
Haiyan	2.74	41	21	489	−549	−30	254	−285	−16	279	−943	−332	145	−490	−173
Haining	3.92	151	113	483	−556	−36	359	−413	−27	274	−951	−338	204	−708	−252
Tongxiang	4.42	128	107	504	−634	−65	424	−533	−55	122	−846	−362	102	−711	−304
Huzhou	6.02	−309	−353	163	−1184	−511	186	−1354	−584	92	−915	−411	106	−1046	−470
Changxing	5.62	42	45	422	−863	−221	450	−922	−236	71	−901	−415	76	−962	−443
Anji	4.18	−66	−52	178	−1025	−423	141	−814	−336	58	−873	−408	46	−694	−324
Deqing	3.11	17	10	379	−684	−152	224	−404	−90	77	−930	−426	46	−550	−252
Yuhang	5.27	−91	−92	344	−371	−13	345	−372	−13	120	−1061	−456	120	−1034	−457
Linan	3.06	−220	−128	7	−381	−187	4	−222	−109	218	−754	−268	127	−439	−156
Minhang	3.49	62	41	294	−239	28	195	−158	18	257	−974	−359	171	−645	−238
Jiading	4.29	238	194	387	−197	95	315	−161	77	281	−940	−329	229	−766	−268
Chuangsha	3.71	188	133	339	−295	22	239	−208	15	286	−925	−319	202	−653	−225
Nanhui	4.11	152	119	189	−279	−45	148	−208	−35	287	−939	−326	224	−734	−255
Qingpu	5.68	−52	−56	432	−1042	−305	467	−1125	−329	2	−756	−377	3	−816	−407
Songjiang	5.90	−287	−322	249	−1010	−381	278	−1131	−426	−60	−816	−438	−67	−914	−491
Jinshan	5.63	−77	−83	213	−1481	−634	228	−1585	−679	275	−945	−335	294	−1011	−359
Fengxian	5.87	9	10	171	−270	−49	191	−301	−55	280	−944	−332	321	−1053	−370
Baoshan	3.13	200	119	393	−92	151	234	−55	90	311	−876	−283	185	−522	−168
Chongming	3.73	195	138	283	47	165	201	34	117	165	−108	28	117	−77	20
Tai-Lake region	232	34	1483	340	−521	−91	14987	−22977	−3994	46	−744	−349	2022	−32790	−15380

WA = Weighted average of annual mean SOC change (kg C ha^{-1} yr^{-1}) by the area of each polygon; ASC = Average annual SOC change (kg C ha^{-1} yr^{-1}); TSC = Total SOC change (Gg C); Max = Maximum value of ASC (or TSC); Min = Minimum value of ASC (or TSC); Ave = Average of maximum and minimum ASC (or TSC).

that from P5, and the average clay content sourced from C5 for 24 counties was also lower than that from P5. Some research showed that high initial SOC content and low clay content is favorable for C losses [22,57,58].

As can be seen from the Table 5, a big number of counties where the average annual-, total SOC change modeled with the C14 and P5 differed greatly. There was only one county that gained SOC from 1982 to 2000, while other 36 counties lost SOC in C14. The SOC losses of C14 ranged from 360 kg C ha^{-1} yr^{-1}

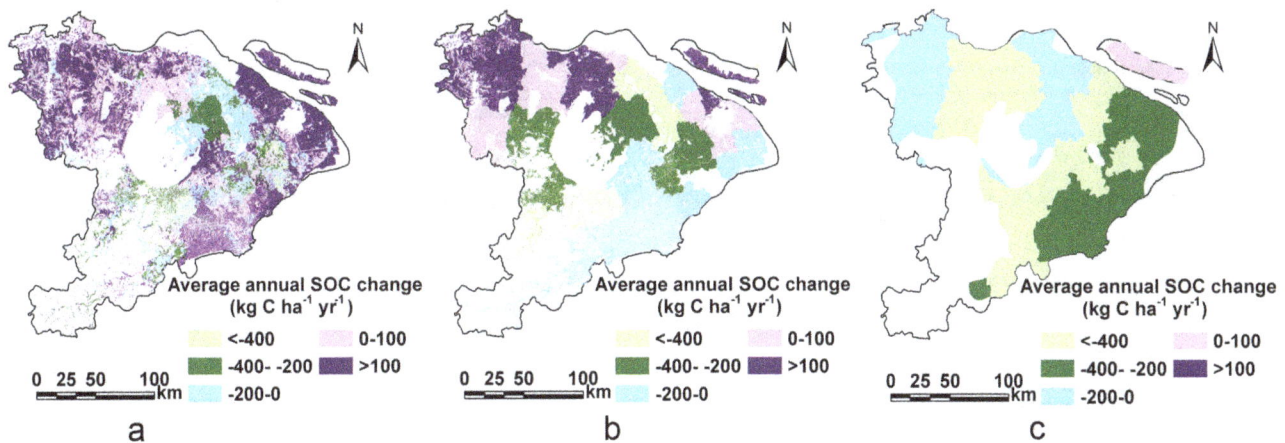

Figure 7. Spatial distribution of average annual SOC change modeled with P5, C5 and C14 in the Tai-Lake region, China (a: P5, b: c5, and c:C14).

to 620 kg C ha^{-1} yr^{-1} in most counties. With the total SOC change as modeled with P5 inputs as the baseline, the relative deviations of counties in Zhangjiagang, Taicang, Kunshan, Wuxi, Jiangyin, Changxing, Deqing, and Fengxian were more than 1000%. Only five counties (Wuxian, Huzhou, Linan, Songjiang, and Chongming) in C14 had relatively low deviation (<100%). The main reasons were that the average pH value of C14 in most counties ranged from 6.5 to 7.5, which were closer to neutral than that from C5 and C14. Moreover, the average initial SOC contents of C14 in most counties were higher than 20 g kg^{-1}, which was also much higher than that from P5 or C5. Therefore, high SOC losses occurred in C14.

The modeled data at county scale in three simulation methods indicated the underestimation with the county-based database was related to its soil data source and simulation unit resolution, especially the coarse soil maps (1:14,000,000) that missed relatively small soil patches containing low or high soil properties (i.e., initial SOC content, pH, and clay content) which were sensitive to SOC change. This would also explain why the precision of soil database plays an important role in elevating the accuracy of modeled SOC change at regional scale.

Conclusions

Using different spatial information, process-based models integrated with GIS databases can play an important role in describing C biogeochemical cycles, such as targeting mitigation efforts to the most beneficial regions. However, SOC models have often been applied to regions with high heterogeneity but limited spatially differentiated soil information and simulated unit resolution.

Simulation results indicate that total SOC change from 1982 to 2000 in the top layer (0–30 cm) of the 2.3 M ha of paddy rice

fields in the Tai-Lake region was +1.48 Tg C for P5. However, discrepancies in the results existed among the three databases, because different soil data and basic simulation units were used. The total SOC changes in the Tai-Lake region were -3.99 Tg C and -15.38 Tg C for C5 (or C14), respectively, from 1982 to 2000. With the total SOC change as modeled with P5 inputs as the baseline, the relative deviation of C5 was lower than C14 due to the more precise soil data. In contrast, the relative deviation of C14 was higher than other databases due to using coarser soil data and low-resolution simulation units. In addition, with the same basic simulation unit, average annual-, total SOC change between C5 and C14 for the Tai-Lake region also had a large discrepancy due to the use of different soil data. The comparison demonstrated that the most sensitive factors (e.g., initial SOC content and pH) for modeling SOC dynamics should be given a high priority during the input data acquisition as they contribute disproportionately to the uncertainties produced during the upscaling process [20]. The results also indicate that improving the performance of the biogeochemical DNDC model is essential in creating accurate models of the soil carbon cycle.

Acknowledgments

We gratefully acknowledge support for this research from the National Natural Science Foundation of China (No. 41001126), and the National Basic Research Program of China (973 Program) (2010CB950702). Sincere thank is also given to Professor Changsheng Li (University of New Hampshire, USA) for his useful advice on DNDC model.

Author Contributions

Conceived and designed the experiments: DSY XZS SHX. Performed the experiments: LMZ SXX. Analyzed the data: LMZ SXX. Contributed reagents/materials/analysis tools: SXX YCZ. Wrote the paper: LMZ.

References

1. Eswaran H, Berg EVD, Reich P (1993) Organic carbon in soil of the world. Soil Science Society of America Journal 57:192–194.

2. Lal R (2006) World soils and greenhouse effect: An overview, in soils and global change. Encyclopedia of Soil Science. doi:10.1081/E-ESS-120042696.

3. Shi XZ, Yang RW, Weindorf DC, Wang HJ, Yu DS, et al. (2010) Simulation of organic carbon dynamics at regional scale for paddy soils in China. Climatic Change 102:579–593.

4. Li QK (1992) Paddy soil of China. Beijing: Science Press. 514 p.

5. Gong ZT (1999) Chinese soil taxonomic classification. Beijing: Science Press. 5–215 p.

6. Liu QH, Shi XZ, Weindorf DC, Yu DS, Zhao YC, et al. (2006) Soil organic carbon storage of paddy soils in China using the 1:1,000,000 soil database and their implications for C sequestration. Global Biogeochemical Cycles 20:GB3024. doi:10.1029/2006GB002731.

7. Jenkinson DS, Rayner JH (1977) The turnover of soil organic matter in some of Rothamsted classical experiments. Soil Science 125:298–305.

8. Smith P, Smith JU, Powlson DS, Arah JRM, Chertov OG, et al. (1997) A comparison of the performance of nine soil organic matter models using datasets from seven long term experiments. Geoderma 81:153–225.

9. Ardö J, Olsson L (2003) Assessment of soil organic carbon in semi-arid Sudan using GIS and the CENTURY model. Journal of Arid Environments 54:633–651.

10. Shirato Y (2005) Testing the suitability of the DNDC model for simulating long-term soil organic carbon dynamics in Japanese paddy soils. Soil Science and Plant Nutrition 51(2):183–192.

11. Tang HJ, Qiu JJ, van Ranst E, Li CS (2006) Estimations of soil organic carbon storage in cropland of China based on DNDC model. Geoderma 134:200–206.

12. Cerri CEP, Easter M, Paustian K, Killian K, Coleman K, et al. (2007) Predicted soil organic carbon stocks and changed in the Brazilian Amazon between 2000 and 2030. Agriculture, Ecosystems and Environment 122:58–72.

13. Huang Y, Yu YQ, Zhang W, Sun WJ, Liu SL, et al. (2009) Agro-C: A biogeophysical model for simulating the carbon budget of agroecosystems. Agricultural and Forest Meteorology 149:106–129.

14. Tang HJ, Qiu JJ, Wang LG, Li H, Li CS, et al. (2010) Modeling soil organic carbon storage and its dynamics in croplands of China. Agricultural Sciences in China 9(5):704–712.

15. Li CS, Frolking S, Frolking TA (1992) A model of nitrous oxide evolution from soil driven by rainfall events: I. Model structure and sensitivity. Journal of Geophysical Research 97:9759–9776.

16. Li CS, Frolking S, Frolking TA (1992) A model of nitrous oxide evolution from soil driven by rainfall events:II. Model applications. Journal of Geophysical Research 97:9777–9783.

17. Tonitto C, David MB, Li CS, Drinkwater LE (2007) Application of the DNDC model to tile-drained Illinois agroecosystems: Model comparison of conventional and diversified rotations. Nutrient Cycling in Agroecosystems 78 (1):65–81.

18. Pathak H, Li CS, Wassmann H (2005) Greenhouse gas emissions from Indian rice fields: calibration and upscaling using the DNDC model. Biogeoscience 2:113–123.

19. Neufeldt H, Schäfe M, Angenendt E, Li CS, Kaltschmitt M, et al. (2006) Disaggregated greenhouse gas emission inventories from agriculture via a coupled economic-ecosystem model. Agriculture, Ecosystems and Environment 112:233–240.

20. Zhang F, Li CS, Wang Z, Wu HB (2006) Modeling impacts of management alternatives on soil carbon storage of farmland in Northwest China. Biogeosciences 3:451–466.

21. Wang LG, Qiu JJ, Tang HJ, Li CS, van Ranst E (2008) Modelling soil organic carbon dynamics in the major agricultural regions of China. Geoderma 147:47–55.

22. Li CS, Mosier A, Wassmann R, Cai ZC, Zheng XH, et al. (2004) Modeling greenhouse gas emissions from rice-based production systems: Sensitivity and upscaling. Global Biogeochemical Cycles 18:GB1043. doi:10.1029/2003 GB002045.

23. Cai ZC, Sawamoto T, Li CS, Kang GD, Boonjawat J, et al. (2003) Field validation of the DNDC model for greenhouse gas emissions in East Asian cropping systems. Global Biogeochemical Cycles 17 (4): GB1107, doi:10.1029/2003 GB002046.

24. Xu Q, Lu YC, Liu YC, Zhu HG (1980) Paddy soil of Tai-Lake region in China. Shanghai: Science Press.

25. Shi XZ, Yu DS, Warner ED, Sun WX, Petersen GW, et al. (2006) Cross-reference system for translating between genetic soil classification of China and Soil Taxonomy. Soil Science Society of America Journal 70:78–83.

26. Soil Survey Staff in USDA (2010) Keys to Soil Taxonomy (11th Edition). Washington: USDA-Natural Resources Conservation Service.

27. Li CS, Frolking S, Harriss R (1994) Modeling carbon biogeochemistry in agricultural soils. Global Biogeochemical Cycles 8 (3):237–254.

28. Li CS, Narayanan V, Harriss R (1996) Model estimates of nitrous oxide emissions from agricultural lands in the United States. Global Biogeochemical Cycles 10 (2):297–306.

29. Li CS, Qiu JJ, Frolking S, Xiao XM, Salas W, et al. (2002) Reduced methane emissions from large-scale changes in water management in China's rice paddies during 1980–2000. Geophysical Research Letters 29 (20):1972, doi:10.1029/2002GL01 5370.

30. Li CS (2007) Quantifying greenhouse gas emissions from soils: Scientific basis and modeling approach. Soil Science and Plant Nutrition 53 (4):344-352.

31. Qiu JJ, Wang LG, Tang HJ, Li H, Li CS (2005) Studies on the situation of soil organic carbon storage in croplands in northeast of China. Agricultural Sciences in China 37 (8):1166–1171.

32. Zhao YC, Shi XZ, Weindorf DC, Yu DS, Sun WX, et al. (2006) Map scale effects on soil organic carbon stock estimation in north China. Soil Science Society of America Journal 70:1377–1386.

33. Institute of Soil Science (1986) The soil atlas of China. Beijing: Institute of Soil Science, Academia Sinica, Cartographic Publishing House.

34. National Soil Survey Office of China (1993–1997) Soils in China (Vol. 1–6). Beijing: Agricultural Publishing House.

35. Zhang LM, Yu DS, Shi XZ, Xu SX, Wang SH, et al. (2012) Simulation soil organic carbon change in China's Tai-Lake paddy soils. Soil and Tillage Research 121:1–9.

36. Zhang LM, Yu DS, Shi XZ, Xu SX, Weindorf DC, et al. (2009) Quantifying methane emissions from rice fields in the Taihu region, China by coupling a detailed soil database with biogeochemical model. Biogeosciences 6:739–749.

37. Xu SX, Zhao YC, Shi XZ, Yu DS, Li CS, et al. (2013) Map scale effects of soil databases on modeling organic carbon dynamics for paddy soils of China. Catena 104:67–76.

38. Yu DS, Yang H, Shi XZ, Warner ED, Zhang LM, et al. (2011) Effects of soil spatial resolution on quantifying CH_4 and N_2O emissions from rice fields in the Tai Lake region of China by DNDC model. Global Biogeochemical Cycles 25:GB2004. doi:10.1029/2010GB003825.

39. Yu DS, Zhang LM, Shi XZ, Warner ED, Zhang ZQ, et al. (2013) Soil assessment unit scale affects quantifying CH_4 emissions from rice fields. Soil Science Society of America Journal 77:664–672.

40. Li CS (2007) Quantifying soil organic carbon sequestration potential with modeling approach. In: Tang HJ, Van Ranst E, Qiu JJ (Eds). Simulation of soil organic carbon storage and changes in agricultural cropland in China and its impact on food security. Beijing: China Meteorological Press. 1–14 p.

41. Gou J, Zheng XH, Wang MX, Li CS (1999) Modeling N_2O emissions from agriculture fields in Southeast China. Advances in Atmospheric Sciences 16 (4):581–592.

42. China Meteorological Administration (2011) China meteorological data daily value. China Meteorological Data Sharing Service System, Beijing, China. http://cdc.cma.gov.cn/index.jsp.

43. Lu RK, Shi TJ (1982) Agricultural chemical manual. Beijing:China Science Press. 142 p.

44. Whitmore AP, Klein-Gunnewiek H, Crocker GJ, Klir J, Körschens M, et al. (1997) Simulating trends in soil organic carbon in long-term experiments using the Verberne/MOTOR model. Geoderma 81:137–151.

45. Admassu Y, Shakoor A, Wells N (2012) Evaluating selected factors affecting the depth of undercutting in rocks subject to differential weathering. Engineering Geology 124:1–11.

46. Leech NL, Barret KKC, Morgan G (2008) SPSS for intermediate statistics. New York: Lawrence Erlbaum Associates. 270 p.

47. Dielman TE (2001) Applied regression analysis for business and economics. California: Duxbury Thomson Learning. 647 p.

48. Arnold RW (1995) Role of soil survey in obtaining a global carbon budget. In: Lal R, Kimble J, Levine E, Stewart BA (Eds.) Advances in Soil Science: Soils and Global Change. Boca Raton, FL: CRC Press. 57–263 p.

49. Zhong B, Xu YJ (2011) Scale effects of geographical soil datasets on soil carbon estimation in Louisiana, USA: a comparison of STATSGO and SSURGO. Pedosphere 21 (4):491–501.

50. Liu YH, Yu ZR, Chen J, Zhang FR, Reiner D, et al. (2006) Changes of soil organic carbon in an intensively cultivated agricultural region: A denitrification-decomposition (DNDC) modelling approach. Science of the Total Environment 372:203–214.

51. Liu Q, Sun B, Jie XL, Li ZP (2009) The spatial-temporal dynamic change and simulation of county-scale paddy soil organic carbon red soil hilly region. Acta pedologica sinica 46 (6):1059–1067.

52. Xu SX, Shi XZ, Zhao YC, Yu DS, Wang SH, et al. (2012) Spatially explicit simulation of soil organic carbon dynamics in China's paddy soils. Catena 92:113–121.

53. Studdert GA, Monterubbianesi GM, Domínguez GF (2011) Use of RothC to simulate changes of organic carbon stock in the arable layer of a Mollisol of the southeastern Pampas under continuous cropping. Soil and Tillage Research 117:191–200.

54. Yu YQ, Huang Y, Zhang W (2012) Modeling soil organic carbon change in croplands of China, 1980-2009. Global and Planetary Change 82–83:115–128.

55. Wu TY, Schoenau JJ, Li FM, Qian PY, Malhi SS (2004) Influence of cultivation and fertilization on total organic carbon and carbon fractions in soils from the Loess Plateau of China. Soil and Tillage Research 77:59–68.

56. Pacey JG, DeGier JP (1986) The factors influencing landfill gas production. In: Energy from landfill gas. Proceeding of a conference jointly sponsored by the United Kingdom Department of Energy and the United States Department of Energy (October 1986). 51–59 p.

57. Burke IC, Lauenroth WK, Conffin DP (1995) Soil organic matter recovery in semiarid grassland: implications for the conservation reserve program. Ecological Monographs 5:793–801.

58. Kay BD (1998) Soil structure and organic carbon: a review. In: Lal R, Kimble JM, Follett RF. Soil Processes and the carbon cycle. Boca Raton, FL: CRC Press.169–198 p.

9

Timing of the Departure of Ocean Biogeochemical Cycles from the Preindustrial State

James R. Christian[1,2]*

1 Canadian Centre for Climate Modelling and Analysis, Victoria, B.C., Canada, **2** Fisheries and Oceans Canada, Institute of Ocean Sciences, Sidney, BC, Canada

Abstract

Changes in ocean chemistry and climate induced by anthropogenic CO_2 affect a broad range of ocean biological and biogeochemical processes; these changes are already well underway. Direct effects of CO_2 (e.g. on pH) are prominent among these, but climate model simulations with historical greenhouse gas forcing suggest that physical and biological processes only indirectly forced by CO_2 (via the effect of atmospheric CO_2 on climate) begin to show anthropogenically-induced trends as early as the 1920s. Dates of emergence of a number of representative ocean fields from the envelope of natural variability are calculated for global means and for spatial 'fingerprints' over a number of geographic regions. Emergence dates are consistent among these methods and insensitive to the exact choice of regions, but are generally earlier with more spatial information included. Emergence dates calculated for individual sampling stations are more variable and generally later, but means across stations are generally consistent with global emergence dates. The last sign reversal of linear trends calculated for periods of 20 or 30 years also functions as a diagnostic of emergence, and is generally consistent with other measures. The last sign reversal among 20 year trends is found to be a conservative measure (biased towards later emergence), while for 30 year trends it is found to have an early emergence bias, relative to emergence dates calculated by departure from the preindustrial mean. These results are largely independent of emission scenario, but the latest-emerging fields show a response to mitigation. A significant anthropogenic component of ocean variability has been present throughout the modern era of ocean observation.

Editor: Vanesa Magar, Centro de Investigacion Cientifica y Educacion Superior de Ensenada, Mexico

Funding: The authors have no support or funding to report.

Competing Interests: The author has declared that no competing interests exist.

* Email: jim.christian@ec.gc.ca

Introduction

The direct effect of anthropogenic CO_2 on ocean chemistry already exceeds the range of natural variability in many locations [18]. However, many aspects of ocean biogeochemistry are forced only indirectly by CO_2, via the effect of atmospheric CO_2 on climate. Detecting an anthropogenic climate change signal in ocean biogeochemical data is difficult due to short data records and high natural variability [10,13,24]. Trends are not monotonic, and even strong anthropogenic forcing is subject to modulation by a variety of physical and biogeochemical processes [14]. In addition, the effects of climate warming are complex and competing processes can offset each other. For example, primary production will tend to increase with increasing temperature, but the same increases in temperature cause increasing stratification that limits the supply of nutrients to the surface ocean [37]. In addition, some fields have opposing trends in different regions [23].

Variability in climate can be divided into forced and unforced components. In model simulations the unforced component is referred to as the model's internal variability, or natural variability. Models never reproduce the timing of natural variability exactly, but a good model will reproduce the statistical characteristics [38]. A "white noise" spectrum implies approximately equal variability across the spectrum of frequencies resolved, while a "red noise" spectrum implies greater variance at lower frequencies (i.e., significant autocorrelation in time). Climate variability is generally considered to have a "red" spectrum [19,25], but modern data records are too short for complete characterization and it is therefore difficult to know with confidence whether the models' 'natural' variability is too weak (or too strong). Nonetheless we know that climate models can accurately reproduce important aspects of observed climate variability [17,40].

There is a relatively well-established, although certainly not uniform, methodology for detection of anthropogenic climate change and attribution of those changes to specific forcing factors [6,9,20,34]. In this literature only a handful of papers to date have dealt with ocean biogeochemistry [1,8,18,24]. Detection normally refers to a demonstration that observed variability in the climate system exceeds the range expected from natural variability at some specified (e.g., 5%) level of significance [5,21]. Detection period refers to the length of a data record required to unequivocally detect an anthropogenic signal, while detection time indicates the point in time at which the signal becomes detectable [35]. The latter is related to the "time of emergence", or the point in time at which the anthropogenic signal emerges from the "historical envelope" of natural variability, but there are subtle differences between the two. In model simulations, detection of even extremely small signals is possible given a sufficiently large ensemble; so one could argue that the term should not be used at all for studies using only model simulations. Emergence implies that thereafter the signal remains consistently outside the envelope of natural variability as defined e.g., by some multiple of the standard deviation of the unforced control simulation [29,31]. In this study I define emergence as the point at which the anthropogenic signal exceeds natural variability (estimated from a preindustrial control simulation) at some specified threshold, and remains continuously in excess of that threshold (except when reversals are clearly attributable to mitigation). Note that this definition differs from that of [29], who include both natural and anthropogenic forcing up to 2005 as part of the envelope of historical variability from which emergence is estimated. It is also important to note that as observing systems inevitably have incomplete coverage, the point of potential detection of an anthropogenic signal is always later than the point of emergence as defined here.

While this sort of analysis does not offer any immediate hope of unambiguously detecting an anthropogenic signal in the real world, it represents a useful measure of the 'true' point of departure from the preindustrial climate, which is surprisingly early in many cases. Having an estimate of the emergence time is useful when examining historical data records, few if any of which are entirely free of anthropogenic influence. It can also be useful to impacts research in that it provides information about the magnitude of the anthropogenic signal relative to the natural variability [9,27].

Modern coupled climate/carbon cycle models provide a homogeneous data set with which to conduct such an experiment, which despite its shortcomings overcomes some of the problems of earlier studies. Some early detection studies were limited by the lack of extended control simulations with models identical to their forced runs, or their control runs were done with ocean-only models for which stochastic forcing had to be employed to generate internal variability [35]. The current data set includes multiple realizations of the forced simulations, and an extended control run (with an identical model) from which the preindustrial variability can be estimated. It can not be known with certainty that the model does not underestimate natural variability, but if such biases exist they are probably small [12].

Methods

Model description

The model used is the Canadian Earth System Model, which has been previously described by [3,4,12]. In the version used here (CanESM2.0), the atmosphere model is run at T63 spectral resolution, a 128×64 horizontal grid. The 256×192 ocean model, with six grid cells (2x, 3y) to each atmosphere grid cell, has a longitude resolution of approximately $1.4°$ and latitude resolution of approximately $0.94°$. The ocean model has 40 vertical levels (increased from 29 in CanESM1, with all of the additional levels in the upper 300 m), with a resolution of 10–15 m in the upper 100 m. The ocean carbon cycle model is based on the Ocean Carbon Model Intercomparison Project II protocols [32] and couples the carbon cycle to an NPZD ecosystem model via a fixed C/N Redfield Ratio and a temperature-dependent rain ratio (ratio of inorganic to organic carbon in vertical flux at the base of the euphotic zone) [12,41].

Simulations used here have specified atmospheric CO_2 concentrations; runs with freely-varying atmospheric CO_2 give similar results. Historical (1850–2005) forcing includes volcanic eruptions and solar variability. Greenhouse gas forcing after 2005 is provided by the Representative Concentration Pathways (RCPs)

[30]. RCPs 8.5, 4.5, and 2.6 are often referred to as the "no mitigation", "moderate mitigation", and "strong mitigation" scenarios respectively; in RCP8.5 emissions continue to increase and atmospheric CO_2 concentration exceeds 900 ppm by 2100 [4].

Nine data fields were considered: sea surface temperature and salinity, mixed layer depth (MLD), surface ocean pCO_2 and air-sea CO_2 flux, surface nitrate concentration, total primary production, organic export flux at 100 m, and dinitrogen fixation. All of these are standard (2D, monthly) data fields of the 5th Coupled Model Intercomparison Project [39]; all data are in the public domain. The choices are somewhat arbitrary, but cover a range of fields commonly measured by oceanographers and used to diagnose the performance of ocean biogeochemical models. MLD was excluded from some analyses because in the historical simulation it was available only for a single realization (see below Figure 1).

Statistics

For global and station means, emergence from the envelope of natural variability was defined as the first year the mean of an ensemble of five realizations of the forced (historical, RCP) run differed from the mean of the unforced preindustrial (1850 greenhouse gas concentrations) control run by n standard deviations (nσ) of the interannual variability of the control run, and remained in excess of nσ thereafter, with n = 3 for global metrics [31] and n = 2 for individual locations. The different criteria for global means and specific locations are arbitrary but yield consistent results across the range of stations considered; for individual locations a 2σ threshold was applied, because in many cases emergence would not occur by 2100 with the more conservative 3σ threshold. The ensemble mean is used to isolate the anthropogenic signal by averaging out the effect of internal variability. Three hundred years of control run were used to calculate the standard deviation. Annual means were used in all cases. Trends were not corrected for drift in the control run, but drift is small as only the surface and euphotic zone are considered, and is also present in the forced runs. Drift will bias the method slightly towards later emergence because it increases the preindustrial variability. In some cases the fields emerged but later fell back within the nσ window under the "strong mitigation" scenario (RCP2.6); these were considered emergent if they remained outside the window for 50 years (in practice, 20–25 years would be sufficient as the longest any fields remained outside the window in an unforced control run was ~15 years).

Spatial 'fingerprints' were calculated for the 14 ocean regions defined by Sarmiento et al. [36] (hereafter S02) and for zonal means of $10°$ latitude bands as in [31]; results are largely insensitive to which of these was used (see below Results). S02 divided the ocean up by both latitude (regions are bounded by $15°$ and $45°$ north and south and exclude areas north of $60°N$) and basin, even in the Southern Ocean which was divided into several 'sectors'. So if there are interbasin differences in trends there will be a small amount of information in this pattern that is not present in simple zonal means.

Prior to spatial averaging, data were normalized as

$$x_n = \left(\frac{x - \overline{x}}{\sigma_x} \right) \qquad (1)$$

Where $\overline{x}$ and σ_x are the mean and standard deviation of all valid data from the preindustrial control run, except for dinitrogen fixation (DNF) where only data equatorward of $40°$ latitude were

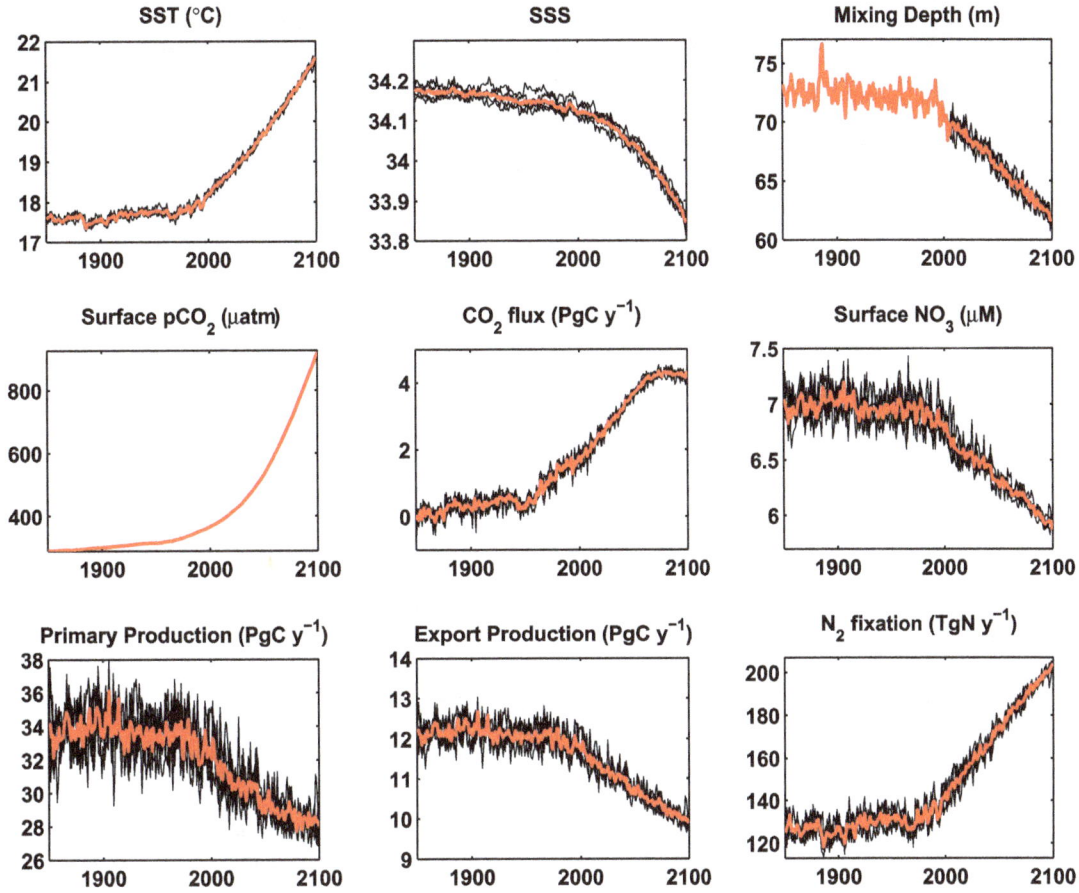

Figure 1. Global means or integrals of ocean surface fields from 1850–2100 under RCP8.5. Five ensemble members (thin black lines) are shown along with the ensemble mean (thick red line). Mixed layer depth has only a single realization for the historical run. Export is at 100 m. Primary production, export, CO_2 flux and N_2 fixation are global integrals; other fields are global mean surface values.

considered. The difference between the means for these regions for 2081–2100 of RCP8.5 and the means for 300 years of unforced control run was considered to be the 'fingerprint' of anthropogenic climate change. The projection of this fingerprint onto each year of the forced (historical + RCP) run, i.e.,

$$Y_i = P + aX \qquad (2)$$

where X is the anthropogenic fingerprint, Y_i is a vector of regional means for the current year of the forced run and P is a vector of regional means for the long-term mean of the unforced control run, was calculated by linear least squares.

$$a = \left(X^t C^{-1} X\right)^{-1} X^t C^{-1} (Y_i - P) \qquad (3)$$

where C is the covariance matrix for the control run. The same method was applied to (300) individual years (P_i) of the control run in order to generate a distribution of values (a_0) from which emergence can be determined. As for the global means, the year of emergence (YOE) was considered to be the year in which a exceeded three times the standard deviation of a_0 (σ_{a0}) and remained consistently in excess of $3\sigma_{a0}$ thereafter [31].

A 'combined' fingerprint experiment was conducted where the time series of regional means of all variables except pCO2, CO2 flux and mixing depth were superimposed, for a total of 92

variables/regions in the S02 case (sea surface nitrate concentration in the tropical Atlantic was excluded because its extremely low variance made the covariance matrix singular). pCO2 and CO2 flux were excluded because their much earlier emergence would dominate the result (pCO2 in particular has a much smaller preindustrial standard deviation than other fields and emerges very early); mixing depth was excluded because there was only a single realization of the historical run available. As noted above, high-latitude regions were excluded for DNF, because it only occurs where temperature exceeds 20°C.

Ocean observing station network

Model simulations were sampled at ten ocean observing stations, eight of the nine used by Moore et al. [28] and Station KNOT in the northeast Pacific (Table 1). The Ross Sea station used by Moore et al. was replaced, because of its shallow depth and proximity to the ice shelf, with a more oceanic location in the South Atlantic (SATL). These stations were chosen to represent most major regions of the world ocean, excluding the Arctic and the marginal seas, and to include actual sampling stations where observations have been made. Four of the stations are located in the tropics and subtropics and three each in the Southern Ocean and the northern midlatitudes (Table 1). Northern midlatitude stations range in latitude from 44–50°N, and Southern Ocean stations from 51–62°S.

Table 1. Locations of stations at which the model simulations were sampled for local emergence.

Station	Latitude	Longitude
BATS	32°N	64°W
HOT	23°N	158°W
PAPA	50°N	145°W
KNOT	44°N	155°E
ARAB	16°N	62°E
EQPAC	0°	140°W
SATL	51°S	19°W
NABE	47°N	19°W
KERFIX	51°S	68°E
PLRFR	62°S	170°W

At each station, the YOE was determined in the same manner as for the global means, except that the criterion for emergence was set to 2σ instead of 3σ. In addition, 20 and 30 year linear trends were calculated for individual ensemble members to identify the point at which these become consistently positive or negative, termed the "Last Zero Crossing" (LZC). In this case individual ensemble members are used because the influence of natural variability must be preserved. For a LZC to be recorded for N year trends, the last sign reversal must occur at least $N/2+10$ years prior to the end of the run for at least 4 of 5 ensemble members; LZC is then averaged over the realizations in which a LZC was recorded.

Results

Global mean trends

Global mean trends in ocean physical and biogeochemical fields show substantial alteration in the 21st century under the no-mitigation scenario (Figure 1), and in many cases these trends are well underway by the end of the 20th (which is scenario-independent). MLD declines by ~3 m by 2000 and 10 m by 2100 (Figure 1). Export production and dinitrogen fixation show more or less monotonic trends that are well underway by 2000 (Figure 1). Ocean CO_2 uptake continues to increase, but the rate of growth declines rapidly near the end of the 20th century (Figure 1). Some biogeochemical fields, such as primary production, have weak trends due to competing influences of e.g., temperature and stratification, as well as offsetting trends in different regions, so that the trend is relatively small compared to natural variability, at least initially (Figure 1).

The emergence of the global mean or integral values of the selected fields from the envelope of natural variability was tested by comparing ensemble means for historical + RCP2.6/4.5/8.5 simulations with the unforced control simulations, for each year of the simulation (1850–2100). Year of emergence was recorded as the first year that the ensemble mean differed from the preindustrial mean by at least 3 standard deviations of the control run and remained in excess of this threshold continuously thereafter. Surface ocean pCO_2 emerges in the 1870s, but air-sea CO_2 flux does not emerge until about a century later (Figure 2). Sea surface temperature (SST) emerges before sea surface salinity (SSS), as is expected for the global mean. SSS has offsetting trends in different regions, because the global surface net freshwater flux is close to zero, but anthropogenic warming increases the local flux in both net evaporative and net precipitation regions [15,23]. Emergence for SST closely approximates the date at which detection of an anthropogenic contribution to ocean heat content change is deemed to have become statistically significant [9]. YOE for MLD was not determined but would be around 2005 (not shown).

Primary production does not emerge in RCPs 2.6 or 4.5, and even in RCP8.5 does not emerge until the 2040s. This is partly due to offsetting increasing and declining trends in different regions (see below sections 3.2 and 3.3). In RCP4.5 global mean primary production exceeds the 3σ threshold for about 10 years in the 2040s, but does not remain there because the overall trend is not monotonic under the mitigation scenarios (not shown). In RCP2.6, air-sea CO_2 flux, surface nitrate concentration, and export production all fall back within the 3σ window after initially emerging from it the early 21st century (not shown); in these cases the fields are recorded as emergent (Figure 2), because they remained outside the 3σ window continuously for >50 years. Note that emergence can occur slightly earlier in a lower emission scenario due to natural variability, which is reduced but not eliminated by using ensemble means (Figure 2). In fact, it is possible for emergence to occur after 2005 in one scenario while occurring before 2005 in the others, even though atmospheric CO_2 is identical up to 2005.

Detecting the spatial 'fingerprint' of anthropogenic change

Because some fields have offsetting increases and decreases in different regions [23], global means are not necessarily a good metric of whether the forced simulation has emerged from the envelope of variability of the unforced control. I have calculated a spatial 'fingerprint' of anthropogenic change by taking the difference between the mean for 2081–2100 of RCP8.5 and the mean of the preindustrial control run (equation 2). The projection of the anthropogenic fingerprint onto individual years of the transient run (a in equations 2 and 3) for individual data fields is shown in Figure 3 for two different choices of the averaging regions. The difference between the two is generally small and emergence from the 3σ window occurs by 2005 in all cases. Historical volcanic eruptions such as Krakatoa (1883), Agung (1963) and Pinatubo (1991) are visible in the trends for some fields (Figure 3). The eruption of Pinatubo occurs near the point of emergence for primary production, export production and dinitrogen fixation and delays emergence by 5–10 years (Figure 3).

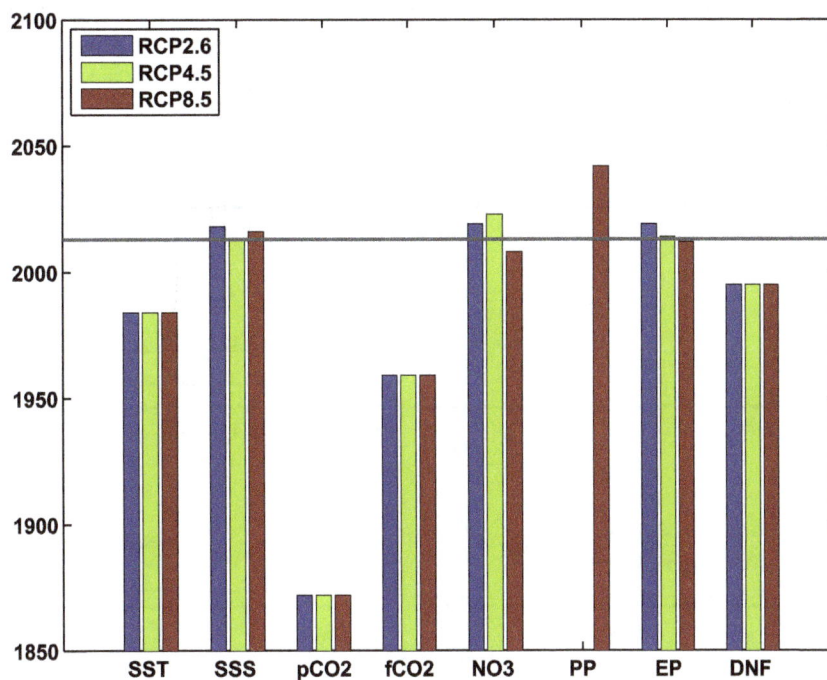

Figure 2. Year of emergence of global mean ocean surface fields from the range of natural variability for RCPs 2.6, 4.5, and 8.5. Emergence is defined as the year that the ensemble mean of 5 forced (historical + RCP) runs exceeds three times the standard deviation (3σ) of 300 years of unforced control run and remains in excess of 3σ thereafter. Horizontal grey line indicates beginning of 2013. SST = sea surface temperature; SSS = sea surface salinity; pCO2 = surface ocean pCO_2; fCO2 = ocean-atmosphere CO_2 flux; NO3 = surface nitrate concentration; PP = primary production; EP = export production at 100 m; DNF = dinitrogen fixation.

Trends in the presence of the anthropogenic fingerprint do not in most cases differ much between the 14 regions of S02 or global zonal means for 10° latitude bands as in [31] (Figure 3). The variables that show the largest difference, i.e. the most sensitivity to basin-specific information, are sea-surface salinity and air-sea CO_2 flux (Figure 3). There are large interbasin differences in evaporation and precipitation [11], so it is not surprising that the spatial 'fingerprint' of anthropogenic warming for SSS has a larger component that is basin-specific than for most other fields. For CO_2 flux the anthropogenic fingerprint shows enhanced net uptake (which may be reduced outgassing in outgassing regions) in the equatorial Pacific upwelling zone, the Gulf Stream and Kuroshio termination regions, the high-latitude North Atlantic and much of the Southern Ocean (not shown). However, there is a fair amount of variation among regions of the Southern Ocean, with the strongest enhancement in the Atlantic sector and a mosaic of positive and negative trends in the Pacific and Indian sectors. By recalculating the fingerprint for a reduced set of regions, with different basins being combined for specific latitude ranges, the latitudes in which basin-specific information is important can be identified (not shown). For CO_2 flux about half of the total effect is in the Southern Ocean, with the balance in the tropics. For SSS the only region where there is sensitivity to basin-specific averaging is the subtropics (i.e., regions of net evaporation).

The combined fingerprint of all non-carbon fields (except MLD) shows early emergence, and is also insensitive to the choice of averaging regions (Figure 4, Table 2). As with some of individual fields, volcanic eruptions can delay emergence. In this case the eruption of Agung (1963) delays emergence by nearly half a century in the zonal means case, whereas the S02 case is unaffected because it remains slightly outside the window following the eruption (Figure 4, Table 2). This illustrates how

sensitive the exact date of emergence can be to the somewhat arbitrary criteria employed, but further serves to illustrate that an anthropogenic signal was present even in the first half of the twentieth century. The biological fields (surface nitrate concentration, primary production, export production and dinitrogen fixation) do not affect emergence time much compared to SST and SSS alone (Table 2), but this results in large part from the particular timing of the eruption of Agung (Figure 4). Inclusion of both sets of fields narrows the window of preindustrial variability substantially relative to either alone (Figure 4). Had the eruption not occurred when it did, the differences in emergence times among these three cases could be much larger.

Local emergence and ocean observing networks

An increasing number of ocean time series data span several decades [7,14]. But since these are largely localized observations, and interannual to interdecadal variability is also present in the data records, how can such data be used to detect a longer-term trend? And how clearly can such trends be associated with anthropogenic forcing? The following analysis of a network of 10 ocean observing stations explores these questions.

Year of emergence at 2σ for the ten station means is shown in Figure 5. YOE's are generally later for individual stations than for the global means, and are quite variable among stations (Figure 5). However, the ranges for different variables are generally consistent with YOE's estimated for global means or spatial fingerprints except for CO_2 flux (Figure 5). Surface ocean pCO_2 emerges much earlier than other fields (prior to ~1960), but the range among stations is almost 100 years. For all fields except pCO_2 and SST, there are some stations at which emergence does not occur by 2100, and a few (MLD, surface nitrate, and export production) emerge at less than half of the stations (Figure 5).

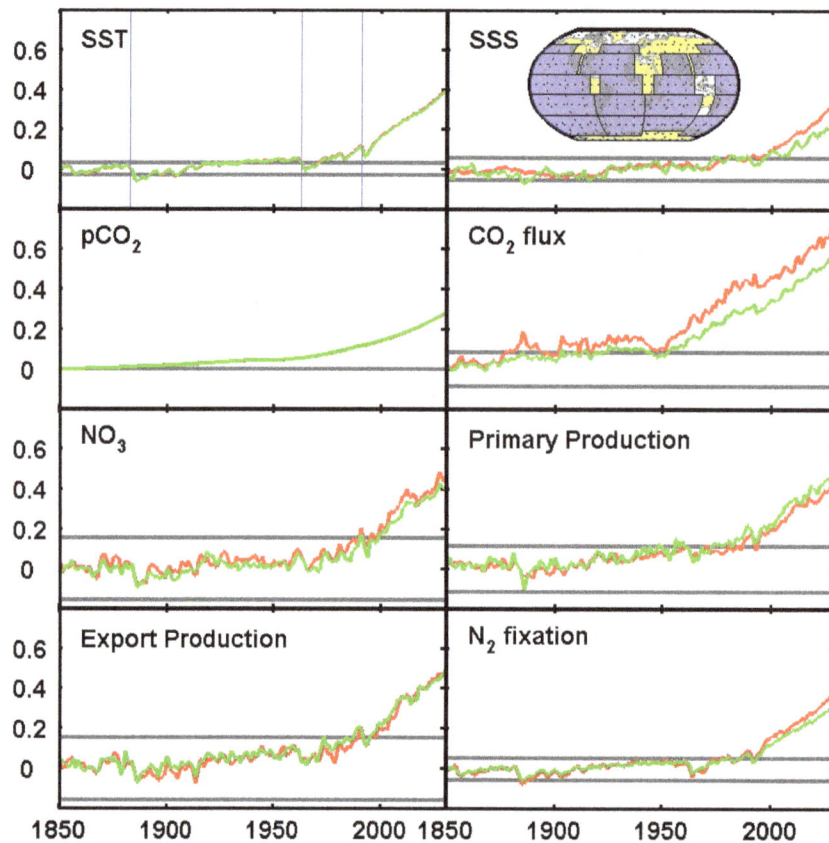

Figure 3. Contribution of 2081–2100 (RCP8.5) anthropogenic fingerprint to difference of current year from preindustrial, for individual fields. Fingerprint is based on areal means for the 14 ocean regions of Sarmiento et al. (2002) (S02, red) or global zonal means of 10° latitude bands (green). Vertical axes are normalized as shown in equation (1). Horizontal black lines are plus or minus three standard deviations of preindustrial values (3σ range is shown for S02 only; values are almost identical for the two methods). Vertical blue lines in first panel indicate eruptions of Krakatoa (1883), Agung (1963), and Pinatubo (1991). Inset map in second panel indicates S02 regions (blue rectangles); Pacific regions are single boxes spanning the basin.

Quasi-linear trends have been shown to occur over extended periods at ocean observing stations, particularly for carbon-related fields, for which the anthropogenic trend is large relative to the natural variability [7]. However, for other fields the time series are as yet too short for trends to be unambiguously associated with anthropogenic forcing. I have calculated 20 and 30 year linear trends from the model solution at the 10 observing stations in Table 1, and compared the time of emergence as estimated above with the time of the last sign reversal, or LZC.

The LZC for various fields and stations is shown in Table 3. The table shows the final year where the sign of the regression coefficient was opposite to the mean for the last 10 intervals. The LZC is not a statistically rigorous estimate of the time at which the anthropogenic trend becomes significant relative to natural variability, but it does give a rough indication (i.e., in the control climate the coefficients are equally often positive and negative), and it is shown below that its relation to the YOE is quite consistent. In quite a few cases sign reversals occur throughout the 20th century (blank entries in Table 3). In other cases the LZC occurs quite early, suggesting that the anthropogenic impact is present in the observations taken at these locations, which in most cases date from the 1980s or 1990s.

Because in this experiment the YOE (based on the ensemble mean and preindustrial standard deviation) is known (Figure 5), one can determine whether the LZC is correlated with the YOE

across different fields and locations and whether it tends to over- or underestimate the YOE. The LZC as an estimate of emergence time is strongly correlated with YOE for both 20 and 30 years trends, with r = 0.91 and 0.92 respectively (Figure 6). It provides consistently conservative estimates (exceeds the YOE) for 20 year trends, but not for 30 year trends, which tend to have an "early emergence" bias (Figure 6). This suggests that the proper trend length for which the LZC will approximate the YOE is ~25 years, but the exact value is sensitive to the characteristics of the specific data sets employed (see Discussion).

Discussion

A central result of this analysis is that if the model is a reasonable representation of the real world, most of the fields have already emerged from the preindustrial range (Figures 2 and 3, Table 2). As a result, the emission scenario does not matter much for the global YOE because it precedes the point where they start to diverge. However, for individual locations YOEs are generally later, sometimes much later (Figure 5) and therefore emergence may not occur, depending on the future trajectory of emissions. Global YOE is affected by mitigation only for primary production (Figure 2).

YOE is strongly affected by historical volcanic eruptions, and in some cases delayed by decades (Figures 3 and 4). The eruption of Mt Pinatubo (1991) delays emergence by about 10 years for

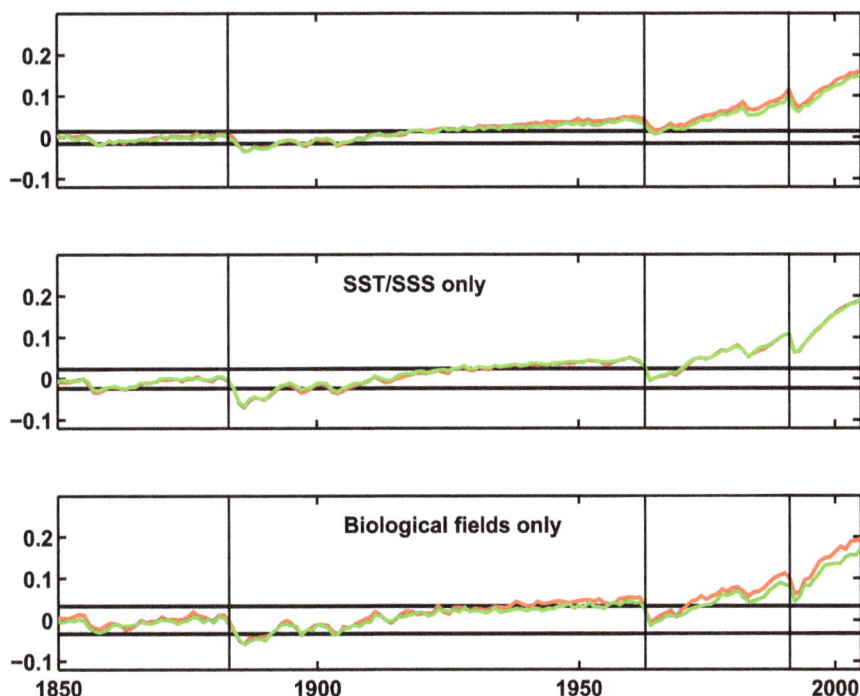

Figure 4. As Figure 3 but for combined fingerprint of all fields except pCO₂, CO₂ flux and mixing depth, for SST + SSS only, and for biological fields (surface nitrate concentration, primary production, export production and dinitrogen fixation) only. Red line indicates fingerprint based on the 14 ocean regions of S02; green line is for global zonal means of 10° latitude bands. Vertical black lines indicate eruptions of Krakatoa (1883), Agung (1963), and Pinatubo (1991).

several fields (Figure 3). Eruption of Agung in 1963 delayed emergence in the combined fingerprint experiment by more than 40 years in the zonal means case, while the S02 case is unaffected because it remains slightly outside the window following the eruption (Figure 4). Note that the downward trend at this time appears to begin prior to the eruption: this may result in part from natural variability, in part from errors in the volcanic aerosol data set, and in part from a small increase in volcanism globally in the years 1960–1963; this downward trend is present in multiple models (see e.g., [22]).

Table 2. Year of emergence by estimation of single or multiple field anthropogenic fingerprint calculated for the 14 ocean regions of Sarmiento et al. (2002) or for global zonal means for 10° latitude bands, for two or three standard deviations of preindustrial values.

Regions	S02	S02	Zonal	Zonal
Emergence threshold	2σ	3σ	2σ	3σ
Individual				
SST	1972	1973	1972	1973
SSS	1976	1990	1997	1997
Surface pCO2	1859	1860	1859	1860
Air-sea CO2 flux	1877	1913	1918	1953
Surface [NO3]	1986	2000	1996	2000
Primary production	1969	1995	1973	1984
Export production	1988	1996	1981	1997
N2 fixation	1994	1995	1994	1995
Combined				
All, except pCO2 and CO2 flux	1919	1923	1967	1968
Biological fields	1971	1972	1971	1972
SST + SSS only	1971	1973	1973	1973

Combined fingerprint excludes pCO₂ and CO₂ flux; biological fields are surface nitrate concentration, primary production, export production and dinitrogen fixation.

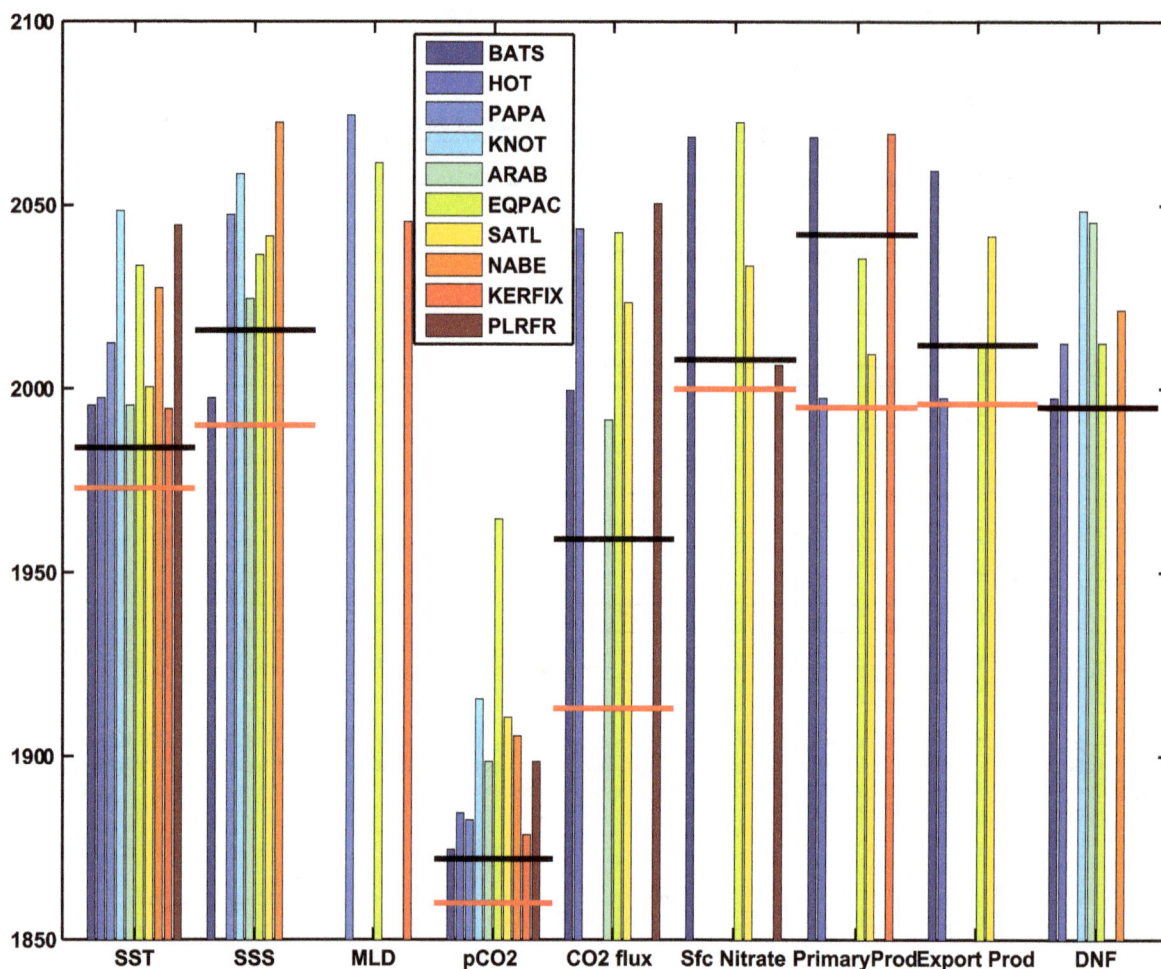

Figure 5. Year of emergence of mean ocean surface fields from the range of natural variability at 2σ for individual locations under RCP8.5 (see Table 1). Horizontal lines indicate YOE's for global means or spatial fingerprints from S02 (see Figures 2 and 3). Black indicates YOE for the global mean, red for the spatial fingerprint. For dinitrogen fixation red bars are not visible because YOE's are the same. No vertical bar indicates that the variable does not emerge by 2100 at that location.

The spatial fingerprint analysis provides a measure of the importance of diverse observations for detection of anthropogenic effects. The analysis presented here does not deal with detection directly, because only model simulations are used, but it is likely that earlier YOE implies earlier detection assuming sufficient observations are available. Spatial fingerprints generate earlier YOE than the global means for all fields except DNF (Figure 5). In many cases basin-specific information provides little additional information above what is provided by global zonal means (Figure 3, Table 2). A combined fingerprint of all fields results in very early emergence (∼1920 were it not for the eruption of Agung in 1963). Inclusion of the biological fields has relatively little effect on time of emergence beyond what is available from SST and SSS alone, but does make the window smaller, i.e., the additional information increases confidence that the projection of the anthropogenic fingerprint on the preindustrial climate is close to zero (Figure 4). This is probably because the SST response to planetary heating is global and more or less instantaneous (disregarding modulation by natural variability), whereas impacts on ocean biology are more indirect, resulting from stratification that derives from surface heating. It is therefore unsurprising that fields like export production emerge later and have relatively little effect on the combined fingerprint YOE. DNF is (in the model) an

approximately linear function of SST, so it has strong trends but contains little information beyond what is present in SST alone. Nonetheless, emergence of the combined fingerprint for biological fields alone is quite similar to the total fingerprint, so there is likely to be an anthropogenic influence on most modern ocean biogeochemistry measurements.

Simplifying assumptions such as a single plankton species with fixed elemental ratios limit variability of the modelled ocean ecosystem. Plankton models with multiple species are better able to reproduce changes in ecosystem structure that occur under different physical forcing regimes [2]. There is some evidence that changes in plankton elemental stoichiometry can also 'amplify' the biological response to relatively small changes in physical forcing [26]. These biases are present in both the control and the forced runs, but they bias the model towards a 'damped' biological response to changing physical forcing and thus induce a bias towards later emergence. Excluding DNF at higher latitudes also potentially biases the method towards later emergence because a small amount of DNF occurs in regions where it is absent in the control run (as SST begins to exceed 20°C), but these rates are very low (not shown).

The use of ensembles to average out internal variability in the transient runs gives a statistically robust estimate of emergence

Table 3. Last year where 20 or 30 year trend had opposite sign to that recorded at the end of the 21st century.

	# years	BATS	HOT	PAPA	KNOT	ARAB	EQPAC	SATL	NABE	KERFIX	PLRFR
SST	20	1979	2027	2011	2041	1983	2058	2019	2049	1999	
	30	1968	1968	1974	1977	1967	1979	1968	2001	1957	2073
SSS	20	1980	2060	2074	2080	2039	2059	2050	2060	2067	
	30	1958		2038	2038	1997	2015	2021	2021	2064	
Mixing Depth	20			2070			2076	2077	2077		2075
	30	2073	2074	2049			2021	2056	2054		2070
Surface pCO$_2$	20	1869	1931	1883	1963	1947	1972	1943	1936	1909	1942
	30	*	1869	*	1940	1873	1941	1885	1867	*	1853
Air-sea CO$_2$ flux	20	2083		2062	2084	2044		2041	2078		2059
	30			2046		1993		1993	2057		2021
Surface [NO$_3$]	20		2081		2082			2053		2074	
	30	2029			2081		2069	2036	2046	2073	
Primary production	20		2052		2082		2085	2051		2069	2079
	30	2030	1968		2072		2047	1976	2046	2047	
Export production	20		2069		2082	2074	2085	2080		2075	2079
	30	2019	1988		2081	2072	2059	2033	2044	2066	2075
N$_2$ fixation	20	2001	2055			2083	2078				
	30	1968	1968			2071	2008				

No data indicates sign reversals continue up until the end of the 21st century, except for N$_2$ fixation which only occurs at the stations where dates are listed.
* indicates that the trend was of a consistent sign from the outset. Mixing depth has only a single realization for the historical simulation but no values <2005 appear.

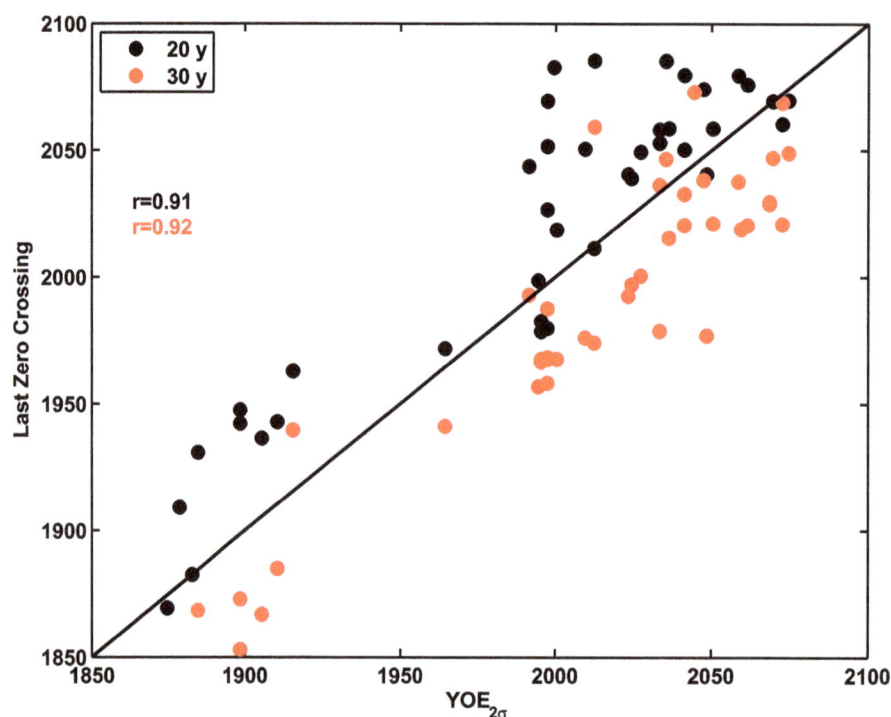

Figure 6. Last-zero-crossing for 20- and 30-year regression coefficients (Table 3) relative to YOE defined as in Figure 5, for all data fields at individual sampling stations where a last-zero-crossing was recorded.

relative to the variability of the control run. The time of emergence in an individual realization would depend strongly on internal variability, because natural variability superimposed on an overall trend will tend to produce periods where the record is flat or has a weak counter-trend (similar to the recent "warming hiatus", cf. [16]), followed by periods of rapid change. If there were no overall trend, the positive and negative trends would be symmetric (as in the unforced control run). But in the presence of a long term trend they are not, and there will be short periods with strong positive trends.

To fully understand the probability of medium-term trends occurring, one would ideally wish to know the frequency spectrum of natural variability. In the model, the trends are clearly forced by anthropogenic greenhouse gases; they do not occur in the unforced control run. But it is not known whether the model ocean underestimates (or overestimates, although that is less likely) natural variability [12,25]. In the forced model runs, the probability of a trend occurring is inversely proportional to the length, i.e., 20 year countertrends appear much more often than 30 year ones (see Table 3). But in a stationary climate it is not always true that shorter countertrends appear more frequently; it depends on the frequency spectrum. The more 'red' the spectrum (i.e., the stronger the autocorrelation), the more likely it is that a longer period with a trend will occur. If the spectrum is sufficiently 'red', longer periods with a consistent trend can occur more frequently than short ones [19]. In the model this does not occur, and there is a clear association of longer trends with external forcing [16], but it is not possible to know with certainty whether the model's internal variability has a less 'red' spectrum than the real ocean. At some locations, such as BATS in the subtropical Atlantic, the model's variance spectrum is red, and longer periods with consistent trends occur more frequently than shorter ones in the unforced control (not shown). In the 250 years of historical + RCP simulation, however, the anthropogenic trend is large

relative to the natural variability, so trends counter to the forced response are rare. This is at the heart of the detection problem: the instrumental record is gradually becoming long enough to resolve interdecadal variability, but there is already a large and growing anthropogenic component. As ocean time-series data accumulate, understanding of the spectrum of variability will increase, but separating out the anthropogenic component will remain extremely challenging.

Using the 2σ YOE estimated here as a benchmark (Figure 5), the last-zero-crossing for a 20-year trend (LZC_{20}) is a conservative estimate of emergence, while LZC_{30} has an early-emergence bias (Figure 6). Whether a particular length scale for this analysis is a conservative criterion depends on the specific data sets employed and can not readily be generalized across fields or emissions scenarios (although in these results there are consistent relationships among LZC_{20}, LZC_{30} and YOE for different data fields). In general, the faster the anthropogenic signal grows, the stronger the tendency for the LZC to give early emergence. This is a simple function of the signal-to-noise ratio with a nonlinearly growing signal ($d^2y/dt^2 > 0$) and a constant amount of noise. A noise-driven countertrend that exceeds the forced component is much more likely to occur early in the experiment, when dy/dt is small. Similar logic explains why the criterion becomes more conservative with redder noise. The greater the probability that a trend will occur over a period of e.g. 30 years, the more likely it is that a short-term trend counter to the anthropogenically forced trend will occur in the forced simulation after the 'true' emergence point has been passed. For the simulations considered here, the appropriate period appears to be about 25 years (Figure 6).

This analysis - using climate model simulations with future emissions scenarios - obviously benefits from hindsight that can never be available in ocean observations. If a 30 year secular trend is observed, and in the 31st year it does not reverse, nor in the 32nd, the observer is inclined to believe that a long-term trend is present. It is not the presence of a positive trend that is diagnostic, but the absence of counter-trends over shorter periods. Because in observations we lack the hindsight available in models, it is impossible to say with certainty how long a time since LZC is required to diagnose emergence, but model simulations can help us to estimate the probability of a counter-trend emerging in the future. In most locations it is unlikely that counter-trends will ever cease for periods less than about 15 years, even under very strong anthropogenic forcing, although the statistics are likely to be a latitude dependent [27]. For the stations and data fields examined here, LZC is a fairly reliable diagnostic of emergence for a period between 20 and 30 years, with the former being too conservative and the latter giving many false positives (Figure 6).

These results suggest that it will be difficult to detect an anthropogenic trend in localized ocean time-series data with confidence. The existence of a trend on timescales longer than interdecadal does not necessarily imply that it is anthropogenic, but as time series extend beyond the interdecadal range opportunities for detection and attribution will arise. Ultimately the case for anthropogenic forcing will have to rest on a mechanistic understanding of the underlying processes [33]. It has only recently become possible to conduct an analysis of the kind presented here even with models, and the models will continue to improve. At the same time, ocean time series will be extended with modern methods that were novel when the time series began but are now mature and operationalized. Increased computational power also makes possible high resolution regional models that can aid in the interpretation of observations and elucidation of mechanisms.

Acknowledgments

Nathan Bindoff, Nathan Gillett, Debby Ianson, Slava Kharin and an anonymous reviewer made useful comments on an earlier draft of this manuscript.

Author Contributions

Conceived and designed the experiments: JC. Analyzed the data: JC. Wrote the paper: JC.

References

1. Andrews O, Bindoff N, Halloran P, Ilyina T, Le Quéré C (2013) Detecting an external influence on recent changes in oceanic oxygen using an optimal fingerprinting method. Biogeosciences 10: 1799–1813.

2. Armstrong R (1994) Grazing limitation and nutrient limitation in marine ecosystems - steady-state solutions of an ecosystem model with multiple food-chains. Limnol Oceanogr 39: 597–608.

3. Arora V, Boer G, Christian J, Curry C, Denman K, et al. (2009) The effect of terrestrial photosynthesis down regulation on the twentieth-century carbon budget simulated with the CCCma Earth System Model. J Clim 22: 6066–6088.

4. Arora V, Scinocca J, Boer G, Christian J, Denman K, et al. (2011) Carbon emission limits required to satisfy future representative concentration pathways of greenhouse gases. Geophys Res Lett 38 doi:10.1029/2010GL046270

5. Barnett T, Pierce D, Schnur R (2001) Detection of anthropogenic climate change in the world's oceans. Science 292: 270–274.

6. Barnett T, Zwiers F, Hegerl G, Allen M, Crowley T, et al. (2005) Detecting and attributing external influences on the climate system: a review of recent advances. J Clim 18: 1291–1314.

7. Bates N (2007) Interannual variability of the oceanic CO_2 sink in the subtropical gyre of the North Atlantic Ocean over the last 2 decades. J Geophys Res 112 doi:10.1029/2006JC003759

8. Beaulieu C, Henson S, Sarmiento J, Dunne J, Doney S, et al. (2013) Factors challenging our ability to detect long-term trends in ocean chlorophyll. Biogeosciences 10: 2711–2724.

9. Bindoff NL, Stott PA, AchutaRao KM, Allen MR, Gillett N, et al. (2013) Detection and attribution of climate change: from global to regional. In: Stocker, TF, Qin D, Plattner G-K, Tignor M, Allen SK, et el., editors. Climate Change 2013: The Physical Science Basis.Contribution of Working Group I to the Fifth Assessment Report of the Intergovernmental Panel on Climate Change. Cambridge: Cambridge University Press. pp. 867–952.

10. Chavez F, Ryan J, Lluch-Cota S, Niquen M (2003) From anchovies to sardines and back: Multidecadal change in the Pacific Ocean. Science 299: 217–221.

11. Chou S, Nelkin E, Ardizzone J, Atlas R, Shie C (2003) Surface turbulent heat and momentum fluxes over global oceans based on the Goddard satellite retrievals, version 2 (GSSTF2). J Clim 16: 3256–3273.

12. Christian J, Arora V, Boer G, Curry C, Zahariev K, et al. (2010) The global carbon cycle in the Canadian Earth System Model (CanESM1): Preindustrial control simulation. J Geophys Res 115 doi:10.1029/2008JG000920

13. Deutsch C, Brix H, Ito T, Frenzel H, Thompson L (2011) Climate-forced variability of ocean hypoxia. Science 333: 336–339.

14. Dore J, Lukas R, Sadler D, Church M, Karl D (2009) Physical and biogeochemical modulation of ocean acidification in the central North Pacific. Proc Natl Acad Sci USA 106: 12235–12240.

15. Durack P, Wijffels S, Matear R (2012) Ocean salinities reveal strong global water cycle intensification during 1950 to 2000. Science 336: 455–458.

16. Easterling D, Wehner M (2009) Is the climate warming or cooling? Geophys Res Lett 36 doi:10.1029/2009GL037810

17. Flato G, Marotzke J, Abiodun B, Braconnot P, Chou SC, et al. (2013) Evaluation of climate models. In: Stocker, TF, Qin D, Plattner G-K, Tignor M, Allen SK, et el., editors. Climate Change 2013: The Physical Science Basis. Contribution of Working Group I to the Fifth Assessment Report of the Intergovernmental Panel on Climate Change. Cambridge: Cambridge University Press. pp. 741–866.

18. Friedrich T, Timmermann A, Abe-Ouchi A, Bates N, Chikamoto M, et al. (2012) Detecting regional anthropogenic trends in ocean acidification against natural variability. Nat Clim Chang 2: 167–171.

19. Hasselmann K (1976) Stochastic climate models. 1. Theory. Tellus 28: 473–485.

20. Hasselmann K (1993) Optimal fingerprints for the detection of time-dependent climate-change. J Clim 6: 1957–1971.

21. Hegerl G, Hasselmann K, Cubasch U, Mitchell J, Roeckner E, et al. (1997) Multi-fingerprint detection and attribution analysis of greenhouse gas, greenhouse gas-plus-aerosol and solar forced climate change. Clim Dyn 13: 613–634.

22. Hegerl G, Zwiers F, Tebaldi C (2011) Patterns of change: whose fingerprint is seen in global warming? Environ Res Lett 6 doi:10.1088/1748-9326/6/4/044025

23. Helm K, Bindoff N, Church J (2010) Changes in the global hydrological-cycle inferred from ocean salinity. Geophys Res Lett 37 doi:10.1029/2010GL044222

24. Henson S, Sarmiento J, Dunne J, Bopp L, Lima I, et al. (2010) Detection of anthropogenic climate change in satellite records of ocean chlorophyll and productivity. Biogeosciences 7: 621–640.

25. Huybers P, Curry W (2006) Links between annual, Milankovitch and continuum temperature variability. Nature 441: 329–332.

26. Karl D, Letelier R, Hebel D, Tupas L, Dore J, et al. (1995) Ecosystem changes in the North Pacific subtropical gyre attributed to the 1991–92 El-Niño. Nature 373: 230–234.

27. Mahlstein I, Knutti R, Solomon S, Portmann R (2011) Early onset of significant local warming in low latitude countries. Environ Res Lett 6 doi:10.1088/1748-9326/6/3/034009

28. Moore J, Doney S, Kleypas J, Glover D, Fung I (2002) An intermediate complexity marine ecosystem model for the global domain. Deep Sea Res Part 2 Top Stud Oceanogr 49: 403–462.

29. Mora C, Frazier A, Longman R, Dacks R, Walton M, et al. (2013) The projected timing of climate departure from recent variability. Nature 502: 183–187.

30. Moss R, Edmonds J, Hibbard K, Manning M, Rose S, et al. (2010) The next generation of scenarios for climate change research and assessment. Nature 463: 747–756.

31. Muir L, Brown J, Risbey J, Wijffels S, sen Gupta A (2013) Determining the time of emergence of the climate change signal at regional scales. CAWCR Res Lett 10: 8–19.

32. Najjar R, Orr J (1998), Design of OCMIP-2 simulations of chlorofluorocarbons, the solubility pump and common biogeochemistry. http://ocmip5.ipsl.fr/documentation/OCMIP/phase2/simulations/Biotic/HOWTO-Biotic.html. Accessed 2014 Sep 19.

33. Rosenzweig C, Karoly D, Vicarelli M, Neofotis P, Wu Q, et al. (2008) Attributing physical and biological impacts to anthropogenic climate change. Nature 453: 353–357.

34. Santer BD, Wigley TML, Barnett TP, Anyamba E (1996) Detection of climate change and attribution of causes. In: Houghton JY, Meira Filho LG, Callander BA, Harris N, Kattenberg A, et al., editors. Climate Change 1995: The Science of Climate Change. Contribution of Working Group I to the Second Assessment Report of the Intergovernmental Panel on Climate Change. Cambridge: Cambridge University Press. pp. 406–443.

35. Santer B, Mikolajewicz U, Bruggemann W, Cubasch U, Hasselmann K, et al. (1995) Ocean variability and its influence on the detectability of greenhouse warming signals. J Geophys Res 100: 10693–10725.

36. Sarmiento J, Dunne J, Gnanadesikan A, Key R, Matsumoto K, et al. (2002) A new estimate of the CaCO$_3$ to organic carbon export ratio. Global Biogeochem Cycles 16 doi:10.1029/2002GB001919

37. Taucher J, Oschlies A (2011) Can we predict the direction of marine primary production change under global warming? Geophys Res Lett 38 doi:10.1029/2010GL045934

38. Taylor K (2001) Summarizing multiple aspects of model performance in a single diagram. J Geophys Res 106: 7183–7192.

39. Taylor K, Stouffer R, Meehl G (2012) An overview of CMIP5 and the experiment design. Bull Am Meteorol Soc 93: 485–498.

40. Wang M, Overland J, Bond N (2010) Climate projections for selected large marine ecosystems. J Mar Syst 79: 258–266.

41. Zahariev K, Christian J, Denman K (2008) Preindustrial, historical, and fertilization simulations using a global ocean carbon model with new parameterizations of iron limitation, calcification, and N$_2$ fixation. Prog Oceanogr 77: 56–82.

10

Temperature and Cyanobacterial Bloom Biomass Influence Phosphorous Cycling in Eutrophic Lake Sediments

Mo Chen[1,2], Tian-Ran Ye[1], Lee R. Krumholz[3], He-Long Jiang[1]*

1 State Key Laboratory of Lake Science and Environment, Nanjing Institute of Geography and Limnology, Chinese Academy of Sciences, Nanjing, China, **2** Graduate University of Chinese Academy of Sciences, Beijing, China, **3** Department of Microbiology and Plant Biology, University of Oklahoma, Norman, Oklahoma, United States of America

Abstract

Cyanobacterial blooms frequently occur in freshwater lakes, subsequently, substantial amounts of decaying cyanobacterial bloom biomass (CBB) settles onto the lake sediments where anaerobic mineralization reactions prevail. Coupled Fe/S cycling processes can influence the mobilization of phosphorus (P) in sediments, with high releases often resulting in eutrophication. To better understand eutrophication in Lake Taihu (PRC), we investigated the effects of CBB and temperature on phosphorus cycling in lake sediments. Results indicated that added CBB not only enhanced sedimentary iron reduction, but also resulted in a change from net sulfur oxidation to sulfate reduction, which jointly resulted in a spike of soluble Fe(II) and the formation of FeS/FeS_2. Phosphate release was also enhanced with CBB amendment along with increases in reduced sulfur. Further release of phosphate was associated with increases in incubation temperature. In addition, CBB amendment resulted in a shift in P from the Fe-adsorbed P and the relatively unreactive Residual-P pools to the more reactive Al-adsorbed P, Ca-bound P and organic-P pools. Phosphorus cycling rates increased on addition of CBB and were higher at elevated temperatures, resulting in increased phosphorus release from sediments. These findings suggest that settling of CBB into sediments will likely increase the extent of eutrophication in aquatic environments and these processes will be magnified at higher temperatures.

Editor: Brett Neilan, University of New South Wales, Australia

Funding: This work was supported by grants from the National Natural Science Foundation of China (51079139), the Innovation Program of the Chinese Academy of Sciences (KZCX2-EW-314), 135 project of Nanjing Institute of Geography and Limnology, CAS (No. NIGLAS2012135008), and Chinese Academy of Sciences visiting professorship for senior international scientists (2011T1Z37). The funders had no role in study design, data collection and analysis, decision to publish, or preparation of the manuscript.

Competing Interests: The authors have declared that no competing interests exist.

* E-mail: hljiang@niglas.ac.cn

Introduction

Due to climate change and anthropogenic carbon and nitrogen runoff, cyanobacterial blooms are becoming more common in freshwater lakes and estuaries throughout the world, threatening the sustainability of aquatic ecosystems [1], [2]. The formation of large mucilaginous cyanobacterial blooms in freshwater lakes restricts light penetration, which depletes oxygen levels, thereby reducing water quality and adversely affecting the ecosystem [1]. These changes can result in reduction in the numbers of submerged plants, death of aquatic animals, and alteration of food web dynamics [3].

As cyanobacterial bloom biomass (CBB) dies, it settles on surface sediments, eventually becoming incorporated into sediments through resuspension and bioturbation. Decomposed CBB can be an important benthic food source [4] and decomposition products can be assimilated by rooted macrophytes [5]. As CBB undergoes decomposition, both nitrogen and phosphorus containing compounds are released. This release results in changes in the nutrient composition of sediments and water and eventually alters the sediment microbial community [6]. While decaying CBB in sediments has been found to strongly influence the bacterial community composition of lake sediments [7], the role of settled CBB in biogeochemical cycling in lakes with seasonal temperature changes have not been well studied.

In lakes, especially shallow lakes, sediment processes dominate the overall metabolic activities [8]. Lake sediments are important in the global carbon cycle, as they can act as both a sink and a source of critical elements [9]. Phosphorous is cycled in lake sediments [10] and excessive phosphorous input often directly causes eutrophication [11]. Phosphorus occurs in lake sediments in both organic and inorganic forms. Inorganic phosphorus typically associates with amorphous and crystalline forms of Fe, Al, Ca, and other elements. Organic phosphorus varies in ease of decomposition, therefore in phosphate bioavailability [12]. Transformation of phosphate compounds in sediments is highly dependent on environmental parameters, with the most important being temperature and redox potential. Typically, an increase in temperature depletes labile organic phosphorus [13]. Also, internal phosphorus cycling in sediments can be enhanced by increased temperatures, leading to more significant eutrophication [14]. Although increased temperatures result in higher levels of P release [15], especially in concert with extended anoxic conditions [16], a substantial fraction of P still remains buried in sediments. The mechanisms behind these shifts in P pools, as well as the impact of seasonal temperature changes and settled CBB on P movement between pools under anaerobic conditions, has not been well studied. Phosphorus cycling is also intertwined with iron and sulfur cycling. Under anaerobic conditions, both iron and sulfate reduction increase phosphorus release from sediments [17]. The lowering of the redox potential allows reduction of Fe (III) in iron oxides to Fe(II), resulting in a decrease in Fe-bound phosphorus [18].

In this study, the effect of settled CBB and seasonal temperature changes on phosphorus cycling in the sediments of a subtropical shallow lake were investigated. CBB was amended into surface sediments taken from a eutrophic lake, Lake Taihu, and sediments were incubated under anaerobic conditions at temperatures ranging from 4 to 32°C. Our results indicate that settling of CBB into sediments and increasing temperatures likely play important roles in influencing Fe/S biogeochemical processes and maintaining the eutrophic status of lakes.

Materials and Methods

Ethics statement

No specific permits were required for the described field studies. The location studied is not privately-owned or protected in any way and our studies did not involve any endangered or protected species.

Sediment sampling

Samples of both sediments and CBB were taken from Lake Taihu. Lake Taihu (31°10′ N, 120°24′ E), the third largest shallow freshwater lake in China, is situated south of the Yangtze River delta. It has a water surface area of 2340 km^2 and mean and maximum depths of 1.9 m and 3.4 m, respectively [19]. Water temperatures are between 0 and 38.0°C, with minimum temperatures occurring in January and maxima in July and August [20]. Increasing nutrient inputs associated with both population and economic growth have led to eutrophication in the lake. The average total phosphorus and sulfate concentrations in lake water are 0.086 and 100 mg L^{-1}, respectively. Since the 1980s, cyanobacterial blooms have occurred with increasing frequency and intensity in Lake Taihu [21].

Sediments were sampled using a gravity core sampler, and CBB samples were harvested by sieving lake surface water through a fine mesh plankton net in May, 2012. CBB samples were immediately stored in polyethylene bottles. Sediments and CBB samples were placed on ice and transported to the laboratory within several hours of collection. Subsequent storage of all samples was at 4°C for less than 24 hours until usage.

Sediment incubations

CBB samples were poured in to medical trays and placed in a Fume hood for several days to air dry. The dried CBB was then scraped into a mortar and ground into powder. Surface sediments (0–5 cm) from 18 sediments cores (10-cm-i.d. Plexiglas tubes) were sliced and homogenized thoroughly in a nitrogen-filled glove bag. Sediments with wet weights of 300 g and air-dried CBB with weights of 2 g were mixed thoroughly in a 500-mL glass jar inside the nitrogen-filled glove bag. Unamended experiments did not contain CBB. Finally, the glass jars containing sediment and CBB mixtures or unamended sediment were sealed tightly with rubber stoppers under pure nitrogen and incubated at 4°C, 15°C, 25°C or 32°C in a dark environment. Sediments were incubated for 41 days. During this period, the glass jars were opened in the nitrogen-filled glove bag for sampling every 2–13 days. Immedi-

ately after sampling, the jars were sealed tightly and incubation was resumed. The initial water content in unamended and bloom-amended sediments were 51% and 50.6%, respectively. Experiments were performed in triplicate.

Determination of iron and sulfate reduction rates

Fe (III) reduction rates were calculated after Fe (II) concentration in the sediments was determined using the method described below. Rates were calculated by regression of Fe (II) concentration vs. time [22]. Similarly, sulfate reduction rates were calculated based on regressions of sulfate concentrations vs. time.

Analytical methods

Water content in sediments was determined as weight loss after drying at 105°C for 12 hours. Total organic carbon (TOC) in sediments was measured with Walkley and Black's rapid titration method [23]. pH in sediments was measured by inserting a glass electrode calibrated with NBS standards (National Bureau of Standards, Gaithersburg, USA) directly into sediments.

Total phosphorus (TP) content in sediments was analyzed as phosphate after acid hydrolysis at 340°C [24]. The presence of different forms of phosphorus was determined through sequential extractions and digestions [25]. Samples were air-dried and extracted sequentially as follows: (a) 1 M NH$_4$Cl at pH 7, (b) 0.11 M NaHCO$_3$/0.11 M Na$_2$S$_2$O$_4$, (c) 0.1 M NaOH, and (d) 0.5 M HCl. The different forms of phosphorus extracted according to the above processes were referred to as (with description in parentheses): (a) NH$_4$Cl-P (Loosely bound or labile-P including soluble reactive phosphate (SRP) in pore water, Labile-P); (b) Fe-adsorbed P (Fe-P); (c) NaOH-rP (Al-adsorbed P, Al-P). (d) HCl-P (Ca-bound P, Ca-P); (e) Residual-P (residual P, Res-P). Res-P was calculated as TP (determined above using acid hydrolysis) minus the total phosphorus measured during the sequential extraction. Organic P (Org-P) was calculated from the difference between total NaOH extracted P (NaOH-Tot P) and NaOH-rP (Al-P) after digestion in the NaOH extraction step c [25].

To determine the dissolved Fe(II) concentration, sediment samples were placed into a 2 mL centrifuge tube in the nitrogen-filled glove bag, leaving no gas phase. The sediment samples were centrifuged at 12000 rpm for 30 sec and 0.1 mL of the supernatant was acidified with 0.1 mL of 1 M HCl (under pure nitrogen). Concentrations of Fe(II) was determined colorimetrically [26]. Sulfate and orthophosphate (PO$_4^{3-}$) concentrations in pore water were measured using ion chromatography (ICS-2000, Dionex, USA).Sediment Amorphous Fe(III) oxide and total Fe(II) concentrations in the sediments were measured by extracting sediments with 0.5 M HCl, followed by determination using the Ferrozine method [26]. Total Fe was reduced with hydroxylamine hydrochloride and Fe(III) concentration was calculated from the difference between total Fe and Fe (II).

Concentrations of acid volatile sulfide (AVS) and chromium reducible sulphur (CRS) were determined according to the procedure described previously [27], with slight modifications. Briefly, 2 g wet sediment was put into a sample flask with a small vial containing 5 mL alkaline Zn solution. The flask was closed with a two-outlet stopper and flushed with nitrogen gas for 5 min, then both outlets of the flask were closed. In order to measure AVS (including dissolved sulfide and FeS), 15 mL of deoxygenated 9 M HCl solution and 2 mL of 1 M ascorbic acid solution were injected into the flask through the syringe stopcock in a nitrogen filled glove bag. The syringe stopcock was closed, and the flask was incubated at room temperature. The flask was opened after 24 hours and the Zn trap retrieved to determine the concentration of precipitated Zn-sulfide. Immediately after the Zn trap was retrieved, the trap (5 mL alkaline Zn solution) was replaced and the flask closed and flushed with nitrogen again for the following CRS (pyrite-S) separation. Cr(II) solution (15 mL) was immediately injected into the flask from the previous AVS procedure, after replacing the small vial containing 5 mL fresh alkaline Zn solution. The sample flask was then closed, allowing the reaction to take place at room temperature. After 48 hours, the flask was opened and the Zn trap was retrieved for Zn-sulfide analysis as described above.

The temperature coefficient (Q10) can be used to express the influence of temperature on the rate of various metabolic processes [28]. In this study, Q10 was used to investigate the influence of temperature on the rate of iron reduction, sulfate reduction, and phosphate release in sediments. Q10 values were calculated from reduction and release rates at 15, 25 and 32°C, using the equation: $Q10 = (R_2/R_1)^{[10/(T_2-T_1)]}$, where R_2 and R_1 are the metabolic rates at two temperatures, T_2 and T_1.

Statistical analyses

Statistical significance of differences was determined by one-way analysis of variance using Origin Pro 7.5 and SPSS 19.0 software. A $P<0.05$ was considered significant.

Results

Concentrations of inorganic elements (Fe, S, and P) in pore-water

The initial concentrations of Fe(II) in pore-water in unamended and CBB-amended sediments were 47.0±6.4 µmol L^{-1} and 75.1±6.2 µmol L^{-1}, respectively (Fig. 1a). Fe(II) concentration in the pore-water of CBB-amended sediments at 32°C increased rapidly to a maximum of 1361.8±43.1 µmol L^{-1} on day 2 (Fig. 1a) and then decreased quickly over the next 5 days. The maximum concentration of Fe(II) in the pore-water in CBB-amended sediments at 4°C (398.7±5.1 µmol L^{-1}), 15°C (1045.9± 43.4 µmol L^{-1}) and 25°C (1075.1±54.5 µmol L^{-1}) appeared on days 28, 17, and 7, respectively (Fig. 1a). In unamended sediments, Fe(II) concentration in pore-water increased to around 100 µmol L^{-1} at higher temperatures (15°C, 25°C, and 32°C), compared to no obvious change of Fe(II) concentration at 4°C (Fig. 1a).

Initial sulfate concentration in the pore-water of unamended and CBB-amended sediments was around 1122–1136 µmol L^{-1}. As shown in Fig. 1b, in sediments without CBB amendment, the sulfate concentration in pore-water increased gradually at 15°C, 25°C and 32°C, with little change at 4°C. Rate of sulfate production was a function of the increase in temperature. In CBB-amended incubations at 32°C, sulfate reduction occurred immediately. Sulfate concentration decreased during the first 7 days and stayed at approximately 26 µmol L^{-1} until the end of the experiment (Fig. 1b). The rate of sulfate reduction was also temperature dependent, with relatively rapid loss of sulfate at 25°C and a marked lag in sulfate reduction at the other two temperatures (4°C and 15°C). At 4°C, sulfate concentration declined slowly until day 41, finally reaching a level similar to those seen in other experiments.

Orthophosphate concentration in pore-water of unamended sediments had an initial concentration of 48.8±4.6 µmol L^{-1} and increased only slightly, reaching peak concentrations of 50.6±6.5 µmol L^{-1} at 4°C, 56.8±5.8 µmol L^{-1} at 15°C, 59.3±8.5 µmol L^{-1} at 25°C, and 69.0±8.6 µmol L^{-1} at 32°C (Fig. 2a). In contrast, CBB amendment resulted in higher pore-water phosphate concentrations at all incubation temperatures. Also, higher incubation temperature was associated with higher

phosphate concentrations in CBB amended sediments (Fig. 2a). Maximum phosphate concentrations were 82.8±19.7 µmol L^{-1} at 4°C, 131.8±20.0 µmol L^{-1} at 15°C, 238.5±78.0 µmol L^{-1} at 25°C, and 371.0±28.4 µmol L^{-1} at 32°C. The effect of CBB amendment on phosphate release was significant at all temperatures ($P=0.024$ at 4°C, 0.022 at 15°C, 0.017 at 25°C, and <0.001 at 32°C).

Iron and sulfate reduction in sediments

The initial Fe(II) levels in unamended and CBB-amended sediments were 27.9±3.1 µmol g^{-1} and 27.1±2.4 µmol g^{-1}, respectively. Peak Fe(II) concentrations ranged from 34.1±1.6 µmol g^{-1} to 59.8±7.3 µmol g^{-1} in unamended sediments and from 40.3±5.2 µmol g^{-1} to 74.6±2.3 µmol g^{-1} in the CBB-amended sediments. Peak concentrations increased with incubation temperature (Table 1). The rate of iron reduction varied from 1.96±1.72 µmol cm^{-3} day^{-1} to 4.51±0.81 µmol cm^{-3} day^{-1} in the unamended sediments and from 5.93±4.00 µmol cm^{-3} day^{-1} to 18.31±2.29 µmol cm^{-3} day^{-1} in the CBB-amended sediments with increasing temperatures from 15 to 32°C. No obvious iron reduction occurred at 4°C in either system.

The CBB-amendment experiments were carried out at four temperatures. Thus, it was possible to calculate the temperature dependence of metabolic processes, which were expressed as Q10 values. Q10 values of iron reduction ranged from 1.19 to 3.54 in CBB-amended incubations based on paired comparisons (Table 2). Interestingly, the highest Q10 value in amended experiments was 3.54, with incubation temperature increasing from 25°C to 32°C, indicating that the rate of iron reduction with CBB amendment increased most at higher temperatures.

While sulfate reduction did not occur in unamended sediments, the rates of sulfate reduction calculated using pore-water sulfate concentrations in the CBB-amended experiments ranged from 0.017±0.001 µmol cm^{-3} day^{-1} at 4°C to 0.152±0.008 µmol cm^{-3} day^{-1} at 32°C (Table 1). Rate of sulfate reduction increased linearly with temperature, with Q10 values ranging from 1.37 to 1.44 in the CBB-amended incubation (Table 2). Q10 values for phosphorus release rates ranged from 1.19 to 1.61 in the CBB-amended incubation, with the highest Q10 value occuring as temperature increased from 25°C to 32°C.

AVS and CRS in sediments

The initial value of AVS in unamended and bloom-amended experiments was about 0.6 µg g^{-1} with initial CRS values of 41.5±0.8 and 41.8±1.1 µg g^{-1}, respectively (Table 3). For unamended sediments, AVS and CRS values did not change much over the entire incubation at all temperatures. However, at the end of the experiments, AVS and CRS values in the CBB-amended sediments had increased significantly. They also became higher with increasing incubation temperature, suggesting that the higher levels of reduced iron and sulfur may interact to form more FeS/FeS$_2$.

TOC and pH in sediments

TOC content of sediments was measured and is shown in Fig. 2b. In the unamended sediments, the initial average TOC was 9.7 mg g^{-1}. TOC levels in sediments did not change much during the incubation (Fig. 2b). For the CBB-amended sediments with an initial average level of TOC of 15.7 mg g^{-1} (Fig. 2b), TOC in sediments increased slightly to around 18.5 mg g^{-1} during the initial four days at all incubation temperatures. Thereafter, TOC levels decreased at 25°C and 32°C within several days, and then fluctuated around 12.5–14 mg g^{-1} until the end of the experi-

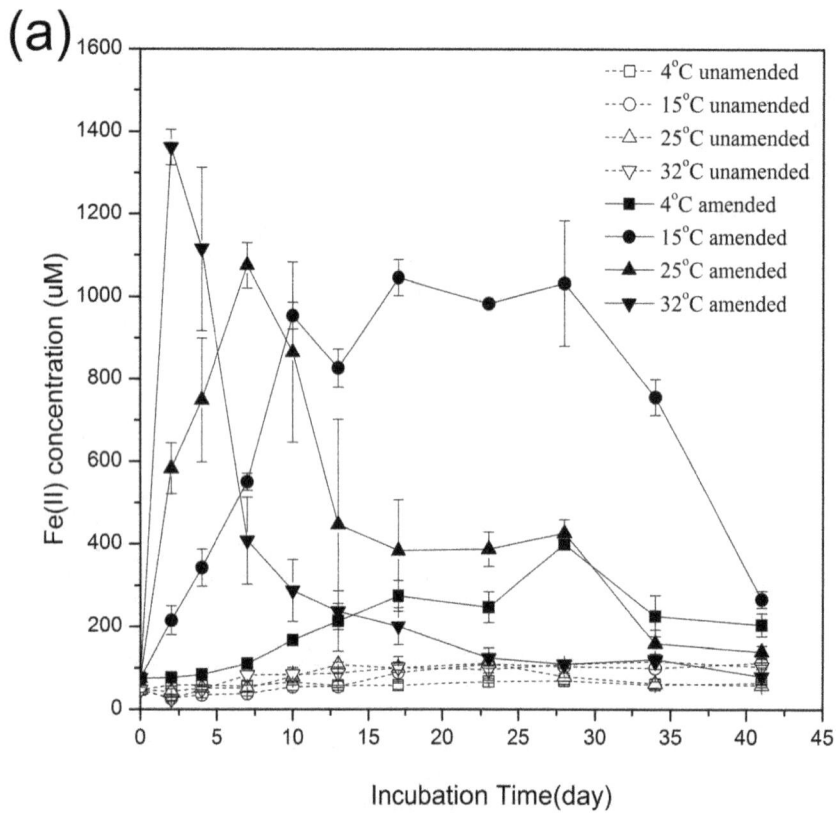

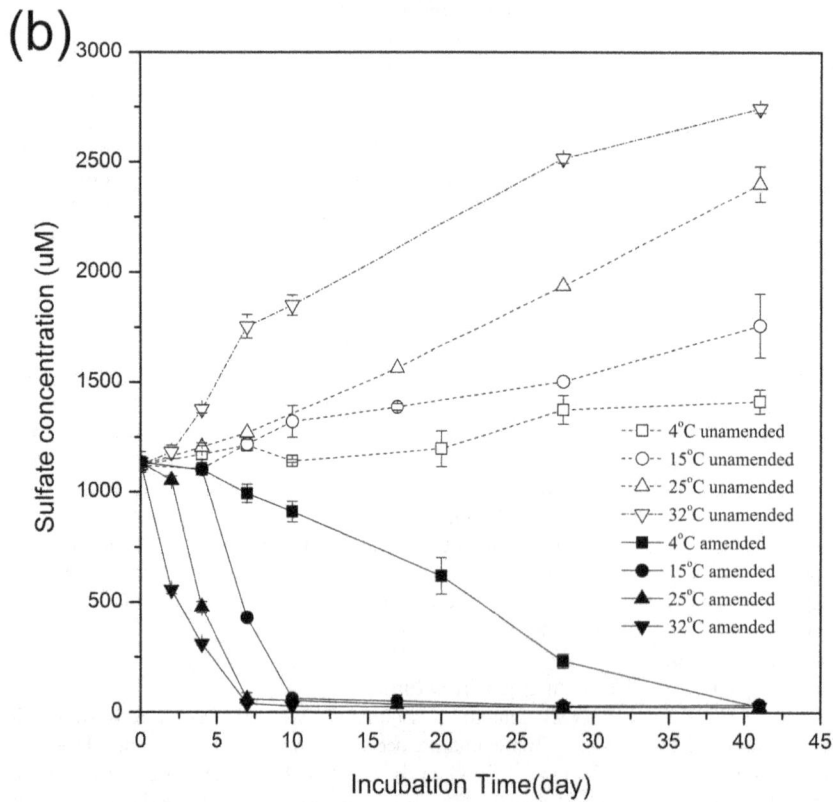

Figure 1. Fe(II) concentrations (a), and sulfate concentrations in sediment pore-water (b).

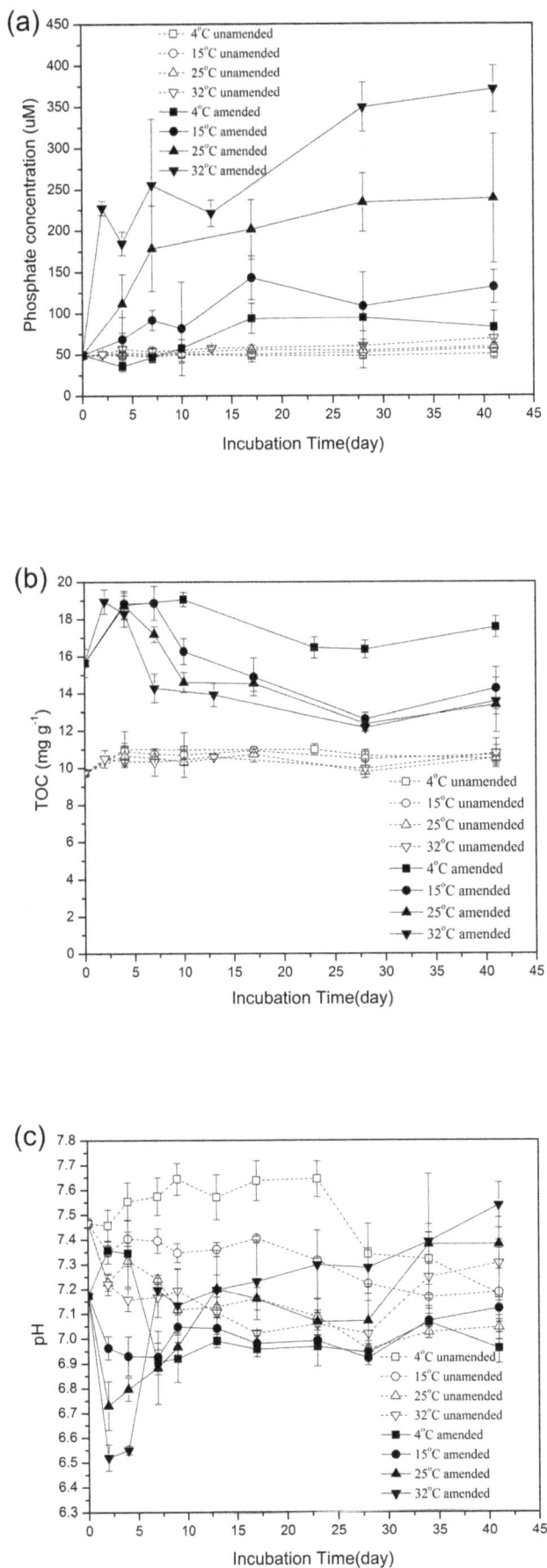

(a)

(b)

(c)

Figure 2. Phosphate concentrations in sediment pore-water(a), total organic carbon (TOC)(b) and pH in sediments(c).

ments. The increase in TOC was likely due to the procedure measuring relatively labile organic carbon more effectively than more refractory compounds [29]. The TOC increase suggests that added bloom biomass enhanced transformation of some of the refractory organic matter or that bloom biomass might contain some refractory compounds.

At 25°C and 32°C, pH values in the CBB-amended sediments first rapidly declined, possibly due to biomass fermentation in the initial of two days, and then increased (Fig. 2c). At the end of the experiments, only pH values in the CBB-amended sediments at 25°C and 32°C were higher than initial pH values in sediments, reaching 7.38 ± 0.02 and 7.54 ± 0.09, respectively (Fig. 2c). The decrease that initially occurred in all sediments was likely associated with production of fermentation products and the subsequent increase in pH may result from iron reduction and sulfate reduction, which can consume protons and fermentation products.

Phosphate fractionation

The initial TP in CBB-amended sediments was 880.7 ± 5.5 µg g^{-1}, only slightly higher than that in the unamended sediments (844.0 ± 2.7 µg g^{-1}) (Table 4). Labile-P levels were low (5–8 µg g^{-1}) and did not show much variation throughout the experiment. The decrease in Fe-P content was greater at higher temperatures in both amended and unamended experiments, and was also affected by the CBB-amendment. Final values of Fe-P decreased by as much as 13.3% in the unamended sediments and 54.5% in CBB-amended sediments during the incubation (Table 4). The decrease in Fe-P was associated with higher levels of iron reduction, with the greatest change in sediments with amendment of bloom biomass. Al-P levels decreased slightly in unamended sediments and high incubation temperatures led to greater reduction in Al-P. In contrast, Al-P levels in the CBB-amended sediments was observed to increase, and concentrations increased more with higher incubation temperatures.

Both Ca-P and organic-P concentrations increased during incubation of unamended and amended sediments with resulting increases of 12.2 to 48.4 µg g^{-1}. However, Ca-P levels were less sensitive to temperature and CBB amendment than other pools. In contrast, the organic-P content of bloom-amended sediments at the end of experiments was much higher than that in unamended experiments at all incubation temperatures. Temperature had a lesser effect on levels of organic-P, with increased levels only observed after the 32°C CBB unamended incubation (Table 4).

At the end of the experiments, Res-P contents of the sediments were less than the initial values (Table 4). Res-P content in unamended sediments did not vary much at the four incubation temperatures, and decreased by about 45 µg g^{-1} by the end of the experiments from an initial value of 305.2 ± 3.0 µg g^{-1}. For biomass-amended sediments, Res-P content decreased as a function of increase in temperature, ranging from 41.9 µg g^{-1} at 4°C to 128.9 µg g^{-1} at 32°C. Phosphate moved from the Fe-P and Res-P pools to the Al-P, Ca-P and organic-P pools in the amended sediments and from the Fe-P, Al-P and Res-P pools to the Ca-P and organic-P pools in the unamended sediments.

Based on our statistical analyses, CBB amendment had a significant influence on all P pools except Labile-P. Fe-P, Al-P, and Res-P levels were significantly different between temperature intervals ($P < 0.05$), indicating that increased temperatures play an important role in regulating P conversions following an influx of

Table 1. Iron reduction and sulfate reduction in sediments.

	15°C		25°C		32°C	
	Unamended	**CBB-amended**	**Unamended**	**CBB-amended**	**Unamended**	**CBB-amended**
Iron reduction rate (μmol cm^{-3} day^{-1})	$1.96\pm1.72^{\triangle}$	5.93 ± 4.00^b	3.00 ± 0.46^a	8.85 ± 2.48^c	4.51 ± 0.8^a	$18.31\pm2.29^{b,c}$
Maximum Fe(II) concentration (μmol cm^{-3})	65.88 ± 9.10 (17)¶	90.67 ± 15.88 (34)	68.77 ± 4.25 (10)	96.65 ± 4.72 (10)	86.32 ± 10.52 (10)	107.74 ± 3.37 (4)
Sulfate reduction rate (μmol cm^{-3} day^{-1})^d	ND*	0.085 ± 0.001	ND	0.122 ± 0.006	ND	0.152 ± 0.008

$^{\triangle}$Data are means $\pm$ standard deviation;
¶Parentheses indicate the time (day) of the maximum iron reduction rate;
*ND, not detected;
^{a}Significant difference for iron reduction at 25°C and 32°C in unamended sediments ($P<0.05$);
^{b}Significant difference for iron reduction at 15°C and 32°C in CBB-amended sediments ($P<0.05$);
^{c}Significant difference for iron reduction at 25°C and 32°C in CBB-amended sediments ($P<0.05$);
^{d}Sulfate reduction rates in amended sediments at any two respective temperatures show significant difference ($P<0.05$).

CBB. Organic P levels were unexpectedly not significantly influenced by temperature in CBB amended samples($P>0.05$). However, levels of error were high for these analyses. In unamended samples, only organic P and Res-P pools were significantly influenced by temperature ($P<0.05$), indicating organic P and Res-P pools are the most temperature-sensitive pools. Moreover, CBB not only enhanced P pool cycling and conversion but also had the most significant impact on concentrations in Fe-P and Al-P pools.

Discussion

In shallow lakes, the majority of metabolic activities, including most of the organic matter mineralization and nutrient cycling, occur in sediments [8]. Here we report that CBB addition strongly influences biogeochemical processes in sediments taken from the shallow Lake Taihu. Aside from the obvious increase in rate and extent of iron reduction, CBB addition to sediments reversed the sulfur transformation process from sulfur oxidation to sulfate reduction.

Results show that TOC levels in unamended sediments did not change much at any temperature (Fig. 2b), suggesting that organic

matter in sediments was somewhat resistant to microbial utilization. Under these conditions, the availability of electron acceptors allowed chemoautotrophic processes to proceed, specifically the anaerobic oxidation of reduced sulfur compounds [30], which resulted in increasing sulfate concentrations. With CBB addition, sulfate reduction occurred and sulfides accumulated. This enhancement of sulfate reduction then affected iron cycling in sediments, as sulfides reduce Fe(III) directly and precipitate Fe(II), resulting in the removal of soluble Fe(II) and the formation and burial of FeS and pyrite [31]. CBB amendment directly increased the Fe loading in sediments (Table 3). Typically, Fe(II) forms ferrous phosphate compounds such as vivianite ($Fe_3(PO_4)_2 \cdot 8H_2O$) [32], [33], however, when Fe(II) is removed from the system by precipitation, phosphate is released into pore water or the overlying waters. Coupled Fe/S cycles after CBB addition to sediments, directly affected the mobilization of phosphorus (Fig. 2a).

While Fe/S cycling, including dissolution of Fe(III) phases and precipitation of FeS/FeS_2, is regarded as a critical step in phosphorus release from estuarine and marine sediments [17], [34], standing pools of sulfate are typically in the low micromolar-range in lake sediments and are generally interpreted to be too

Table 2. Temperature effects on calculated Q10 of metabolic processes in freshwater systems.

Temperature interval	Iron reduction	Sulfate reduction	Phosphate release	Sample source	Reference	Trophic status
15–25°C	1.19	1.44	1.19	Lake Taihu	This study	Hyper-eutrophic
25–32°C	3.54	1.37	1.61	Lake Taihu	This study	Hyper-eutrophic
5–15°C		2.25		Lake Constance	[57]	Meso-eutrophic
6.3–20°C		1.07*		Lake Kizaki	[58]	Meso-eutrophic
20–30°C		1.61*		Lake Kizaki		
4–37°C		2.9		Lake Mendota	[59]	Eutrophic
15–25°C			1.03	Cootes Paradise Marsh	[60]	
10–20°C			1.21	Cootes Paradise Marsh		
12–19°C		1.9		Lake Arreskov Littoral sediment	[61]	Eutrophic
20–35°C		1.18		SP-1,Jamsil Dam area,Han River	[62]	

*Q10 values were calculated from the data in the reference.

Table 3. Acid volatile sulfide (AVS) and chromium reducible sulphur (CRS) content in sediments.

| | Initial values (µg g⁻¹) | | Final values at the end of experiments (µg g⁻¹) | | | | | | | |
| | | | 4°C | | 15°C | | 25°C | | 32°C | |
	AVS	CRS	AVS	CRS	AVS	CRS	AVS	CRS	AVS	CRS
Unamended	0.6±0.1△	41.5±0.8	1.0±0.2	41.1±3.9	0.9±0.2	43.9±3.0	0.7±0.0	40.2±1.8	1.0±0.0	41.7±3.0
CBB-amended	0.6±0.0	41.8±1.1	1.7±0.2^a,b,c	50.9*±4.7^A,B	10.9*±0.3^a,d	55.3±3.1^C	11.0*±0.3^b,e	65.7*±3.9^A,D	13.2*±0.3^c,d,e	72.1*±1.0^B,C,D

△Data are means ± standard deviation.
*Significant differences between initial values and final values at the respective temperatures ($P<0.05$).
a-eThe same letter represents a significant difference for AVS at two different temperatures ($P<0.05$).
A-DThe same letter represents a significant difference for CRS at two different temperatures ($P<0.05$).
doi:10.1371/journal.pone.0093130.t003

Table 4. Changes in concentrations in individual phosphorus pools during sediment incubations.

| | Initial value (µg g⁻¹) | | Final values at the end of experiments (µg g⁻¹) | | | | | | | |
| | | | 4°C | | 15°C | | 25°C | | 32°C | |
	Unamended	Amended	Unamended	Amended	Unamended	Amended	Unamended	Amended	Unamended	Amended
Labile-P	5.6±0.2△	5.5±0.3	7.4±2.0	6.1±0.3	6.1±0.6	4.1±1.9	7.0±0.3	5.7±0.8	7.6±0.4	7.5±4.5
Fe-P	173.3±4.6	200.0±3.4	178.8±6.1	187.8±17.2^A	174.1±17.2	132.5±17.1	166.7±9.2	130.9*±40.8	150.3*±7.3	91.0*±3.5^A
Al-P	148.3±1.4	155.6±6.9	142.7±0.9	152.0±1.3^B,C,D	136.7±0.0	208.3*±2.3^B,E	133.9±15.0	200.7*±22.6^C,F	120.5±0.8	269.0*±8.0^D,E,F
Ca-P	151.1±2.5	158.7±5.5	173.6±21.2	170.8±0.3	165.8±8.2	183.3*±2.8	170.3*±7.0	207.1±12.6	178.5*±3.5	203.1*±30.7
Org-P	60.5±1.2	71.3±9.9	93.2*±6.4^a	116.1*±18.7	98.2*±4.1^b	108.1*±18.6	97.6*±6.3^c	141.9*±25.9	115.0*±6.7^a,b,c	151.4±36.8
Res-P	305.2±3.0	289.7±6.7	248.6*±2.8^d,e	247.8*±6.5^G,H	257.1*±4.3	245.4*±0.4^I,J	269.6*±4.5^d	197.5±0.7^G,I,K	269.2*±6.1^e	160.8*±1.1^H,J,K
TP	844.0±2.7	880.7±5.5	844.4±25.4	880.7±31.6	838.0±25.8	881.7±32.9	845.0±14.8	883.7±31.5	841.1±15.6	882.7±3.9

△Data are means ± standard deviation.
*Significant differences between initial values and final values at the respective temperatures ($P<0.05$).
a-eThe same letter represents a significant difference at two different temperatures in unamended sediments ($P<0.05$).
A-KThe same letter represents a significant difference in CBB-amended sediments ($P<0.05$)

small for sustained levels of sulfate reduction. Therefore, previous work assumed that direct reduction of FeOOH in freshwater sediments was predominately responsible for phosphorus release [35]. Our work indicates that the role of sulfate reduction in the biogeochemistry of freshwater ecosystems has likely been underestimated [36].

Recently, human practices, including the combustion of coal and other fossil fuels and the application of agricultural fertilizer, have increased sulfate input into aquatic ecosystems [37]. Thus, the impact of sulfate input is increasingly being seen as an issue in the management of inland aquatic ecosystems. Cumulative evidence from low sulfate lakes supports the presence of a hidden sulfur cycle [38]. Currently, sulfate levels in Lake Taihu are much higher than those typically observed in freshwater lakes [39], reaching about 1 mM. The effect of this high level of sulfate and the role of the sulfur cycle has not been previously considered in eutrophic freshwater lakes like Lake Taihu. Previously, the roles of Fe, phosphorus, and nitrogen were emphasized [11], [40]. This study indicates that sulfate transformations must be considered, especially when a large organic input occurs into lake sediments. Sulfate reduction, increasing water temperature, settled CBB, and high sulfate concentrations all dramatically increase the release of P in sediments, which can allow development of algal blooms and other serious eutrophication problems.

Also, amended CBB had an impact on the fractionation of phosphorus in sediments, with increases in Al-P, Ca-P, and organic P pools and decreases in Fe-P and the Residual-P pools. The Fe-P pool was reduced as a result of Fe(III) reduction and subsequent dissolution of Fe compounds. The greatest increase was seen in the Al-P pool after CBB amendment. Previous work has reported that Fe(III) reduction caused a transformation of Fe-P to Al-P [25]. In this study, variation in Al-P levels in sediments might be partially due to variation in sediment pH. Decline in pH during the initial period was likely due to decomposition of algae, which could lead to the dissolution of Al compounds in sediments and the release of soluble Al species [41]. Al binds phosphate most effectively under weakly alkaline conditions [42]. Also, recently formed Al oxyhydroxides bind much more orthophosphate than their diagenetically altered forms [41]. Another pool that was increased during this study was Ca-P (Table 4). Other groups have observed the transformation of Fe-P to Ca-P [43]. In comparison to Fe-P, Ca-P and Al-P are not sensitive to redox changes. Thus, incremental increases in Al-P and Ca-P lead to an increase in the level of phosphorus burial and a lower risk of phosphorus mobilization from sediments [44].

However, transformation of other forms of phosphorus also needs to be considered when evaluating phosphorus burial. The organic-P pool was observed to increase in all CBB amended samples (Table 4). Although release of phosphate from organic P likely occurs during CBB decomposition, the addition of labile organic matter such as CBB can stimulate microbial growth and anaerobic metabolism in sediments [45]. Microorganisms can then incorporate dissolved inorganic phosphorus into the cellular constituents and synthesize organic-P compounds such as phosphonate monoesters and phospholipids [46]. Increased organic P burial due to enhanced burial of organic matter has been previously observed [47].

Under normal conditions, the majority of organic-P in surface sediments in Lake Taihu is stable [48]. But, following organic input, an increase in organic P in sediments can occur (Table 4), accompanied by a further increase as a result of the input of cyanbacterial biomass formed during eutrophication. This increase in organic P indicates that bioavailable organic matter may maintain the eutrophic status of lakes after external sources of P

have been controlled [49]. Others have also argued that in shallow lakes, an infusion of oxygen caused by sediment resuspension could result in rapid mineralization of organic P, further exacerbating eutrophication [50].

As with Fe-P, the Res-P pool decreased significantly in CBB amended sediments (Table 4). Although biotic or abiotic transformations of Res-P are possible and depend on a range of environmental factors, Res-P is generally regarded as nonreactive phosphorus [51]. The conversion of Res-P to more reactive forms of P as a result of the CBB amendment could also lead to phosphorous release and affect pathways of P flow. This effect on P regeneration and burial in lakes has not previously been investigated, and should be considered in restoration of lake ecosystems.

Many factors influence phosphorus pools in sediments, such as sediment resuspension, alum application, sediment microorganisms, E_h, and pH [25], [42], [52], [53]. In this study, we focused on the influence of CBB and temperature on phosphorus pools. Results indicated that phosphate moved from the Fe-P and Res-P pools to Al-P, Ca-P, and organic-P in the amended sediments and from Fe-P, Al-P, and Res-P pools to Ca-P and organic-P pools in the unamended sediments (Table 4, Figure S1).

Global surface temperature has increased by about 0.2°C per decade in the past 30 years [54], with major increases (0.55°C) occurring during that time. These increases will likely influence iron, sulfur, and phosphorus cycling in aquatic ecosystems [15], [55], [56]. Our Q10 values indicate that temperature increases have the greatest effect on iron reduction, with a more linear effect on phosphate release and sulfate reduction (Table 2). Climate change affects biogeochemical cycling in lakes and the results presented here indicate that iron reduction will likely increase in importance relative to sulfate reduction as temperatures warm. When Q10 data from this work is compared to other, mainly eutrophic freshwater systems (Table 2), similar results are observed, indicating that this data for phosphorous release and the effects of increasing temperatures could likely be extrapolated to other systems. The dramatic effect of temperature on iron reduction between 25 and 32°C indicates the significant effect that climate change could have on the system and the potential for further phosphate release as a result of warming. Along with their impact on individual processes, higher temperatures have important effects on the cycling of P between pools, resulting in increases in insoluble P, Al-P, Ca-P, and organic-P. Therefore, higher temperatures might change the pathways of P flow and lead to increased amounts of phosphorous release due to direct increases in rates of release processes as well as effects on the levels of CBB production. Ultimately, the result will likely be the maintenance of the eutrophic status of the system.

Conclusion

The CBB amendment in sediments not only enhanced iron reduction, but also reversed sulfur oxidation to sulfate reduction, which jointly led to the spike in Fe in the sediments. Coupled Fe/S cycling further led to the mobilization of phosphorus in sediments. Considering the critical role of sulfate reduction in phosphorus release from sediments, special attention should be paid to sulfate pollution in freshwater lakes. The amount of phosphorus released from sediments increased after amending CBB, especially at higher temperatures. CBB amendment at high temperatures also caused a shift from Fe-P and the Res-P pools to Al-P, Ca-P, and organic-P in the sediments. The decrease in inert Res-P in sediments led to an increase in levels of more reactive phosphorus. Therefore, settling of cyanobacterial bloom biomass strongly influences Fe/S biogeochemical processes, specifically resulting in the release of Fe the sediment and will likely increase the extent of eutrophication in aquatic environments at higher temperatures.

Acknowledgments

We thank Aleze Krumholz for editorial assistance.

Author Contributions

Conceived and designed the experiments: HLJ MC. Performed the experiments: MC TRY. Analyzed the data: HLJ MC LRK. Contributed reagents/materials/analysis tools: HLJ MC. Wrote the paper: HLJ MC LRK.

References

1. Carey CC, Ibelings BW, Hoffmann EP, Hamilton DP, Brookes JD (2012) Ecophysiological adaptations that favour freshwater cyanobacteria in a changing climate. Water Res 46: 1394–1407.
2. Michalak AM, Anderson EJ, Beletsky D, Boland S, Bosch NS, et al. (2013) Record-setting algal bloom in Lake Erie caused by agricultural and meteorological trends consistent with expected future conditions. Proc Natl Acad Sci USA 110: 6448–6452.
3. Turner AM, Chislock MF (2010) Blinded by the stink: nutrient enrichment impairs the perception of predation risk by freshwater snails. Ecol Appl 20: 2089–2095.
4. Karlson AM, Nascimento FJ, Elmgren R (2008) Incorporation and burial of carbon from settling cyanobacterial blooms by deposit-feeding macrofauna. Limnol Oceanogr 53: 2754–2758.
5. Li K, Liu Z, Gu B (2010) The fate of cyanobacterial blooms in vegetated and unvegetated sediments of a shallow eutrophic lake: a stable isotope tracer study. Water Res 44: 1591–1597.
6. Handley KM, VerBerkmoes NC, Steefel CI, Williams KH, Sharon I, et al. (2012) Biostimulation induces syntrophic interactions that impact C, S and N cycling in a sediment microbial community. ISME J 7: 800–816.
7. Shao K, Gao G, Chi K, Qin B, Tang X, et al. (2013) Decomposition of Microcystis blooms: Implications for the structure of the sediment bacterial community, as assessed by a mesocosm experiment in Lake Taihu, China. J Basic Microbiol 53: 549–554.
8. Pace ML, Prairie YT (2005) Respiration in lakes. In: Del Giorgio PA, le B. Williams PJ, editors. Respiration in aquatic ecosystems. New York: Oxford University Press Inc. pp. 103–121.
9. Cole J, Prairie Y, Caraco N, McDowell W, Tranvik L, et al. (2007) Plumbing the global carbon cycle: integrating inland waters into the terrestrial carbon budget. Ecosystems 10: 172–185.
10. Correll DL (1998) The role of phosphorus in the eutrophication of receiving waters: A review. J Environ Qual 27: 261–266.
11. Conley DJ, Paerl HW, Howarth RW, Boesch DF, Seitzinger SP, et al. (2009) Controlling eutrophication: nitrogen and phosphorus. Science 323: 1014–1015.
12. Reddy K, Kadlec R, Flaig E, Gale P (1999) Phosphorus retention in streams and wetlands: a review. Environ Sci Technol 29: 83–146.
13. Spears BM, Carvalho L, Perkins R, Kirika A, Paterson DM (2007) Sediment phosphorus cycling in a large shallow lake: spatio-temporal variation in phosphorus pools and release. Hydrobiologia 584: 37–48.
14. Genkai-Kato M, Carpenter SR (2005) Eutrophication due to phosphorus recycling in relation to lake morphometry, temperature, and macrophytes. Ecology 86: 210–219.
15. Jiang X, Jin X, Yao Y, Li L, Wu F (2008) Effects of biological activity, light, temperature and oxygen on phosphorus release processes at the sediment and water interface of Taihu Lake, China. Water Res 42: 2251–2259.
16. Wilhelm S, Adrian R (2008) Impact of summer warming on the thermal characteristics of a polymictic lake and consequences for oxygen, nutrients and phytoplankton. Freshwater Biol 53: 226–237.
17. Rozan TF, Taillefert M, Trouwborst RE, Glazer BT, Ma S, et al. (2002) Iron-sulfur-phosphorus cycling in the sediments of a shallow coastal bay: Implications for sediment nutrient release and benthic macroalgal blooms. Limnol Oceanogr 47: 1346–1354.
18. Chacon N, Silver WL, Dubinsky EA, Cusack DF (2006) Iron reduction and soil phosphorus solubilization in humid tropical forests soils: the roles of labile carbon pools and an electron shuttle compound. Biogeochemistry 78: 67–84.
19. Song N, Yan Z-S, Cai H-Y, Jiang H-L (2013) Effect of temperature on submerged macrophyte litter decomposition within sediments from a large shallow and subtropical freshwater lake. Hydrobiologia 714: 131–144.
20. Hu W, Jørgensen SE, Zhang F (2006) A vertical-compressed three-dimensional ecological model in Lake Taihu, China. Ecol Modell 190: 367–398.
21. Qin B, Zhu G, Gao G, Zhang Y, Li W, et al. (2010) A drinking water crisis in Lake Taihu, China: linkage to climatic variability and lake management. Environ Manage 45: 105–112.
22. Roden EE, Wetzel RG (1996) Organic carbon oxidation and suppression of methane production by microbial Fe (III) oxide reduction in vegetated and unvegetated freshwater wetland sediments. Limnol Oceanogr 41: 1733–1748.
23. Walkley A, Black IA (1934) An examination of the Degtjareff method for determining soil organic matter, and a proposed modification of the chromic acid titration method. Soil Sci 37: 29–38.
24. Murphy J, Riley J (1962) A modified single solution method for the determination of phosphate in natural waters. Anal Chim Acta 27: 31–36.
25. Rydin E, Welch EB (1998) Aluminum dose required to inactivate phosphate in lake sediments. Water Res 32: 2969–2976.
26. Lovley DR, Phillips EJ (1987) Rapid assay for microbially reducible ferric iron in aquatic sediments. Appl Environ Microbiol 53: 1536–1540.
27. Hsieh Y, Shieh Y (1997) Analysis of reduced inorganic sulfur by diffusion methods: improved apparatus and evaluation for sulfur isotopic studies. Chem Geol 137: 255–261.
28. Conant RT, Drijber RA, Haddix ML, Parton WJ, Paul EA, et al. (2008) Sensitivity of organic matter decomposition to warming varies with its quality. Global Change Biol 14: 868–877.
29. Strosser E (2010) Methods for determination of labile soil organic matter: an overview. Journal of Agrobiology 27: 49–60.
30. Jost G, Martens-Habbena W, Pollehne F, Schnetger B, Labrenz M (2010) Anaerobic sulfur oxidation in the absence of nitrate dominates microbial chemoautotrophy beneath the pelagic chemocline of the eastern Gotland Basin, Baltic Sea. FEMS Microbiol Ecol 71: 226–236.
31. Taillefert M, Bono A, Luther G (2000) Reactivity of freshly formed Fe (III) in synthetic solutions and (pore) waters: voltammetric evidence of an aging process. Environ Sci Technol 34: 2169–2177.
32. Gächter R, Müller B (2003) Why the phosphorus retention of lakes does not necessarily depend on the oxygen supply to their sediment surface. Limnol Oceanogr 48: 929–933.
33. Roden EE, Edmonds J (1997) Phosphate mobilization in iron-rich anaerobic sediments: microbial Fe (III) oxide reduction versus iron-sulfide formation. Arch Hydrobiol 139: 347–378.
34. Kraal P, Burton ED, Rose AL, Cheetham MD, Bush RT, et al. (2013) Decoupling between water column oxygenation and benthic phosphate dynamics in a shallow eutrophic estuary. Environ Sci Technol 47: 3114–3121.
35. Gunnars A, Blomqvist S (1997) Phosphate exchange across the sediment-water interface when shifting from anoxic to oxic conditions an experimental comparison of freshwater and brackish-marine systems. Biogeochemistry 37: 203–226.
36. Pester M, Knorr K-H, Friedrich MW, Wagner M, Loy A (2012) Sulfate-reducing microorganisms in wetlands-fameless actors in carbon cycling and climate change. Front Microbiol 3: 72.
37. Smith SJ, Pitcher H, Wigley T (2005) Future sulfur dioxide emissions. Climatic Change 73: 267–318.
38. Holmkvist L, Ferdelman TG, Jørgensen BB (2011) A cryptic sulfur cycle driven by iron in the methane zone of marine sediment (Aarhus Bay, Denmark). Geochim Cosmochim Acta 75: 3581–3599.
39. Holmer M, Storkholm P (2001) Sulphate reduction and sulphur cycling in lake sediments: a review. Freshwater Biol 46: 431–451.
40. North R, Guildford S, Smith R, Havens S, Twiss M (2007) Evidence for phosphorus, nitrogen, and iron colimitation of phytoplankton communities in Lake Erie. Limnol Oceanogr 52: 315–328.
41. McLaughlin JR, Ryden JC, Syers JK (1981) Sorption of inorganic phosphate by iron - and aluminum - containing components. J Soil Sci 32: 365–378.
42. Peng J-f, Wang B-z, Song Y-h, Yuan P, Liu Z (2007) Adsorption and release of phosphorus in the surface sediment of a wastewater stabilization pond. Ecol Eng 31: 92–97.
43. Van Cappellen P, Berner RA (1988) A mathematical model for the early diagenesis of phosphorus and fluorine in marine sediments: apatite precipitation. Am J Sci 288: 289–333.
44. Gonsiorczyk T, Casper P, Koschel R (1998) Phosphorus-binding forms in the sediment of an oligotrophic and an eutrophic hardwater lake of the Baltic Lake District (Germany). Water Sci Technol 37: 51–58.
45. Vahtera E, Conley DJ, Gustafsson BG, Kuosa H, Pitkänen H, et al. (2007)Internal ecosystem feedbacks enhance nitrogen-fixing cyanobacteria blooms and complicate management in the Baltic Sea. Ambio 36: 186–194.
46. Ternan NG, Mc Grath JW, Mc Mullan G, Quinn JP (1998) Review: organophosphonates: occurrence, synthesis and biodegradation by microorganisms. World J Microb Biot 14: 635–647.
47. Jilbert T, Slomp C, Gustafsson BG, Boer W (2011) Beyond the Fe-P-redox connection: preferential regeneration of phosphorus from organic matter as a key control on Baltic Sea nutrient cycles. Biogeosciences 8: 655–706.

48. Bai X, Ding S, Fan C, Liu T, Shi D, et al. (2009) Organic phosphorus species in surface sediments of a large, shallow, eutrophic lake, Lake Taihu, China. Environ Pollut 157: 2507–2513.

49. Zhu Y, Wu F, He Z, Guo J, Qu X, et al. (2013) Characterization of Organic Phosphorus in Lake Sediments by Sequential Fractionation and Enzymatic Hydrolysis. Environ Sci Technol 47: 7679–7687.

50. Tang X, Gao G, Qin B, Zhu L, Chao J, et al. (2009) Characterization of bacterial communities associated with organic aggregates in a large, shallow, eutrophic freshwater lake (Lake Taihu, China). Microb Ecol 58: 307–322.

51. Condron LM, Newman S (2011) Revisiting the fundamentals of phosphorus fractionation of sediments and soils. J Soil Sediment 11: 830–840.

52. Da-Peng L, Yong H (2010) Sedimentary phosphorus fractions and bioavailability as influenced by repeated sediment resuspension. Ecol Eng 36: 958–962.

53. Gächter R, Meyer JS (1993) The role of microorganisms in mobilization and fixation of phosphorus in sediments. Hydrobiologia 253: 103–121.

54. Hansen J, Sato M, Ruedy R, Lo K, Lea DW, et al. (2006) Global temperature change. Proc Natl Acad Sci USA 103: 14288–14293.

55. Bullock AL, Sutton-Grier AE, Megonigal JP (2013) Anaerobic Metabolism in Tidal Freshwater Wetlands: III. Temperature Regulation of Iron Cycling. Estuar Coast 36: 482–490.

56. Robador A, Brüchert V, Jørgensen BB (2009) The impact of temperature change on the activity and community composition of sulfate-reducing bacteria in arctic versus temperate marine sediments. EnvironMicrobiol 11: 1692–1703.

57. Bak F, Pfennig N (1991) Microbial sulfate reduction in littoral sediment of Lake Constance. FEMS Microbiol Lett 85: 31–42.

58. Li J-h, Takii S, Kotakemori R, Hayashi H (1996) Sulfate reduction in profundal sediments in Lake Kizaki, Japan. Hydrobiologia 333: 201–208.

59. Ingvorsen K, Zeikus J, Brock T (1981) Dynamics of bacterial sulfate reduction in a eutrophic lake. Appl Environ Microbiol 42: 1029–1036.

60. Kelton N, Chow-Fraser P (2005) A simplified assessment of factors controlling phosphorus loading from oxygenated sediments in a very shallow eutrophic lake. Lake Reserv Manage 21: 223–230.

61. Andersen FØ, Ring P (1999) Comparison of phosphorus release from littoral and profundal sediments in a shallow, eutrophic lake. Hydrobiologia 408: 175–183.

62. Kim L-H, Choi E, Gil K-I, Stenstrom MK (2004) Phosphorus release rates from sediments and pollutant characteristics in Han River, Seoul, Korea. Sci Total Environ 321: 115–125.

11

Linking Stoichiometric Homeostasis of Microorganisms with Soil Phosphorus Dynamics in Wetlands Subjected to Microcosm Warming

Hang Wang[1], HongYi Li[1], ZhiJian Zhang[1]*, Jeffrey D. Muehlbauer[2], Qiang He[3], XinHua Xu[1], ChunLei Yue[4], DaQian Jiang[5]

1 College of Natural Resource and Environmental Sciences, China Academy of West Development, Zhejiang University, Hangzhou, Zhejiang Province, China, 2 Curriculum for the Environment & Ecology, University of North Carolina at Chapel Hill, Chapel Hill, North Carolina, United States of America, 3 Department of Civil & Environmental Engineering, University of Tennessee, Knoxville, Tennessee, United States of America, 4 Institute of Ecology, Zhejiang Forestry Academy, Hangzhou, China, 5 Department of Earth and Environmental Engineering, Henry Krumb School of Mines, Columbia University, New York, New York, United States of America

Abstract

Soil biogeochemical processes and the ecological stability of wetland ecosystems under global warming scenarios have gained increasing attention worldwide. Changes in the capacity of microorganisms to maintain stoichiometric homeostasis, or relatively stable internal concentrations of elements, may serve as an indicator of alterations to soil biogeochemical processes and their associated ecological feedbacks. In this study, an outdoor computerized microcosm was set up to simulate a warmed (+5°C) climate scenario, using novel, minute-scale temperature manipulation technology. The principle of stoichiometric homeostasis was adopted to illustrate phosphorus (P) biogeochemical cycling coupled with carbon (C) dynamics within the soil-microorganism complex. We hypothesized that enhancing the flux of P from soil to water under warming scenarios is tightly coupled with a decrease in homeostatic regulation ability in wetland ecosystems. Results indicate that experimental warming impaired the ability of stoichiometric homeostasis (H) to regulate biogeochemical processes, enhancing the ecological role of wetland soil as an ecological source for both P and C. The potential P flux from soil to water ranged from 0.11 to 34.51 mg m^{-2} d^{-1} in the control and 0.07 to 61.26 mg m^{-2} d^{-1} in the warmed treatment. The synergistic function of C-P acquisition is an important mechanism underlying C:P stoichiometric balance for soil microorganisms under warming. For both treatment groups, strongly significant ($p<0.001$) relationships fitting a negative allometric power model with a fractional exponent were found between n-$H_{C:P}$ (the specialized homeostatic regulation ability as a ratio of soil highly labile organic carbon to dissolved reactive phosphorus in porewater) and potential P flux. Although many factors may affect soil P dynamics, the n-$H_{C:P}$ term fundamentally reflects the stoichiometric balance or interactions between the energy landscape (i.e., C) and flow of resources (e.g., N and P), and can be a useful ecological tool for assessing potential P flux in ecosystems.

Editor: Roeland M. H. Merks, Centrum Wiskunde & Informatica (CWI) & Netherlands Institute for Systems Biology, Netherlands

Funding: The authors wish to thank the National Natural Science Foundation of China (41373073) and Zhejiang Science and Technology Program (2011F20025) for providing the financial support for this project. The funders had no role in study design, data collection and analysis, decision to publish, or preparation of the manuscript.

Competing Interests: The authors have declared that no competing interests exist.

* E-mail: zhangzhijian@zju.edu.cn

Introduction

The principle of stoichiometric homeostasis states that organisms maintain relatively stable levels of biologically-relevant elements (e.g., carbon, nitrogen, and phosphorous) over time. This concept is based on ecological stoichiometry and is primarily applied to trophic interactions [1]. By modulating organism responses to key environmental drivers (e.g., nutrient fertilization), stoichiometric homeostasis is also a major mechanism responsible for the structure, function, and stability of ecosystems. Therefore, deviations from stoichiometric homeostasis can serve as indicators of environmental fluctuations that may be useful in ecological evaluation. The homeostatic regulation coefficient (H), which may be accessed through a regression model (equation 1, in the Materials and Methods) is a measure of homeostatic regulation ability and is commonly used as an indicator of these ecological changes [2].

The average global surface temperature has increased by 0.74°C since 1850 and is likely to increase an additional 1.1–6.4°C by the end of this century [3,4]. Temperature is one of the primary determinants affecting the metabolic rates of organisms, from cells to global-scale ecosystems, based on the metabolic theory of ecology [5,6]. Even a slight increase in temperature can alter energy flow and resource cycles in ecosystems and may affect environmental quality [3,7]. For instance, global soil respiration increased by 0.1 Pg yr^{-1} between 1989 and 2008 due to a high temperature anomaly [8]. Understanding the response of stoichiometric homeostasis to global warming and its associated ecological feedback to the biosphere is gaining an increasing degree of attention worldwide.

Studies of stoichiometric homeostasis have strongly emphasized physiological variation in elemental composition in macroscopic aquatic species [9]. To a lesser extent, some studies have also quantified the degree of stoichiometric homeostasis in terrestrial plants [2,10,11]. Stoichiometric homeostasis investigations in soil microorganisms subjected to environmental changes, however, are rare. Biogeochemical processes at the soil-water interface play a fundamental role in overall soil development, and they are the primary driving force behind key ecosystem functions such as community productivity and water quality [12,13]. Although stoichiometric homeostasis reflects the net outcome of many underlying physiological and biochemical adjustments as organisms respond to their surroundings [14], to our knowledge, few studies have linked any change in stoichiometric homeostasis to biogeochemical processes in a soil ecosystem affected by global warming.

Wetlands are one of the most productive and biologically diverse ecosystem types on the planet, where the chemical and, in particular, nutrient composition of shallow groundwater and surface water can be altered by a range of biogeochemical processes [15]. Many of the elements and nutrients that cycle through wetlands have ecosystem and global-scale effects; for example, phosphorous (P), is the key element affecting eutrophication in aquatic ecosystems [16], limiting coastal ecosystem processes [17], and in regulating allometry in soil food webs [18]. However, global warming may impact the inherent biogeochemical balance of P and other major elements. Diffusive models predict that elemental fluxes of P and nitrogen (N) to receiving waters are more than 2 times higher during summer (i.e., when temperatures are warmer) than in other seasons [19,20]. Further, an 18-month microcosm investigation using a minute and seasonal scale temperature manipulation verified that warming induced substantial mobilization of P from wetland sediment to water [21,22]. In other studies, excessive P loading under high temperature accelerated the risk of eutrophication in receiving waters [15,20,23] and altered carbon (C) assimilation and N accumulation in aquatic plants [24]. These studies suggest that warming or increasing temperature disturbs P equilibrium between sediment and water. However, the link between stoichiometric homeostasis and P dynamics in wetland soil subjected to global warming is poorly understood, which impedes the development of ecological regulatory strategies for coping with global warming scenarios. Additionally, wetlands play a vital role in the global C cycle and may respond strongly to climate change [25]. This is important because the flow of C energy and material also controls the biogeochemical cycling of many other elements in Earth's ecosystems [5,26,27,28]. Thus, integrating C:P stoichiometry is vital to understanding the connection between soil P dynamics and the regulation of stoichiometric homeostasis in wetland ecosystems.

The objective of our study was to shed new light on P dynamics in response to global warming, from the perspective of C:P stoichiometric homeostasis in six subtropical wetlands located in the Yangtze River delta in southeast China. We simulated global warming using a novel, in situ deployment of a microcosm device that mimics a warming scenario of 5°C above ambient temperature. The link between potential P flux from soil to water and the homeostatic regulation ability of soil microorganisms was also examined. We hypothesized that enhancing P flux from soil to water under warming scenarios is tightly coupled with a decrease in homeostatic regulation ability in wetland ecosystems.

Materials and Methods

This work is unrelated to any ethics issues. No specific permit was required for the described field study because the sampling locations were not located in protected areas or private land. The experimental field studies did not involve endangered or protected species.

Study sites

The study sites (120°41′31″E, 30°53′55″N to 120°33′32″E, 30°01′58″N) were located in the southern region of the Taihu Lake Basin and the NingShao Plain within the Yangtze River delta in southeast China. The climate in this area is subtropical monsoon with an annual average rainfall of 1350 mm and an annual average temperature of 26°C in summer and 4°C in winter.

The Supporting Information contains details about the six sites related to geographical position, hydrological parameters, and dominant vegetation during sampling (Table S1), as well as the physico-chemical properties of the six collected soils (Table S2). In brief, YaTang riverine wetland (YT) is in an advanced state of eutrophication and is classified as eutrophic, with the highest P and C total soil content among the six study sites. Soils sampled from XiaZhuhu wetland (XZ), XiXi National Wetland Park (XX), BaoYang riverine wetland (BY), and JinHu wetland (JH) are in a meso-eutrophic state, while the lowest soil C and P content among the sites was found in ShiQiu multipond wetland (SQ). Because these selected study sites represented the range of typical soil C and P conditions across subtropical wetland ecosystems in southeast China, we did not design any additional C or P treatments, which would have required artificial manipulations of the soil concentrations of these elements.

Microcosm configuration and sampling regime

A custom-built, novel microcosm (Fig. S1) simulating climate warming was developed under both present-day ambient temperature conditions (Control) and simulated warming conditions of 5°C above ambient temperature (Warmed). Details on the configuration of this microcosm system and its operation were either described previously [21] or can be found in Text S1. Compared to fixed-temperature laboratory incubations, this novel microcosm offers higher temporal resolution temperature control (on a minute scale), which simulates more realistic warming conditions.

Details on the establishment of wetland columns and water-soil samplings during the microcosm investigation are provided in Text S2. Briefly, each wetland column consisted of a prefabricated PVC pipe assembly (45.0 cm in height and 10.0 cm in internal diameter) that was designed to hold 20 cm of fresh soil and 20 cm of corresponding overlying water. Columns were installed in May 2008, and three replicates were placed inside each of the two incubation boxes. In this study, three samplings of water (overlying water and porewater) and 0–5 cm of topsoil were carried out in July and November 2010 and March 2011, which were used to illustrate the link between soil microorganism stoichiometric homeostasis and soil P dynamics in wetlands subjected to experimental warming.

Water and soil analysis

All water and soil samples were frozen at $-15°C$ prior to analysis. After thawing, water was filtered through a 0.45-μm filter. Phosphorous in filtered samples was measured using a continuous flow analyzer (Autoanalyzer III, Bran+Luebbe, Germany) with a spectrophotometer set at 880 nm (Murphy and Riley, 1962). This is the absorbance of dissolved reactive phosphorus (DRP), which represents the bioavailable P fraction for microorganisms and/or algae in surface water. Porewater samples for dissolved organic carbon (DOC) analysis were first acidified (10% HCl) and purged with inert gas to remove any inorganic carbon, then analyzed using a Shimadzu TOC 5000 analyzer (Shimadzu Scientific Instruments, Columbia, Maryland, USA).

After thawing, soil samples were analyzed for C and P in microbial biomass (Table 1) using the fumigation-extraction method described by Wu et al. [29] and Brookes et al. [30]. In brief, soil C and P in fumigated and non-fumigated soil samples was extracted using solutions of 0.5 mol L^{-1} K_2SO_4 and 0.5 mol L^{-1} $NaHCO_3$. The differences in extractable C and P between fumigated and non-fumigated soil were assumed to be released from lysed soil microbes, i.e., microbial biomass C and P (MB-C, MB-P). By utilizing the susceptibility of organic C to $KMnO_4$ oxidation [31], three fractions of labile organic C in these wetland columns; namely, highly labile organic C (HLOC), mid-labile organic C (MLOC) and labile organic C (LOC), were determined using 33 mmol L^{-1}, 167 mmol L^{-1}, and 333 mmol L^{-1} of $KMnO_4$, respectively.

Calculation of homeostatic regulation ability of soil microorganisms

The homeostatic regulation ability of soil microorganisms was estimated by calculating the homeostatic regulation coefficient (H) according to the following equation [1]:

$$y = c \cdot x^{(1/H)} \qquad (1)$$

where y is the microbial C or P concentration (mg g^{-1} or mg kg^{-1} of dry soil weight) or the C:P molar ratio of soil microorganisms; x is the organic C concentration (including HLOC, MLOC, and LOC) in soil, DRP or DOC concentration in porewater, or the C:P molar ratio among the soil organic C and porewater DRP; and c is a constant. H is defined as the homeostatic regulation ability of soil microorganisms; as the value of H increases, the ecosystem trends toward a higher degree of ecological stability and resilience. For example, higher values of H_N and H_P in grasses receiving field N and P fertilizations were significantly associated with higher species dominance and stability [2]. According to ecological allometry [5,32], H may be classified as an equilibrium constant with respect to functions of organism metabolic and stoichiometric processes.

The first derivative of regression equation (1), reflecting the response rate of y to x, is given as follows:

$$\frac{dy}{dx} = c \cdot (1/H) \cdot \left(x^{(1/H-1)} \right) = \left(c \cdot x^{1/H} \right) \cdot (1/x) \cdot (1/H) \qquad (2)$$

By substituting in variables from equation (1), equation (2) can be transformed as:

$$\frac{dy}{dx} = y \cdot (1/x) \cdot (1/H) = (y/x) \cdot (1/H) \qquad (3)$$

Here, the value of (y/x) is defined as the specific variable for $1/H$, which may be transformed into (x/y) responding to H. Therefore, we define n-H_i as the specialized homeostatic regulation ability for each of the tested wetland soil columns:

$$n - H_i = (x_i/y_i) \cdot H_i \qquad (4)$$

where x_i is the measured value of the organic C concentration (including HLOC, MLOC, and LOC) in soil, DRP or DOC concentration in porewater, or the C:P molar ratio among the soil organic C and porewater DRP for each soil sample; y_i is the microbial C or P concentration (mg g^{-1} or mg kg^{-1} of dry soil weight) or the C:P molar ratio of soil microorganisms for each soil sample; i is the type of parameter (form of HLOC, MLOC, LOC, DRP, DOC, or ratio of C:P for soil or soil microbes); and H refers to the homeostatic regulation coefficient calculated by equation (1) according to the respective for each of the two treatments (Control and Warmed) in terms of the i parameter.

Potential P flux (F_P, mg m^{-2} d^{-1}) was used to evaluate the potential P transfer out of porewater under experimental warming according to the following equation:

$$F_p = (C_{in} - C_{out}) \times V/(S \times T) \qquad (5)$$

where $C_{in}C_{in}$ is the DRP concentration in the porewater (mg L^{-1}), C_{out} is the P concentration in the overlying water (mg L^{-1}), V is the volume of overlying water (mL), S is the area of the wetland column (m^2) and T is the interval between two sampling dates (d). Due to the relatively static hydrological conditions in the wetlands (Table S1), DRP was the most important form for P exchange between soil and water in the study. Soil microorganisms are also more sensitive to DRP than to TP due to its bioavailability. Therefore, values of F_P (and n-H_i) were evaluated by the form of DRP rather than TP. The potential P flux defined here provides a good indicator for assessing P concentration gradients between overlying water and porewater, with high values suggesting a high risk of P transfer from soil to overlying water.

In this study, aquatic macrophytes (e.g., *Phragmites communis*) were temporarily excluded from soil columns, but small benthic organisms (e.g., *Margarya melanoide*) and plankton (e.g., *Spirogyra*) were preserved. Data collected from water and soil samples were expressed as the mean plus standard error. Paired Student's t-tests were used to compare the effects of warming on soil microorganism and C fractions, P concentrations in porewater and overlying water, and potential P flux. Two-way analysis of variance (repeated ANOVA) tests were carried out using SPSS software (version 15.0) to examine the effects of experimental warming, sampling time, and their interaction on soil biochemical traits and potential P flux. Statistical test results were considered significant at the $p < 0.05$ level.

Results

Soil carbon and phosphorus potential flux in response to experimental warming

After >2.5 y of incubation (in Nov 2010), experimental warming increased soil microbial biomass, soil labile C, and porewater DOC (Table 1); however, warming decreased concentrations of total soil C at all wetlands except site SQ. Meanwhile, although fluctuations in DRP concentrations were found in the overlying water, DRP in porewater increased in the warmed treatment by 52% to 137% compared to the ambient treatment, except in the SQ wetland (Table 2). On average, potential P flux in the six wetlands ranged from 0.11 to 34.51 mg m^{-2} d^{-1} in the control and 0.07 to 61.26 mg m^{-2} d^{-1} in the warmed treatment (Table 2). Two-way analysis of variance indicated that experimental warming significantly increased potential P flux, except in the SQ wetland. There were also significant differences in soil C pools and potential P flux between Warmed and Control treatments (Table 3). A significant interaction between treatment and season (sampling date) was found for soil C pools and potential

Table 1. Concentrations of soil microbial biomass (MB; MB-C: microbial carbon; MB-P: microbial phosphorus), soil carbon (TOC: total organic carbon; HLOC: highly labile organic carbon; MLOC: mid-labile organic carbon; and LOC: labile organic carbon), and carbon in porewater (DOC: dissolved organic carbon) in wetland columns in the microcosm experiment (Control: ambient temperature; Warmed: ambient temperature +5°C).

Treatment		Soil microbial biomass				Soil carbon				DOC in porewater (mg kg⁻¹)
		Total mass (mg g⁻¹)	MB-C (mg g⁻¹)	MB-P (mg kg⁻¹)	Microbial C:P stoichiometric ratio	TOC (g kg⁻¹)	HLOC (mg g⁻¹)	MLOC (mg g⁻¹)	LOC (mg g⁻¹)	
JH	Control	0.402±0.023	0.386±0.022	15.0±2.3	68.6:1	25.6±1.4	2.28±0.09	2.75±0.66	3.75±0.71	13.2±2.3
	Warmed	0.493±0.010	0.477±0.008	16.1±2.1	79.0:1	22.3±2.1	2.31±0.37	3.16±0.70	5.48±1.31	16.6±3.1
p		0.064	**0.050**	0.568		**0.028**	0.899	0.498	0.113	0.182
XZ	Control	1.12±0.02	1.08±0.02	25.9±2.4	111:1	54.8±5.9	5.78±0.31	8.55±0.29	9.93±0.43	17.3±0.9
	Warmed	1.47±0.03	1.42±0.04	33.4±2.3	113:1	49.2±2.6	6.35±0.06	8.67±0.13	9.93±0.32	21.6±2.2
p		**<0.001**	**0.009**	**0.005**		0.054	0.055	0.575	0.990	**0.038**
YT	Control	1.42±0.01	1.37±0.01	36.4±2.9	100:1	133±10	9.13±0.23	13.0±1.0	18.9±4.2	20.2±3.3
	Warmed	1.73±0.06	1.66±0.04	46.1±5.7	96.0:1	117±3	10.9±0.1	18.6±0.9	20.5±4.3	22.7±9.9
p		**0.001**	**0.035**	**0.045**		**0.024**	**<0.001**	**0.002**	0.681	0.701
XX	Control	1.01±0.03	0.964±0.024	33.4±4.4	77.0:1	35.6±1.7	2.14±0.07	3.12±0.23	3.70±0.43	16.2±3.0
	Warmed	1.31±0.08	1.25±0.11	42.0±2.3	79.4:1	31.8±1.3	3.39±0.06	3.65±0.19	4.32±0.83	24.3±8.7
p		**0.003**	0.067	**0.028**		**0.011**	**<0.001**	**0.026**	**0.039**	0.208
BY	Control	1.07±0.06	1.04±0.06	34.5±2.4	80.4:1	37.6±1.9	2.90±0.12	3.98±0.05	4.50±0.09	18.6±1.5
	Warmed	1.13±0.07	1.10±0.06	36.6±2.6	80.1:1	32.4±2.1	3.59±0.36	5.91±0.16	7.06±0.33	21.8±1.0
p		0.299	0.317	0.435		**0.043**	**0.033**	**<0.001**	**<0.001**	**0.027**
SQ	Control	0.183±0.018	0.179±0.017	4.25±0.87	112:1	14.2±2.3	0.556±0.025	0.889±0.673	1.73±0.33	8.14±0.96
	Warmed	0.469±0.011	0.456±0.013	13.5±2.7	90.1:1	15.3±1.7	2.61±0.05	3.74±0.15	4.23±0.16	12.3±1.2
p		**<0.001**	**<0.000**	**0.005**		0.443	**<0.001**	**0.002**	**<0.001**	**0.006**

p-values refer to significant differences between two treatments. Sampling date is November 2010.

Table 2. Phosphorus (P) concentrations in overlying water and porewater, and the potential flux of dissolved reactive P in wetland columns in the microcosm experiment (Control: ambient temperature; Warmed: ambient temperature +5°C).

Wetland column	Treatments	Overlying water (mg L^{-1})	Porewater (mg L^{-1})	Potential P flux (mg m^{-2} d^{-1})
	July 2010			
JH	Control	0.066±0.038	0.063±0.005	0.21±0.07
	Warmed	0.017±0.009	0.095±0.001	0.32±0.09
p		**0.022**	0.004	**0.018**
XZ	Control	0.070±0.026	0.266±0.003	0.90±0.03
	Warmed	0.024±0.024	0.612±0.063	2.50±0.38
p		**0.012**	**0.039**	**0.003**
YT	Control	1.507±1.22	2.334±0.703	15.30±3.56
	Warmed	0.909±0.51	4.537±1.510	35.26±4.61
p		0.304	**0.002**	**<0.000**
XX	Control	0.106±0.076	0.330±0.015	2.14±0.17
	Warmed	0.038±0.031	0.570±0.033	3.65±0.10
p		0.091	**0.023**	**0.004**
BY	Control	0.063±0.061	0.104±0.001	0.61±0.02
	Warmed	0.051±0.079	0.231±0.001	0.98±0.24
p		0.783	**<0.000**	**0.017**
SQ	Control	0.016±0.008	0.088±0.015	0.11±0.02
	Warmed	0.012±0.010	0.021±0.009	0.08±0.01
p		0.431	0.025	0.412
	March 2011			
JH	Control	0.018±0.000	0.182±0.102	0.63±0.37
	Warmed	0.016±0.000	0.354±0.048	1.07±0.23
p		0.467	**0.016**	**0.004**
XZ	Control	0.029±0.014	0.720±0.286	2.10±0.13
	Warmed	0.018±0.000	1.704±0.189	4.36±0.78
p		0.187	**0.013**	**0.006**
YT	Control	0.042±0.008	5.83±1.23	34.51±10.36
	Warmed	0.047±0.012	8.34±0.41	61.26±6.12
p		0.572	**0.011**	**<0.000**
XX	Control	0.016±0.006	1.17±0.18	3.38±0.71
	Warmed	0.013±0.004	2.53±0.34	11.72±2.10
p		0.320	**0.004**	**<0.000**
BY	Control	0.016±0.000	0.390±0.067	1.06±0.12
	Warmed	0.022±0.000	0.653±0.182	2.38±0.53
p		**0.012**	**0.039**	**0.034**
SQ	Control	0.012±0.001	0.068±0.019	0.16±0.09
	Warmed	0.016±0.000	0.044±0.040	0.07±0.03
p		0.490	0.193	0.387

p-values refer to significant differences between two treatments.

P flux as well, although there were no significant differences in these variables by season alone.

Microorganism stoichiometric homeostasis in response to experimental warming

The relationships between soil microorganism biomass and three forms of soil carbon concentrations (HLOC, MLOC, and LOC) sampled in Mar. 2011 were rigorously and significantly described by the stoichiometric homeostasis model (Fig. 1,

$p<0.01$). The values of H in the warmed treatment were lower than those in the control, with H values of 1.22 and 1.73 for HLOC, 1.48 and 1.75 for MLOC, and 1.09 and 2.20 for LOC in warmed and control treatments, respectively (Fig. 1). Results from the other two samplings (data not shown) were fundamentally similar to the Mar. 2011 sampling. There was no consistent pattern in the homeostasis model fit between DOC and microorganism C among the three samplings (Table 4). However, strongly significant and positive correlations between DRP in porewater and soil microorganism biomass P fit a power model for

Table 3. Results of two-way analysis of variance (repeated ANOVA) showing the p-values for soil microorganisms and soil carbon and potential phosphorus (P) flux under experimental warming and sampling time. MB-C: microbial carbon, MB-P: microbial phosphorus, TOC: total organic carbon, HLOC: highly labile organic carbon, MLOC: mid-labile organic carbon, and LOC: labile organic carbon.

Factor	Soil microorganisms			Soil carbon				Potential P flux (mg m^{-2} d^{-1})
	Total (mg g^{-1})	MB-C (mg g^{-1})	MB-P (mg kg^{-1})	TOC (g kg^{-1})	HLOC (mg kg^{-1})	MLOC (mg kg^{-1})	LOC (mg kg^{-1})	
Treatment	**0.007**	**0.005**	**0.008**	**0.013**	**0.011**	**0.038**	**0.007**	**0.033**
Time	0.285	0.321	0.308	0.079	0.563	0.439	0.407	0.185
Treatment×Time	**0.017**	**0.016**	**0.009**	**0.006**	**0.019**	0.076	**0.016**	**0.047**

p-values smaller than 0.05 are in bold to indicate statistical significance.

these three consecutive samplings (Fig. 2), with lower values of H under the warmed treatment (1.79 to 2.49) compared to the control (3.08 to 4.24). Although no significant difference between these two treatments was found for the C:P stoichiometric ratios of soil microorganisms (paired t-test $p = 0.607$, Table 1), strongly significant correlations existed between soil C:P ratios of HLOC to DRP and soil microorganism C:P for these three samplings under the two treatments (Fig. 3). More importantly, the C:P ratio-related H values (i.e., $H_{C:P}$) obtained from the warmed group were consistently lower than those from the control group.

No significant relationship was found between porewater ratios of DOC:DRP and soil microbial C:P (Table 4). Relationships were observed for other comparisons of soil LOC to porewater DRP and soil MLOC to porewater DRP vs soil microorganism C:P ratios (Table 4); however, among the three samplings, corresponding $H_{C:P}$ values showed no consistent trends between the two treatment groups. This is in contrast to soil HLOC to porewater DRP results, in which a consistent increase in $H_{C:P}$ was observed from control to warmed treatments across all three samplings (Fig. 3). Specific n-$H_{C:P}$ values for each wetland column and soil treatment (Warmed or Control) were subsequently calculated using the $H_{C:P}$ values derived from these HLOC:DRP regressions, using C:P ratios (HLOC:DRP) in porewater as the 'x_i' variable and soil microbial C:P ratios as the 'y_i' variable in equation (4). Again, strongly significant ($p < 0.001$) linkages that fit a power model between n-$H_{C:P}$ and potential P flux (i.e., allometric relationships [5]) were found for both treatment groups (Fig. 4). Notably, these two fitted curves strongly overlapped, and the regression equation for the two datasets combined ($y = 5.261 \times^{-1.7509}$, $R^2 = 0.5893$) was almost identical to the equations derived from each dataset individually. The data for potential P flux and n-$H_{C:P}$ values were concentrated in ranges less than 20 mg m^{-2} d^{-1} and 3.0, respectively (Fig. 4).

Discussion

Does homeostasis relate to the ecological role of soil in C and P cycling?

The tight linkages in this study between different pools of soil labile C and soil microbial C (Fig. 1), and porewater DRP and soil microbial P (Fig. 2) fit the stoichiometric homeostasis model. This supports the view that stoichiometric homeostasis is an important mechanism underpinning the dynamics of C and P resource availability at the soil-microbe interphase in wetland ecosystems. It confirms that the composition of C and P in microbial assemblages is closely related to resource availability in ecosystems [33,34], as

soil microorganism C and P increases were related to the increases in soil C and P availability observed in the warmed treatment in this study (Table 1 & 2). Based on the principle of homeostatic regulation ability [1] and its prior application to a grassland ecosystem [2], decreased H values for soil labile C and DRP in warmed columns indicate that experimental warming impaired the ability of soil organisms to regulate C and P biogeochemical turnover in the studied wetland ecosystems. This loss of homoeostatic regulation ability is consistent with the observed decrease in soil TOC (Table 1) and the increase in potential P flux (Table 2). Comparison of the data for the decrease in soil TOC and the increase in both soil labile C and microorganism C (Table 1) further indicates that losses of total organic C in warmed soils were mainly derived from non-active C pools; i.e., the recalcitrant organic C. Recalcitrant organic fractions generally represent a large proportion of the total C pool and have relatively slow turnover rates [35,36], and generation of recalcitrant organic matter is regarded as a vital process for C sequestration [36]. Excluding P that is assimilated into microbial biomass (Table 1), warming subjects the remaining P pool to increases in enzymatic biodegradation of organic phosphorus [21], P desorption [20,37], and inorganic P solubilization [38], which could enhance the strength of potential P flux (Table 2) as well as increase DRP in porewater. The ecological role of wetland soil as a source for C and P, coupled with lower H values observed in the warming treatment compared to the control, suggests that H may yield valuable insight into the ecological role of soil in C and P biogeochemical turnover and its response to global warming.

Does homeostasis relate to stoichiometric balance?

Although experimental warming did affect the soil microbial C:P stoichiometric ratios for particular samples to some degree (Table 1), no overall significant difference was found between the two (Warmed and Control) groups. However, C:P stoichiometric ratios in the soil microbial community did seem to vary according to geographical (biotic and abiotic) features [39] across the six wetlands (Table 1 and Fig. 3). Because macronutrients are coupled with various biochemical and cellular constituents, the ratios of these elements remained relatively constant compared to the potentially decoupled resource stoichiometry changes in the surrounding environment that were induced by warming [1,39].

In contrast, tight and significant associations between ratios of soil C:P and microorganism C:P were obtained with the fitted curves of the power model (Fig. 3), which seems contradictory to the results described above. This may be explained by soil microorganisms regulating the equilibrium between the elemental

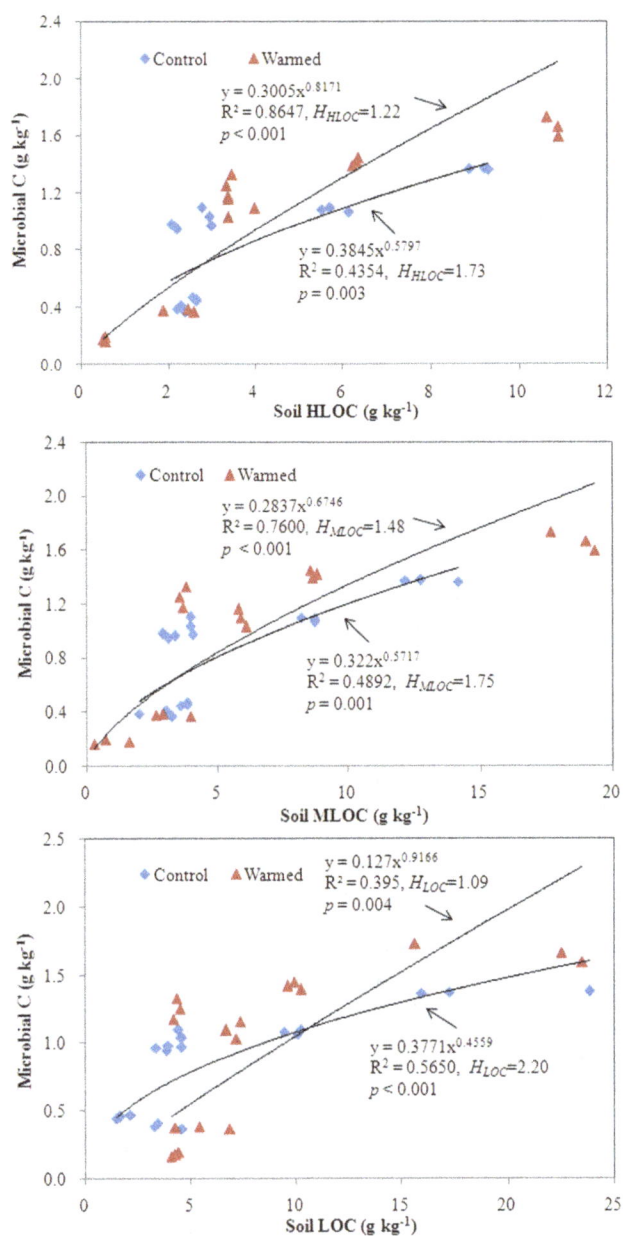

Figure 1. Relationships between soil microbial carbon (MB-C) and soil organic carbon (in forms of HLOC: highly labile organic carbon, MLOC: mid-labile organic carbon, and LOC: labile organic carbon). Control and Warmed represent treatments of ambient temperature and ambient temperature +5°C, respectively. Sampling date is March 2011. Coefficients fitting the data to the equation for stoichiometric homeostasis are given along with an estimate of H_C.

composition of their biomass and the efficiencies of microbial nutrient assimilation and growth through ecoenzymatic interactions [26]. Further, simultaneous N and P enrichment (by photosynthetic biota) produces strongly positive, synergistic responses in ecosystems [16]. This synergistic effect can aid in maintaining "Redfield-like" ratios in soil microorganisms [39], and may also have served to regulate the stoichiometric balance of soil microbial C and P acquisition for the warmed wetland soils in this study. This is also in line with general ecological theory on the elemental stoichiometries of microbial biomass and environmental

availability relative to microbial nutrient assimilation and growth [1,32]. For example, increases in N concentration in biomass are usually accompanied by increases in P uptake for vascular plants [11]. This suggests that the synergistic function of C-P acquisition is an important mechanism underlying the C:P stoichiometric balance for soil microorganisms in wetland ecosystems subjected to warming.

Field N and P fertilization in another study showed that higher values of $H_{N:P}$ in grasses were significantly associated with higher species dominance and stability [2]. In this study, the homeostatic regulation coefficient for C:P stoichiometry in the soil microorganism community vs soil HLOC:DRP ratios (i.e., $H_{C:P}$) was lower under the warmed treatment than the ambient treatment (Fig. 3). In ecoenzymatic stoichiometry theory [26,32], the relative availability of hydrolyzed P via ecoenzymes (such as phosphatase) in soils relative to C in the microbial community composition was higher under experimental warming than in the control treatments. Ecosystem function and stability may have been impacted in the warmed soil columns in this study, with a negative allometric relationship to soil microorganism $H_{C:P}$. Our previous research has shown that a significant increase in soil microorganism biomass measured by total phospholipid fatty acids (PLFAs) occurred under warming [21], which is line with the data in Table 1; however, bacterial abundance in the soil tended to decrease along with an increase in the ratio of fungi:bacteria [21,40,41]. Furthermore, relatively high abundances of fungi in soil ecosystems increase the secretion of P-hydrolyzing enzymes in soils [21,42,43]. Meanwhile, functional gene array data verified that soil P utilization genes (such as polyphosphate kinase and exopolyphosphate) were also enhanced by warming [27], leading to increases in soil microbial P pools (Table 1) and speeding up nutrient cycling processes.

In a 13 d laboratory incubation test of YT wetland sediment (Text S3), lower dissolved oxygen (DO) concentrations in overlying water (Fig. S2) and higher ferrous iron (Fe^{2+}) concentrations in sediment (Fig. S3) were found for sediment sampled from the warmed column compared to the control treatment. This indicates that the redox-enhanced sediment under the warmed treatment could increase dissimilatory reduction of P-bound metal oxides and the decomposition of organic matter, liberating P from sediment aggregates. This may explain the relatively high levels of P in porewater and potential P flux under warming (Table 2). In addition to stimulating net primary production, warming could significantly increase both soil respiration and the associated C-degrading microbial genes [27,44], increasing the C flux from terrestrial ecosystems to the atmosphere in the global C cycle [8].

Results from our previous investigation [21], coupled with the data reported here, lend further support to the use of $H_{C:P}$, H_C, and H_P as ecological 'indicators' (Fig. 3, 1, & 2, respectively). In other words, $H_{C:P}$ values in the soil microorganism community in this study were positively correlated with C-P-related ecological function and stability. Because C:P stoichiometric homeostasis is associated with synergistic C-P acquisitions (as discussed above), a decreased $H_{C:P}$ value under warming conditions (Fig. 4) means that C biodegradation and P mineralization exceeded microorganism utilization of these elements, leading to increases in both C loss and P export. This has a "double-negative" effect on climate change and water quality.

We did not find a consistent relationship between $H_{C:P}$ values in warmed vs. control treatments in terms of LOC:DRP and MLOC:DRP ratios (Table 4). Soil C losses in this study were traced to the recalcitrant fraction, while pools of labile C were not depleted (Table 1). Carbon availability from HLOC for the soil microorganism assemblage and metabolism is generally higher

Table 4. Homeostatic regulation coefficient (*H*) in soils collected from wetland columns in the microcosm experiment (Control: ambient temperature; Warmed: ambient temperature +5°C).

Variable	Dependent variable	Sampling data	Control group			Warmed group		
			H	R^2	*p*	*H*	R^2	*p*
Porewater DOC	Microbial biomass C	Jul 2010	−5.00	0.084	0.203	4.26	0.085	0.219
		Nov 2010	2.76	0.274	**0.004**	3.00	0.257	**0.008**
		Mar 2011	1.71	0.386	**0.000**	−20.0	0.019	0.452
Porewater DOC:DRP	Microbial C:P	Jul 2010	500	0.014	0.672	100	0.005	0.953
		Nov 2010	250	0.000	–	71.4	0.002	0.976
		Mar 2011	55.6	0.007	0.872	43.4	0.016	0.553
Soil LOC: porewater DRP	Microbial C:P	Jul 2010	6.76	0.303	**0.004**	9.43	0.463	**0.000**
		Nov 2010	12.9	0.275	**0.003**	13.7	0.237	**0.010**
		Mar 2011	12.6	0.154	**0.013**	7.46	0.545	**0.000**
Soil MLOC: porewater DRP	Microbial C:P	Jul 2010	8.26	0.340	**0.003**	6.90	0.391	**0.000**
		Nov 2010	16.1	0.268	**0.008**	22.2	0.071	0.237
		Mar 2011	12.8	0.213	**0.013**	5.18	0.461	**0.000**

p-values refer to significant differences between two treatments. DOC and DRP refer to dissolved organic carbon and dissolved reactive phosphorus in porewater while MLOC and LOC refer to mid-labile organic carbon and labile organic carbon in soil, respectively.

than for the rest of the C sources [31]; therefore the index of HLOC is relatively sensitive to C dynamics for microorganism acquisition. Further mechanistic investigations are needed to establish direct links between the destabilization of soil organics and soil microorganism community dynamics in order to better assess the relationship between the availability of different forms of C to C:P stoichiometric dynamics and their ultimate effects on $H_{C:P}$.

Does homeostasis link to soil phosphorus flux?

Efforts to predict P flux or P dynamics from the soil biosphere to aquatic ecosystems are becoming increasingly important in light of concerns related to eutrophication, climate change, and other anthropogenic impacts [19,20,45,46]. Statistics-based modeling integrates various parameters, such as soil physicochemical properties, regional meteorology, and anthropogenic activities [19,20,23], and can be used to predict P flux under IPCC climate scenarios [4]. However, these models are insufficiently related to the ecological forcing of microorganisms in the soil micro-ecosystem, even though such microorganisms essentially mediate and drive Earth's biogeochemical cycles [13,27]. We found that potential P flux (Table 2) was significantly related to n-$H_{C:P}$ according to a negative allometric model (Fig. 4), verifying the hypothesis that a n-$H_{C:P}$ index may be useful as an ecological tool to predict P flux. Although C:P stoichiometric dynamics alone can provide some insight into P dynamics in an ecosystem [5,39], the n-$H_{C:P}$ term (as outlined in equation 4) comprehensively incorporates both soil C and P availability and soil microorganism nutrient acquisition under the principle of stoichiometric homeostasis [1]. The n-$H_{C:P}$ response mirrors mechanisms related to the metabolic theory of ecology [5,6] in that the soil micro-environment of wetland ecosystems responds positively (albeit via a negative allometric relationship) to temperature increases.

According to the derived models of potential P flux to n-$H_{C:P}$ (Fig. 4), lower n-$H_{C:P}$ values in this study indicate higher P flux, which is fundamentally consistent with the ecological interpretation of H_C (Fig. 1), H_P (Fig. 2), and $H_{C:P}$ (Fig. 3) between the two treatments. Although regression models differed, the ecological role of n-$H_{C:P}$ in this study is essentially identical in its nature to the linear and positive relationships between community H and ecological production and stability in a Mongolian grassland [2]. The consistency of these results suggests that although diverse factors may affect P dynamics, n-$H_{C:P}$ may be a useful ecological tool for assessing potential P flux. Notably, the mathematic formulations between n-$H_{C:P}$ and potential P flux were identical between the two experimental treatments (Fig. 4), suggesting that the feasibility of n-$H_{C:P}$ to quantitatively assess potential P flux might be mainly determined by the features of soil C-P stoichiometric homeostasis itself.

It should be noted that both n-$H_{C:P}$ and potential P flux are calculated using one of the same parameters; namely, soil porewater DRP, and are therefore inherently interrelated to some extent. However, n-$H_{C:P}$ (Eq 4) is computed using a multifactor equation linking soil C:P ratio (in form of soil HLOC: soil porewater DRP), soil microorganism biomass C:P ratio, and the associated homeostatic coefficient ($H_{C:P}$). Similarly, changes in potential P flux (Eq 5) were jointly determined by differences in the concentration of DRP in soil porewater vs. overlying water, by the quantity of overlying water and the area of topsoil in the wetland column, and by the overlying water replacement time interval. Moreover, values of these two indices were distinctly influenced by experimental warming (Table 2, Fig. 3). Therefore, the strong, negative allometric relationship observed between these indices (Fig. 4) is indeed an indication of the potential ecological management role of n-$H_{C:P}$ in assessing potential P flux, and is not be expected outright based only on co-variation arising from their common incorporation of a soil porewater DRP parameter.

The stoichiometric parameter $S_{C/P}$ (or $S_{C/N}$) has been defined as a scalar for the relative availability of hydrolyzed P (or N) in relation to microbial community composition [32,47]. Utilizing this $S_{C/P}$ (or $S_{C/N}$) metric, an allometric biogeochemical equilibrium model was developed to predict microbial growth efficiency (GE) from elemental C:N and C:P ratios in biomass ($B_{C/N}$ and $B_{C/P}$, respectively) and environmental substrate sources

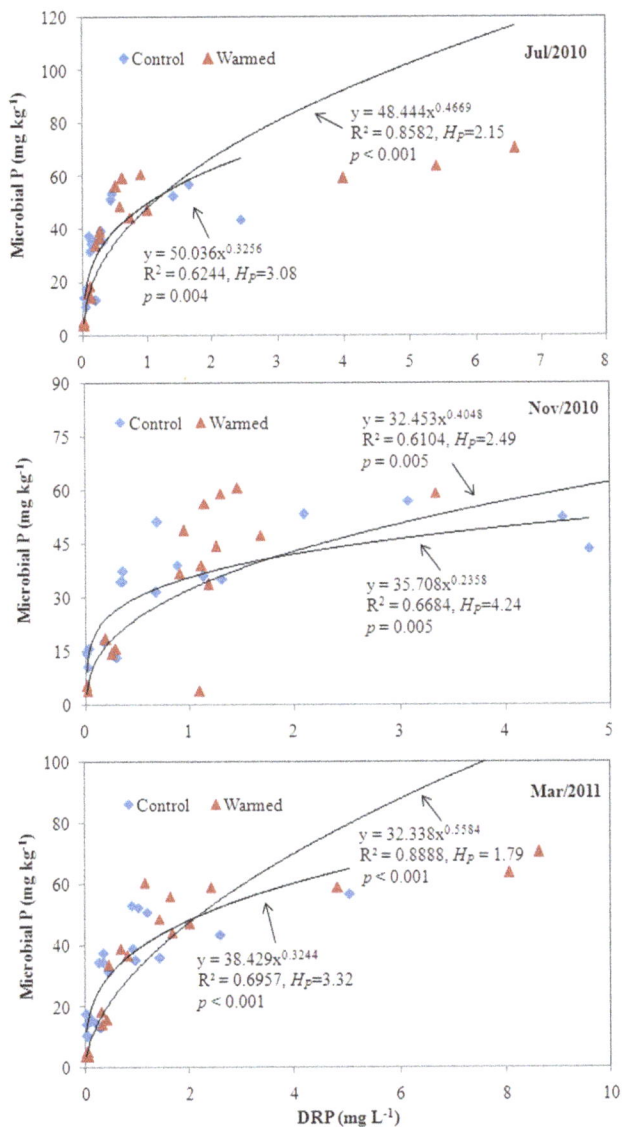

Figure 2. Relationships between soil microbial phosphorus (MB-P) content and dissolved reactive phosphorus (DRP) in porewater. Control and Warmed represent treatments of ambient temperature and ambient temperature +5°C, respectively. Coefficients fitting the data to the equation for stoichiometric homeostasis are given along with an estimate of H_P.

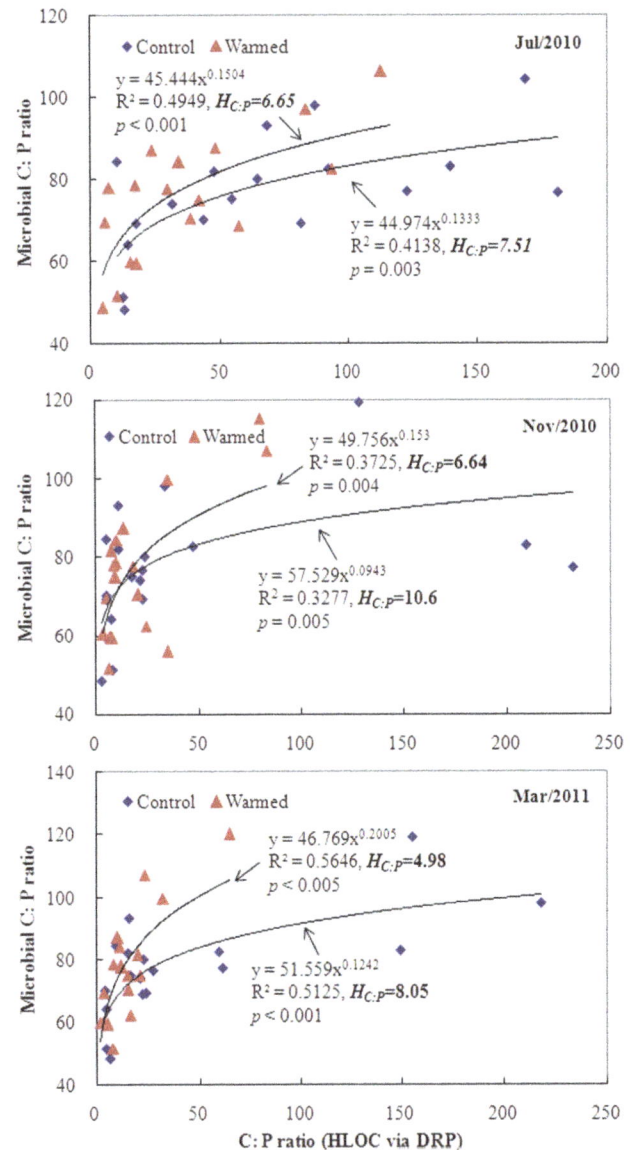

Figure 3. Relationships between the molar ratios of carbon to phosphorus (C:P) in soil microbial biomass and the C:P ratios of soil highly labile organic carbon (HLOC) to dissolved reactive phosphorus (DRP) in porewater. Control and Warmed represent treatments of ambient temperature and ambient temperature +5°C, respectively. Coefficients fitting the data to the equation for stoichiometric homeostasis are given along with an estimate of $H_{C:P}$.

($L_{C/N}$, $L_{C/P}$), integrated with ratios of ecoenzymatic activities (EEAs) that mediate C, N, and P acquisition (EEA$_{C/N}$, EEA$_{C/P}$) [32]. Although the intended prediction outcomes of these approaches differ, the nature of the relationship between microbial n-$H_{C:P}$ and sediment potential P flux is fundamentally similar to the use of the $S_{C/P}$ indictor, because both scalars reflect how stoichiometric balance/interaction is regulated by environmental signals. Further, they can also both be viewed in terms of an energy landscape (i.e., C) that directs the availability and flow of resources (e.g., N and P). Therefore, our experimental data and the measurement of soil n-$H_{C:P}$ might be practical for comparative investigations of soil P dynamics across a variety of wetland ecosystems. This n-$H_{C:P}$ method offers a new alternative for the measurement and understanding of soil P biogeochemical cycling, comparable to current methodologies that utilize biochemistry

[37,42], molecular biology [21,27,43,46], isotopic tracing [48], and mechanistic models [19,23,45]. To the best of our knowledge, this is the first study of integrated C-P stoichiometric homeostasis to clearly demonstrate the linkage between soil n-$H_{C:P}$ and potential P flux in a freshwater wetland ecosystem.

Undoubtedly, aquatic plants are another primary pool for P biogeochemical cycling, because 12%–85% of P in water may be immobilized by wetland plants [48,49]. However, addressing the influence of aquatic plants on n-$H_{C:P}$ and other parameters was beyond the purpose of this study, and thus they were temporarily excluded from the soil columns during the course of the experiment. The involvement of aquatic plants in influencing soil

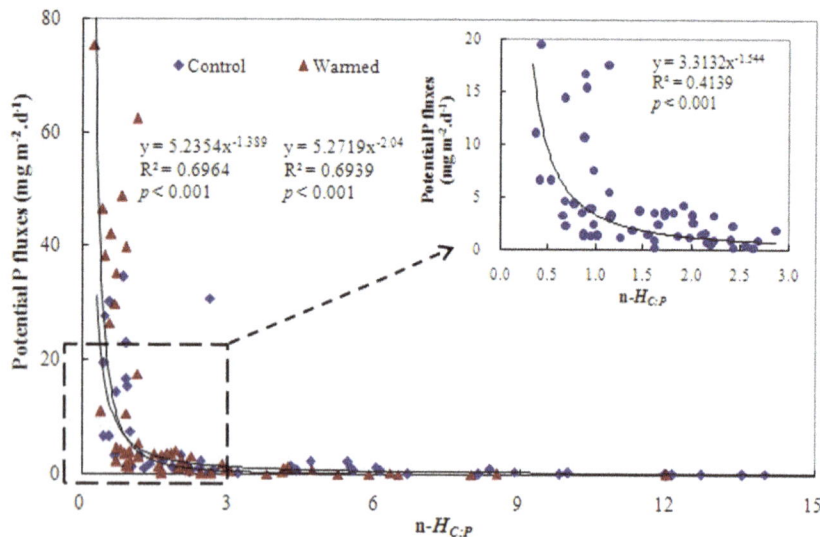

Figure 4. Relationship between potential phosphorus (P) flux and the specialized ability of stoichiometric homeostasis (n-$H_{C:P}$) as a molar ratio of soil highly labial organic carbon (HLOC) to dissolved reactive phosphorus (DRP) in each of the wetland soil columns. Control and Warmed represent treatments of ambient temperature and ambient temperature +5°C, respectively. The data for P flux and n-$H_{C:P}$ less than 20 mg m^{-2} d^{-1} and 3.0, respectively, are plotted in the figure inset.

C:P stoichiometric homeostasis and associated variations in n-$H_{C:P}$ thus requires further investigation.

Although soil respiration at the global scale responds positively to air temperature, in the long term warming may also mobilize older stored carbon, potentially resulting in higher carbon inputs to the soil biosphere [8]. If so, labile P forms would be 'returned' to the soil-microorganism complex under warming according to principles of ecoenzymatic stoichiometry [26] and mechanisms of synergistic C-P acquisition. Such an outcome would be somewhat inconsistent with the near-term decrease in soil C stock (Table 1) and increase in potential P flux (Table 2) we observed in the current microcosm warming experiment. Thus, it will also be necessary to further explore the linkage between n-$H_{C:P}$ and P flux under long-term warming.

In summary, microcosm warming impaired the stoichiometric homeostatic ability (H) of soil microorganisms to regulate P and C biogeochemical processes. This resulted in a "double negative" effect for wetland ecosystem services: increasing their potential for P export and enhancing their role as a recalcitrant organic C source. Moreover, the specialized homeostatic regulation ability (n-$H_{C:P}$, as a function of the C:P ratio in the form of HLOC to DRP in soil porewater relative to the C:P ratio in soil microbial biomass) of soil microorganisms was inversely linked to potential P flux at the soil-water interface. Based on our results, we advocate further use of homeostatic regulation indices (H) as novel ecological tools for assessing potential P flux and other biogeochemical dynamics in the face of global climate change.

Supporting Information

Figure S1 The design of the experimental wetland microcosm system setup by using independently monitored water bath jackets under the current climate condition (Left: Ambient temperature, Control) and the warming climate condition (Right: Ambient temperature +5°C, Warmed treatment).

Figure S2 Dynamics of dissolved oxygen (DO) concentration in overlying water during 12 d of sediment incubation. Sediment samples were collected in July 2010 from YaTang riverine wetland (YT) under control (ambient temperature) and warmed (ambient temperature +5°C) treatments in microcosm experiment. Error bars show ± SD. The differences between control and warmed treatments were tested by Student's t-test for each sampling point, indicated by * $p<0.05$, ** $p<0.01$.

Figure S3 Dynamics of ferric iron (Fe^{3+}) and ferrous iron (Fe^{2+}) concentration in sediment measured during a 13-d laboratory incubation for YaTang riverine wetland (YT) sediment samples under control (ambient temperature) and warmed (ambient temperature +5°C) treatments. Error bars show ± SD. The differences between control and warmed treatments were tested by Student's t-test for each sampling point, indicated by * $p<0.05$, ** $p<0.01$.

Table S1 Details of the six selected wetland sites used in the study.

Table S2 Select physico-chemical parameters of 20-cm depth soil samples collected from the JinHu wetland (JH), XiaZhuhu wetland (XZ), YaTang riverine wetland (YT), XiXi national wetland park (XX), BaoYang riverine wetland (BY), and ShiQiuyang multipond wetland (SQ) in May 2008.

Text S1 Microcosm configuration.

Text S2 Wetland column preparation and field sampling.

Text S3 Laboratory incubation for sediment oxygen demand and reducing capability.

Author Contributions

Conceived and designed the experiments: ZZ JDM XX QH. Performed the experiments: HW CY. Analyzed the data: HW HL DJ. Wrote the paper: HW JDM ZZ. Contributed the experimental materials: ZZ.

References

1. Sterner RW, Elser JJ (2002) Ecological Stoichiometry: The Biology of Elements From Molecules to the Biosphere. Princeton University Press, Princeton.
2. Yu Q, Chen QS, Elser JJ, He NP, Wu HH, et al. (2010) Linking stoichiometric homoeostasis with ecosystem structure, functioning and stability. Ecology Letters 13: 1390–1399.
3. Hansen J, Sato M, Ruedy R, Lo K, Lea DW, et al. (2006) Global temperature change. Proceedings of the National Academy of Sciences of the United States of America 103: 14288–14293.
4. IPCC (2001) Climate Change 2001:The Scientific Basis. Intergovernmental Panel on Climate Change. Cambridge University Press, Cambridge.
5. Allen AP, Gillooly JF (2009) Towards an integration of ecological stoichiometry and the metabolic theory of ecology to better understand nutrient cycling. Ecology Letters 12: 369–384.
6. Gillooly JF, Brown JH, West GB, Savage VM, Charnov EL (2001) Effects of size and temperature on metabolic rate. Science 293: 2248–2251.
7. Englund G, Ohlund G, Hein CL, Diehl S (2011) Temperature dependence of the functional response. Ecology Letters 14: 914–921.
8. Bond-Lamberty B, Thomson A (2010) Temperature-associated increases in the global soil respiration record. Nature 464: 579–U132.
9. Rhee GY (1978) Effects of N: P atomic ratios and nitrate limitation on algal growth, cell compostion and nitrate uptake. Limnology and Oceanography 23: 10–25.
10. Elser JJ, Fagan WF, Kerkhoff AJ, Swenson NG, Enquist BJ (2010) Biological stoichiometry of plant production: metabolism, scaling and ecological response to global change. New Phytologist 186: 593–608.
11. Yu QA, Elser JJ, He NP, Wu HH, Chen QS, et al. (2011) Stoichiometric homeostasis of vascular plants in the Inner Mongolia grassland. Oecologia 166: 1–10.
12. Totsche KU, Rennert T, Gerzabek MH, Kogel-Knabner I, Smalla K, et al. (2010) Biogeochemical interfaces in soil: The interdisciplinary challenge for soil science. Journal of Plant Nutrition and Soil Science 173: 88–99.
13. Falkowski PG, Fenchel T, Delong EF (2008) The microbial engines that drive Earth's biogeochemical cycles. Science 320: 1034–1039.
14. Hessen DO, Agren GI, Anderson TR, Elser JJ, De Ruiter PC (2004) Carbon, sequestration in ecosystems: The role of stoichiometry. Ecology 85: 1179–1192.
15. Verhoeven JTA, Arheimer B, Yin CQ, Hefting MM (2006) Regional and global concerns over wetlands and water quality. Trends in Ecology & Evolution 21: 96–103.
16. Elser JJ, Bracken MES, Cleland EE, Gruner DS, Harpole WS, et al. (2007) Global analysis of nitrogen and phosphorus limitation of primary producers in freshwater, marine and terrestrial ecosystems. Ecology Letters 10: 1135–1142.
17. Sundareshwar PV, Morris JT, Koepfler EK, Fornwalt B (2003) Phosphorus limitation of coastal ecosystem processes. Science 299: 563–565.
18. Mulder C, Elser JJ (2009) Soil acidity, ecological stoichiometry and allometric scaling in grassland food webs. Global Change Biology 15: 2730–2738.
19. Serpa D, Falcao M, Duarte P, da Fonseca LC, Vale C (2007) Evaluation of ammonium and phosphate release from intertidal and subtidal sediments of a shallow coastal lagoon (Ria Formosa-Portugal): A modelling approach. Biogeochemistry 82: 291–304.
20. Nicholls KH (1999) Effects of temperature and other factors on summer phosphorus in the inner Bay of Quinte, Lake Ontario: implications for climate warming. Journal of Great Lakes Research 25: 250–262.
21. Zhang ZJ, Wang ZD, Holden J, Xu XH, Wang H, et al. (2012) The release of phosphorus from sediment into water in subtropical wetlands: a warming microcosm experiment. Hydrological Processes 26: 15–26.
22. Wang H, Holden J, Spera K, Xu XH, Wang ZD, et al. (2013) Phosphorus fluxes at the sediment-water interface in subtropical wetlands subjected to experimental warming: A microcosm study. Chemosphere 90: 1794–1804.
23. Jeppesen E, Kronvang B, Meerhoff M, Sondergaard M, Hansen KM, et al. (2009) Climate change effects on runoff, catchment phosphorus loading and lake ecological state, and potential Adaptations. Journal of Environmental Quality 38: 1930–1941.
24. Cheng WG, Sakai H, Matsushima M, Yagi K, Hasegawa T (2010) Response of the floating aquatic fern Azolla filiculoides to elevated CO(2), temperature, and phosphorus levels. Hydrobiologia 656: 5–14.
25. Hilbert DW, Roulet N, Moore T (2000) Modelling and analysis of peatlands as dynamical systems. Journal of Ecology 88: 230–242.
26. Sinsabaugh RL, Hill BH, Shah JJF (2009) Ecoenzymatic stoichiometry of microbial organic nutrient acquisition in soil and sediment. Nature 462: 795-U117.
27. Zhou JZ, Xue K, Xie JP, Deng Y, Wu LY, et al. (2012) Microbial mediation of carbon-cycle feedbacks to climate warming. Nature Climate Change 2: 106–110.
28. Bradford MA, Fierer N, Reynolds JF (2008) Soil carbon stocks in experimental mesocosms are dependent on the rate of labile carbon, nitrogen and phosphorus inputs to soils. Functional Ecology 22: 964–974.
29. Wu J, Joergensen RG, Pommerening B, Chaussod R, Brookes PC (1990) Measurement of soil microbial biomass C by fumigation extraction - an automated procedure. Soil Biology & Biochemistry 22: 1167–1169.
30. Brookes PC, Powlson DS, Jenkinson DS (1982) Measurement of microbial biomass phosphorus in soil. Soil Biology & Biochemistry 14: 319–329.
31. Loginow W, Wisniewski W, Strony WM (1987) Fractionation of organic carbon based on susceptibility to oxidation Polish. Journal of Soil Science 20.
32. Sinsabaugh RL, Follstad Shah JJF (2012) Ecoenzymatic stoichiometry and ecological theory. In: Futuyma DJ, editor. Annual Review of Ecology, Evolution, and Systematics, Vol 43. Palo Alto: Annual Reviews. pp. 313–343.
33. Small GE, Pringle CM (2010) Deviation from strict homeostasis across multiple trophic levels in an invertebrate consumer assemblage exposed to high chronic phosphorus enrichment in a Neotropical stream. Oecologia 162: 581–590.
34. Marichal R, Mathieu J, Couteaux MM, Mora P, Roy J, et al. (2011) Earthworm and microbe response to litter and soils of tropical forest plantations with contrasting C:N:P stoichiometric ratios. Soil Biology & Biochemistry 43: 1528–1535.
35. Olk DC, Gregorich EG (2006) Overview of the symposium proceedings, "Meaningful pools in determining soil carbon and nitrogen dynamics". Soil Science Society of America Journal 70: 967–974.
36. Schmidt MWI, Torn MS, Abiven S, Dittmar T, Guggenberger G, et al. (2011) Persistence of soil organic matter as an ecosystem property. Nature 478: 49–56.
37. Liikanen A, Murtoniemi T, Tanskanen H, Vaisanen T, Martikainen PJ (2002) Effects of temperature and oxygen availability on greenhouse gas and nutrient dynamics in sediment of a eutrophic mid-boreal lake. Biogeochemistry 59: 269–286.
38. Rinu K, Pandey A (2010) Temperature-dependent phosphate solubilization by cold- and pH-tolerant species of Aspergillus isolated from Himalayan soil. Mycoscience 51: 263–271.
39. Cleveland CC, Liptzin D (2007) C: N: P stoichiometry in soil: is there a "Redfield ratio" for the microbial biomass? Biogeochemistry 85: 235–252.
40. Zhang W, Parker KM, Luo Y, Wan S, Wallace LL, et al. (2005) Soil microbial responses to experimental warming and clipping in a tallgrass prairie. Global Change Biology 11: 266–277.
41. Feng XJ, Simpson MJ (2009) Temperature and substrate controls on microbial phospholipid fatty acid composition during incubation of grassland soils contrasting in organic matter quality. Soil Biology & Biochemistry 41: 804–812.
42. Newman S, McCormick PV, Backus JG (2003) Phosphatase activity as an early warning indicator of wetland eutrophication: problems and prospects. Journal of Applied Phycology 15: 45–59.
43. Allison VJ, Condron LM, Peltzer DA, Richardson SJ, Turner BL (2007) Changes in enzyme activities and soil microbial community composition along carbon and nutrient gradients at the Franz Josef chronosequence, New Zealand. Soil Biology & Biochemistry 39: 1770–1781.
44. Wang H, He ZL, Lu ZM, Zhou JZ, Van Nostrand JD, et al. (2012) Genetic linkage of soil carbon pools and microbial functions in subtropical freshwater wetlands in response to experimental warming. Applied and Environmental Microbiology 78: 7652–7661.
45. Alvarez-Cobelas M, Sanchez-Carrillo S, Angeler DG, Sanchez-Andres R (2009) Phosphorus export from catchments: a global view. Journal of the North American Benthological Society 28: 805–820.
46. He ZL, Deng Y, Van Nostrand JD, Tu QC, Xu MY, et al. (2010) GeoChip 3.0 as a high-throughput tool for analyzing microbial community composition, structure and functional activity. Isme Journal 4: 1167–1179.
47. Sinsabaugh RL, Van Horn DJ, Shah JJF, Findlay S (2010) Ecoenzymatic Stoichiometry in Relation to Productivity for Freshwater Biofilm and Plankton Communities. Microbial Ecology 60: 885–893.
48. Noe GB, Scinto LJ, Taylor J, Childers DL, Jones RD (2003) Phosphorus cycling and partitioning in an oligotrophic Everglades wetland ecosystem: a radioisotope tracing study. Freshwater Biology 48: 1993–2008.
49. White JR, Reddy KR, Majer-Newman J (2006) Hydrologic and vegetation effects on water column phosphorus in wetland mesocosms. Soil Science Society of America Journal 70: 1242–1251.

12

Modelling the Influence of Major Baltic Inflows on Near-Bottom Conditions at the Entrance of the Gulf of Finland

Gennadi Lessin[1]*, Urmas Raudsepp[2], Adolf Stips[3]

1 Plymouth Marine Laboratory, Prospect Place, The Hoe, Plymouth, United Kingdom, **2** Marine Systems Institute, Tallinn University of Technology, Tallinn, Estonia, **3** European Commission, Joint Research Centre, Institute for Environment and Sustainability, Water Research Unit, Ispra, Italy

Abstract

A coupled hydrodynamic-biogeochemical model was implemented in order to estimate the effects of Major Baltic Inflows on the near-bottom hydrophysical and biogeochemical conditions in the northern Baltic Proper and the western Gulf of Finland during the period 1991–2009. We compared results of a realistic reference run to the results of an experimental run where Major Baltic Inflows were suppressed. Further to the expected overall decrease in bottom salinity, this modelling experiment confirms that in the absence of strong saltwater inflows the deep areas of the Baltic Proper would become more anoxic, while in the shallower areas (western Gulf of Finland) near-bottom average conditions improve. Our experiment revealed that typical estuarine circulation results in the sporadic emergence of short-lasting events of near-bottom anoxia in the western Gulf of Finland due to transport of water masses from the Baltic Proper. Extrapolating our results beyond the modelled period, we speculate that the further deepening of the halocline in the Baltic Proper is likely to prevent inflows of anoxic water to the Gulf of Finland and in the longer term would lead to improvement in near-bottom conditions in the Baltic Proper. Our results reaffirm the importance of accurate representation of salinity dynamics in coupled Baltic Sea models serving as a basis for credible hindcast and future projection simulations of biogeochemical conditions.

Editor: Inés Álvarez, University of Vigo, Spain

Funding: These authors have no support or funding to report.

Competing Interests: The authors have declared that no competing interests exist.

* Email: gle@pml.ac.uk

Introduction

The Baltic Sea is a brackish inland water body having a limited water exchange with the North Sea through narrow and shallow Danish Straits. It receives large freshwater runoff and riverine nutrient loads, which in 2006 comprised 638,000 t of total nitrogen and 28,370 t of total phosphorus [1]. After the collapse of the Soviet Union and consequent socio-economic changes in the region in the 1990s, a considerable reduction of nutrient discharge to the sea from agricultural runoff and industrial pollution took place. The total decrease was approximately 35% for both phosphorus and nitrogen [2]. However, eutrophication resulting from direct and indirect input of nutrients is still considered one of the major environmental problems in the sub-basins of the Baltic Sea [3,4], with the Gulf of Finland being the most eutrophied of them [5,6]. The Gulf of Finland, a relatively narrow (50–135 km) basin in the eastern part of the Baltic Sea, is about 400 km long. Its maximum depth decreases from 80–100 m at the entrance to 20–30 m in the eastern part which receives the Neva river discharge. The Gulf of Finland receives about 2 times larger nitrogen and 3 times larger phosphorus inputs than the Baltic Sea in relation to the surface area [7]. Despite the considerable reduction in nutrient discharge to the Baltic Sea as a whole, there have been repeated reportings of the occurrence of near-bottom anoxic conditions and elevated phosphate concentrations in the Gulf of Finland since the beginning of the 2000s [8–11].

The hydrophysical and biogeochemical status of the Baltic Sea and its sub-basins is to a large extent shaped by external forcing, comprising direct interaction with the atmosphere, freshwater runoff and nutrient discharge from the surrounding land as well as interactions between the sub-basins and interactions at the open boundary [12]. The latter is of a special importance because episodic barotropic inflows – termed Major Baltic Inflows (MBIs) – of highly saline and oxygenated water are an important mechanism of deep water renewal and displacement in the Baltic Sea [13]. Although inflows of different intensity take place more or less regularly, according to the classification based on the duration of inflow and mean vertical salinity as proposed by [14] only 7 MBIs that were observed during the period between 1880–2007 are classified as "very strong" [15–17;13]. One of these very strong inflows occurred in 1993, terminating an unusually lengthy stagnant period that lasted for more than a decade. This breakdown of stagnant conditions took place at the beginning of the period of significant reduction of nutrient loads to the Baltic Sea, which raises the question as to what extent the occurred MBIs are responsible for the changes in biogeochemical conditions in the Gulf of Finland during the following decades.

A major part of the research on the influence of MBIs on hydrophysical and biogeochemical conditions in the Baltic Sea concentrated on the deep basins of the Baltic Proper, e.g. [18–20]. In the Gulf of Finland, research on the effect of MBIs was mainly focused on relating the observed physical and biogeo-chemical/ecological conditions in the Gulf to the timing and extent of inflows to the Baltic Proper. For instance, the influence of changes in salinity and density stratification on oxygen concentrations in the second half of the 20th century was assessed based on long-term measurement data [21,22]. A combination of satellite imagery and *in situ* data was used to propose that the MBI of 1993 was responsible for the expansion of *Nodularia spumigena* blooms to the central and eastern Gulf of Finland after 1994 [8]. This type of connection between a physical process and the ensuing ecosystem changes is of special importance, since harmful algae blooms remain one of the major environmental problems for the whole Baltic Sea and for the Gulf of Finland in particular [23,24].

In the present study, a coupled three-dimensional hydrodynamic-biogeochemical model was applied to study the changes in salinity, nutrients and oxygen in the northern part of the Baltic Proper and the western Gulf of Finland in relation to the MBIs that had occurred since the beginning of the 1990s. The main goals of this modelling study were:

a) to realistically simulate the changes in near-bottom salinity, nutrients and oxygen dynamics in the northern Baltic Proper and the western Gulf of Finland during 1991–2009 by means of a 3D hydrodynamic-biogeochemical model;

b) to estimate the role of MBIs in shaping the hydrophysical and biogeochemical conditions in the northern Baltic Proper and the western Gulf of Finland by comparing the realistic results to an experimental run where MBIs were suppressed.

Model Description and Setup

Model simulations were performed using the hydrodynamic model GETM (General Estuarine Transport Model, www.getm. eu, accessed 2014 Oct 24) coupled with the ERGOM (Ecological Regional Ocean Model, www.ergom.net, accessed 2014 Oct 24) biogeochemical model. GETM is a three-dimensional free-surface hydrodynamic model, which solves the primitive equations of water dynamics with a mode splitting technique on the Arakawa C-grid using Boussinesq and hydrostatic approximations [25,26]. GOTM (General Ocean Turbulence Model, www.gotm.net, accessed 2014 Oct 24) is coupled to GETM to resolve vertical mixing using the k-ε turbulence closure scheme [27]. The ERGOM model version applied in this study contains 12 state variables: three phytoplankton groups (diatoms, flagellates and nitrogen-fixing cyanobacteria), nitrate, ammonium, phosphate, bulk zooplankton, detritus, dissolved oxygen, sediment detritus, iron-bound phosphorus in water and in the sediments. ERGOM uses nitrogen as a model currency. Nitrate, ammonium and phosphate are taken up by phytoplankton in accordance with Redfield nitrogen to phosphorus ratio 16:1. It is assumed that cyanobacteria are able to fix atmospheric nitrogen and are limited only by availability of phosphate. Ammonium and phosphate are released by respiration, excretion and detritus mineralisation. In the presence of oxygen, part of the ammonium is converted to nitrate through the process of nitrification. Under anaerobic conditions, and in the presence of nitrate, detritus is oxidised by reducing nitrate to dinitrogen gas which leaves the system. Under anaerobic conditions and depleted nitrate, hydrogen sulphide is produced through microbial use of oxygen bound in sulphate. The hydrogen sulphide concentration is counted as negative oxygen. In the case of oxic near-bottom conditions, a fixed portion of nitrogen recycled in the sediments is removed from the system through consecutive nitrification and denitrification. The model accounts

for the oxygen-dependent dynamics of phosphate in sediments: under oxygenated conditions, part of the mineralised phosphate is forming iron-phosphate complexes which are stored in the sediments, whereas in anoxic conditions the previously stored phosphate is liberated to the overlying water. Detailed description and formulation of the model is given in [28–30].

The model domain covers the entire Baltic Sea area with an open boundary in the northern Kattegat (Fig. 1). Bathymetry was interpolated to a 2×2 nm (3704×3704 m) model grid from the digital topography of the Baltic Sea [31]. 25 layers were applied in the vertical, using adaptive coordinates. Adaptive coordinates are based on a vertical optimization of the layer distribution which depends on vertical density and velocity gradients and the distance to surface and bottom [20]. The time step implemented is 30 s for the barotropic and 600 s for the baroclinic mode. The period modelled is 01.01.1990–31.12.2009. During the first year of the simulation only hydrodynamics was modelled as a spin-up for the coupled hydrodynamic-biogeochemical simulation.

Initial distributions of water temperature and salinity for January 1990 were interpolated to the model grid from the monthly climatological data set [32]. Initial distributions of nitrate, ammonium, phosphate and dissolved oxygen were reconstructed from a limited amount of available measurement data covering the winter of 1991 [33] and interpolated to the model grid. All the other biogeochemical model variables were given uniform initial distributions over the model domain based on previously reported typical winter values. Prescribed salinity and temperature distributions at the open boundary were interpolated using monthly climatological data [32]. Hourly sea level fluctuations at the open boundary were interpolated from gauge measurements at Kattegat.

The model was forced with European Centre for Medium Range Weather Forecasting (ECMWF) ERA-Interim reanalysis meteorological data. The ERA-Interim configuration uses a 30 min time step and has a spectral T255 horizontal resolution, which corresponds to approximately 79 km spacing on a reduced Gaussian grid [34]. The original data on air temperature, dew point temperature, air pressure, cloud cover, wind speed and wind direction, were interpolated to a regular Gaussian grid corresponding to approximately 50 km spacing with 6-hourly temporal resolution. The model took into account land-based runoff and nutrient loads which had been incorporated into 20 major rivers [30]. Atmospheric deposition of nutrients was taken constant over the entire modelled period.

The setup of the experimental run was identical to the reference run as described above. The only difference being that sea level fluctuations at the open boundary were not prescribed, thus suppressing one of the major factors driving the barotropic inflows to the Baltic Sea [35].

Results

Validation of the reference run

The performance of the model was validated at locations corresponding to HELCOM monitoring stations BY15 (57.32°N, 20.05°E, depth 238 m), LL17 (59.03°N, 21.08°E, depth 171 m), LL12 (59.48°N, 22.9°E, depth 82 m), LL7 (59.85°N, 24.84°E, depth 100 m) and LL3A (60.07°N, 26.35°E, depth 68 m), representing a transition path from the Baltic Proper (Gotland Deep) to the central-eastern parts of the Gulf of Finland.

To summarize the model performance over the whole period, the modelled and measured salinity, temperature, nitrate, ammonium, phosphate and dissolved oxygen at the sea surface and at depths of 20 m, 50 m and (if applicable) 100 m as well as near the

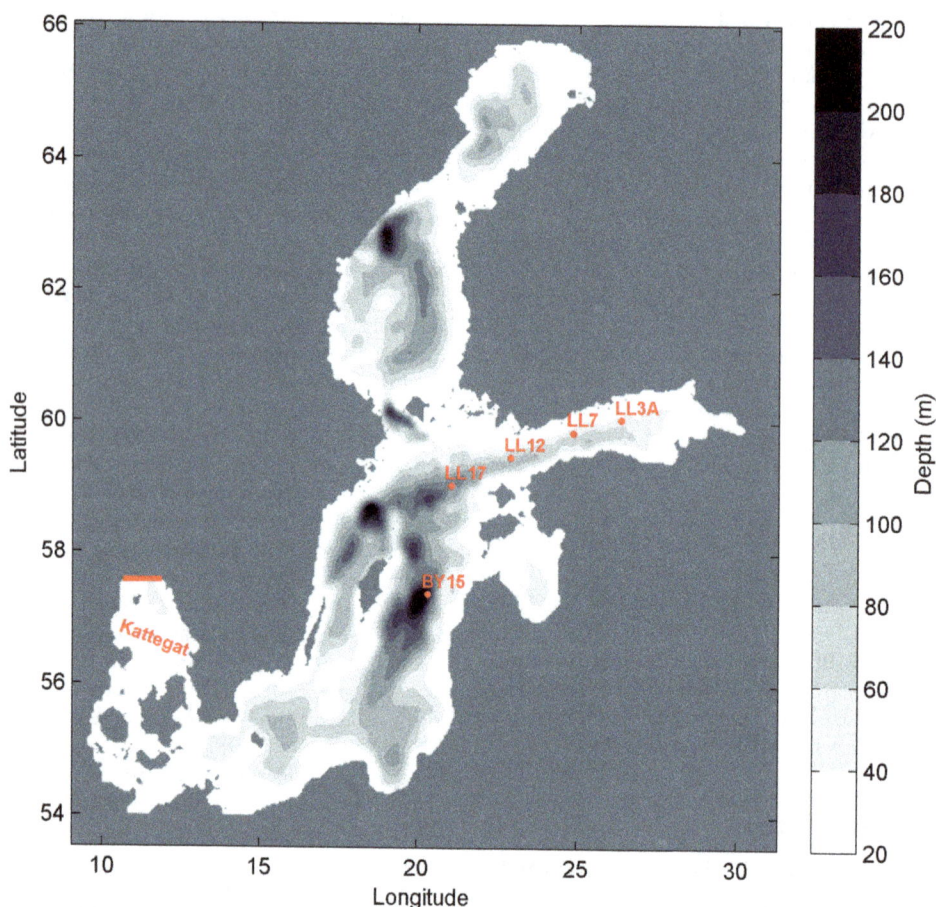

Figure 1. Bathymetric map of the model domain. Locations of monitoring stations BY15 (57.32°N, 20.05°E, depth 238 m), LL17 (59.03°N, 21.08°E, depth 171 m), LL12 (59.48°N, 22.9°E, depth 82 m), LL7 (59.85°N, 24.84°E, depth 100 m) and LL3A (60.07°N, 26.35°E, depth 68 m) are shown. Location of the model open boundary in Kattegat is indicated by red line.

bottom were compared at all stations using a cost function, formulated as CF = |(M–D)/SD|, where the bias (M–D) of the model mean (M) relative to the mean of observations (D) is normalized to the standard deviation (SD) of the observations. Cost function values 0–1 indicate a good match between the model results and measurements, 1–2 indicate a reasonable match and values above 2 indicate a poor match [36]. Among the parameters compared, modelled temperature follows measured values most accurately, followed by oxygen and inorganic nutrients, while salinity has slightly higher cost function values in the upper layers (Table 1).

Agreement of the temporal dynamics of modelled salinity, temperature, oxygen, nitrate and phosphate with corresponding measurement data for the near-bottom layer (ranging 0–2 m from the sea floor) of the northern Baltic Proper (station LL17) and the western Gulf of Finland (station LL12) was investigated.

At station LL17 (Fig. 2) near-bottom salinity was modelled most accurately during the years 1991–1996. An increase of around 1 PSU caused by the MBIs of 1993–1994 was properly reproduced. However, the effect of the subsequent inflow events was underestimated by the model, which by the end of 2008 led to a modelled salinity of about 1 PSU lower than in the measurements. In general, the model underestimated deep water temperature by a margin of up to 0.5°C, with the exception of the year 2003 when the difference between the measured and modelled temperatures was 1°C. Temporal course of the near-

bottom temperature was simulated rather well, with a decreasing trend lasting up until 1997, an increasing one from 1997 to 2005 and a stable temperature in the final period of the model. Both model and measurement oxygen data confirmed that at this station either anoxic or hypoxic (oxygen concentrations 0–2 ml/l, see [22]) conditions were dominant near the bottom. While model results showed anoxic conditions during 1991–1995, measurements indicate a presence of oxygen for this time period, which might be explained by uncertainty in the initial distributions of the biogeochemical variables in the model. It must be noted that negative oxygen concentrations shown in the model results mean that there is a presence of H_2S and in the current context represent the severity of anoxia. In the measurement data the absence of oxygen is indicated by zero oxygen concentrations. The temporal evolution of near-bottom nitrate was in general well-reproduced: it is present most of the time up to the year 2000 and absent thereafter. Near-bottom nitrate dynamics are closely dependent on oxygen concentrations, and therefore the mismatch between modelled and measured nitrate (i.e. too low nitrate during 1991–1997 and too high around 2005) is caused by mismatch between modelled and measured oxygen concentrations. The model accurately reproduced phosphate dynamics, where after the onset of anoxic conditions it showed an increase in phosphorus concentrations caused by the release of phosphate which was previously stored in the sediments.

Table 1. Cost function values for salinity (PSU), temperature (°C), nitrate (mmol N/m³), phosphate (mmol P/m³), oxygen (ml/l) and ammonium (mmol N/m³) at the surface (S), 20 m, 50 m and 100 m depths and near the bottom (B) at stations LL3A, LL7, LL12, LL17 and BY15.

		S	20	50	100	B
Salinity	LL3A	1.82	0.61	0.19		0.52
	LL7	0.93	1.63	0.26		0.62
	LL12	1.35	2.46	1.02		0.32
	LL17	1.27	1.42	0.27	0.65	0.93
	BY15	1.42	1.20	0.98	0.70	0.32
Temperature	LL3A	0.24	0.26	0.07		0.10
	LL7	0.19	0.01	0.12		0.14
	LL12	0.19	0.00	0.11		0.33
	LL17	0.17	0.31	0.09	0.65	0.90
	BY15	0.16	0.47	0.22	0.75	0.79
Nitrate	LL3A	0.34	0.01	1.89		1.36
	LL7	0.29	0.09	0.60		0.67
	LL12	0.06	0.14	0.07		0.74
	LL17	0.07	0.22	0.79	0.06	0.23
	BY15	0.03	0.35	1.21	0.61	1.24
Phosphate	LL3A	1.26	1.42	1.47		0.68
	LL7	0.89	0.86	1.70		0.08
	LL12	0.81	0.56	0.45		0.50
	LL17	0.02	0.03	0.09	0.30	0.09
	BY15	0.32	0.23	0.33	0.63	0.90
Oxygen	LL3A	0.29	0.81	0.84		0.86
	LL7	0.04	0.42	0.81		0.21
	LL12	0.10	0.58	0.67		0.75
	LL17	0.11	0.85	0.84	0.34	0.50
	BY15	0.17	1.08	1.55	0.06	0.89
Ammonium	LL3A	0.20	0.57	1.70		0.99
	LL7	0.33	0.15	0.35		0.53
	LL12	0.55	0.11	0.56		0.68
	LL17	0.50	0.09	0.54	0.84	1.13
	BY15	0.63	0.50	0.47	0.30	1.08

Cost function values 0–1 indicate a good result, 1–2 indicate a reasonable result and >2 indicate a poor result.

At station LL12 in the western part of the Gulf of Finland (Fig. 3) the model accurately reproduced the steady increase of near-bottom salinity and its variability during the course of the investigated period. Near-bottom temperature was rather accurately simulated by the model, capturing intra-annual variability. Simulated oxygen followed the measurement data rather closely, reproducing the shift from well-oxygenated (concentrations of up to about 8 ml/l) to mostly hypoxic and anoxic near-bottom conditions that started from 1995 as well as the increase in variability since 2004. Near-bottom nitrate concentrations were high during the first half of the modelled period, while in the second half concentrations of nitrate showed high fluctuations and nitrate-depleted conditions were present frequently. Compared to the measurement data, the model somewhat underestimated nitrate concentrations in the mid-1990s. Near-bottom nitrate variability clearly reflects its dependence on oxygen dynamics. The modelled phosphate matched the measurement data rather accurately, as an increase over the modelled period

and small-scale variability in concentrations were correctly reproduced.

Validation and detailed analysis of the results of the model for the central Gulf of Finland (HELCOM station LL7) are given in [37].

Results of the experimental run

The effect of eliminating highly saline water inflows on the near-bottom conditions at stations LL17 and LL12 was analysed by subtracting the daily mean values of the reference run from the results of the experimental run (Fig. 4). Over the entire modelled period the salinity was lower in the experimental run than in the reference run. Temporal dynamics was similar at both stations, but at LL17 the mean difference between the two runs was slightly higher and showed less small-scale variability than at LL12. In the absence of saltwater inflows, near-bottom oxygen at LL17 decreased until the difference between the two runs reached about 5 ml/l at the end of the modelled period. At LL12 the experimental run showed a general improvement of oxygen

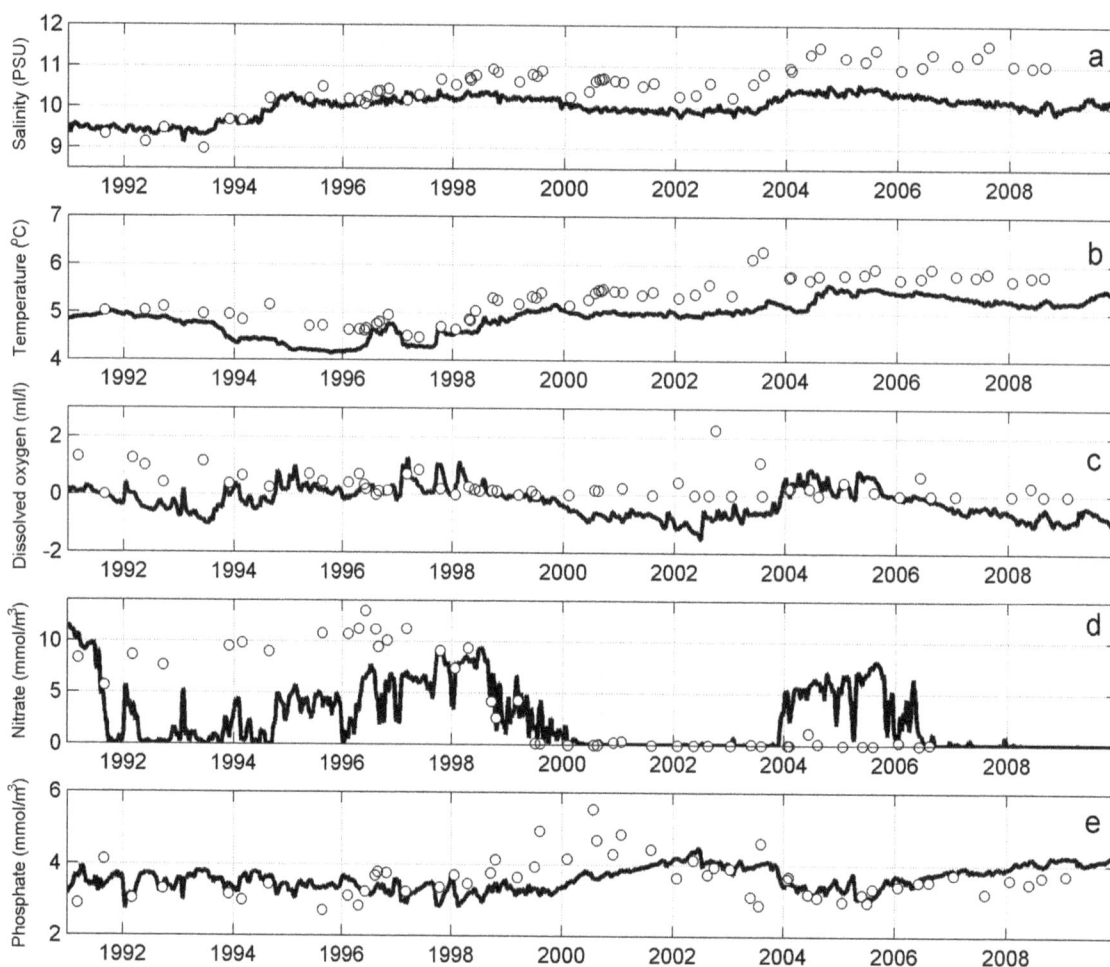

Figure 2. Comparison of modelled (lines) and monitored (circles) near-bottom salinity (a), temperature (b), dissolved oxygen (c), nitrate (d) and phosphate (e) at station LL17.

conditions, but the variability was high and there were short periods when the oxygen concentrations were lower than in the reference run. This occurred more often during the last third of the modelled period. At station LL17 near-bottom nitrate in the experimental run was close to or lower than in the reference run, reflecting both faster consumption of nitrate in the first half of the modelled period and the fact that as a result of stronger anoxia nitrate was later completely depleted. At station LL12 the opposite situation occurred: in the experimental run nitrate was usually more abundant, especially during the second half of the modelled period, although the difference was variable due to frequent changes in oxygen conditions. The difference in near-bottom phosphate at station LL17 increased steadily over the modelled period, reaching about 2.5 mmol/m3 higher concentration in the experimental run. At LL12 the experimental run showed a decrease in mean phosphate until the year 2002, thereafter variability of phosphate increased (the difference varying approximately from −2 to 2 mmol/m3) and, while the mean was still lower in the experimental run, higher phosphate concentrations than in the reference run were often present.

Near-bottom N:P ratio

To summarize the effect of the absence of MBIs and the resulting lower water salinity on near-bottom nutrient content, yearly mean inorganic nitrogen to phosphate (N:P) ratios (N

representing the sum of nitrate and ammonium) for near-bottom layers in both the reference and experimental runs were calculated (Fig. 5). At station LL17 N:P ratio in the reference run was up to 2–3 times higher than in the experimental run, except for the last four years, when N:P ratios in both runs were very close to each other. In 2006 and 2007 N:P ratio was even slightly higher in the experimental run. N:P ratio decreases during the first half of the modelled period in the experimental run and subsequently increases due to higher near-bottom ammonium concentrations under anoxic conditions.

At station LL12 N:P ratio was always higher for the experimental run. The ratio was relatively high up to the year 1998 in the reference run and up to 2000 in the experimental run and in both cases decreased thereafter due to the onset of anoxic conditions near the bottom. Due to the sporadic character of short-lasting periods of anoxia, N:P ratio in the experimental run remained slightly higher until the end of the modelled period.

Comparison of variability of near-bottom oxygen and salinity

The dynamics of the variability of near-bottom oxygen concentrations in relation to salinity variations at stations LL17 and LL12 were analysed for the reference and the experimental runs. Table 2 summarizes the correlation coefficients of daily

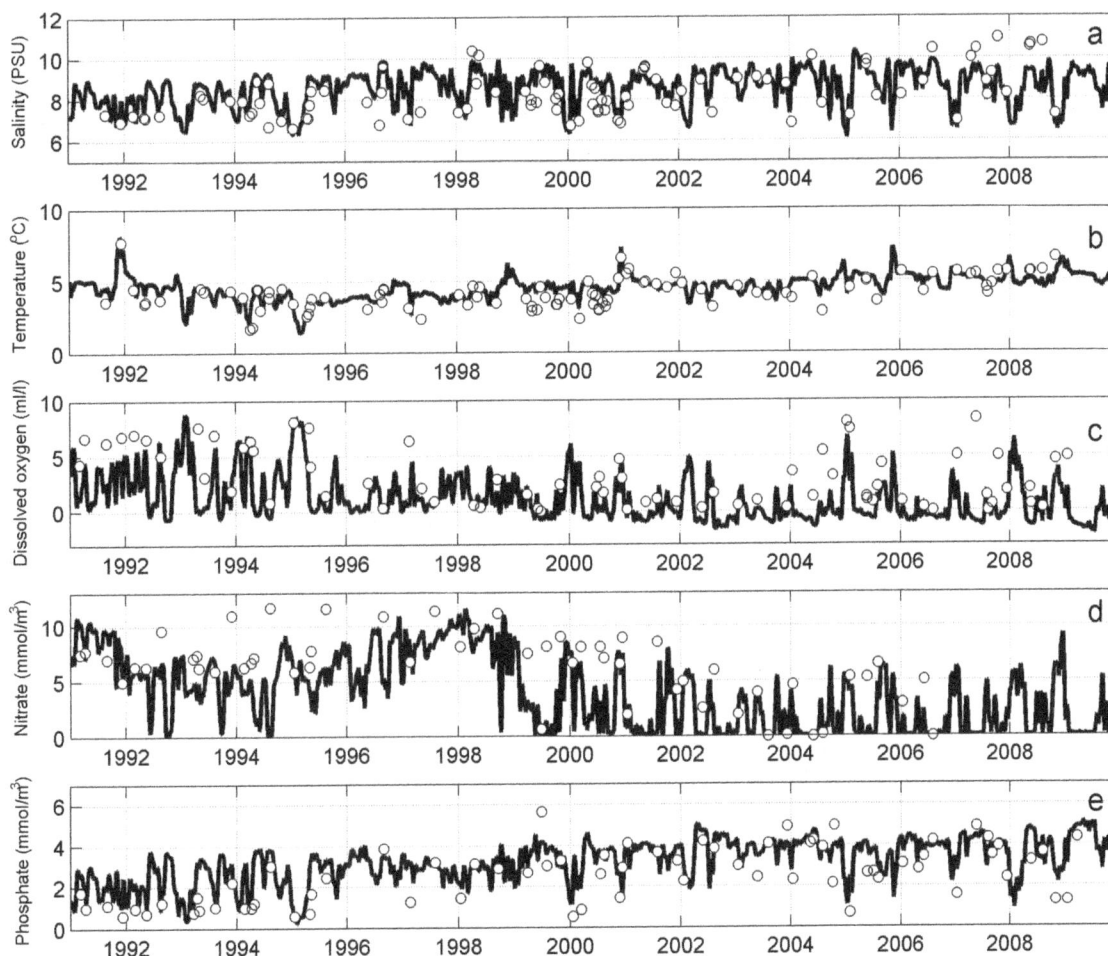

Figure 3. Comparison of modelled (lines) and monitored (circles) near-bottom salinity (a), temperature (b), dissolved oxygen (c), nitrate (d) and phosphate (e) at station LL12.

mean salinity and oxygen for each modelled year and for the total simulation period calculated at 95% confidence level.

At station LL17 salinity and oxygen variability in the reference run was lower than in the experimental run (Fig. 6a). Before the occurrence of the first MBI in 1993, salinity was mostly in the range between 9 to 10 PSU, while oxygen concentration was between −1 and 1 ml/l. As a consequence of the MBIs, the following years were characterized by a shift towards slightly higher salinity, up to approximately 10.5 PSU, and slightly increased oxygen concentrations. On the other hand, during the following more stagnant years a slight salinity decrease (down to 9.5 PSU) and lowering of oxygen content were evident. Nevertheless, oxygen always stayed in the range of −2 to 2 ml/l. High positive correlation coefficients were characteristic for the periods of MBIs (e.g. 1993–1994) and stagnant years (e.g. 1999–2000). In the first case they were caused by the inflow of saline and relatively oxygenated water and in the second case by steady decrease of salinity and loss of oxygen. Years with relatively stable conditions (e.g. 1998, 2007) had a lower correlation coefficient of salinity and oxygen concentrations. If we consider the entire modelled period as one data set, then the correlation coefficient between salinity and oxygen, similarly to the correlation coefficients of individual years, is positive.

In the experimental run at station LL17 there was a more significant drop in salinity (from around 9.5 to 7.6 PSU) and a shift

towards more severe oxygen depletion (from slightly above 0 to around −7 ml/l) over the course of the modelled period (Fig. 6b). The correlation coefficient between salinity and oxygen was above 0.5 only during the first two simulation years, due to the decrease of both salinity and oxygen. While there were some years with low but positive correlation coefficient values, most of the years were characterized by a negative correlation coefficient explained by the supply of relatively more saline but oxygen-depleted water from the deeper areas of the Baltic Proper and its subsequent mixing with less saline and relatively oxygenated (or less oxygen-depleted) water from the overlying layers. Since in this run salinity in the Baltic Proper is not compensated by inflows through the Danish Straits, the system experiences yearly shifts towards lower salinity and stronger anoxia. Therefore, strong positive correlation (0.74) exists for the entire modelled period, although correlation coefficients for most of the individual years were negative.

In the reference run, station LL12 was characterized by variations of salinity from around 6 to 10.5 PSU, and of oxygen from around −2.5 to 9 ml/l (Fig. 7a). Compared to the Northern Baltic Proper area, there was a high intra-annual variability in both parameters. Every individual year was characterized by a strong negative correlation between salinity and oxygen. High salinity is associated with low oxygen or anoxia, while lower salinity is connected to the occurrence of higher oxygen concentrations at the near-bottom. However, at salinity higher

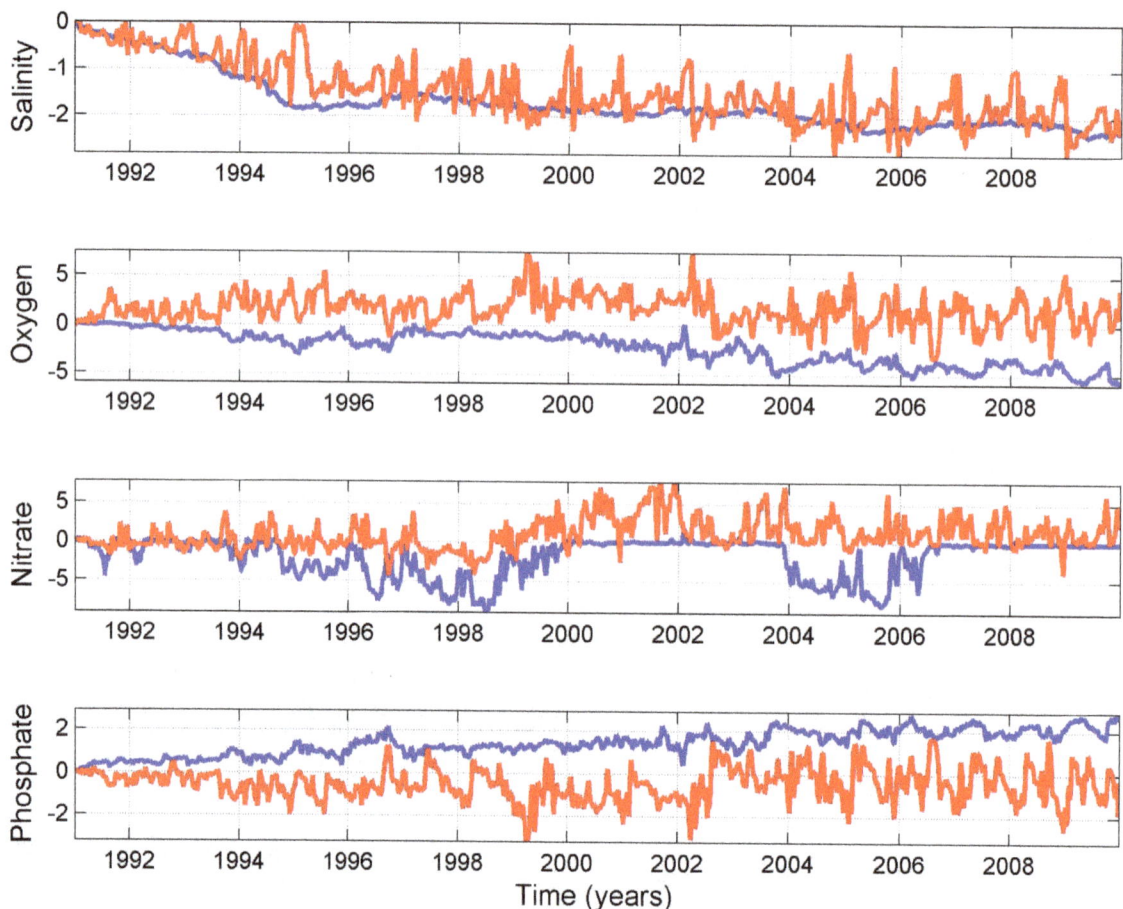

Figure 4. Difference between results of reference and experimental run for near-bottom salinity (PSU), oxygen (ml/l), nitrate (mmol N/m³) and phosphate (mmol P/m³) at station LL17 (blue lines) and LL12 (red lines).

than 10 PSU no decrease in oxygen was seen, indicating that in case of favourable conditions intense water inflows are able to import oxygenated water to the western Gulf of Finland from deeper areas of the Baltic Sea. The correlation coefficients between salinity and oxygen during individual years were close to the correlation coefficient for the entire period (-0.84).

In the experimental run, salinity ranged from about 5 to 9 PSU and oxygen from -4 to almost 10 ml/l at station LL12 (Fig. 7b). Similarly to the reference run, individual years were characterized by negative correlations of salinity and oxygen. However, in comparison to the reference run, decreased saltwater supply from the Baltic Proper leads to a decreased stratification of the water column. As a consequence, the salinity becomes gradually lower over the course of the entire modelled period, while oxygen values remain relatively stable. This results in weaker correlation for the entire modelled period (-0.6), than for each individual year (~ -1).

Discussion

Validation of the reference run showed that the model rather accurately reproduced hydrodynamic and biogeochemical conditions in the study area. Eilola et al [38] evaluated biogeochemical cycles in three state-of-the art numerical models of the Baltic Sea and presented long-term (1970–2005) cost-function values at 6 stations, which included 2 of the stations evaluated in the present paper, namely BY15 (Gotland Deep) and LL7 (central Gulf of

Finland). Although the modelled period was shorter in the present study, the quality of the results reproduced is comparable to those presented in [38].

The mismatch between the modelled parameters and observed values can be explained by uncertainties in the setup, initial conditions and forcing of the hydrodynamic model as well as from the uncertainties in parameterization of biogeochemical processes. Meier et al [39] showed that a horizontal resolution of 2 nm was necessary for the model to reproduce the January 1993 saltwater inflow to the Baltic Sea. Near-bottom salinity in the deep areas of the Baltic Sea after the first inflow events was underestimated in our model. This issue could be solved by further refining the setup of the hydrodynamic model. A robust example of the influence of uncertainties in hydrodynamic modelling on biogeochemical cycling is the presence of substantial concentrations of nitrate in the near-bottom layers at station LL17 in the model results for 2004–2006. This was caused by propagation of less dense but oxygenated water along the bottom, which triggered the nitrification of ammonium. Comparison of time-series of near-bottom variables at LL17 and LL12 leads to the conclusion that uncertainties in setup and forcing have higher effect at deeper stations due to longer water residence time and weaker response to the variability of external forcing.

Accurate representation of hydrodynamical processes is a prerequisite for credible simulations of biogeochemical cycling in a coupled 3D ecosystem-physical model. This is especially valid for

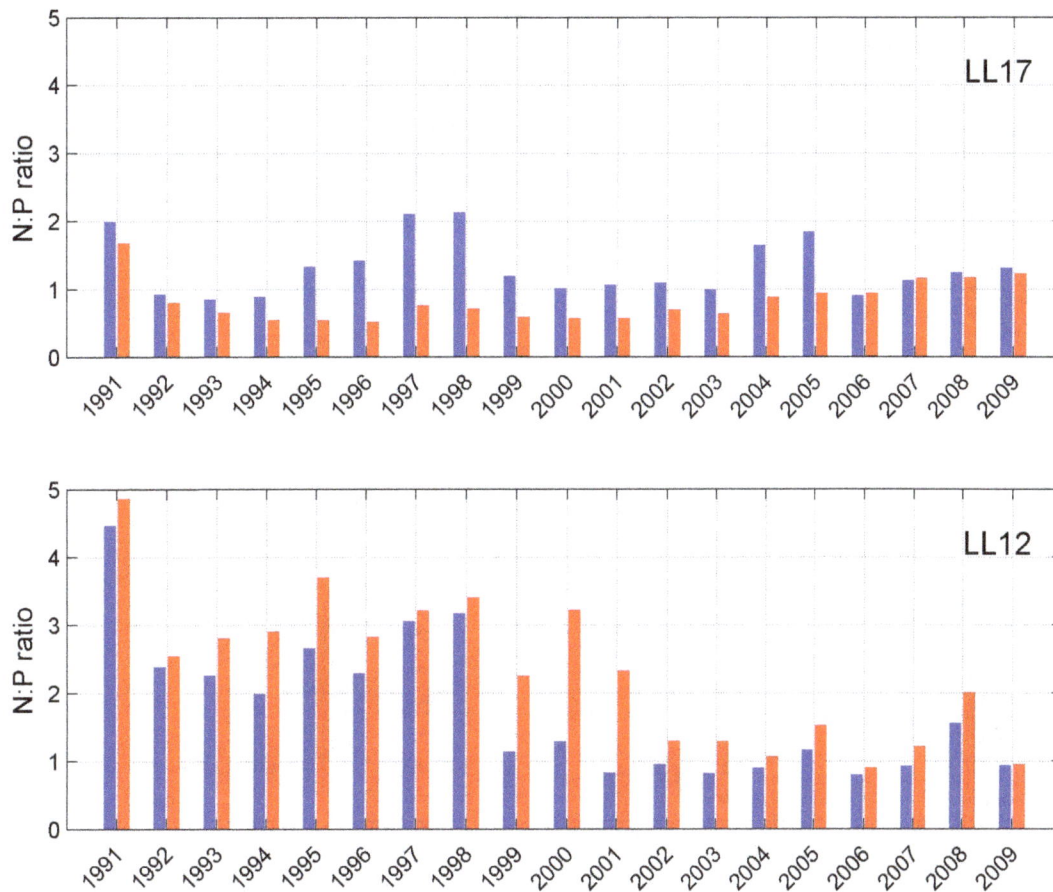

Figure 5. Yearly mean N:P ratios at stations LL17 and LL12 in the reference run (blue bars) and experimental run (red bars). N represents the sum of nitrate and ammonium concentrations.

the Baltic Sea, where ecosystem dynamics are strongly controlled by physical processes. Due to the differences in depth and vertical density stratification between central Baltic Sea and its sub-basins, the biogeochemical response of these regions to decreasing salinity can differ significantly.

Results of the experimental run showed decreasing near-bottom salinity at both stations. However, the response of biogeochemical variables was different in the north-eastern (NE) Baltic Proper and in the western Gulf of Finland. In the absence of strong inflows the near-bottom layer at station LL17 became increasingly anoxic as time passed, while at the shallower station LL12 mean near-bottom oxygen level increased due to weaker stratification and consequent stronger mixing throughout the water column. On the other hand, starting from the year 2002, more severe anoxia in the Baltic Proper was also manifested at LL12 in the form of frequent but short-lasting events of oxygen depletion. Inflows of saline and oxygen-depleted water from NE Baltic Proper to the Gulf of Finland usually take place due to the typical estuarine circulation in the Gulf. Liblik et al [40] observed that the deep water salt wedge in winter 2012 originated from the Baltic Proper at a depth range of 110–115 m and concluded that deterioration of deep layer oxygen conditions was solely related to the advective transport of hypoxic water from the NE Baltic Proper. In the case of a weaker and deeper halocline in the Baltic Proper, oxygen conditions in the Gulf of Finland improve. As confirmed by the results of experimental run, the mean near-bottom oxygen concentration in the western Gulf increases in the absence of

MBIs, but after several years of stagnant conditions anoxia in the deeper layers of the Baltic Proper becomes more severe and estuarine transport leads to the emergence of short-lasting events of deterioration of water quality (lower oxygen or anoxia and higher phosphate concentrations) at the near-bottom of the western Gulf of Finland. Extrapolating our results beyond the modelled period, further deepening of the halocline in the Baltic Proper is likely to prevent the inflows of anoxic water to the Gulf of Finland and will lead to improved near-bottom conditions in the long term.

This is in accordance with the study of Gustafsson et al [41] who modelled the influence of several engineering measures which aimed to reduce the effects of eutrophication of the Baltic Sea. Some of the measures were connected to the management of the flow capacity of the Danish Straits, e.g. closing Öresund at the Drogden Sill. This scenario showed the presence of a long (lasting for more than 30 years) transitional period of stagnation in the Baltic Proper, during which hypoxia increased in deeper waters. After this period the water quality improved and salinity was extremely reduced [42]. Our modelling experiment confirms that within the time frame of simulation deep areas of the Baltic Proper would become more anoxic in the absence of strong saltwater inflows, while in the shallower areas (western Gulf of Finland) near-bottom conditions would improve on average. Thus, in most cases the improvement of near-bottom oxygen conditions follow-ing the decrease in saltwater inflows through the Danish Straits initially takes place in the shallower areas of the Baltic and then

Table 2. Correlation coefficients of mean near-bottom salinity and oxygen for each modelled year and the total simulation period at stations LL17 and LL12 in the reference run (ref) and experimental run (exp).

Year	LL17 ref	LL17 exp	LL12 ref	LL12 exp
1991	0.27	0.69	−0.92	−0.89
1992	0.10	0.64	−0.94	−0.95
1993	0.50	−0.03	−0.97	−0.97
1994	0.53	−0.21	−0.91	−0.87
1995	0.42	−0.78	−0.97	−0.94
1996	0.43	0.17	−0.90	−0.84
1997	0.48	0.29	−0.79	−0.78
1998	0.16	0.03	−0.89	−0.88
1999	0.61	−0.25	−0.94	−0.89
2000	0.73	0.22	−0.95	−0.91
2001	0.42	−0.43	−0.87	−0.84
2002	0.66	−0.54	−0.96	−0.98
2003	0.63	−0.86	−0.82	−0.76
2004	0.47	−0.47	−0.82	−0.87
2005	0.72	−0.22	−0.90	−0.96
2006	0.74	−0.50	−0.83	−0.95
2007	0.33	−0.29	−0.88	−0.93
2008	0.79	−0.05	−0.88	−0.94
2009	0.74	−0.24	−0.83	−0.95
total	0.40	0.74	−0.84	−0.60

propagates to the deeper areas as water becomes fresher and stratification becomes weaker.

Experimental run results showed a rapid response in near-bottom dynamics of nitrate and phosphate to the variability in oxygen. At LL17 nitrate was consumed faster than in the reference run, while at LL12 mean nitrate concentrations increased, yet variability was high with sporadic periods of depletion in response to the rapidly changing oxygen conditions. The lack of difference between nitrate in the experimental and reference runs during 2000–2004 can be explained by the absence of nitrate in both cases, and the difference in 2004–2006 is an artefact of overestimated nitrate in the reference run during the same period. Compared to the reference run, near-bottom phosphate at LL17 increased over the course of the modelling period due to the lack of phosphate binding to the sediments under anoxic conditions. At station LL12 better oxygen conditions led to decreasing phosphate concentrations in the water column. After 2002, mean phosphate concentrations in the experimental run slightly increased due to the enhanced frequency of short-lasting events of anoxic water inflow from the Baltic Proper which caused the release of phosphate from sediments into the overlying water. Therefore, in the time frame of the simulated period, the absence of strong saltwater inflows and the consequent freshening of the Baltic Sea led to opposite responses in near-bottom inorganic nutrients in the Baltic Proper and the Gulf of Finland. However, despite the decrease and depletion of nitrate and increase of phosphate, near-bottom N:P ratios in the experimental and the reference runs were rather similar and even increased slightly towards the end of the simulation period. Although there was more phosphate at LL17 in the experimental run, nitrogen concentration (in the form of ammonium) was also higher than in the reference run. In the western Gulf of Finland N:P ratio started to decrease from the year

1999 in the reference run and from 2001 in the experimental run. This is clearly correlated to the change in deep oxygen conditions. However, since 2002 N:P ratio in the experimental run only slightly exceeded the reference run values. This can be attributed to the emergence of inflows of anoxic water from the Baltic Proper starting from that year. Phosphate release from sediments occurs rapidly after sediments become anoxic, but phosphate binding takes place at a slower pace following the deposition and the subsequent mineralization of organic matter. Kahru et al [8] reported the decrease of N:P ratio starting from 1995 in the Gulf of Finland, which was possibly triggered by the major saline water inflow in 1993. Results of our experiment show that the decrease in near-bottom N:P ratio takes place also in the absence of Major Baltic Inflows. However, in that case it is caused by inflowing anoxic water from NE Baltic Proper to the Gulf of Finland, releasing phosphate previously bound to the sediments and the simultaneous decreasing nitrate due to denitrification.

Nutrient dynamics in the deep layers of the Baltic Sea depends to a large extent on the dynamics of salt and oxygen. Analysis of correlation of the two provides a better understanding of the origin of water masses and their impact on the governing biogeochemical conditions. Salinity-oxygen scatter plots for station LL17 show that near-bottom oxygen conditions in that area improve as a consequence of oxygen-rich saltwater inflows. In the case of stagnant conditions together with decreasing salinity also more severe anoxia is expected. Although stratification becomes weaker, mixing processes are still not intensive enough to improve near-bottom oxygen conditions. The situation is different in the shallower western Gulf of Finland area. Vertical mixing as well as estuarine transport reversals are able to increase near-bottom oxygen concentrations to 8–9 ml/l, which is typical in surface water [40,43]. This process is more easily achieved in the

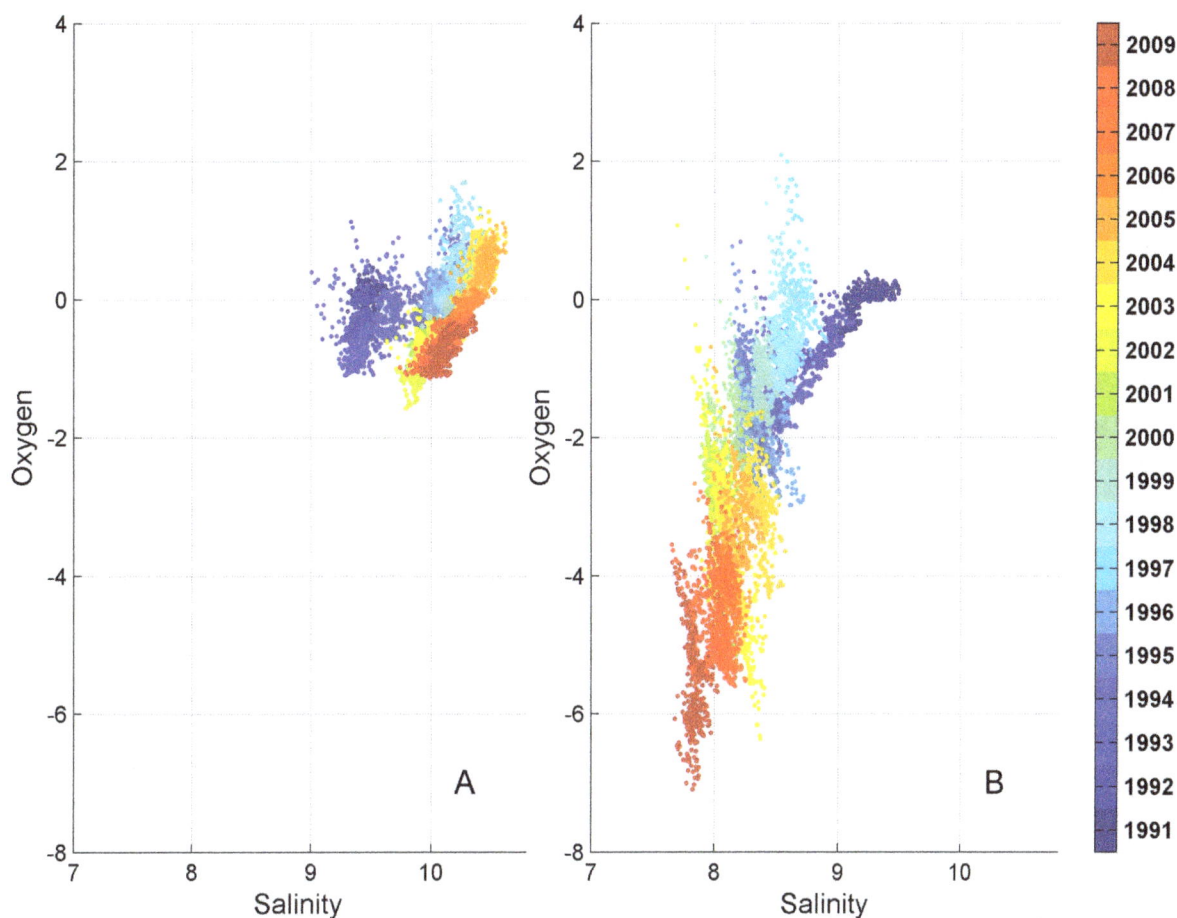

Figure 6. Scatter plots of near-bottom salinity vs oxygen at station LL17 in the reference (A) and experimental (B) runs. Individual years are represented by distinct colours.

conditions of the less stratified water column of the experimental run. In the reference run, during the usual estuarine transport in the Gulf of Finland, near-bottom anoxia becomes slightly less pronounced after the inflows of highly saline deep water (>9 PSU), which bring oxygen originating from the water masses of the Major Baltic Inflows. In the experimental run this time period is characterized by the most severe anoxic conditions despite lower salinity and density stratification.

The dynamic nature of external forcing, leading to a spatially variable biogeochemical response, makes the Baltic Sea unique amongst regions with the occurrence of deep water hypoxia and/ or anoxia. For instance, in case of the northern Gulf of Mexico it was found that chemical-biological processes were mainly responsible for maintaining anoxic near-bottom conditions [44]. Similarly, water column models simulating biogeochemical cycles in regions where hypoxic conditions are permanently present (for example, the highly stratified Black Sea), indicate the importance of organic matter supply and decomposition on the structure of the redox interface and the corresponding processes [45]. In contrast, the occurrence of large oxygen minimum zones is considered to be primarily caused by weak ocean ventilation through subsurface currents [46].

Since ocean warming and increased stratification caused by climate change will likely reduce deep-water oxygen concentrations in the future, Peña et al [47] stressed urgency to develop models coupling realistic physics and biogeochemistry at appro-

priate scales. For the Baltic Sea region it is expected that climate change will decrease frequency of MBIs and increase run-off due to higher precipitation. Decreasing salinity stratification will have important implications for the ecosystem of the entire Baltic Sea, as shown in our modelling study.

We propose the approach of comparing results of a hindcast scenario (reference run) and its modified versions with controlled alterations in e.g. external forcing, initial conditions or process rates (experimental run). Such an approach can be credibly used to further improve our understanding of marine system functioning, and its response patterns to global and regional changes and management measures. In contrast to future scenario modelling, this method is more substantiated because the results of hindcast simulations are validated against observational data so that robust quantitative comparisons between default and perturbed cases can be made.

Conclusions

A coupled three-dimensional hydrodynamic-biogeochemical model was set up and applied for the Baltic Sea area with an aim to simulate the changes in salinity, nutrients and oxygen dynamics during 1991–2009, focussing on the northern Baltic Proper and the western Gulf of Finland. Using available measurement data, validation showed that the model performance was rather accurate. In order to investigate the effects of MBIs on

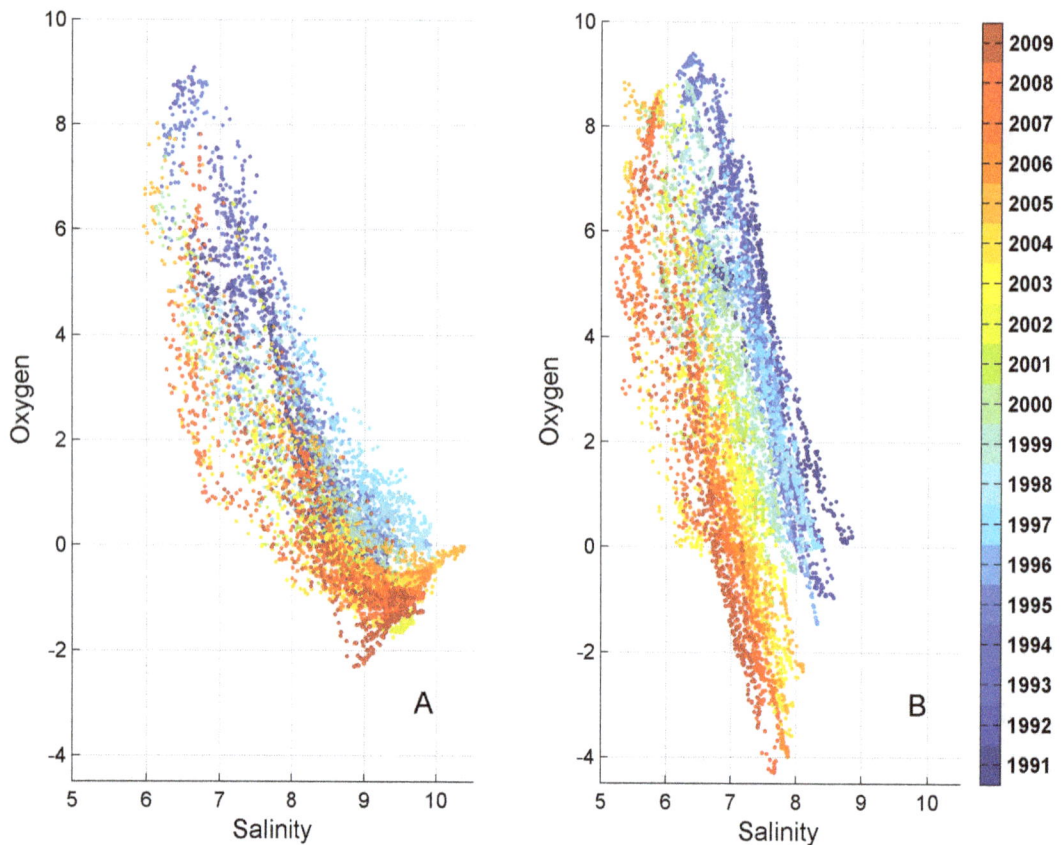

Figure 7. Scatter plots of near-bottom salinity vs oxygen at station LL12 in the reference (A) and experimental (B) runs. Individual years are represented by distinct colours.

hydrophysical and biogeochemical conditions in the focus area, we compared the results of the reference run with the results of the experimental run where inflows of highly saline water to the Baltic Sea from the North Sea were suppressed. Results of the experimental run showed decreasing near-bottom salinity at both stations LL17 and LL12. In the course of the modelled period anoxia became stronger at deep station LL17, while shallower station LL12 showed an average increase in near-bottom oxygen levels due to weaker stratification and intensified water mixing. On the other hand, compared to the reference run at station LL12, emergence of short-lasting events of lower oxygen were associated with the estuarine transport of anoxic water masses from the Baltic Proper. Consequently, during the second half of the modelled period near-bottom N:P ratios decreased in the absence of MBIs both in the stagnant Baltic Proper and in the western Gulf of

Finland. Our results confirmed that accurate representation of MBIs in hydrodynamic models is important for realistic simulations of biogeochemical properties of both deep central areas and shallower sub-basins of the Baltic Sea.

Acknowledgments

The authors thank the GETM and ERGOM developers groups for actively improving the models' code, and the ECMWF for the atmospheric forcing datasets.

Author Contributions

Conceived and designed the experiments: GL. Performed the experiments: GL. Analyzed the data: GL UR AS. Wrote the paper: GL UR AS.

References

1. HELCOM (2011) The Fifth Baltic Sea Pollution Load Compilation (PLC-5) Balt. Sea Environ. Proc. No. 128.
2. Lääne A, Pitkänen H, Arheimer B, Behrendt H, Jarosinski W, et al. (2002) Evaluation of the implementation of the 1988 ministerial declaration regarding nutrient load reductions in the Baltic Sea catchment area. Technical report, Finnish Environment Institute.
3. Schiewer U, editor (2008) Ecology of Baltic Coastal Waters. Ecological Studies 197, Springer-Verlag Berlin Heidelberg.
4. Hong B, Swaney DP, Mörth CM, Smedberg E, Hägg HE, et al. (2012) Evaluating regional variation of net anthropogenic nitrogen and phosphorus inputs (NANI/NAPI), major drivers, nutrient retention pattern and management implications in the multinational areas of Baltic Sea basin. Ecological Modelling, 227: 17–135. DOI: 10.1016/j.ecolmodel.2011.12.002.
5. Lundberg C, Lönnroth M, von Numers M, Bonsdorff E (2005) A multivariate assessment of coastal eutrophication. Examples from the Gulf of Finland,

northern Baltic Sea. Marine Pollution Bulletin, 50(11): 1185–1196, ISSN 0025-326X, DOI: 10.1016/j.marpolbul.2005.04.029.
6. Lehtoranta J, Ekholm P, Pitkänen H (2008) Eutrophication-driven sediment microbial processes can explain the regional variation in phosphorus concentrations between Baltic Sea sub-basins. J. Mar. Syst. 74(1–2): 495–504, DOI: 10.1016/j.jmarsys.2008.04.001.
7. Pitkänen H, Lehtoranta J, Peltonen H (2008) The Gulf of Finland. In: U Schiewer, editor. Ecology of Baltic Coastal Waters. Ecological Studies 197. Springer-Verlag Berlin Heidelberg, 285–308.
8. Kahru M, Leppänen JM, Rud O, Savchuk OP (2000) Cyanobacteria blooms in the Gulf of Finland triggered by saltwater inflow in the Baltic Sea. Mar. Ecol. Progr. Ser., 207: 13–18.
9. Kauppila P, Lepistö L (2001) Changes in phytoplankton. In: Kauppila P, Beck S, editors. The state of Finnish coastal waters in the 1990s. The Finnish Environment, 472: 61–70.

10. Raateoja M, Seppälä J, Kuosa H, Myrberg K (2005) Recent changes in trophic state of the Baltic Sea along SW coast of Finland. Ambio 34: 188–191.

11. Lessin G, Stips A (2012) Modeling of eutrophication processes in the Gulf of Finland during 1991–2010, in: Baltic International Symposium (BALTIC), 2012 IEEE/OES. 1–7.

12. Stigebrandt A, Gustafsson BG (2003) Response of the Baltic Sea to climate change – theory and observations. J. Sea Res., 49: 243–256. DOI: 10.1016/S1385-1101(03)00021-2.

13. Matthäus W, Nehring D, Feistel R, Nausch G, Mohrholz V, et al. (2008) The inflow of highly saline water into the Baltic Sea. In: Feistel R, Nausch G, Wasmund N, editors. State and evolution of the Baltic Sea, 1952–2005: a detailed 50-year survey of meteorology and climate, physics, chemistry, biology, and marine environment. Hoboken, N.J.: John Wiley & Sons, Inc. 265–309.

14. Franck H, Matthäus W, Sammler R (1987) Major inflows of saline water into the Baltic Sea during the present century. Gerlands Beitr. Geophys. 96: 517–531.

15. Matthäus W, Franck H (1992) Characteristics of major Baltic inflows – a statistical analysis. Cont. Shelf Res. 12: 1375–1400.

16. Fischer H, Matthäus W (1996) The importance of the Drogden Sill in the Sound for major Baltic inflows. J. Mar. Syst. 9: 137–157.

17. Matthäus W (2006) The History of Investigation of Salt Water Inflows Into the Baltic Sea: From Early Beginning to Recent Results. Institut für Meereskunde Warnemhunde.

18. Heiser U, Neumann T, Scholten J, Stüben D (2001) Recycling of manganese from anoxic sediments in stagnant basins by seawater inflow: a study of surface sediments from the Gotland Basin, Baltic Sea. Mar. Geol. 177: 151–166.

19. Yakushev EV, Kuznetsov IS, Podymov OI, Burchard T, Neumann T, et al. (2011) Modeling the influence of oxygenated inflows on the biogeochemical structure of the Gotland Sea, central Baltic Sea: changes in the distribution of manganese. Comput. geosci. 37: 398–409, DOI: 10.1016/j.cageo.2011.01.001.

20. Hofmeister R, Beckers JM, Burchard H (2011) Realistic modelling of the exceptional inflows into the central Baltic Sea in 2003 using terrain-following coordinates. Ocean Modelling 39 (3–4): 233–247, DOI: 10.1016/j.ocemod.2011.04.007.

21. Laine AO, Andersin AB, Leiniö S, Zuur AF (2007) Stratification-induced hypoxia as a structuring factor of macrozoobenthos in the open Gulf of Finland (Baltic Sea). Journal of Sea Research 57: 65–77.

22. Conley DJ, Bjorck S, Bonsdorff E, Carstensen J, Destouni G, et al. (2009) Hypoxia-related processes in the Baltic Sea. Environ. Sci. Technol. 43 (10), DOI: 10.1021/es802762a.

23. Vahtera E, Conley DJ, Gustafsson BG, Kuosa H, Pitkänen H, et al. (2007) Internal ecosystem feedbacks enhance nitrogen-fixing cyanobacteria blooms and complicate management in the Baltic Sea. Ambio, 36: 186–194.

24. Lessin G, Raudsepp U, Maljutenko I, Laanemets J, Passenko J, et al. (2014) Model study on present and future eutrophication and nitrogen fixation in the Gulf of Finland, Baltic Sea. Journal of Marine Systems 129: 76–85, DOI: 10.1016/j.jmarsys.2013.08.006.

25. Burchard H, Bolding K (2002) GETM – A General Estuarine Transport Model, Scientific documentation, Tech. Rep. EUR 20253 EN. European Commission, 220.

26. Burchard H, Bolding K, Villarreal MR (2004) Three-dimensional modelling of estuarine turbidity maxima in a tidal estuary. Ocean Dynamics 54: 250–265.

27. Umlauf L, Burchard H (2005) Second-order turbulence closure models for geophysical boundary layers. A review of recent work. Cont. Shelf Res., 25: 795–827.

28. Neumann T (2000) Towards a 3D-ecosystem model of the Baltic Sea. J. Mar. Sys. 25: 405–419.

29. Neumann T, Fennel W, Kremp C (2002) Experimental simulations with an ecosystem model of the Baltic Sea: A nutrient load reduction experiment. Glob. Biogeochem. Cycles 16(3). DOI: 10.1029/2001GB001450.

30. Neumann T, Schernewski G (2008) Eutrophication in the Baltic Sea and shifts in nitrogen fixation analyzed with a 3D ecosystem model. J. Mar. Syst. 74(1–2): 592–602.

31. Seifert T, Tauber F, Kayser B (2001) A high resolution spherical grid topography of the Baltic Sea – 2nd edition, Baltic Sea Science Congress, Stockholm 25–29. November 2001, Poster #147, Available: www.io-warnemuende.de/iowtopo. Accessed 2014 Oct 24.

32. Janssen F, Schrum C, Backhaus J (1999) A climatological data set of temperature and salinity for the Baltic Sea and the North Sea. Ocean Dynamics 51: 5–245.

33. FIMR monitoring of the Baltic Sea environment. – Annual report (2007) Olsonen R (ed): MERI – Report Series of the Finnish Institute of Marine Research. No. 62, 2008.

34. Dee DP, Uppala SM, Simmons AJ, Berrisford P, Poli P, et al. (2001) The ERA-Interim reanalysis: configuration and performance of the data assimilation system. Q.J.R. Meteorol. Soc. 137 (656). John Wiley & Sons, Ltd. DOI: 10.1002/qj.828.

35. Feistel R, Nausch G, Wasmund N, editors (2008) State and evolution of the Baltic Sea, 1952–2005: a detailed 50-year survey of meteorology and climate, physics, chemistry, biology, and marine environment. Hoboken, N.J.: John Wiley & Sons, Inc.

36. Eilola K, Meier HEM, Almroth E (2009) On the dynamics of oxygen, phosphorus and cyanobacteria in the Baltic Sea: A model study. J. Mar. Syst. 75: 163–184.

37. Lessin G, Stips A (2012) Modeling of eutrophication processes in the Gulf of Finland during 1991–2010. In: IEEE/OES Baltic 2012 International Symposium: May 8–11, 2012, Klaipeda, Lithuania, Proceedings: Baltic International Symposium (BALTIC), 2012 IEEE/OES, Klaipeda, Lithuania, 8–10 May 2012.

38. Eilola K, Gustafsson BG, Kuznetsov I, Meier HEM, Neumann T, et al. (2011) Evaluation of biogeochemical cycles in an ensemble of three state-of-the-art numerical models of the Baltic Sea. J. Mar. Syst. 88(2): 267–284.

39. Meier HEM, Döscher R, Faxen T (2003) A multiprocessor coupled ice-ocean model for the Baltic Sea: application to salt inflow. J Geophys Res 108(C8): 3273.

40. Liblik T, Laanemets J, Raudsepp U, Elken J, Suhhova I (2013) Estuarine circulation reversals and related rapid changes in winter near-bottom oxygen conditions in the Gulf of Finland, Baltic Sea. Ocean Sci. Discuss., 10: 727–762, DOI: 10.5194/osd-10-727-2013.

41. Gustafsson BG, Meier HEM, Eilola K, Savchuk OP, Axell L, et al. (2008) Simulation of some engineering measures aiming at reducing effects from eutrophication of the Baltic Sea; Report C82; Earth Sciences Centre, Göteborg University: Göteborg, Sweden.

42. Conley DJ, Bonsdorff E, Carstensen J, Destouni G, Gustafsson BG, et al. (2009) Tackling hypoxia in the Baltic Sea: Is engineering the solution? Environ. Sci. Technol. 43 (10), doi10.1021/es8027633.

43. Elken J, Raudsepp U, Laanemets J, Passenko J, Maljutenko I, et al. (2014) Increased frequency of wintertime stratification collapse events in the Gulf of Finland since the 1990s. J. Mar. Syst, 129: 47–55. DOI: 10.1016/j.jmarsys.2013.04.015.

44. Bierman VJ, Hinz SC, Dong-Wei Z, Wiseman WJ, Rabalais NN, et al. (1994) A preliminary mass balance model of primary productivity and dissolved oxygen in the Mississippi River Plume/Inner Gulf Shelf Region. Estuaries 17: 886–899.

45. Yakushev EV, Pollehne F, Jost G, Kuznetsov I, Schneider B, et al. (2007) Analysis of the water column oxic/anoxic interface in the Black and Baltic seas with a numerical model. Mar. Chem. 107, 388–410.

46. Resplandy L, Lévy M, Bopp L, Echevin V, Pous S, et al. (2012) Controlling factors of the oxygen balance in the Arabian Sea's OMZ. Biogeosciences 9: 5095–5109. DOI:10.5194/bg-9-5095-2012.

47. Peña MA, Katsev S, Oguz T, Gilbert D (2010) Modeling dissolved oxygen dynamics and hypoxia. Biogeosciences 7: 933–957.

13

Carbon and Nitrogen Isotopes from Top Predator Amino Acids Reveal Rapidly Shifting Ocean Biochemistry in the Outer California Current

Rocio I. Ruiz-Cooley[1]*, Paul L. Koch[2], Paul C. Fiedler[3], Matthew D. McCarthy[1]

1 Ocean Sciences Department, University of California Santa Cruz, Santa Cruz, California, United States of America, **2** Earth and Planetary Sciences Department, University of California Santa Cruz, Santa Cruz, California, United States of America, **3** Southwest Fisheries Science Center, National Marine Fisheries Service, National Oceanic and Atmospheric Administration, La Jolla, California, United States of America

Abstract

Climatic variation alters biochemical and ecological processes, but it is difficult both to quantify the magnitude of such changes, and to differentiate long-term shifts from inter-annual variability. Here, we simultaneously quantify decade-scale isotopic variability at the lowest and highest trophic positions in the offshore California Current System (CCS) by measuring $\delta^{15}N$ and $\delta^{13}C$ values of amino acids in a top predator, the sperm whale (*Physeter macrocephalus*). Using a time series of skin tissue samples as a biological archive, isotopic records from individual amino acids (AAs) can reveal the proximate factors driving a temporal decline we observed in bulk isotope values (a decline of ≥ 1 ‰) by decoupling changes in primary producer isotope values from those linked to the trophic position of this toothed whale. A continuous decline in baseline (i.e., primary producer) $\delta^{15}N$ and $\delta^{13}C$ values was observed from 1993 to 2005 (a decrease of ~4‰ for $\delta^{15}N$ source-AAs and 3‰ for $\delta^{13}C$ essential-AAs), while the trophic position of whales was variable over time and it did not exhibit directional trends. The baseline $\delta^{15}N$ and $\delta^{13}C$ shifts suggest rapid ongoing changes in the carbon and nitrogen biogeochemical cycling in the offshore CCS, potentially occurring at faster rates than long-term shifts observed elsewhere in the Pacific. While the mechanisms forcing these biogeochemical shifts remain to be determined, our data suggest possible links to natural climate variability, and also corresponding shifts in surface nutrient availability. Our study demonstrates that isotopic analysis of individual amino acids from a top marine mammal predator can be a powerful new approach to reconstructing temporal variation in both biochemical cycling and trophic structure.

Editor: Wei-Chun Chin, University of California, Merced, United States of America

Funding: Funding was provided by Marine Mammal and Turtle Division, Southwest Fisheries Science Center-National Oceanographic Atmospheric Administration for data collection and isotope analysis and National Science Foundation(Division of Ocean Sciences(OCE)-1155728, and OCE-0623622) for analysis of amino acids and data. Funding for Open Access provided by the University of California, Santa Cruz, Open Access Fund. The funders had no role in study design, data collection and analysis, decision to publish, or preparation of the manuscript.

Competing Interests: The authors have declared that no competing interests exist.

* Email: rcooley@ucsc.edu

Introduction

The California Current System (CCS) contains one of the five major coastal upwelling zones in the world's oceans, and hosts a great diversity and abundance of marine life [1]. The oceanographic state of this large ecosystem is dynamic. Natural climate variation and anthropogenic stressors alter biochemical cycling, food web dynamics, and the fitness of species [1–3]. Known interannual and decadal changes are related both to the El Niño-Southern Oscillation (ENSO) and to basin-scale processes associated with the Pacific Decadal Oscillation (PDO) [1].The latter, is an index of interannual sea surface temperature (SST) variability in the North Pacific, that is related to physical and biochemical variations and influences community changes in plankton, fish and other taxa [4,5]. In addition to this natural variability, humans have perturbed climate by increasing atmospheric CO_2 concentrations, which have increased ocean temperatures, water column stratification, hypoxia, and water column

anoxia and have decreased surface ocean pH [6,7]. These environmental factors may negatively impact populations of species, increasing mortality and decreasing reproductive success due to habitat compression and metabolic constraints [8]. Other anthropogenic pressures, such as intensive fisheries and the past whaling industry (which principally targeted sperm whales, *Physeter macrocephalus*) might have triggered top-down effects. Given the lack of detailed proxy records to trace simultaneously biochemical baselines and length of food webs, assessing the extent to which biogeochemical cycling and community structure in pelagic ecosystems have changed over the past century is difficult, as is attributing change to natural cycles versus anthropogenic disturbances.

The isotopic values of marine primary producers are sensitive to environmental variation, such as change in temperature, and CO_2 or nitrate concentrations, as well as biological differences such as physiology and growth rate [9–11]. Hence, the carbon and nitrogen isotope values ($\delta^{13}C$ and $\delta^{15}N$ values, respectively) of primary producers, also known as "baseline isotope values", vary in space and time as a function of these fundamental ecosystem properties [12]. Baseline isotope values are then integrated into consumers' tissues through diet, typically with metabolic fractionation leading to enrichment in the heavier isotope (especially ^{15}N) in consumers [13,14]. Therefore, isotopic values of marine consumers could be used to reconstruct changes in diet and/or ecosystem biogeochemistry. The $\delta^{13}C$ and $\delta^{15}N$ values from a resident animal can potentially provide an integrated record of the biogeochemical characteristics of its habitat, as well as its trophic position [15]. However, because multiple factors influence the bulk $\delta^{13}C$ and $\delta^{15}N$ values ultimately recorded in consumer tissues, it is often difficult to disentangle the effects of changing trophic position from shifts in baseline values.

Studies in different ocean basins have shown that bulk tissue $\delta^{13}C$ or $\delta^{15}N$ values have declined over the last century, but interpretations of these trends have varied widely [16]. For example, declining bulk tissue $\delta^{15}N$ values are sometimes attributed to a drop in consumer trophic level [17,18] or to baseline shifts due to either changes in foraging zone or biogeochemical cycles [16]. In particular, two recent studies in the Pacific have revealed pervasive declines in $\delta^{15}N$ values in the offshore Central Pacific [18] and North Pacific Subtropical Gyre (NPSG) [19], but offered diametrically opposing interpretations as to underlying mechanism. In the highly productive CCS, despite accumulating evidence for oceanographic changes since the 1950s [2], isotopic data from plankton species have been contradictory. Bulk $\delta^{15}N$ values from three zooplankton species have exhibited no long-term trends, whereas data for a specialized zooplankton feeder decreased by approximately 3‰ [20,21]. Declines in $\delta^{13}C$ values over the 20th century are expected due to the combustion of fossil fuels (i.e., the Suess effect), and have been observed in many records and ecosystems [22]. However, variability in the magnitude and timing of $\delta^{13}C$ declines has suggested that other factors, such as declining primary productivity, could also contribute in some regions [16]. In the offshore CCS, there are currently no $\delta^{13}C$ time series for organic or inorganic material.

Isotopic analysis of individual amino acids (AAs) can effectively separate trophic effects from shifts in baseline isotope values [23,24]. Regardless of an animal's trophic position, the original $\delta^{15}N$ and $\delta^{13}C$ values from primary producers are relatively well preserved within the group of 'source-AAs' for nitrogen [25] and the 'essential-AAs' for carbon [26]. In contrast, isotopic values from the 'trophic-AAs' for nitrogen, and 'non-essential-AAs' for carbon, undergo significant metabolic fractionation, and vary in association with a consumer's diet [23,24], tissue turnover rates, and possibly metabolism [27]. Hence, isotopic analysis of amino acids from apex marine mammal predators offers a unique opportunity to simultaneously investigate temporal variation at the lowest and highest trophic levels of their food web. Sperm whales are top predators of the mesopelagic ocean. Mark-recapture studies, morphology, and acoustic analysis indicate that female sperm whales forage within the same oceanic region year round [28]. Consequently, they can function as natural biological samplers, broadly integrating biogeochemical information from their home ecosystem. In this study, we use sperm whale skin as a novel biological archive of time series data. Our data combine bulk tissue and AA isotope analysis to examine temporal variation in baseline values (reflecting ecosystem biogeochemistry) and whale trophic position (indicating trophic structure) from offshore waters of the California Current ecosystem.

Results and Discussion

Foraging zone of sperm whales sampled in CCS

In the CCS off the US west coast, sperm whales are found in oceanic waters from California to Washington [29]. Their habitat therefore excludes the coastal upwelling system that exhibits strong latitudinal isotopic gradients [30]. Mitochondrial and nuclear markers reveal that the CCS whales are an independent population and a single genetic stock [31]. Our isotopic data from skin biopsies (Figure 1) indicate that whales fed homogenously within the offshore northern and central CCS. First, the variation in bulk isotope values (n = 18; SD = 1.2‰ for $\delta^{13}C$ and 1.2‰ for $\delta^{15}N$) is similar to the variation observed in other sperm whale populations that are considered to be resident (i.e., Gulf of Mexico and Gulf of California, SD≤0.8‰ for both $\delta^{15}N$ and $\delta^{13}C$ [15]; SE Pacific, SD = 3.5‰ for $\delta^{15}N$ and 0.7‰ for $\delta^{13}C$ [32]). In addition, the $\delta^{15}N$ values for phenylalanine (Phe; n = 12; mean (SD) = 10.9‰ (0.9)) are relatively consistent with expected nitrate and particulate organic matter $\delta^{15}N$ values from the oceanic northern CCS (~6 to 10 ‰) [12], and also with published Phe $\delta^{15}N$ values from muscle of the jumbo squid (Dosidicus gigas; potential prey of sperm whales) [33]. Phe $\delta^{15}N$ values are a proxy for primary producer values [25] as they exhibit only minor ^{15}N-enrichment with trophic transfer [23]. In top predators (such as sperm whales), this likely results in slightly higher Phe $\delta^{15}N$ values versus baseline inorganic N sources. Lastly, because latitudinal trends in the $\delta^{15}N$ values from predator source-AAs can indicate their geographic residency [24,33], the lack of any latitudinal variation in Phe $\delta^{15}N$ values ($r^2 = 0$; n = 12) strongly suggests that the individual sperm whales sampled here were not foraging in different localized regions, but rather foraged over a broad latitudinal range within the northern and central CCS. While the isotopic incorporation rate for extremely large animals like whales is not well known, the thick skin of sperm whales likely integrates information for at least three and possibly more than six months prior to sampling [34]. Our data set encompasses information mainly from the fall and winter, except for the samples collected in 2001 and 2003, which also integrate information from the summer.

Coupled decadal declines in $\delta^{15}N$ and $\delta^{13}C$ values

Bulk $\delta^{13}C$ and $\delta^{15}N$ values in whale skin decreased from 1993 to 2005 by 1.1‰ and 1.7‰, respectively. These decreases were statistically significant at the alpha = 0.05 level. Inclusion of a single sample available from 1972 further suggests possible longer-term temporal declines for both $\delta^{13}C$ and $\delta^{15}N$ values by ≥4‰ and >3‰, that are also statistically significant (Table 1). Together, these coupled time-series declines in bulk $\delta^{13}C$ and $\delta^{15}N$ values suggest coincident biogeochemical or trophic system perturbation (Table 1, Figure 2). In particular, the rate of decrease for bulk $\delta^{15}N$ values since the 1970's is at least five times greater than the rate for the long-term $\delta^{15}N$ decrease recently documented in the central Pacific from proteinaceous corals (2.3‰ in 150 years; annual decrease calculated at 0.015‰) [19], and it is more similar to the rate of change observed for a single zooplankton $\delta^{15}N$ record from southern California (~3‰ in 50 years) [21].

To disentangle the factors driving the declines in bulk isotope values, we analyzed individual AA isotope values, focusing on AAs that have been demonstrated to track baseline changes (as noted above, essential AA for $\delta^{13}C$ values, source AA for $\delta^{15}N$ values). Linear regression models for average $\delta^{13}C$ and $\delta^{15}N$ values from the most accurately measured essential- and source-AAs both exhibited strong negative temporal trends across all the data (i.e. for both 1972 and 1993 to 2005; Table 1, Figure 2), with drops of

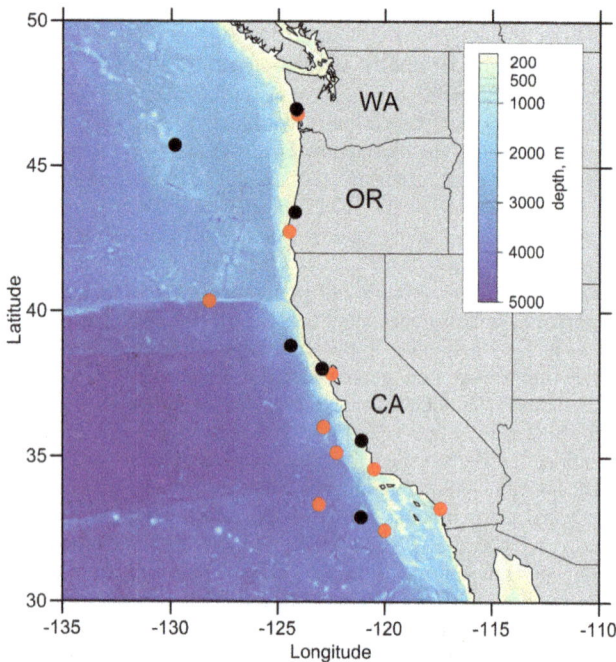

Figure 1. Sperm whales are distributed year-round in offshore deep waters (~>150 km off the US west coast [29]). Skin samples (○) from free-ranging sperm whales were collected together with skin from stranded individuals. Tissue samples were used for bulk (in black) and amino acid (in red) stable isotope analysis.

$\geq 3‰$ and $>4‰$ respectively indicated by compound-specific isotope data. Residuals for all regressions exhibited a random pattern. In contrast, average $\delta^{15}N$ values for the trophic-AAs were much variable, resulting in a lower r^2, but overall they paralleled the source-AA trend (Table 1, Figure 2A). These results are not consistent with any significant drop in sperm whale trophic level as the primary driver of decreases in bulk isotope ratios, and instead strongly implicate coupled changes in baseline $\delta^{15}N$ and $\delta^{13}C$ values.

These negative trends in baseline $\delta^{15}N$ and $\delta^{13}C$ values might relate to changes in biochemical cycling, rates of primary production, or primary producer species composition. In particular, the decline in average essential-AA $\delta^{13}C$ values (Figure 2B), which are a direct proxy for primary producers, is far too high to be explained solely by the Seuss effect (~0.2 ‰ per decade since 1960 [35]), and it also coincides with the decline in average source-AA $\delta^{15}N$ values. This suggests that the mechanism explaining a drop in primary producer $\delta^{15}N$ values should be consistent with a concurrent large decline in $\delta^{13}C$ values. One possiblity, which would represent a direct analogy to changes in other ocean regions, would be a shift towards more oligotrophic conditions for the outer CCS. This explanation would be consistent with coupled declines in both isotopes, linked to decreased primary production and a shift in species composition that is typically associated with warmer and more stratified ocean conditions [36]. Oligotrophy in the world ocean is increasing due to climate shifts [37] and is projected to continue increasing in the North Pacific [38]. Recent isotopic records from deep sea proteinaceous corals, for example, provide strong support for such linked trends associated with warming of the NPSG [39]. The nitrogen isotope record from deep sea coral indicate that the long-term declines in baseline $\delta^{15}N$ values are likely linked to progressive increases in seasonal gyre

extent, leading to steady increases in N contribution from diazotrophy [19]. Therefore, an analogous explanation would imply that oceanographic conditions in the offshore CCS region (which have conditions more similar to the open ocean and represent the base of sperm whales' food web) might have shifted toward more "gyre-like" conditions, driving baseline isotope values toward those more typical of the oligotrophic open ocean.

However, to our knowledge, there is currently no evidence for substantially increasing SST and diazotrophy in the CCS itself. Instead, recent analyses suggest largely the opposite: overall, the thermocline weakened and shoaled in the offshore CCS between 1950 and 1993 [7], possibly increasing nutrient availability in the euphotic zone despite increased stratification [40]. Additionally, the offshore CCS has cooled (not heated) since the early 1990s (Figure 3), and this trend is also reflected in the present "cool" PDO regime. Furthermore, the generalization that global warming will universally increase stratification and thus decrease surface nutrient supply has been recently challenged for some regions including the CCS [41]. For example, one recent model projects increases in nitrate supply and productivity in the CCS during the 21st century despite increases in stratification and limited change in wind-driven upwelling [42]. In the southern CCS, coastal surface nutrients have increased possibly linked to a general shoaling of the nutricline [43]. In the Southern California Bight, the most intensively monitored region of the CCS, nutrients in source waters have also increased over the last three decades, but the N:P and Si:N ratios were greatly reduced, possibly shifting phytoplankton species composition and abundance [44]. Whether or not these trends in nutrient dynamics extend to other regions of the CCS is unclear, because the oceanographic state of this ecosystem varies regionally [1,45].

In particular, shifts in offshore and onshore oceanographic conditions appear to be decoupled. Coastal upwelling has recently increased, as expected for enhanced alongshore winds [46], but has decreased offshore where upwelling is driven by wind-stress curl [47]. Since 1997, trends in satellite chlorophyll estimates, an index of phytoplankton biomass, have been positive in coastal upwelling waters but tend to be zero or negative in offshore waters [48]. Together, this current evidence indicates cooling, but not increases in productivity, in the offshore CCS concurrent with the observed 1993–2006 trends in sperm whale $\delta^{15}N$ and $\delta^{13}C$ values. Lower temperatures increase the solubility of CO_2 and change the fractionation associated with carbon fixation, often resulting in lower phytoplankton $\delta^{13}C$ values [49]; lower temperatures might have also changed phytoplankton growth and species composition. If surface nitrate also increased along the outer CCS region sampled by these whales, then the degree of nitrate utilization by primary producers (and so their $\delta^{15}N$ values) could have also changed, since phytoplankton preferentially assimilate $^{14}NO_3^-$ [50]. In general, proportional nitrate utilization is lower where surface NO_3^- concentrations are higher [50]. Therefore, lower NO_3^- utilization during seasonal upwelling might also be expected to depress the $\delta^{15}N$ values of primary producers, propagating the ^{15}N-depleted signal into food webs during their most productive periods. At present, there simply are not enough detailed data on nutrient concentrations and other oceanographic factors in the outer CCS to deduce a mechanism. However, the observed declining baseline values revealed by sperm whales do indicate a recently progressive shift in primary producer dynamics, likely associated with changes in SST, average state of surface nutrients and/or primary production.

Table 1. Temporal variation in $\delta^{15}N$ and $\delta^{13}C$ values from the offshore California Current System in sperm whale skin samples.

Time period	Tracer	Linear Regression	n	r^2	p-value	Isotopic shift (‰)	Annual decrease
1993–2005	$\delta^{15}N$						
	Bulk	y = 302–0.143 * year	17	0.25	<0.05	1.7	0.14
	Mean Source-AA	y = 717–0.354 * year	11	0.52	= 0.01	4.2	0.35
	Mean Trophic-AA	y = 311–0.143 * year	11	0.12	>0.05	1.7	
1972–2005	$\delta^{15}N$						
	Bulk	y = 218–0.101 * year	18	0.37	<0.05	3.3	0.10
	Mean Source-AA	y = 298–0.145 * year	12	0.39	<0.05	4.7	0.14
	Mean Trophic-AA	y = 88–0.031 * year	12	0.03	>0.05	1.0	
1993–2005	$\delta^{13}C$						
	Bulk	y = 174–0.095 * year	17	0.24	<0.05	1.1	0.09
	Mean Essential-AA	y = 474–0.250 * year	8	0.62	<0.05	3.0	0.25
1972–2005	$\delta^{13}C$						
	Bulk	y = 242–0.129 * year	18	0.67	<0.01	4.2	0.12
	Mean Essential-AA	y = 184–0.105 * year	9	0.58	<0.05	3.4	0.10

For mean calculations: Source-AAs are phenylalanine, glycine, lysine, tyrosine; Trophic-AA: glutamic acid, alanine, isoleucine, leucine, proline; Essential-AA: phenylalanine, valine, leucine. Isotopic shifts were calculated using the corresponding linear regression equations listed in this table. The annual decrease was calculated for shifts that exhibited a p-value ≤ 0.05.

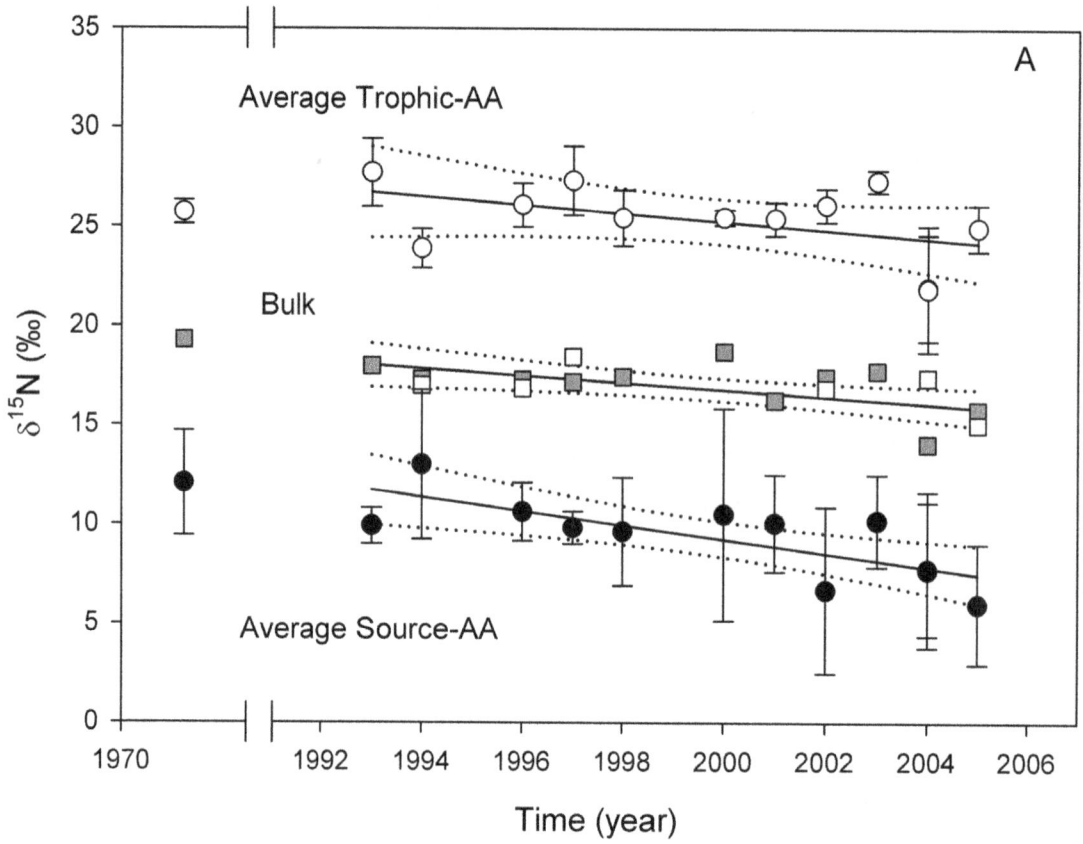

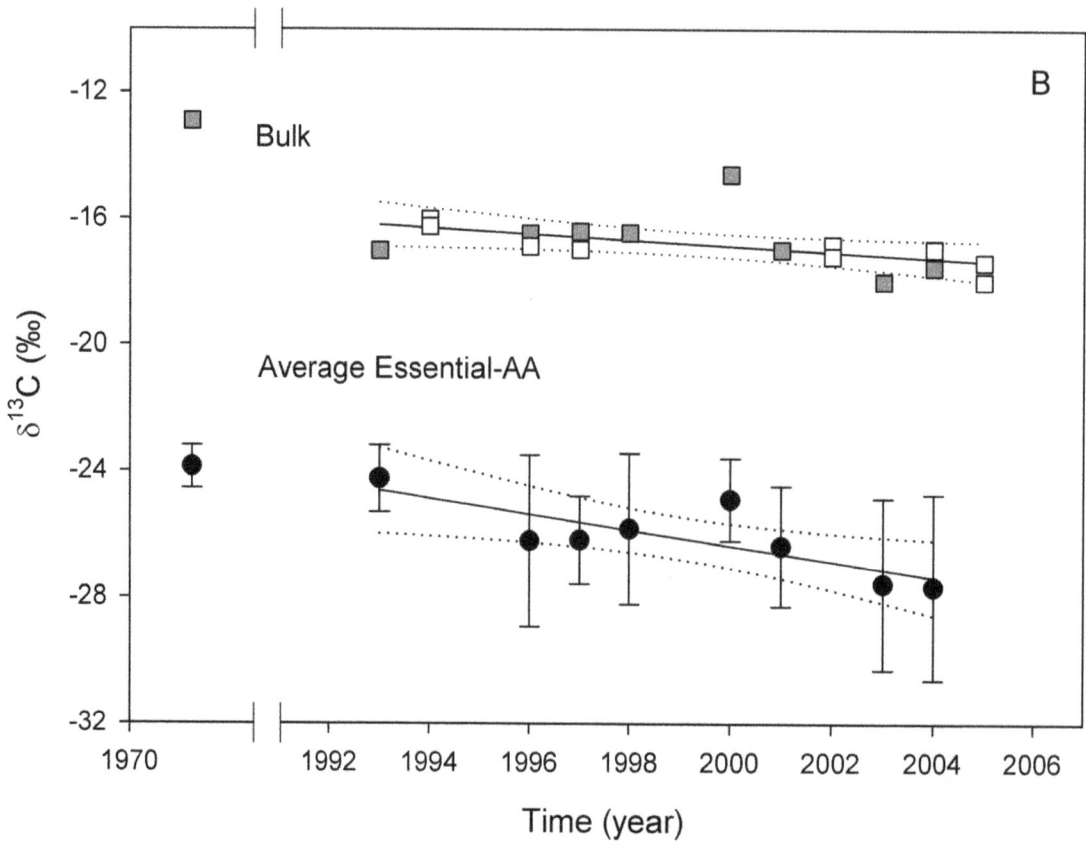

Figure 2. Time series of isotopic data from sperm whale skin. (A) $\delta^{15}N$ values from bulk skin, average source-AAs and average trophic- AAs ($\pm$ SD); and (B) $\delta^{13}C$ values from bulk skin and average essential-AAs ($\pm$SD). Bulk isotope data are plotted with a square symbol ($\square$), filled grey squares indicate the samples that were also analyzed for amino acid stable isotope analysis. The corresponding linear regression equations are provided in Table 1, as are the amino acids included within each AA-group.

Implications of Rapid Change for offshore CCS Biogeochemistry

Although our time series data are limited for both elements, the compound-specific AA data identify a parallel decline in both baseline $\delta^{15}N$ and $\delta^{13}C$ values in the outer CCS from 1992 to 2005, likely indicative of major recent shifts in biochemical cycling. At the same time, however, the overall similarity in whale trophic position signifies that the broad trophic structure is realtively unaffected. We note that in comparison with the recent deep sea coral data from the gyre offshore of this region [19], our data suggest that both the rate and scale of biochemical change on the CCS margin may be far greater than in the open Pacific Ocean. The coral record from the NPSG indicates a fairly steady $\delta^{15}N$ annual decrease of ~0.015‰ over the last 150 years with a total drop of 2.3 ‰ in exported primary production $\delta^{15}N$ values over that period. In contrast, our molecular-level proxies for $\delta^{15}N$ values at the base of the food chain (the source AAs) indicate more rapid annual declines of 0.35 ‰ since the 1990's. The independent molecular proxies for primary production $\delta^{13}C$ values (the essential AAs) indicate relatively similar declines.

Together with the CCS observations discussed above, the contrast with the NPSG coral data (while not directly comparable in terms of time scale), suggests that despite the fact that baseline $\delta^{15}N$ declines are observed in both data sets, different biogeochemical mechanisms may underlie the changes in these very different oceanographic regions. Climate variability likely affects the biochemistry of ecosystems differently depending on the oceanographic properties, microbial and phytoplankton communities, and species assemblages. In the eastern Pacific Ocean, the structure of the pycnocline varies strongly among the known biogeochemical provinces [51]. This likely influences geographic variation in surface nutrient availability, and therefore stable isotope ratios in POM, primary producers [12] and consumers [14]. Temporal trends in pycnocline depth, SST, stratification, and mixed layer depths also differ between these biogeochemical provinces [40]. For example, while SST decreased overall since 1958 in many parts of the California Current, SST increased in

the easternmost southern subtropical gyre and equatorial Pacific [40]. Ultimately, more detailed data that couple integrated measures of ecosystem baseline with oceanographic state will be required to understand the substantial biogeochemical changes our data indicate.

Our work highlights that detailed time-series of biochemical baseline and trophic structure records among different ecosystems will be crucial to identify rapid ecosystem shifts in response to climate change. In particular, in the face of uncertain coupling of natural and anthropogenic climate forcing, understanding the timing, extent and especially the mechanistic basis for baseline shifts now represents an urgent challenge. However, despite many efforts to unravel the linkage and feedback controls between the carbon and nitrogen cycles, and the effect of their variability on primary production and food-web dynamics, they are still not well understood. This study has demonstrated the great potential in coupling molecular isotopic tools with the unique bioarchive of sperm whales (or other top predators), as sentinels of offshore ecosystems. This may allow, for the first time, decoding of the factors that underlie temporal trends in bulk isotopic records, while simultaneosly monitoring changes at both the highest and lowest trophic levels. We suggest that integrating this approach with detailed oceanographic data will be a major new tool to identify the effects of natural climate variability versus anthropogenic global warming on ecosystem biochemistry and primary production. Elucidating such patterns from this and other ocean margin regions, in particular their relationships with oceanographic and climatic variations and shifts in primary production, will be an essential part of the critical task of predicting future trends in both ecosystem biochemistry and trophic dynamics.

Material and Methods

A total of 18 skin samples (Figure 1) were analyzed for bulk stable isotope analysis. Skin tissue samples with enough material (3.5 mg) were selected for CSIA-AA. Data from 12 samples were obtained for individual AA $\delta^{15}N$ values, and 9 samples for $\delta^{13}C$ values. The Southwest Fisheries Science Center/Pacific Islands

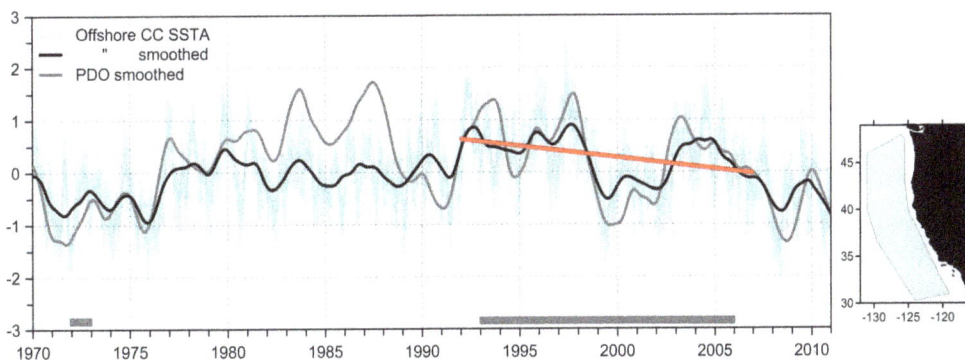

Figure 3. Time series data of sea surface temperature anomaly (SSTA) from the offshore California Current (inset map) and the Pacific Decadal Oscillation (PDO). Monthly SSTA was computed in 0.5-deg fields from the Simple Ocean Data Assimilation version 2.2.4 reanalysis (http://coastwatch.pfeg.noaa.gov/erddap/griddap/hawaii_d90f_20ee_c4cb.html), and then averaged in the offshore area (the plot shows $\pm$1sd). Monthly SSTA (°C) and PDO values were smoothed with a 25-month lowess smooth. The linear fit is for 1992–2006 (red line, slope -0.044°C y-1). Sample periods are indicated along the time axis.

Fisheries Science Center Institutional Animal Care and Use Committee (IACUC) approved the original animal work that produced the samples. Sex was determined genetically using qPCR sexing assay by the PRD-Genetic Lab at NOAA [52]. These samples consisted of 5 females, 2 males and 2 unidentified individuals possibly corresponding to females or juvenile males. Large adult males were not included. Bulk isotope values were analyzed by continuous flow isotope ratio mass spectrometry (IRMS; Thermo Finnigan) and standardized relative to Vienna-Pee Belemnite (V-PDB) for carbon and atmospheric N_2 for nitrogen. Results are expressed in part per thousand (‰) and standard notation: $\delta^H X = [(R_{sample}/R_{standard}) - 1] \times 1000$, where H is the mass number of the heavy isotope, X is either C or N, and R_{sample} and $R_{standard}$ are the ratio of $^{13}C/^{12}C$ or $^{15}N/^{14}N$ in the sample and standard, respectively.

We hydrolyzed and prepared approximately 3.5 mg of skin as well as a control (Cyanno; bacteria tissue) [53] to quantify $\delta^{15}N$ values from source- and trophic-AAs and $\delta^{13}C$ values from essential- and non-essential-AAs. All derivatives were injected with an AA control, N-leucine, to verify accuracy during each run, and analyzed via gas chromatography-IRMS to obtain $\delta^{15}N$ and $\delta^{13}C$ values from individual AAs. Each sample was run 3–4 times to maximize accuracy among chromatograms. The associated analytical error among replicates was <1.0 ‰. For all samples, $\delta^{15}N$ values were obtained from a total of four source-AAs (phenylalanine, glycine, lysine, tyrosine), and five trophic-AAs (glutamic acid, alanine, isoleucine, leucine, proline) (Figure S1A). For $\delta^{13}C$ values, the essential-AAs that we consistently determined were phenylalanine, valine and leucine, and the non-essential-AA were alanine, proline, aspatic acid, glutamic acid and tyrosine (Figure S1B).

The relative pattern of AA $\delta^{15}N$ and $\delta^{13}C$ values was highly consistent with past work from other organisms and tissues [23,25,54]. We grouped data as source- or trophic-AAs for $\delta^{15}N$ values, and essential- or non-essential-AAs for $\delta^{13}C$ values to increase power in the analysis and evaluate temporal variation. We calculated average values for each AA group and they are reported in Table S1. Regression analyses were conducted to evaluate linear relationship between time and each isotopic tracer for both bulk and individual-AA $\delta^{15}N$ and $\delta^{13}C$ values (Table 1).

There was a weak correlation between average source-AA and trophic-AA ($r^2 = 0.13$; $p = 0.67$), indicating that trophic-AA $\delta^{15}N$ values could not be predicted by the variability in source-AAs, and vice versa. However, the correlation between average essential-AA and non-essential-AA $\delta^{13}C$ values was moderate ($r^2 = 0.63$, $p = 0.06$). Since the controls on isotopic patterns for non-essential-AA $\delta^{13}C$ values are complex and dependent on diet quality and quantity, including *de novo* synthesis and routing of AAs from diet-to-tissue, this group was not considered in the linear regression analysis.

Supporting Information

Figure S1 Stable isotope values of individual amino acids (AAs) in skin samples of sperm whales (*Physeter macrocephalus*). (A) Four $\delta^{15}N$ Source-AAs: phenylalanine (phe), glycine (gly), lysine (lys), tyrosine (tyr), and five Trophic-AAs: glutamic acid (glx), alanine (ala), isoleucine (ile), leucine (leu), proline (Pro); and (B) Three $\delta^{13}C$ essential-AAs: phe, leu, and valine (val).

Table S1 Average values and one standard deviations (SD) were calculated for Source-AAs (phenylalanine, glycine, lysine, tyrosine), Trophic-AAs (glutamic acid, alanine, isoleucine, leucine, proline) and Essential-AAs (phenylalanine, valine, leucine).

Acknowledgments

We thank L. T. Ballance, J. Barlow, K. Robertson (SWFSC/NMFS/NOAA) and J. Calambokidis (Cascadia Research) for facilitating the use of tissues samples, and the genetic SWFSC lab for molecular whale sex identification.

Author Contributions

Conceived and designed the experiments: RIRC MDM. Performed the experiments: RIRC MDM. Analyzed the data: RIRC PCF. Contributed reagents/materials/analysis tools: MDM. Contributed to the writing of the manuscript: RIRC PLK PCF MDM.

References

1. Checkley JDM, Barth JA (2009) Patterns and processes in the California Current System. Prog Oceanogr 83: 49–64.
2. Bograd SJ, William JS, Barlow J, Booth A, Brodeur RD, et al. (2010) Status and trends of the California Current region, 2003–2008. PICES Special Publication. 106–141 p.
3. McGowan JA, Bograd SJ, Lynn RJ, Miller AJ (2003) The biological response to the 1977 regime shift in the California Current. Deep Sea Res II 50: 2567–2582.
4. Brinton E, Townsend A (2003) Decadal variability in abundances of the dominant euphausiid species in southern sectors of the California Current. Deep Sea Res II 50: 2449–2472.
5. Chavez FP, Ryan J, Lluch-Cota SE, Ñiquen CM (2003) From anchovies to sardines and back: Multidecadal change in the Pacific Ocean. Science 299: 217–221.
6. Chan F, Barth JA, Lubchenco J, Kirincich A, Weeks H, et al. (2008) Emergence of anoxia in the California Current Large Marine Ecosystem. Science 319: 920.
7. Palacios DM, Bograd SJ, Mendelssohn R, Schwing FB (2004) Long-term and seasonal trends in stratification in the California Current, 1950–1993. J Geophy Res 109: C10016.
8. Bograd SJ, Castro CG, Di Lorenzo E, Palacios DM, Bailey H, et al. (2008) Oxygen declines and the shoaling of the hypoxic boundary in the California Current. Geophys Res Lett 35: L12607.
9. Farrell JW, Pedersen TF, Calvert SE, Nielsen B (1995) Glacial-interglacial changes in nutrient utilization in the equatorial Pacific Ocean. Nature 377: 514–517.
10. Rau GH, Sweeney RE, Kaplan IR (1982) Plankton ^{13}C:^{12}C ratio changes with latitude: differences between northern and southern oceans. Deep Sea Res 29: 1035–1039.
11. Goericke R, Fry B (1994) Variations of marine plankton $\delta^{13}C$ with latitude, temperature, and dissolved CO_2 in the World Ocean. Global Biogeochem Cy 8: 85–90.
12. Somes CJ, Schmittner A, Galbraith ED, Lehmann MF, Altabet MA, et al. (2010) Simulating the global distribution of nitrogen isotopes in the ocean. Global Biogeochem Cy 24.
13. Peterson BJ, Fry B (1987) Stable Isotopes in Ecosystem Studies. Annu Rev Ecol Evol Syst 18: 293–320.
14. Ruiz-Cooley RI, Gerrodette T (2012) Tracking large-scale latitudinal patterns of $\delta^{13}C$ and $\delta^{15}N$ along the eastern Pacific using epi-mesopelagic squid as indicators. Ecosphere 3: 63.
15. Ruiz-Cooley R, Engelhaupt D, Ortega-Ortiz J (2012) Contrasting C and N isotope ratios from sperm whale skin and squid between the Gulf of Mexico and Gulf of California: effect of habitat. Mar Biol: 1–14.
16. Schell DM (2001) Carbon isotope ratio variations in Bering Sea biota: The role of anthropogenic carbon dioxide. Limnol Oceanogr Methods 46: 999–1000.
17. Emslie SD, Patterson WP (2007) Abrupt recent shift in $\delta^{13}C$ and $\delta^{15}N$ values in Adélie penguin eggshell in Antarctica. Proc Natl Acad Sci USA 104: 11666–11669.
18. Wiley AE, Ostrom PH, Welch AJ, Fleischer RC, Gandhi H, et al. (2013) Millennial-scale isotope records from a wide-ranging predator show evidence of recent human impact to oceanic food webs. Proc Natl Acad Sci USA 110: 8972–8977.
19. Sherwood OA, Guilderson TP, Batista FC, Schiff JT, McCarthy MD (2013) Increasing subtropical North Pacific Ocean nitrogen fixation since the Little Ice Age. Nature 505: 78–81.

20. Rau GH, Ohman MD, Pierrot-Bults A (2003) Linking nitrogen dynamics to climate variability off central California: a 51 year record based on $^{15}N/^{14}N$ in CalCOFI zooplankton. Deep Sea Res Part II 50: 2431–2447.

21. Ohman MD, Rau GH, Hull PM (2012) Multi-decadal variations in stable N isotopes of California Current zooplankton. Deep Sea Res Part I 60: 46–55.

22. Sonnerup RE, Quay PD, McNichol AP, Bullister JL, Westby TA, et al. (1999) Reconstructing the oceanic ^{13}C Suess Effect. Global Biogeochem Cy 13: 857–872.

23. Chikaraishi Y, Ogawa NO, Kashiyama Y, Takano Y, Suga H, et al. (2009) Determination of aquatic food-web structure based on compound-specific nitrogen isotopic composition of amino acids. Limnol Oceanogr Methods 7 740–750.

24. Popp BN, Graham BS, Olson RJ, Hannides CCS, Lott MJ, et al. (2007) Insight into the trophic ecology of yellowfin tuna, Thunnus albacares, from compound-specific nitrogen isotope analysis of proteinaceous amino acids. In: Dawson TD, Siegwolf, R. T W., editor. Stable isotopes as indicators of ecological change. New York: Elsevier Academic Press. pp. 173–190.

25. McClelland JW, Montoya JP (2002) Trophic relationships and the nitrogen isotopic composition of amino acids in plankton. Ecology 83: 2173–2180.

26. O'Brien DM, Fogel ML, Boggs CL (2002) Renewable and nonrenewable resources: Amino acid turnover and allocation to reproduction in Lepidoptera. Proc Natl Acad Sci USA 99: 4413–4418.

27. Germain LR, Koch PL, Harvey JT, McCarthy MD (2013) Nitrogen isotopic fractionation of amino acids in harbor seals (Phoca vitulina): Differential trophic enrichment factors based on ammonia vs. urea excretion. Mar Ecol Prog Ser 482: 265–277.

28. Default S, Whitehead H, Dillon M (1999) An examination of the current knowledge on the stock structure of sperm whales (Physeter macrocephalus) worldwide. J Cetac Res Manage 1: 1–10.

29. Carretta JV, Forney KA, Lowry MS, Barlow J, Baker J, et al. (2010) U.S. Pacific marine mammal stock assessments: 2009. California, USA. 336 p.

30. Sigman DM, Casciotti KL (2001) Nitrogen Isotopes in the Ocean. In: Editor-in-Chief: John HS, editor. Encyclopedia of Ocean Sciences. Oxford: Academic Press. pp. 1884–1894.

31. Mesnick SL, Taylor BL, Archer FI, Martien KK, TreviÑO SE, et al. (2011) Sperm whale population structure in the eastern and central North Pacific inferred by the use of single-nucleotide polymorphisms, microsatellites and mitochondrial DNA. Mol Ecol Resour 11: 278–298.

32. Marcoux M, Whitehead H, Rendell L (2007) Sperm whale feeding variation by location, year, social group and clan: Evidence from stable isotopes. Mar Ecol Prog Ser 333: 309–314.

33. Ruiz-Cooley RI, Ballance LT, McCarthy MD (2013) Range expansion of the jumbo squid in the NE Pacific: $\delta^{15}N$ decrypts multiple origins, migration and habitat Use. PLoS ONE 8: e59651.

34. Ruiz-Cooley RI, Gendron D, Aguiniga S, Mesnick S, Carriquiry JD (2004) Trophic relationships between sperm whales and jumbo squid using stable isotopes of C and N. Mar Ecol Prog Ser 277: 275–283.

35. Francey RJ, Allison CE, Etheridge DM, Trudinger CM, Enting IG, et al. (1999) A 1000-year high precision record of $\delta^{13}C$ in atmospheric CO_2. Tellus B 51: 170–193.

36. Karl DM, Bidigare RR, Letelier RM (2001) Long-term changes in plankton community structure and productivity in the North Pacific Subtropical Gyre: The domain shift hypothesis. Deep Sea Res Part II 48: 1449–1470.

37. Polovina JJ, Howell EA, Abecassis M (2008) Ocean's least productive waters are expanding. Geophys Res Lett 35: L03618.

38. Polovina JJ, Dunne JP, Woodworth PA, Howell EA (2011) Projected expansion of the subtropical biome and contraction of the temperate and equatorial upwelling biomes in the North Pacific under global warming. ICES J Mar Sci 68: 986–995.

39. Guilderson TP, McCarthy MD, Dunbar RB, Englebrecht A, Roark EB (2013) Late Holocene variations in Pacific surface circulation and biogeochemistry inferred from proteinaceous deep-sea corals. Biogeosciences 10: 3925–3949.

40. Fiedler PC, Mendelssohn R, Palacios DM, Bograd SJ (2012) Pycnocline Variations in the Eastern Tropical and North Pacific, 1958–2008. J Climate 26: 583–599.

41. Dave AC, Lozier MS (2013) Examining the global record of interannual variability in stratification and marine productivity in the low-latitude and mid-latitude ocean. J Geophysi Res-Oceans 118: 3114–3127.

42. Rykaczewski RR, Dunne JP (2010) Enhanced nutrient supply to the California Current Ecosystem with global warming and increased stratification in an earth system model. Geophys Res Lett 37: L21606.

43. Aksnes DL, Ohman MD (2009) Multi-decadal shoaling of the euphotic zone in the southern sector of the California Current System. Limonol Oceanogr 54: 1272–1281.

44. Bograd SJ, Buil MP, Lorenzo ED, Castro CG, Schroeder ID, et al. (2014) Changes in source waters to the Southern California Bight. Deep Sea R Part II. Available: http://dx.doi.org/10.1016/j.dsr2.2014.04.009.

45. McClatchie S (2013) Regional fisheries oceanography of the California Current System: the CalCOFI Program. Dordrecht: Springer. 253 p.

46. García-Reyes M, Largier J (2010) Observations of increased wind-driven coastal upwelling off central California. J Geophy Res 115.

47. Jacox MG, Moore AM, Edwards CA, Fiechter J (2014) Spatially resolved upwelling in the California Current System and its connections to climate variability. Geophysl Res Lett 41: 3189–3196.

48. Kahru M, Kudela RM, Manzano-Sarabia M, Greg Mitchell B (2012) Trends in the surface chlorophyll of the California Current: Merging data from multiple ocean color satellites. Deep Sea Res Part II 77–80: 89–98.

49. Rau GH, Takahashi T, Marais DJD (1989) Latitudinal variations in plankton $\delta^{13}C$: implications for CO_2 and productivity in past oceans. Nature 341: 516–518.

50. Wada E, Hattori A (1991) Nitrogen in the sea: forms, abundances, and rate processes. Boca Raton: CRC Press. 208 p.

51. Longhurst AR (2007) Ecological Geography of the Sea; Press. EA, editor. 542 p.

52. Morin PA, Nestler A, Rubio-Cisneros NT, Robertson KM, Mesnick S (2005) Interfamilial characterization of a region of the ZFX and ZFY genes facilitates sex determination in cetaceans and other mammals. Mol Ecol 14: 3275–3286.

53. McCarthy MD, Benner R, Lee C, Fogel M (2007) Amino acid nitrogen isotopic fractionation patterns as indicators of heterotrophy in plankton, particulate, and dissolved organic matter. Geochim Cosmochim Acta 71: 4727–4744.

54. Sherwood OA, Lehmann MF, Schubert CJ, Scott DB, McCarthy MD (2011) Nutrient regime shift in the western North Atlantic indicated by compound-specific $\delta^{15}N$ of deep-sea gorgonian corals. Proc Natl Acad Sci USA.

14

Unmanned Aerial Survey of Fallen Trees in a Deciduous Broadleaved Forest in Eastern Japan

Tomoharu Inoue[1]*, Shin Nagai[1], Satoshi Yamashita[2], Hadi Fadaei[1], Reiichiro Ishii[1], Kimiko Okabe[2], Hisatomo Taki[2], Yoshiaki Honda[3], Koji Kajiwara[3], Rikie Suzuki[1]

1 Department of Environmental Geochemical Cycle Research, Japan Agency for Marine-Earth Science and Technology (JAMSTEC), Yokohama, Japan, 2 Department of Forest Entomology, Forestry and Forest Products Research Institute (FFPRI), Tsukuba, Ibaraki, Japan, 3 Center of Environmental Remote Sensing, Chiba University, Chiba, Japan

Abstract

Since fallen trees are a key factor in biodiversity and biogeochemical cycling, information about their spatial distribution is of use in determining species distribution and nutrient and carbon cycling in forest ecosystems. Ground-based surveys are both time consuming and labour intensive. Remote-sensing technology can reduce these costs. Here, we used high-spatial-resolution aerial photographs (0.5–1.0 cm per pixel) taken from an unmanned aerial vehicle (UAV) to survey fallen trees in a deciduous broadleaved forest in eastern Japan. In nine sub-plots we found a total of 44 fallen trees by ground survey. From the aerial photographs, we identified 80% to 90% of fallen trees that were >30 cm in diameter or >10 m in length, but missed many that were narrower or shorter. This failure may be due to the similarity of fallen trees to trunks and branches of standing trees or masking by standing trees. Views of the same point from different angles may improve the detection rate because they would provide more opportunity to detect fallen trees hidden by standing trees. Our results suggest that UAV surveys will make it possible to monitor the spatial and temporal variations in forest structure and function at lower cost.

Editor: Dale A. Quattrochi, NASA Marshall Space Flight Center, United States of America

Funding: This study was funded by the Environment Research and Technology Development Fund (S-9) of the Ministry of the Environment of Japan. The funders had no role in study design, data collection and analysis, decision to publish, or preparation of the manuscript.

Competing Interests: The authors have declared that no competing interests exist.

* Email: tomoharu@jamstec.go.jp

Introduction

Fallen trees are an ecologically relevant indicator of forest biodiversity [1], since they provide habitat for many species, such as small animals (e.g., birds, mammals, and insects) and fungi [2–4]. In addition, the decomposition of fallen trees is an important mechanism driving biogeochemical cycles such as nutrient and carbon cycling [5–6]. Thus, information about their spatial distribution is of use in indicating species distribution, biodiversity, and biogeochemical cycles in forest ecosystems [7–8]. Fallen trees are typically assessed by field surveys [9]. However, because field surveys are time consuming and labour intensive [10], it is expensive to assess the spatial distribution of fallen trees over a wide area. One way to reduce costs is to use remote-sensing technologies such as airborne and satellite imagery (e.g., [10–14]). For example, some studies have utilized the remote-sensing technologies for assessment of post-hurricane forest damage (e.g., [15–16]).

Recent advances in the technology of unmanned aerial vehicles (UAVs) have made UAVs ideal for remote sensing [17–20]. Equipped with sensors such as digital cameras, UAVs can gather aerial photographs with fine spatial and temporal resolution [20].

They also have greater flexibility in flying height and schedule and lower operating costs than manned aircraft [18,19,21]. For these reasons, aerial surveys using UAVs have been suggested for low-cost ecological research (e.g., vegetation monitoring and wildlife surveys) [19,20,22–26].

In this study, we photographed a deciduous broadleaved forest from a UAV, compared the positions of fallen trees identified in the images by eye with those recorded on the ground, and compared the number of the fallen trees detected between the original images and an orthorectified mosaic. The purpose of this study was to show the applicability and limits of this technique.

Materials and Methods

Study site

The study was conducted in a 300-m × 200-m plot in the Ogawa Forest Reserve (OFR) in Kitaibaraki, Ibaraki, Japan (36°56′10″N, 140°35′18″E, 650–700 m above sea level; Figs. 1, 2). The forest was dominated by deciduous broadleaved trees such as oak (*Quercus serrata*) and beech (*Fagus japonica* and *F. crenata*) [27], with a patchy distribution of dwarf bamboos on the forest floor [28]. The OFR is described in detail by Nakashizuka and

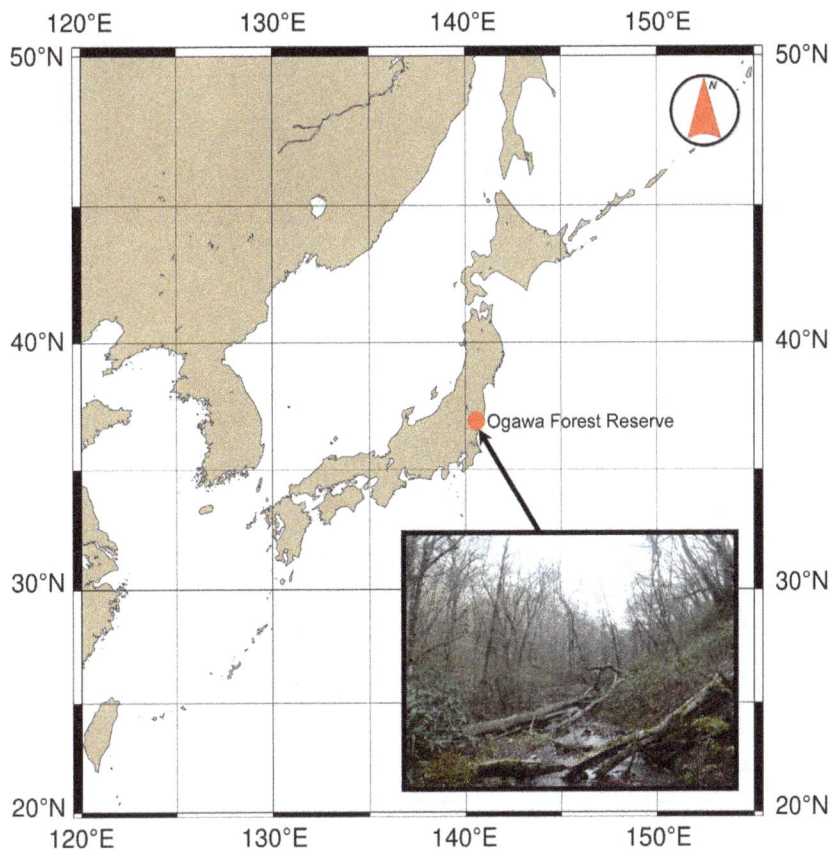

Figure 1. Location of the Ogawa Forest Reserve (OFR).

Matsumoto [29]. This study was conducted including a national forest under the permission of the Ibaraki district forest office of the Forestry Agency in Japan.

Collection of low-altitude aerial photographs from a UAV

On 29 November 2011, when the trees were bare, we flew a UAV (RMAX-G1 helicopter, Yamaha-Motor Co. Ltd., Shizuoka, Japan; Fig. 3; a model commonly used in Japan (e.g., [30–32])) over the OFR plot at 30 to 70 m above the ground in a north–south orientation at 3 m s^{-1}. A consumer-grade digital camera with a 35-mm lens (EOS Kiss X5, Canon, Tokyo, Japan; image sensor 14.9 mm × 22.3 mm) mounted beneath the UAV pointing straight down took images (5184 × 3456 pixels) every 5 s. The

UAV was also equipped with a global positioning system (GPS) detector that recorded altitude, latitude, and longitude every 0.1 s. The GPS position with the timestamp closest to that of each photograph was used as the position of the UAV.

Generation of a digital elevation model (DEM)

The UAV was also equipped with a laser range finder (LRF; SkEyesBOX MP-1, SkEyes Unlimited Corp., Washington, PA, USA), which assembled a 3D point cloud. We divided the plot (300 m × 200 m) into 1-m × 1-m cells, and the lowest elevation in each cell was used in the creation of a digital elevation model (DEM) of the site.

Table 1. Relationship of maximum diameter between ground-surveyed and visually identified fallen trees.

Maximum diameter* of fallen tree (m)	Number of ground-surveyed fallen trees	Number of visually identified fallen trees	Identification rate (%)
0.30≤ x	8	7	88
0.20≤ x <0.30	7	2	29
0.10≤ x <0.20	28	2	7
0.05≤ x <0.10	1	0	0

*Maximum of diameters at each end and middle (see Table S1).

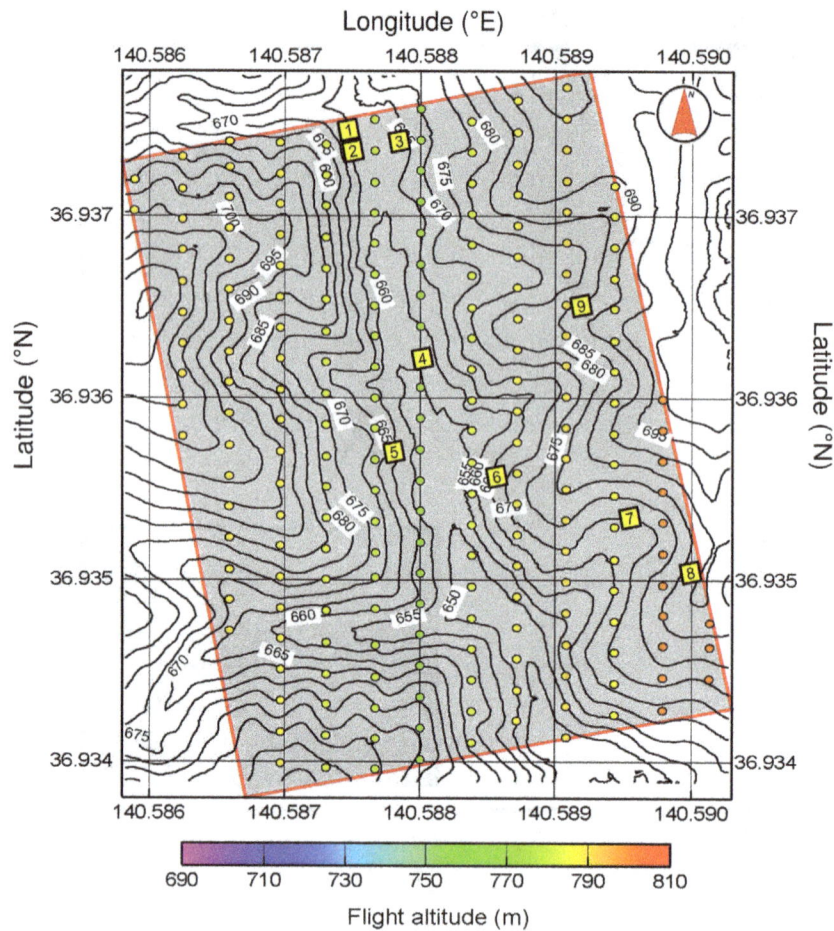

Figure 2. OFR plot (grey area). Yellow squares mark nine sub-plots for ground survey. Circles indicate UAV photograph points. Circle colour indicates flight altitude.

Orthorectification and mosaic of aerial photographs

The aerial photographs were orthorectified according to the DEM and the position (longitude and latitude) and attitude (pitch, roll, and heading) of the UAV, and assembled into one mosaic image with the help of "tie points" in overlapping photographs.

Ground survey

On 31 May 2012 in nine 10-m × 10-m sub-plots (Fig. 2), ground observers recorded the position, stem diameters (each end and midpoint) and length of all fallen trees with a diameter of >

5 cm. The ground survey was conducted in leafy season because it was easier to check the tree species in the study site when there were leaves on the trees, as opposed to a leaf-off season.

Detection of fallen trees in aerial photographs

Fallen trees were identified by eye in the original aerial photographs on a computer monitor. Positions were compared with those of fallen trees mapped in the ground survey and those identified by eye in the orthorectified mosaic.

Table 2. Relationship between lengths of ground-surveyed and visually identified fallen trees.

Length of fallen tree (m)	Number of ground-surveyed fallen trees	Number of visually identified fallen trees	Identification rate (%)
$10 \leq x$	9	7	78
$5 \leq x < 10$	15	3	20
$0 \leq x < 5$	20	1	5

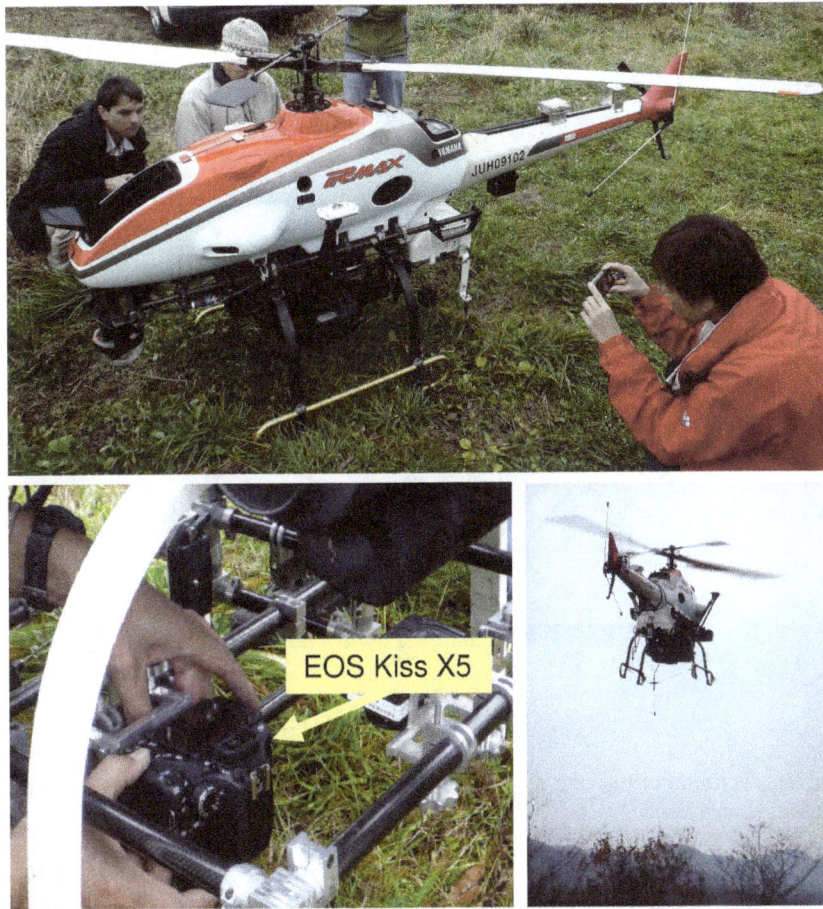

Figure 3. UAV (RMAX-G1) equipped with a digital camera (EOS KISS X5).

Figure 4. Close-up image of part of the forest floor. Two fallen trees are detectable.

Results

The OFR plot was covered by 211 aerial photographs (Fig. 2) with a spatial resolution of 0.5 to 1.0 cm per pixel. Since the trees were bare in late November, fallen trees were visible (Fig. 4). A DEM of the plot was generated from a cloud of 5445612 points (Fig. 5). The DEM showed a valley with an elevation range of 640 m (in the south) to 720 m (in the north). From the DEM and the position of the UAV, an orthorectified mosaic image was generated (Fig. 6).

In the ground survey, we found a total of 44 fallen trees in the sub-plots (Table S1). By eye, however, we identified only 11 fallen trees in the sub-plots on the original images (Table S1). We identified 80% to 90% of fallen trees which were >30 cm in diameter or >10 m in length, but few that were thinner or shorter (Tables 1, 2). Over the whole plot, we detected 244 fallen trees on the original photographs and 209 on the orthorectified mosaic (Fig. 6).

Discussion

Because fallen trees are generally defined as being >2.5 cm in diameter [2], we assumed that the high spatial resolution of our aerial photographs (0.5–1.0 cm per pixel) would allow us to detect

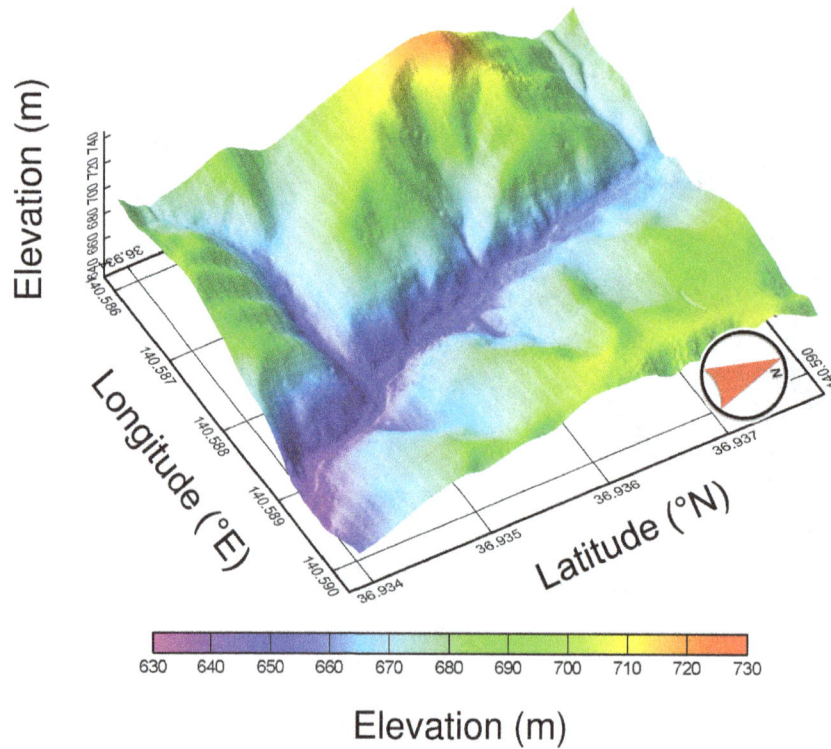

Figure 5. Digital elevation model (DEM) of the OFR plot. This DEM was used for orthorectification of the aerial photographs.

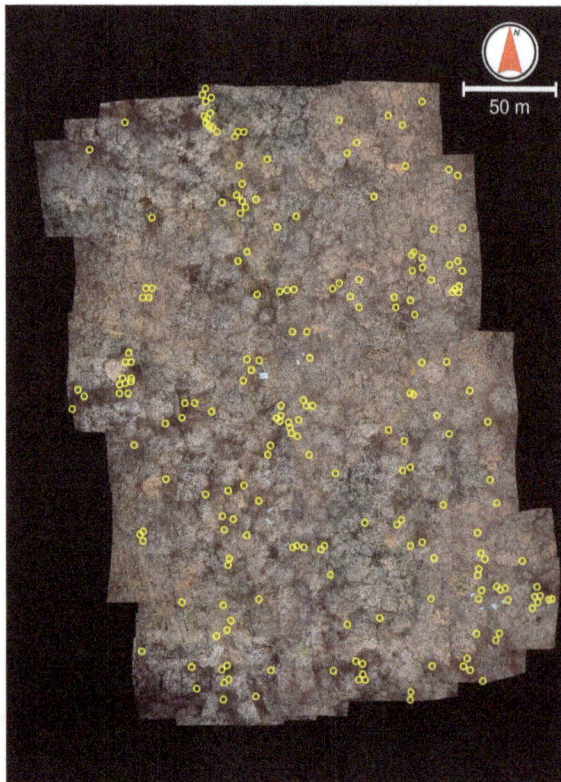

Figure 6. Fallen trees (yellow circles) detected by eye in the orthorectified mosaic.

them. However, our results suggest difficulty in identifying narrow or short fallen trees (Tables 1, 2). This failure may be due to the similarity of fallen trees to trunks and branches of standing trees, and masking of fallen trees by the branches of standing trees and forest floor vegetation. Hodgson et al., who surveyed marine mammals by UAV in Australia, reported the usefulness of overlaps between photographs for detecting animals that are masked by sun glitter [24]. For a similar reason, views of the same point from different angles may provide more opportunity to detect fallen trees hidden by standing trees (Fig. 7). The memory capacity of the camera [24] will determine the balance between coverage and overlap. The optimal degree of overlap will also depend on flying speed, flight altitude, and camera specifications (e.g., frames per second).

Another factor contributing to the poor rate of visual identification of fallen trees might be ambiguity in colour. One of the clues we used in detecting fallen trees was the colour of mosses growing on them. Trees with mosses are easy to detect visually, but freshly fallen trees with no mosses might be confused with tree branches or fallen leaves.

The orthorectification and mosaicking of aerial photographs according to the DEM require much labour and time, even for experts, and can also require specialized and expensive software. However, we identified fewer fallen trees from the orthorectified mosaic than from the original photographs. Thus, non-orthorectified photographs of the same point from different angles would allow better identification of fallen trees, at no additional cost, than orthorectified mosaics.

Conclusions

We showed the applicability of aerial photographs captured from a UAV for the detection of large fallen trees in a deciduous

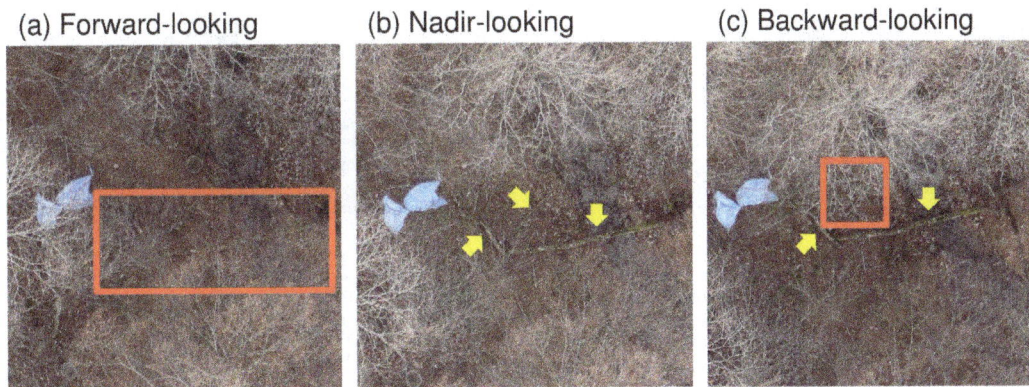

Figure 7. Example of overlapping images of the same point taken from different angles. In this example, the visual detection of three fallen trees from the forward- and backward-looking images may be difficult owing to masking by tree branches (within red boxes). The fallen trees are clearly visible in the nadir-looking image (yellow arrows).

broadleaved forest in eastern Japan. Because much tree death is episodic and irregular [33], high-frequency monitoring at multiple points is necessary for the detection of newly fallen trees and for the understanding of species distribution, biodiversity, and nutrient and carbon cycling in forest ecosystems. UAVs now permit high-frequency monitoring at low cost. This approach has great potential for forest ecology, especially for measuring temporal and spatial variations in forest structure and functioning. Furthermore, as UAVs are advancing and the payload of them is increasing, new sensors for forest monitoring on an UAV will become more common in the future [26]. Installation of a multi-angle photographing system, like a PRISM (Panchromatic Remote-sensing Instrument for Stereo Mapping) on-board the Japanese satellite ALOS (Advanced Land Observing Satellite), on an UAV would potentially help to find more opportunity for detection of fallen trees hidden by standing trees and it may increase the accuracy of fallen tree identification.

Acknowledgments

The authors are thankful to the administrators of the OFR site of the Forestry and Forest Products Research Institute (FFPRI) for their cooperation during the field data collection. We also thank the anonymous reviewers and editor for their helpful and constructive comments.

Author Contributions

Conceived and designed the experiments: RS SY KK HT SN RI HF KO YH. Performed the experiments: RS SY KK HT SN RI HF KO YH. Analyzed the data: KK RS TI SN HF. Contributed to the writing of the manuscript: TI SN RS.

References

1. Biała K, Condé S, Delbaere B, Jones-Walters L, Torre-Marín A (2012) Streamlining European biodiversity indicators 2020: Building a future on lessons learnt from the SEBI 2010 process. EEA Tech Rep No.11/2012. doi: 10.2800/55751.
2. Harmon ME, Franklin JF, Swanson FJ, Sollins P, Gregory SV, et al. (1986) Ecology of coarse woody debris in temperate ecosystems. Adv Ecol Res 15: 133–302. doi: 10.1016/S0065-2504(08)60121-X.
3. Bunnell FL, Houde I (2010) Down wood and biodiversity – implications to forest practices. Environ Rev 18: 397–421. doi: 10.1139/A10-019.
4. Stokland JN, Siitonen J, Jonsson BG (2012) Biodiversity in dead wood. Cambridge University Press, Cambridge.
5. Herrmann S, Prescott CE (2008) Mass loss and nutrient dynamics of coarse woody debris in three Rocky Mountain coniferous forests: 21 year results. Can J For Res 38: 125–132. doi: 10.1139/X07-144.
6. Ohtsuka T, Shizu Y, Hirota M, Yashiro Y, Shugang J, et al. (2014) Role of coarse woody debris in the carbon cycle of Takayama forest, central Japan. Ecol Res 29: 91–101. doi: 10.1007/s11284-013-1102-5.
7. Edman M, Jonsson BG (2001) Spatial pattern of downed logs and wood-decaying fungi in an old-growth *Picea abies* forest. J Veg Sci 12(5): 609–620. doi: 10.2307/3236900.
8. Wu JB, Guan DX, Han SJ, Zhang M, Jin C (2005) Ecological functions of coarse woody debris in forest ecosystem. J For Res 16: 247–252. doi: 10.1007/BF02856826.
9. Ståhl G, Ringvall A, Fridman J (2001) Assessment of coarse woody debris: A methodological overview. Ecol Bull 49: 57–70.
10. Bütler R, Schlaepfer R (2004) Spruce snag quantification by coupling colour infrared aerial photos and a GIS. For Ecol Manage 195: 325–339. doi: 10.1016/j.foreco.2004.02.042.
11. Meentemeyer RK, Rank NE, Shoemaker DA, Oneal CB, Wickland AC, et al. (2007) Impact of sudden oak death on tree mortality in the Big Sur ecoregion of California. Biol Invasions 10: 1243–1255. doi: 10.1007/s10530-007-9199-5.
12. Chambers JQ, Fisher JI, Zeng H, Chapman EL, Baker DB, et al. (2007) Hurricane Katrina's carbon footprint on U.S. Gulf Coast forests. Science 318: 1107. doi: 10.1126/science.1148913.
13. Kupfer JA, Myers AT, McLane SE, Melton GN (2008) Patterns of forest damage in a southern Mississippi landscape caused by Hurricane Katrina. Ecosystems 11: 45–60. doi: 10.1007/s10021-007-9106-z.
14. Pasher J, King DJ (2009) Mapping dead wood distribution in a temperate hardwood forest using high resolution airborne imagery. For Ecol Manage 258: 1536–1548. doi: 10.1016/j.foreco.2009.07.009.
15. Wang F, Xu YJ (2010) Comparison of remote sensing change detection techniques for assessing hurricane damage to forests. Environ Monit Assess 162: 311–326. doi: 10.1007/s10661-009-0798-8.
16. Wang W, Qu JJ, Hao X, Liu Y, Stanturf JA (2010) Post-hurricane forest damage assessment using satellite remote sensing. Agric For Meteorol 150: 122–132. doi: 10.1016/j.agrformet.2009.09.009.
17. Valavanis KP (ed) (2007) Advances in unmanned aerial vehicles: State of the art and the road to autonomy. Springer, Dordrecht, Netherlands.
18. Watts AC, Perry JH, Smith SE, Burgess MA, Wilkinson BE, et al. (2010) Small unmanned aircraft systems for low-altitude aerial surveys. J Wildl Manag 7: 1614–1619. doi: 10.1111/j.1937-2817.2010.tb01292.x.
19. Watts AC, Ambrosia VG, Hinkley EA (2012) Unmanned aircraft systems in remote sensing and scientific research: Classification and considerations of use. Remote Sens 4: 1671–1692. doi: 10.3390/rs4061671.
20. Anderson K, Gaston KJ (2013) Lightweight unmanned aerial vehicles will revolutionize spatial ecology. Front Ecol Environ 11: 138–146. doi: 10.1890/120150.
21. Rango A, Laliberte A, Herrick JE, Winters C, Havstad K, et al. (2009) Unmanned aerial vehicle-based remote sensing for rangeland assessment, monitoring, and management. J Appl Remote Sens 3: 033542. doi: 10.1117/1.3216822.
22. Koh LP, Wich SA (2012) Dawn of drone ecology: low-cost autonomous aerial vehicles for conservation. Trop Conserv Sci 5: 121–132.

23. Sardà-Palomera F, Bota G, Viñolo C, Pallarés O, Sazatornil V, et al. (2012) Fine-scale bird monitoring from light unmanned aircraft systems. Ibis 154: 177–183. doi: 10.1111/j.1474-919X.2011.01177.x.

24. Hodgson A, Kelly N, Peel D (2013) Unmanned aerial vehicles (UAVs) for surveying marine fauna: a dugong case study. PLoS One 8(11): e79556. doi: 10.1371/journal.pone.0079556.

25. Vermeulen C, Lejeune P, Lisein J, Sawadogo P, Bouché P (2013) Unmanned aerial survey of elephants. PLoS One 8(2): e54700. doi: 10.1371/journal.pone.0054700.

26. Getzin S, Nuske RS, Wiegand K (2014) Using unmanned aerial vehicles (UAV) to quantify spatial gap patterns in forests. Remote Sens 6: 6988–7004. doi: 10.3390/rs6086988.

27. Masaki T, Suzuki W, Niiyama K, Iida S, Tanaka H, et al. (1992) Community structure of a species-rich temperate forest, Ogawa Forest Reserve, central Japan. Vegetatio 98: 97–111. doi: 10.1007/BF00045549.

28. Nakashizuka T, Iida S, Tanaka H, Shibata M, Abe S, et al. (1992) Community dynamics of Ogawa Forest Reserve, a species rich deciduous forest, central Japan. Vegetatio 103: 105–112. doi: 10.1007/BF00047696.

29. Nakashizuka T, Matsumoto Y (eds.) (2002) Diversity and interaction in a temperate forest community: Ogawa Forest Reserve of Japan. Ecological Studies vol. 158, Springer, Tokyo.

30. Kaneko T, Koyama T, Yasuda A, Takeo M, Yanagisawa T, et al. (2011) Low-altitude remote sensing of volcanoes using an unmanned autonomous helicopter: an example of aeromagnetic observation at Izu-Oshima volcano, Japan. Int J Remote Sens 32: 1491–1504. doi: 10.1080/01431160903559770.

31. Koyama T, Kaneko T, Ohminato T, Yanagisawa T, Watanabe A, et al. (2013) An aeromagnetic survey of Shinmoe-dake volcano, Kirishima, Japan, after the 2011 eruption using an unmanned autonomous helicopter. Earth Planets Sp 65: 657–666. doi: 10.5047/eps.2013.03.005.

32. Sanada Y, Kondo A, Sugita T, Nishizawa Y, Yuuki Y, et al. (2014) Radiation monitoring using an unmanned helicopter in the evacuation zone around the Fukushima Daiichi nuclear power plant. Explor Geophys 45: 3–7. doi: 10.1071/EG13004.

33. Franklin JF, Shugart HH, Harmon ME (1987) Tree death as an ecological process: the causes, consequences, and variability of tree mortality. BioScience 37(8): 550–556. doi: 10.2307/1310665.

15

Defining Mediterranean and Black Sea Biogeochemical Subprovinces and Synthetic Ocean Indicators Using Mesoscale Oceanographic Features

Anne-Elise Nieblas[1]*, Kyla Drushka[2], Gabriel Reygondeau[3], Vincent Rossi[4], Hervé Demarcq[5], Laurent Dubroca[6], Sylvain Bonhommeau[1]

1 Unité Mixte Recherche Ecosystèmes Marins Exploités 212, Institut Français de Recherche pour l'Exploitation de la Mer (IFREMER), Sète, France, **2** Applied Physics Laboratory, University of Washington, Seattle, Washington, United States of America, **3** Center for Macroecology, Evolution and Climate, National Institute for Aquatic Resources, Technical University of Denmark (DTU Aqua), Charlottenlund, Copenhagen, Denmark, **4** Instituto de Física Interdisciplinar Sistemas Complejos, Institute for Cross-Disciplinary Physics and Complex Systems, (CSIC-UIB), Campus Universitat de les Illes Balears, Palma de Mallorca, Spain, **5** Unité Mixte de Recherche Ecosystèmes Marins Exploités 212, Institut de Recherche pour le Développement (IRD), Sète, France, **6** European Commission, Joint Research Center, Institute for Environment & Sustainability, Water Resources, Ispra, Italy

Abstract

The Mediterranean and Black Seas are semi-enclosed basins characterized by high environmental variability and growing anthropogenic pressure. This has led to an increasing need for a bioregionalization of the oceanic environment at local and regional scales that can be used for managerial applications as a geographical reference. We aim to identify biogeochemical subprovinces within this domain, and develop synthetic indices of the key oceanographic dynamics of each subprovince to quantify baselines from which to assess variability and change. To do this, we compile a data set of 101 months (2002–2010) of a variety of both "classical" (i.e., sea surface temperature, surface chlorophyll-a, and bathymetry) and "mesoscale" (i.e., eddy kinetic energy, finite-size Lyapunov exponents, and surface frontal gradients) ocean features that we use to characterize the surface ocean variability. We employ a k-means clustering algorithm to objectively define biogeochemical subprovinces based on classical features, and, for the first time, on mesoscale features, and on a combination of both classical and mesoscale features. Principal components analysis is then performed on the oceanographic variables to define integrative indices to monitor the environmental changes within each resultant subprovince at monthly resolutions. Using both the classical and mesoscale features, we find five biogeochemical subprovinces for the Mediterranean and Black Seas. Interestingly, the use of mesoscale variables contributes highly in the delineation of the open ocean. The first axis of the principal component analysis is explained primarily by classical ocean features and the second axis is explained by mesoscale features. Biogeochemical subprovinces identified by the present study can be useful within the European management framework as an objective geographical framework of the Mediterranean and Black Seas, and the synthetic ocean indicators developed here can be used to monitor variability and long-term change.

Editor: Silvia Mazzuca, Università della Calabria, Italy

Funding: A. -E. N. was supported by a joint grant from France Filière Pêche (http://www.francefilierepeche.fr/) and Institut Français de Recherche pour l'Exploitation de la Mer (http://wwz.ifremer.fr/). V. R. acknowledges support from Ministerio de Ciencia e Innovación (http://www.idi.mineco.gob.es/portal/site/MICINN/) and FEDER (http://www.europe-en-france.gouv.fr/Configuration-Generale-Pages-secondaires/FEDER) through the ESCOLA project (CTM2012-39025-C02-01). S. B. was supported by the Agence nationale de la recherche SEAS-ERA MERMAID project (ANR-12-SEAS-0003) and the French National Research project INSU-MERMEX. The funders had no role in study design, data collection and analysis, decision to publish, or preparation of the manuscript.

Competing Interests: The authors have declared that no competing interests exist.

* Email: anne.elise.nieblas@gmail.com

Introduction

Growing pressure on the European marine environment has led to an increasing demand for comprehensive evaluation and monitoring programs [1–4]. The Mediterranean and Black Seas are ecologically- and economically-important semi-enclosed seas characterized by highly specific biogeochemcial, oceanographic, and environmental conditions that have resulted in pronounced endemism of exploited marine species [5–7]. The Mediterranean Sea is commonly divided into two basins, east and west, which each have specific hydrological conditions and marked seasonal cycles [8]. Recently, the International Panel on Climate Change has designated the Mediterranean as one of the most perturbed marine ecosystems of the global ocean, as both deep and surface environments show significant change in the open seas, coastal, benthic and neritic areas [9–11]. In addition, it is undergoing increasing anthropogenic pressure, including pollution, overfishing, and habitat loss via coastal development [1,7,12].

In this context, the European Union has recently adopted the Integrated Maritime Policy framework for the protection of European Seas; the primary objective of which is to achieve environmentally healthy waters by 2020 [1–4]. The first step toward the goal of healthy waters and the aim of this study is to identify an objective spatial partitioning in the Mediterranean and Black Seas, where environmental conditions are homogeneous, to act as a framework for marine zoning [13,14], for ecological management [15,16], as well as to determine baseline conditions which can then be used to effectively monitor variability and change.

Marine bioregionalization aims to identify unique and homogeneous biogeochemical partitions delineated by observable frontiers, such as frontal structures. This discipline, recently redefined by [17], is based on objective statistical methodologies and has been applied in several regions of the global ocean at several different scales [18–21]. However, owing to the complex hydrodynamics [22,23] and the important influence of mesoscale activity on biogeochemcial processes [24–26], bioregionalization of the Mediterranean and Black Seas remains difficult.

Marine bioregionalizations are classically performed on oceanographic features that are thought to be representative of the oceanographic and biogeochemical structure of a region; for example, sea surface temperature (SST), bathymetry and surface chlorophyll-*a* (chl) [20,27,28]. However, mesoscale processes must also be important for defining biogeochemical partitions as they are known to impact ocean productivity, including the spatial distribution and stocks of chlorophyll-*a* [29,30]; and basin-scale circulation and its mesoscale variability have been shown to be crucial in delineating hydrodynamical regions [31].

Previous studies have used single or multivariate analyses to derive regions of similar features in the Mediterranean Sea, including classical oceanographic indicators (i.e., chl [8]; SST, chl, sea surface salinity, and bathymetry [32]), bio-physical indicators, such as Ekman pumping, nutrient concentration, euphotic depth, and stratification [32]; and exploited fish distributions and biodiversity [7]. Recently, the Mediterranean Sea was subdivided into several hydrodynamical provinces delineated by multi-scale oceanic frontal structures in order to assess the ecological connectivity of the whole basin [31].

In this study, we derive objective biogeochemical subprovinces (sensu [33]) of the Mediterranean and Black Seas based on multivariate analyses of classical oceanographic features (SST, chl, and bathymetry), mesoscale features (eddy kinetic energy (EKE), SST and chl surface fronts, finite-size Lyapunov exponents (FSLE), and the Okubo-Weiss (OW) parameter), and a combination of both classical and mesoscale features. We also quantify the stability of the boundaries between the biogeochemical subprovinces in time and space. Synthetic oceanographic indices for the subprovinces are then extracted to act as baseline indicators, using principal components analysis (PCA), similar to the multivariate ocean-climate indices recently developed by [34]. Finally, we examine the temporal variability of these indices and their relationships with large-scale climate indices. The biogeochemical subprovinces identified in this study and the time series of their synthetic indicators could become important tools within the European management framework for assessing the environmental variability and change within the Mediterranean Sea.

Materials and Methods

Data

We used daily 4-km, version 5 Advanced Very High Resolution Radiometer pathfinder SST (1982–2012) available at http://www.nodc.noaa.gov/sog/pathfinder4km/. Chl data were taken from the National Aeronautics and Space Administration's daily 4-km level-3 Moderate Resolution Imaging Spectroradiometer daily data set (2002–2010) available at http://oceancolor.gsfc.nasa.gov/. We extracted weekly 1/3° (i.e., about 33 km at these latitudes) Ssalto/Duacs sea level anomalies and geostrophic velocity anomalies (u,v) computed and distributed by Aviso (1992–2012), with support from the Centre National d'Études Spatiales (http://www.aviso.oceanobs.com/duacs/). The bathymetry of the Mediterranean basin was extracted from the ETOPO1 database hosted on the National Oceanic and Atmospheric Administration's website at ~4 km resolution using the *getNOAA.bathy* function (marmap package, http://cran.r-project.org/). For consistency between variables, analyses were performed for data between May 2002 and November 2010, totaling 101 months.

Oceanographic indices

Several features were derived by further processing the remotely-sensed data. SST and chl fronts were computed with the gradient method, using a common sobel operator (e.g., [35,36]). These continuous values indicate the frontal intensity between water masses.

Using geostrophic velocity anomalies, we calculated several indicators of mesoscale ocean features. These data were used to derive backward-calculated FSLEs, which measure the horizontal mixing and dispersion in the ocean [37] and help to detect mesoscale Lagrangian coherent structures of ecological significance (e.g., [38]). FSLEs are defined as $\lambda(\mathbf{x},t,\delta_0,\delta_f) = (1/\tau)\log(\delta_0/\delta_f)$ where $\lambda(\mathbf{x},t,\delta_0,\delta_f)$ is the FSLE at position $\mathbf{x}$ and time t with an initial separation distance from $\mathbf{x}$ of δ_0 and final separation distance from $\mathbf{x}$ of δ_f. Here, we assign δ_0 to be 0.04 degrees and δ_f to be 0.6 degrees, and a time interval, τ, of 200 days, following the FSLE parameterizations of the Center for Topographic studies of the Ocean and Hydrosphere (http://ctoh.legos.obs-mip.fr/products/submesoscale-filaments/fsle-description), allowing us to detect mesoscale structures of <100 km, an appropriate scale for these seas [37]. Geostrophic velocity anomalies were also used to compute EKE, $(u^2+v^2)/2$, which is an indicator of the intensity of the eddy activity. Finally, we use geostrophic velocity anomalies to compute the OW parameter [39,40], $W = s^2_n+s^2_s+\omega^2$, where s_n and s_s are the normal and shear components of strain, and ω is the relative vorticity. This parameter is used to identify regions of high vorticity (W<0), which are likely related to the cores of mesoscale ocean eddies [41].

Monthly mean time series (2002–2010) were derived for all variables (except bathymetry) for the region including the Mediterranean and the Black Seas (30°N to 47°N, −6°E to 42°E) and regridded at 4 km resolution. We natural log transformed chl, chl fronts and bathymetry in order to stabilize their variance as their values can span several orders of magnitude.

Spatio-temporal multivariate k-means cluster analysis

Multivariate arrays were created from time-averages of the monthly scaled oceanographic indices (Figure 1) and combined into "classical" (i.e., SST, chl, and bathymetry), "mesoscale" (i.e., FSLE, OW, EKE, and SST and chl surface fronts), and "full" arrays (i.e., all features). After initial tests, k-means (*kmeans*, stats package, http://cran.r-project.org/; [42]) was determined to be the most robust cluster analysis algorithm to objectively classify

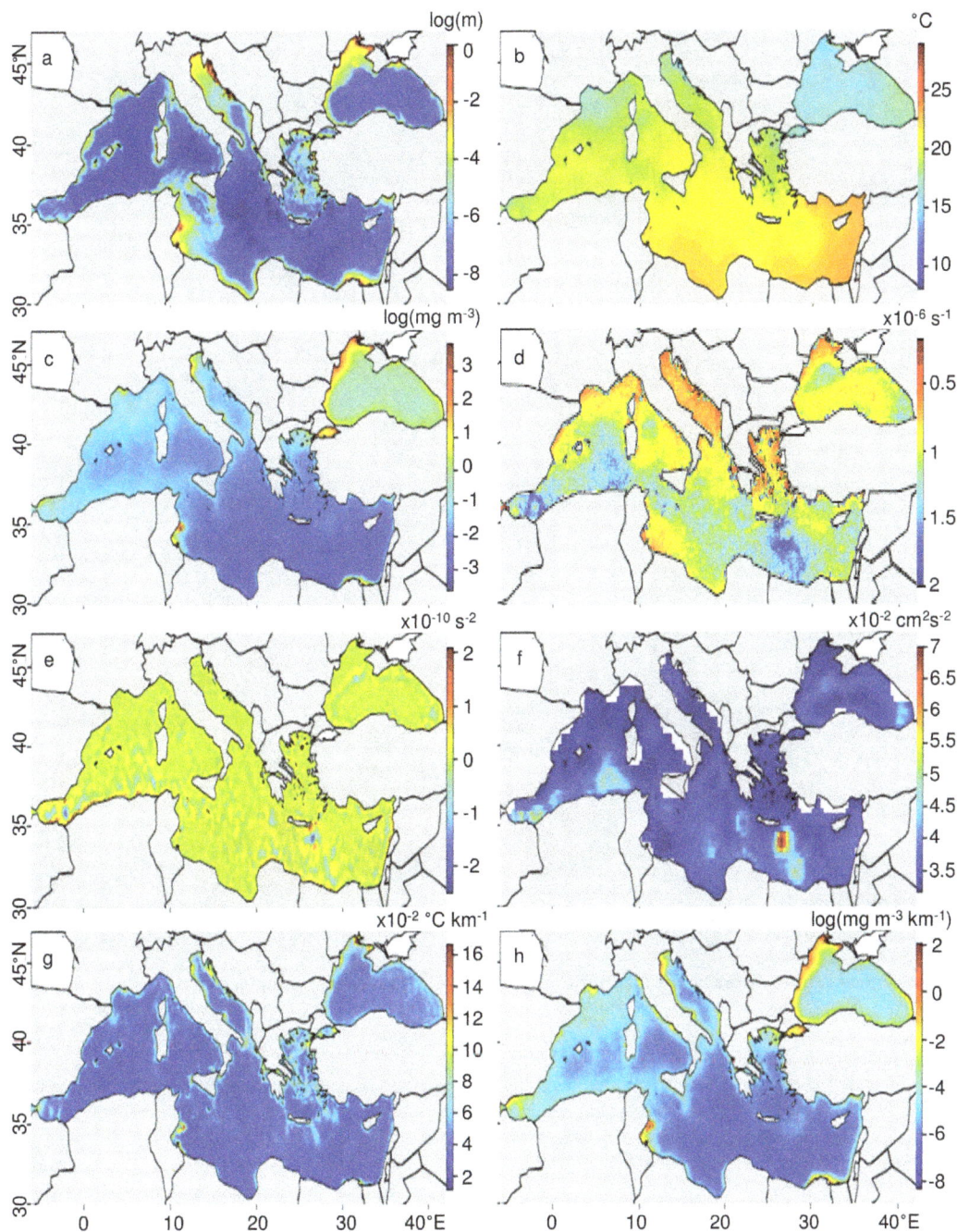

Figure 1. Time-averages of all oceanographic variables collected for the Mediterranean Sea. Variables include a) natural log-transformed bathymetry, b) sea surface temperature (SST), c) natural log-transformed chlorophyll-*a* (chl), d) finite-size Lyapunov exponents (FSLE), e) Okubo-Weiss parameter (OW), f) eddy kinetic energy (EKE), g) SST surface frontal gradients, and h) natural log-transformed chl surface frontal gradients.

biogeochemical subprovinces. This partitioning method, using Euclidean distances, assigns data points to k clusters and minimizes the sum of squares between the data points to cluster centre. With this algorithm, k must be defined *a priori*. In order to define k, we bootstrap (1000 times) k between 2 and 30. The between-clusters sum of squares is then divided by the total sum of squares to find the explained sum of squares. Arbitrary 1% and 5% thresholds are defined (Figure S1 in File S1), which we used to define the optimal k for the three multivariate arrays (Table 1), whereby the explained sum of squares for each additional k increases by less than 1% and 5%, respectively. K-means analyses were then

performed on each array using the optimal k for both threshold levels (1%; Figure S2 in File S1 and 5%; Figure 2). The resultant clusters were defined as the biogeochemical subprovinces of the Mediterranean and Black Seas as a subdivision of the Mediterranean provinces defined by [33].

To investigate the spatial stability of the subprovinces through time, we used the optimal k values found for each of the three multivariate arrays for both the 1% and 5% threshold levels, and performed a k-means analysis on each of the multivariate arrays for every month of the data set (n = 101 months). Then, based on an adaptation of the effectiveness test implemented by [43], the

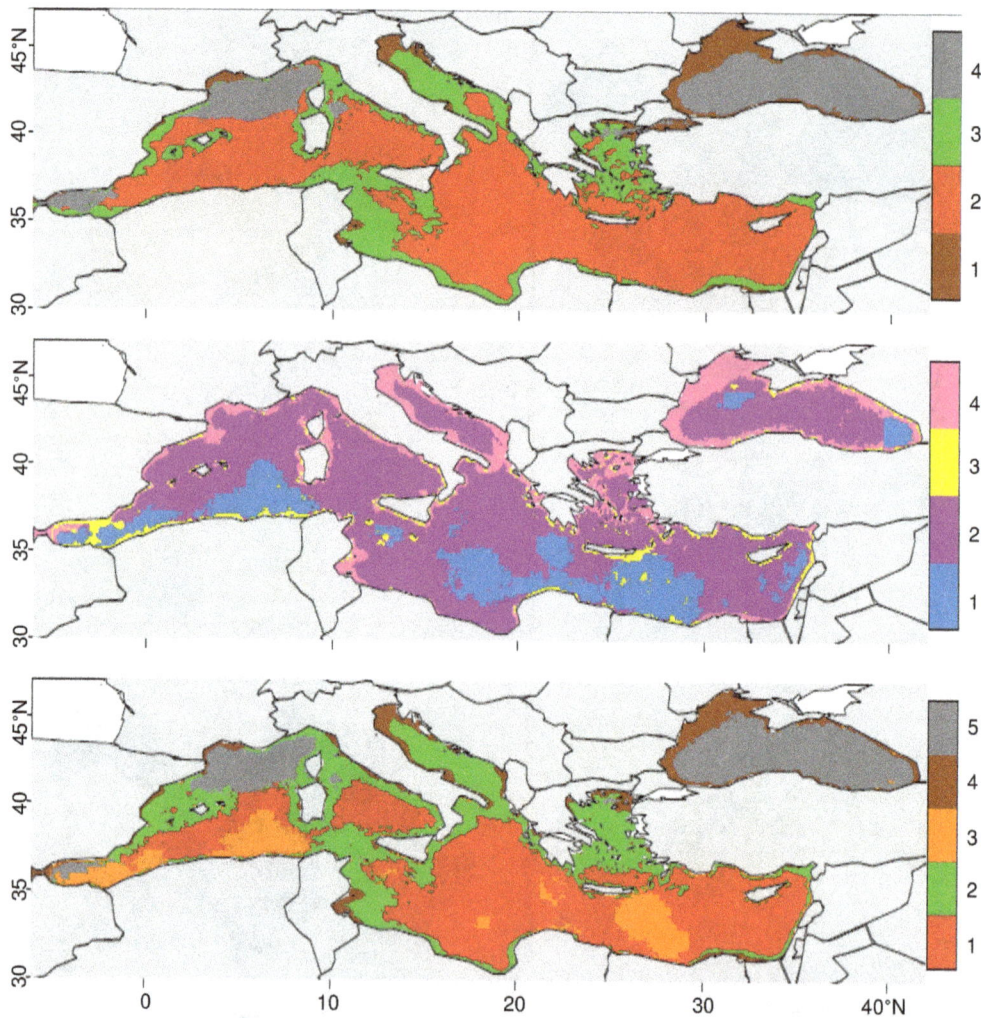

Figure 2. Biogeochemical subprovinces of the Mediterranean and Black Seas. Subprovinces for the (a) "classical", (b) "mesoscale", and (c) "full" multivariate arrays using a 5% threshold for the explained sum of squares to define the optimal number of subprovinces (see text).

temporal stability of each geographical cell is computed as the percentage of time that a geographical cell is considered as a boundary between two clusters at each temporal step (Figure 3, Figure S3 in File S1).

Development of synthetic indices through PCA

In order to develop synthetic indices of the oceanographic indicators for each biogeochemical subprovince, we extracted the scaled and centered monthly time series of each oceanographic

variable (except bathymetry) for each pixel within each biogeochemical subprovince. Although bathymetry is important for determining the biogeochemical subprovinces, it does not vary in time and was not included in the PCA. The strong seasonal cycle observed in all time-series was removed before performing the PCA as this signal swamps both the lower- and higher-frequencies of the time series (e.g., [44]). We then performed a PCA for each biogeochemical subprovince with an individual being the monthly value of each oceanographic variable for each pixel. We used the

Table 1. Optimal number of clusters, k, for each multivariate array for the 1% and 5% threshold levels obtained by bootstrapping 1000 times the k-means analysis on k between 2 and 30.

Array	Threshold	
	1%	5%
Classical	9	4
Mesoscale	14	4
Full	13	5

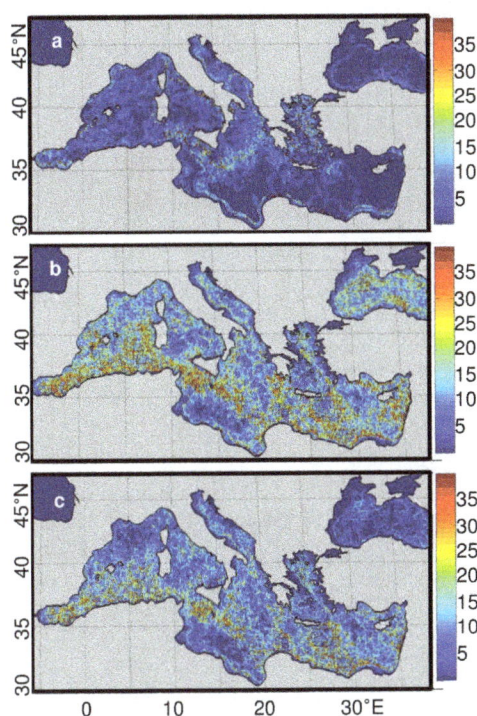

Figure 3. Spatial stability of the borders of biogeochemical subprovinces. Stability plots for the (a) classical, (b) mesoscale, and (c) full multivariate arrays. K-means analyses, using the k found in the time-averaged analyses (Table 1), are performed on the multivariate arrays at monthly time steps for the 101 months of the data set and using a 5% threshold of the explained sum of squares to define the optimal number of subprovinces (see text). Spatial stability is represented as the percentage of time that a boundary of the biogeochemical subprovinces is found at a particular pixel over the 101 months of the data set. Red colors indicate stable borders.

common cutoff of eigenvalues >1 to retain the unrotated principal components (PCs) (Table S1). We then took the monthly mean of the retained PCs over all the pixels, and used these as the synthetic indices of each biogeochemical subprovince.

Finally, we investigated the mode of temporal variability of these synthetic indices. Spectra were calculated to show the variability of each time-series. Lagged correlations were then investigated between time-series and monthly anomalies of four independent large-scale climate indices known to influence Mediterranean Sea dynamics [45,46]: North Atlantic Oscillation (NAO), the East Atlantic pattern (EA), the East Atlantic-West Russia pattern (EAWR), and the Scandinavian pattern (SCAND). These indices were computed by the National Oceanic and Atmospheric Administration/Climate Prediction Center.

Results

At the 5% threshold level, we find four subprovinces for the classical oceanographic features, four subprovinces for the mesoscale features and five subprovinces for the full combination of features (Table 1, Figure 2). Overall, the "classical" array (i.e., defined using the set of classical features) has the most stable boundaries in time and space (Figure 3a), while the boundaries for the "mesoscale" array (i.e., defined using the set of mesoscale features only) are highly variable (Figure 3b), as are the boundaries of the "full" array (i.e., defined using both classic and mesoscale features) (Figure 3c). This indicates that the apparent stability

found for the classical array is not representative of the "true" *in-situ* variability of the oceanic environment. The 5% threshold identifies fewer biogeochemical subprovinces than the 1% threshold (Table 1; Figure 2; Figure S2 in File S1), which, in addition to higher stability (Figure 3; Figure S3 in File S1), makes them easier to monitor in a management context. The full array at the 5% threshold was finally determined to be the most useful for management purposes, as the subprovinces are realistic and inclusive of both classical and mesoscale features (Figure 2, 3, 4). We perform a PCA for the subprovinces defined for the 5% threshold of the full array to derive synthetic time series (Table S1, Figure 5). The first two PCs are retained for full subprovinces 1–3 (explaining up to 39% and 21% of the variance for PC1 and PC2, respectively), and the first three PCs are retained for full subprovinces 4–5 (explaining up to 27%, 20%, and 18% of the variance for PC1, PC2 and PC3, respectively). We find strong low frequency (interannual) energy for the first two PCs in all full subprovinces (Figure 6), with the PCs of subprovince 5 being particularly different from the rest, especially PC2. We do not find any meaningful correlations to large-scale climate indices for any of the retained PCs (Table S2).

The stability of the boundaries for a given biogeochemical subprovince indicates whether the subprovince is representative of the local hydrological and biogeographical conditions, with higher stability giving greater credence. We found that the stability of the borders of the biogeochemical subprovinces defined by the 5% threshold (Figure 3) were more stable in terms of time and space than those of the 1% threshold (Figure S3 in File S1). The boundaries of the biogeochemical subprovinces of the classical array are very clearly defined and spatially stable through time (Figure 3a), with the greatest stability of boundaries found at the coastal isobaths (Figure 1a). Greater variability in the boundaries occurs through the Strait of Sicily and in the Aegean Seas, which correspond to high spatial bathymetric variability. The variability in biogeochemical subprovince boundaries observed in the Ionian Sea appears to correspond to a latitudinal transition of SST (Figure 1b). The boundaries for the mesoscale array are extremely variable (Figure 3b), with boundaries shifting significantly for each month of the data set. Especially high variability occurs in the southern parts of both the eastern and western basins, including the Alboran Sea, Algerian Basin, through the Strait of Sicily and south of Malta, where mesoscale activity is particularly high (Figure 1d,e,f).

Despite this spatial variability, we find that the subprovinces derived from the classical and mesoscale features are often grouped in the same manner. For example, the coastal Gulf of Lions, the northern Adriatic Sea, the Aegean Sea, the coastal Gulf of Gabes, the coast near the Suez Canal, and the western Black Sea are consistently grouped together in all three multivariate arrays (classical subprovince 1, mesoscale subprovince 4, full subprovince 4; Figure 2). These subprovinces are characterized by shallow bathymetry, generally low but variable SSTs, and high and variable chl (Figure 4a, c; Table 2). In terms of mesoscale features, these subprovinces have relatively consistently high FSLE, indicating low horizontal mixing, low EKE, and high SST and chl frontal gradients (Figure 4b, c; Table 2). Offshore of the Gulf of Lions and the offshore Black Sea are also commonly grouped by the three multivariate arrays (classical subprovince 4, mesoscale subprovince 2, full subprovince 5), and are characterized by deep bathymetry, low and relatively constant SSTs, and high chl (Figure 4a, c; Table 2). This subprovince has high FSLE, again indicating low horizontal mixing, low EKE (Figure 4b, c; Table 2), and high chl frontal gradients for the full subprovince 5 (Figure 4c; Table 2). The Strait of Gibraltar/Alboran Sea is

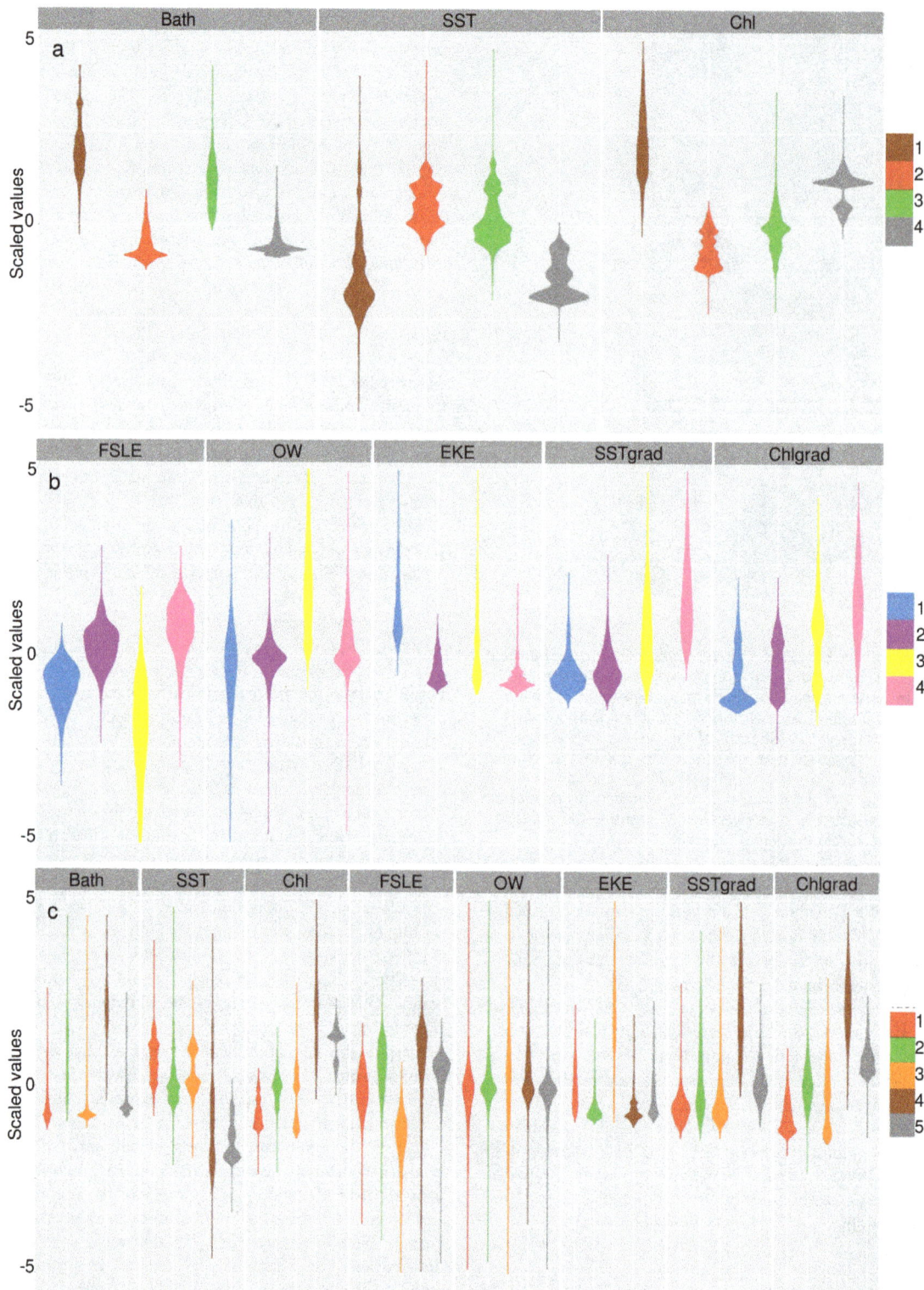

Figure 4. Violin plots of the scaled values for each oceanographic variable. Violin plots for the (a) classical, (b) mesoscale, and (c) full multivariate arrays. Colors indicate biogeochemical subprovinces. The mean of each variable is represented by the bulge in the violin and its variability is indicated by the tails. Here, positive bathymetry is shallow and negative bathymetry is deep. Variable abbreviations are as follows: sea surface temperature (SST), chlorophyll-*a* concentration (chl), finite-sized Lyapunov exponents (FSLE), the Okubo-Weiss parameter (OW), eddy kinetic energy (EKE), and SST and chl frontal gradients (SSTgrad, and chlgrad, respectively). Colorbars represent subprovinces.

grouped mainly by classical subprovinces 3 and 4 (Figure 2a, c), but is more variably characterized by mesoscale features (Figure 1, Figure 4). The majority of the southern parts of the eastern and western basins are grouped together (classical subprovince 2, mesoscale subprovince 2, full subprovince 1), and are characterized by deeper bathymetry, higher SSTs, low chl, and low

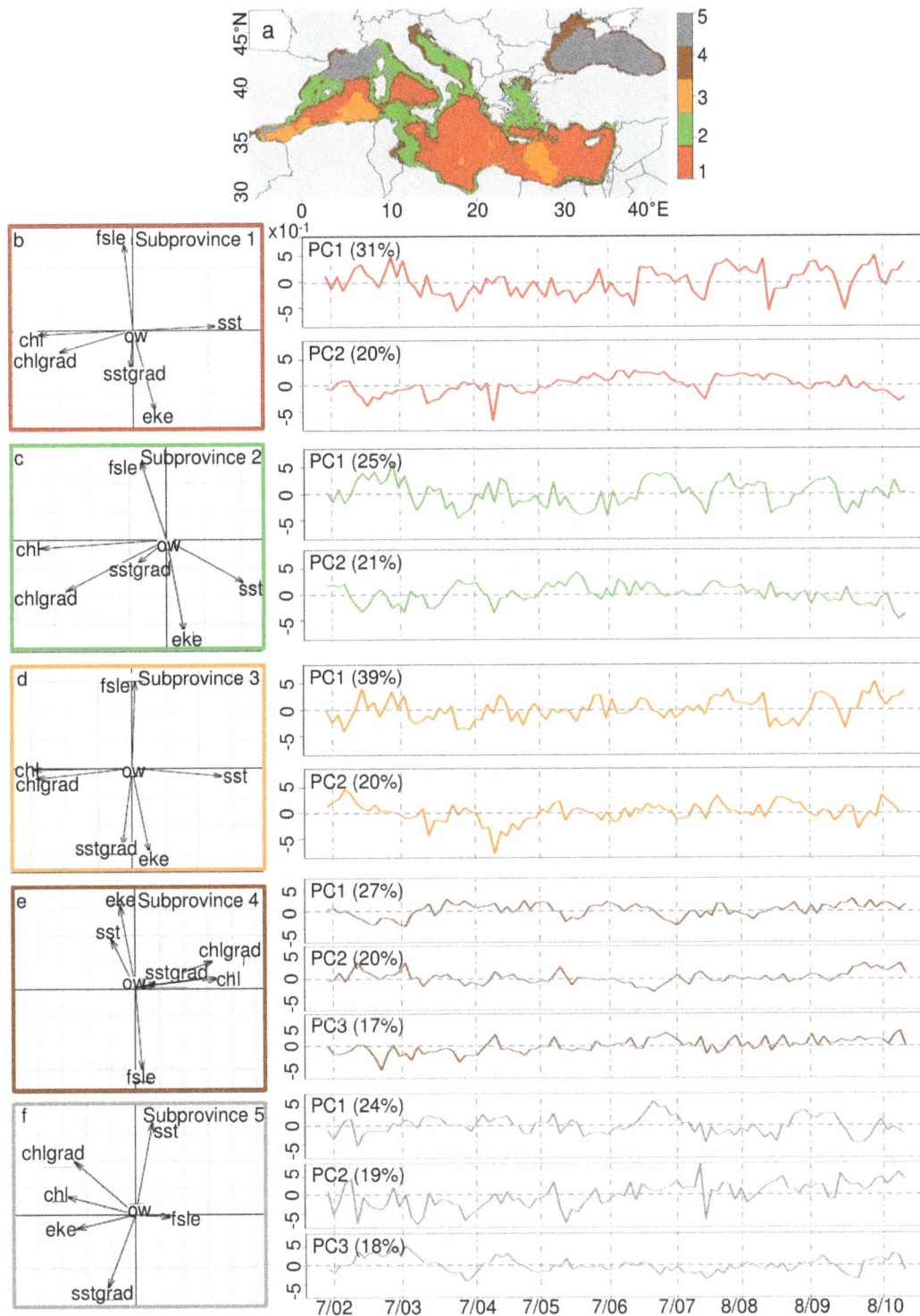

Figure 5. Principal component (PC) analysis for the biogeochemical subprovinces of the Mediterranean Sea. The (a) biogeochemical subprovinces of the Mediterranean Sea, as defined by the 5% threshold for the full multivariate array, and the (b–f) PC analysis arrow plots and time series of the retained PCs for each of the subprovinces, derived from the mean of all pixels in the subprovince for each month of the data set. Arrows that align well with an axis are well-explained by that axis. The longer the arrow, the more it contributes to explaining the variability of an axis. Variable abbreviations are as in Figure 4.

mesoscale activity (Figure 4; Table 2). The eastern and western basin are divided by the Strait of Sicily for the classical and full subprovinces (Figure 2a, c) due to relatively shallow bathymetry (Figure 4a,c; Table 2). Mesoscale subprovince 1 appears related to

full subprovince 3, which both occupy the southern basin, and are characterized by low FSLE (strong mixing), low and variable OW (indicative of eddy cores), high EKE, and low SST and chl frontal gradients (Figure 4b; Table 2). This high EKE is especially

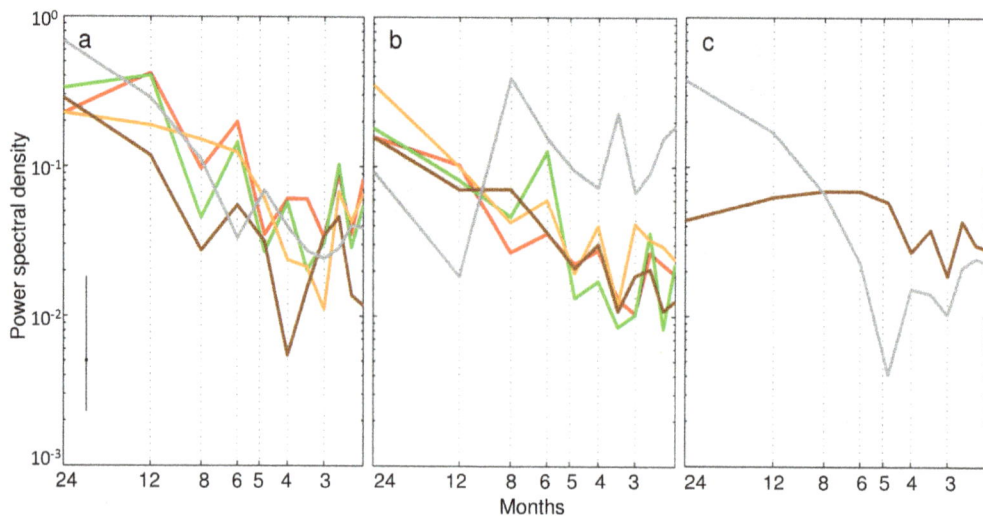

Figure 6. Spectral energy plots. Spectral energy plots for (a) principal component (PC) 1, (b) PC2, and (c) PC3 as retained for the different biogeochemical subprovinces computed for 101 months, having removed the seasonal signal. Peaks are indicated for all PCs at interannual frequencies. The error bar in the bottom-left indicates the 95% significance level.

apparent in full subprovince 3, south of Crete in the Levantine Sea (e.g., the wind-driven Ierapetra anticyclonic gyre), in the Algerian Basin, and in the Alboran Sea (Figure 1f, 2c).

In general, we find that classical variables for the full subprovinces primarily explain the first PC, and mesoscale variables for the full subprovinces generally explain the second PC (Figure 5b–f arrow plots, left panels; Table 3), with full subprovince 5 exhibiting much different patterns for all PCs than the other full subprovinces. For each PC, the variables having a correlation value >0.5 with a particular PC time series are deemed significant (a necessarily subjective cutoff value relevant only to this data set; Table 3). We find that the first two axes are typically divided between classical (chl and SST, as well as chl frontal gradients; PC1), and mesoscale features (typically FSLE and EKE, and SST frontal gradients for subprovince 3; PC2) (Figure 5b–d arrow plots), while OW does not explain any axis (Table 3). Chl and chl frontal gradients are highly correlated (r = 0.92), explaining the alignment of this mesoscale feature with classical features on the first axis. FSLE and EKE are also highly negatively correlated for each subprovince (Figure 5b–d arrow plots). PC1 for subprovinces 1, 2, and 3 are significantly correlated (p<0.001), with relatively high correlation coefficients (p<0.05, r = 0.61 to 0.8; Table S2), indicating that their classical features vary in the same manner. PC2 for subprovinces 1 and 2 are also significantly correlated (p<0.05, r = 0.58; Table S2).

Three axes are retained for subprovinces 4 and 5, as the eigenvalues for their PC3 were >1 (Figure 5e, f). The classical and mesoscale features are not as clearly divided by the retained axes for subprovinces 4 and 5 as they are for the first three subprovinces. The first axis for subprovince 4 is primarily explained by chl and chl frontal gradients and the second axis is explained by FSLE and EKE, similar to subprovinces 1–3, but the third axis is primarily explained by SST (Table 3). Subprovince 5 is even more different as the first and second axes are explained by a mix of classical and mesoscale features: PC1 is primarily explained by chl, chl frontal gradients and EKE, PC2 is explained by SST and SST frontal gradients, and PC3 is explained by FSLE.

The PC time series represent synthetic indices of oceanographic dynamics for each full subprovince (Figure 5b–f, right panels). After removing the dominant seasonal cycle, power spectra reveal

that low-frequency (interannual) variability dominates PC1 for all subprovinces (Figure 6a) as well as PC2 for all subprovinces except subprovince 5 (Figure 6b). Subprovince 5 has distinctive spectral characteristics, with no high-frequency peak for PC1 (Figure 6a) and no decrease in energy at high frequencies for PC2. Instead, PC2 has a strong peak at a period of 8 months (Figure 6b). Subprovince 5 also has a strong low-frequency signal for PC3, but no significant signals are found for PC3 of subprovince 4 (Figure 6c).

We find six out of 48 significant correlations between the retained PCs and large-scale climate indices (r = −0.2 to 0.27 with p-value <0.05). However, this could be due to multiple testing of time series. Since the correlations were low, we did not correct the p-values for multiple testing and we consider that there is no major links between the PCA axes and climatic indices. Lagged correlations between the PCs and the climate indices were also considered but did not render stronger relationships (Table S2).

Discussion

Our results synthetically characterize the hydrodynamics of the Mediterranean and Black Seas, complex and variable oceanic systems that function on multiple temporal and spatial scales [47]. Mesoscale features in the Mediterranean are an important source of variability [48], and we find that they are an important component to include in this bioregionalization. Indeed, the omission of mesoscale features and their temporal variability in such spatial analyses are misleading, as already suggested by [20], who promoted the use of dynamical biogeochemical subprovinces instead of their static equivalents. Classical features here are stable and are clearly representative of the biogeochemistry of the subprovince that they describe. Though highly variable, mesoscale features enable us to further discriminate additional regions. This is especially useful in the open ocean, which appears homogenous when considering only classical variables.

Overall, the biogeochemical subprovinces defined for the full array of variables (Figure 3c) can be organized into four broad categories that compare relatively well with previous studies. Subprovinces 1 and 2 represent the open ocean regions of the southern basin. These two subprovinces are highly correlated in

Table 2. Mean and standard deviations of the environmental variables for each biogeochemical subprovince defined by the 5% threshold of the classical, mesoscale, and full multivariate arrays over the 101 months of the data set, including sea surface temperature (SST, °C), chlorophyll-a concentration (chl; mg m⁻³), finite-sized Lyapunov exponents (FSLE; s⁻¹), the Okubo-Weiss parameter (OW; s⁻²), eddy kinetic energy (EKE; cm² s⁻²), and SST and chl frontal gradients (SSTgrad; °C km⁻¹, and chlgrad; mg m⁻³ km⁻¹, respectively).

Subprovince	SST	Chl	Bath	EKE	FSLE	OW	SSTgrad	Chlgrad
Classical								
1	17.04±6.66	0.19±0.24	−56.75±73.12					
2	20.85±4.41	0.15±0.12	−2133.17±945.04					
3	20.19±4.55	0.24±0.26	−203.08±178.70					
4	16.98±6.01	0.65±0.59	−1786.03±658.50					
Mesoscale								
1				0.042±0.012	−1.32e-06±7.01e-07	−7.36e-12±8.69e-11	0.038±0.024	0.005±0.015
2				0.035±0.004	−1.00e-06±4.91e-07	1.44±3.59e-11	0.039±0.025	0.006±0.017
3				0.037±0.008	−1.84e-06±1.47e-06	4.99e-11±8.33e-11	0.052±0.035	0.026±0.094
4				0.034±0.003	−8.69e-07±5.67e-07	9.94e-12±3.34e-11	0.065±0.043	0.086±0.210
Full								
1	21.13±4.38	0.13±0.09	−2152.13±1021.32	0.036±0.005	−1.12e-06±5.5e-07	−7.62e-13±4.87e-11	0.03±0.02	0.003±0.004
2	19.95±4.57	0.25±0.005	−609.96±548.89	0.034±0.003	−9.39e-07±5.6e-07	8.09e-12±3.74 e-11	0.046±0.029	0.008±0.018
3	20.54±4.10	0.20±0.25	−2210.62±962.95	0.044±0.014	−1.57e-06±1.05e-06	4.00e-12±1.078e-10	0.040±0.026	0.008±0.026
4	17.80±6.48	1.91±2.84	−77.36±132.71	0.034±0.003	−8.95e-07±6.95e-07	1.18e-11±3.53e-11	0.072±0.047	0.155±0.279
5	16.87±6.09	0.65±0.49	−1774.19±722.23	0.035±0.005	−9.69e-07±4.66e-07	−3.76e-13±3.82e-11	0.045±0.030	0.017±0.045

Table 3. Correlation matrix of the first four principal components (PC) analyzed for the full multivariate array, as defined by the 5% threshold, including sea surface temperature (SST), chlorophyll-*a* concentration (chl), eddy kinetic energy (EKE), finite-sized Lyapunov exponents (FSLE), the Okubo-Weiss parameter (OW), and SST and chl frontal gradients (SSTgrad and chlgrad, respectively).

Subprovince	PC	Variance explained (%)	Eigenvalues of environmental variables						
			SST	Chl	EKE	FSLE	OW	SSTgrad	Chlgrad
1	1	31	**0.56**	**−0.64**	0.15	−0.06	0	−0.02	**−0.5**
	2	20	0.03	−0.04	**−0.64**	**0.68**	0	−0.31	−0.18
	3	16	−0.1	0.03	0.24	−0.2	0	−0.94	−0.02
2	1	25	0.42	**−0.69**	0.09	−0.14	0	−0.16	**−0.54**
	2	21	−0.31	−0.06	**−0.65**	**0.56**	0	−0.18	−0.37
	3	16	0.1	0.19	0.12	0.01	0	−0.96	0.14
3	1	39	**0.55**	**−0.6**	0.1	0.02	0	−0.06	**−0.58**
	2	20	−0.06	−0.01	**−0.58**	**0.6**	0	**−0.54**	−0.07
	3	15	0	0.04	0.12	−0.59	0	−0.79	0.04
4	1	27	−0.21	**0.69**	−0.13	0.06	0	0.19	**0.65**
	2	20	0.39	0.08	**0.64**	**−0.62**	0	0.05	0.21
	3	17	**0.8**	−0.15	−0.18	0.4	0	0.24	0.28
5	1	24	0.14	**−0.57**	**−0.5**	0.3	0	−0.24	**−0.51**
	2	19	**0.7**	0.14	−0.11	−0.02	0	**−0.56**	0.4
	3	18	0.23	−0.38	0.41	**−0.73**	0	−0.08	−0.3

Variables with correlation values >0.5 (bold) are deemed to significantly contribute to each PC.

terms of their PC1 and PC2 (Table S2), indicating that both their classical and mesoscale features vary similarly and appear to differ mostly in their bathymetry. Subprovince 3 is representative of regions of particularly high mesoscale activity. This is clearly the case for the Alboran Sea, the Algerian Basin, the Strait of Sicily and the Ierapetra, Rhodes and Mersa-Matrouh gyres, as confirmed in other studies (e.g., [8,22,37,47,49,50]). Mesoscale activity is also particularly high for the phenological regions in the Alboran Sea and the Strait of Sicily found by [8], who defined seven regions based on satellite ocean color. They found no apparent bloom pattern in these regions, which coincide primarily with full subprovince 3 (characterized by highly variable chl; Table 2). Subprovince 4 represents the coastal regions with both narrow and wide continental shelves. Finally, subprovince 5 represents oceanic gyres at high latitude, including the Lion Gyre, though the South Adriatic Gyre [50] is not included in this subprovince, as might be expected. The biodiversity hot spots identified by [7] using exploited fish distributions also show good spatial agreement with our subprovinces. They highlight the importance of the western Mediterranean shelves, the Alboran Sea, the Adriatic Sea, and the Levantine Basin, which coincide with our coastal and high mesoscale activity subprovinces. These subprovinces are characterized by high and highly variable chl, an indicator of primary productivity [51], which may be associated with the high ecological productivity and biodiversity here [52]. The coastal regions and the large oceanic gyres revealed in this study are consistent with the subdivisions found by [31] using only dynamical criteria, e.g., advection and dispersion schemes due to surface currents. This suggests that the horizontal circulation potentially explains a significant part of the basin-scale distribution of oceanic tracers such as SST and chl. The mesoscale variability of the oceanic circulation also controls their smaller-scale patterns (through the formation of SST and chl frontal gradients) and is thus responsible for the lower stability of the boundaries in regions where mesoscale features are particularly ubiquitous.

The synthetic indices developed in this study via PCA show temporal variability at seasonal and interannual time scales. The seasonal signal of oceanographic conditions in the Mediterranean that is the dominant signal in all PCs is generally related to changes of heat and momentum fluxes, which also vary at seasonal time scales driven by synoptic weather patterns [53]. However, interannual variability, as shown to be strong for PC1 and 2, is often more complex and puzzling. Our attempt to explain the interannual signal that we found in PCs 1 and 2 showed no clear relationship to large-scale climate indices, which supposedly represent external atmospheric forcing over the basin [45,46]. Though it is possible that our indices are not sufficiently long (101 months) to reveal any relationship to low-frequency signals, other potential drivers may help explain the interannual variability that we find. Among these are internal nonlinear ocean dynamics (e.g., unstable mesoscale eddy fields) that alter water properties and movements, the region and timing of deep-water formation, the extreme events and long-term variability of atmospheric forcing (e.g., winds, solar radiation, precipitation) that impact both the surface and sub-surface circulation, or interannual variations in the Gibraltar inflow [50].

The oceanographic variables defined here were included to capture as much of the ocean variability and dynamics as possible; however, our analysis indicated that some variables may be redundant. As noted, chl and chl frontal gradients are highly positively correlated (r = 0.92) and are almost always aligned with the same axis (Figure 5b–d arrow plots, Table S1, Table 3). For PC2, FSLE and EKE could also be redundant in that they are consistently aligned with the same axis for subprovinces 1–4 and

are significantly negatively correlated (r = −0.53, p<0.001). This is consistent with the compact relationship (power-law) found between FSLE and EKE in the global ocean [54], although it appears less robust in the Mediterranean [55]. OW does not appear to be a useful variable to include in this bioregionalization, as it does not significantly contribute to the explanation of any axis (Table 3). This may be because OW is related to EKE and FSLE [55] and may be redundant in this analysis. In addition, it may exist at a lower spatial resolution than that detected by the 5% threshold level. At the 1% threshold level, it appears that OW may play a more important role in the bioregionalization (Figure S2b, c in File S1). Finally, frontal structures themselves indicate the boundaries between two different water masses [56] and between macro- or meso-provinces [33], making these variables potentially redundant to a partitioning analysis. However, this is not obvious *a priori*, as there is clearly a complex relationship between chl and chl fronts as indicated by their high correlation. We find that the contribution of SST and chl fronts is not strong or consistent in the Mediterranean or Black Seas.

In this study, we characterize the biogeographical conditions of the Mediterranean Sea in a simple and synthetic approach. We develop objective biogeochemical subprovinces that agree with spatial characterizations made in previous studies and are specifically adapted to be useful as geographic reference points in a management context. Our results highlight the importance of mesoscale features to help delineate further regions in the seemingly homogeneous open ocean. We suggest that these subprovinces could be relevant for defining pelagic habitats for marine protected areas that are dynamic, yet predictable enough to be important for foraging and breeding aggregations [57]. In addition, the synthetic indices developed here represent a baseline from which variability and future changes of important classical and mesoscale features can be assessed. An understanding of the fundamental ocean processes of these heavily-impacted bodies of water has important implications for climatic studies, anthropogenic impact mitigation, and marine resources management.

Acknowledgments

We would like to thank NASA, NOAA, and AVISO for the freely available remotely-sensed data. We would also like to thank Francesco d'Ovidio for his expertise and FSLE code and Giorgio Caramanna for his help improving the manuscript.

Supporting Information

File S1   Supporting figures and tables. Figure S1 in File S1: Boxplots of the bootstrapped (1000 times) between-clusters sum of squares divided by the total sum of squares (i.e., y-axis represents the proportion of the explained sum of squares) for k between 2 and 30 for k-means analyses performed on the (a) "classical", (b) "mesoscale" and (c) "full" multivariate arrays. To identify the most appropriate k for each multivariate array, we define thresholds whereby the explained sum of squares for each additional k increases by less than 5% (red line) or less than 1% (blue line). Table S1 in File S1: The eigenvalues of each axis for the "full" multivariate array for biogeochemical subprovinces defined by the 5% threshold. To determine which principal components (PC) to retain, we used the common cutoff of eigenvalues ≥1. Figure S2 in File S1: Biogeochemical subprovinces of the Mediterranean Sea for the (a) "classical", (b) "mesoscale", and (c) "full" multivariate arrays using a 1% threshold on the explained sum of squares to define the optimal number of subprovinces (see text). Figure S3 in File S1: Spatial stability of the borders of biogeochemical subprovinces for the (a) classical, (b) mesoscale, and (c) full multivariate arrays. K-means analysis, using the k found in the time-averaged analyses (Table 1), are performed on the multivariate arrays at monthly time steps for the 101 months of the data set and using a 1% threshold on the explained sum of squares to define the optimal number of subprovinces (see text). Spatial stability is represented as the percentage of time that a boundary of the biogeochemical subprovinces is found at a particular pixel over the 101 months of the data set. Red colors indicate stable borders. Table S2 in File S1: Correlation coefficients between the retained principal components (PC) for each of the full biogeochemical subprovinces and the monthly anomalies of the large-scale climate indices: North Atlantic Oscillation (NAO), the East Atlantic pattern (EA), the East Atlantic-West Russia pattern (EAWR), and the Scandinavian pattern (SCAND). Only correlations above the 95% significance level are included. Table S3 in File S1: Correlation coefficients between the retained principal components (PC) for each of the full biogeochemical subprovinces. Significance levels are represented as p<0.001 '***', p<0.05 '*', not significant ' '.

Author Contributions

Conceived and designed the experiments: AEN SB LD. Performed the experiments: AEN SB. Analyzed the data: AEN SB KD GR. Contributed reagents/materials/analysis tools: HD. Wrote the paper: AEN KD GR VR HD LD SB.

References

1. Durrieu de Madron X, Guieu C, Sempéré R, Conan P, Cossa D, et al. (2011) Marine ecosystems' responses to climatic and anthropogenic forcings in the Mediterranean. Prog Oceanogr 91: 97–166. doi:10.1016/j.pocean.2011.02.003.

2. Fraschetti S (2012) Mapping the marine environment: advances, relevance and limitations in ecological research and marine conservation and management. Biol Mar Mediterr 19: 79–83.

3. Micheli F, Levin N, Giakoumi S, Katsanevakis S, Abdulla A, et al. (2013) Setting Priorities for Regional Conservation Planning in the Mediterranean Sea. PLoS One 8: e59038. doi:10.1371/journal.pone.0059038.

4. Portman ME, Notarbartolo-di-Sciara G, Agardy T, Katsanevakis S, Possingham HP, et al. (2013) He who hesitates is lost: Why conservation in the Mediterranean Sea is necessary and possible now. Mar Policy 42: 270–279. doi:10.1016/j.marpol.2013.03.004.

5. Abdulla A, Gomei M, Hyrenbach D, Notarbartolo-di-Sciara G, Agardy T (2009) Challenges facing a network of representative marine protected areas in the Mediterranean: prioritizing the protection of underrepresented habitats. ICES J Mar Sci 66: 22–28.

6. Bianchi CN (2007) Biodiversity issues for the forthcoming tropical Mediterranean Sea. Hydrobiologia 580: 7–21.

7. Coll M, Piroddi C, Steenbeek J, Kaschner K, Lasram FBR, et al. (2010) The biodiversity of the Mediterranean Sea: estimates, patterns, and threats. PLoS One 5: e11842. doi:10.1371/journal.pone.0011842.

8. d'Ortenzio F, Ribera d'Alcalà M (2009) On the trophic regimes of the Mediterranean Sea: a satellite analysis. Biogeosciences 6: 139–148. doi:10.5194/bg-6-139-2009.

9. Parry ML, Canziani O, Palutikof J, van der Linden P, Hanson C (2007) Climate change 2007: impacts, adaptation and vulnerability: contribution of Working Group II to the fourth assessment report of the Intergovernmental Panel on Climate Change. Cambridge: Cambridge University Press. 982 p.

10. Pachauri RK, Reisenger A (2007) Climate change 2007 Synthesis report: Contribution of Working Groups I, II and III to the fourth assessment report of the Intergovernmental Panel on Climate Change. Geneva: Intergovernmental Panel on Climate Change.

11. Lejeusne C, Chevaldonne P, Pergent-Martini C, Boudouresque CF, Perez T (2012) Climate change effects on a miniature ocean: the highly diverse, highly impacted Mediterranean Sea, Trends Ecol Evol 25: 250–260. doi:10.1016/j.tree.2009.10.009.

12. Halpern BS, Walbridge S, Selkoe KA, Kappel CV, Micheli F, et al. (2008) A global map of human impact on marine ecosystems. Science 319: 948–952.

13. Spalding MD, Fox HE, Allen GR, Davidson N, Ferdaña ZA, et al. (2007) Marine Ecoregions of the World: A Bioregionalization of Coastal and Shelf Areas. BioScience 57: 573–583. doi:10.1641/B570707.

14. Rice J, Gjerde KM, Ardron J, Arico S, Cresswell I, et al. (2011) Policy relevance of biogeographic classification for conservation and management of marine biodiversity beyond national jurisdiction, and the GOODS biogeographic classification. Ocean Coast Manage 54: 110–122.

15. Gabrié C, Lagabrielle E, Bissery C, Crochelet E, Meola B, et al. (2012) The status of marine protected areas in the Mediterranean Sea. MedPAN and RAC/SPA. 256 p. Available: http://www.medpan.org/documents/10180/0/The+Status+of+the+Marine+Protected+Areas+in+the+Mediterranean+Sea+2012/069bb5c4-ce3f-4046-82cf-f72dbae29328. Accessed 2014 October 7.

16. Coll M, Cury P, Azzurro E, Bariche M, Bayadas G, et al. (2013) The scientific strategy needed to promote a regional ecosystem-based approach to fisheries in the Mediterranean and Black Seas. Rev Fish Biol Fish 23: 415–434.

17. Vierros M, Bianchi G, Skjoldal HR (2008) The Ecosystem approach of the convention on biological diversity. In: Bianchi G, Skjoldal, editors. The Ecosystem approach to fisheries. pp. 39–46.

18. Edgar GJ, Moverley J, Barrett NS, Peters D, Reed C (1997) The conservation-related benefits of a systematic marine biological sampling programme: the Tasmanian reef bioregionalisation as a case study. Biol Conserv 79: 227–240.

19. Grant S, Constable A, Raymond B, Doust S (2006) Bioregionalisation of the Southern Ocean: report of experts workshop (Hobart, September 2006). Sydney: WWF-Australia and ACE CRC. 44 p. Available: http://awsassets.wwf.org.au/downloads/mo007_bioregionalisation_of_the_southern_ocean_8sep06.pdf. Accessed 2014 October 7.

20. Reygondeau G, Longhurst A, Martinez E, Beaugrand G, Antoine D, et al. (2013) Dynamic biogeochemical provinces in the global ocean. Global Biogeochem Cycles 27: 1046–1058.

21. Kavanaugh MT, Hales B, Saraceno M, Spitz YH, White AE, et al. (2014) Hierarchical and dynamic seascapes: A quantitative framework for scaling pelagic biogeochemistry and ecology. Prog Oceanogr 120: 291–304.

22. Millot C, Taupier-Letage I (2005) Circulation in the Mediterranean Sea. In: Saliot A, editor, The Mediterranean Sea, Handbook of Environmental Chemistry. Berlin Heidelberg: Springer, pp. 29–66.

23. Giakoumi S, Sini M, Gerovasileiou V, Mazor T, Beher J, et al. (2013) Ecoregion-based conservation planning in the Mediterranean: dealing with large-scale heterogeneity. PLoS One 8: e76449. doi:10.1371/journal.pone.0076449.

24. Millot C (1991) Mesoscale and seasonal variabilities of the circulation in the western Mediterranean. Dyn Atmospheres Oceans 15: 179–214. doi:10.1016/0377-0265(91)90020-G.

25. Millot C (1999) Circulation in the Western Mediterranean Sea. J Mar Syst 20: 423–442. doi:10.1016/S0924-7963(98)00078-5.

26. Robinson AR, Malanotte-Rizzoli P, Hecht A, Michelato A, Roether W, et al. (1992) General circulation of the Eastern Mediterranean. Earth-Science Reviews 32: 285–309. doi:10.1016/0012-8252(92)90002-B.

27. Sherman K, Alexander LM (1989) Biomass yields and geography of large marine ecosystems. Boulder: Westview Press. 493 p.

28. Sherman K, Sissenwine M, Christensen V, Duda A, Hempel G (2005) A global movement toward an ecosystem approach to management of marine resources: Politics and socio-economics of ecosystem-based management of marine resources. Mar Ecol Prog Ser 300: 275–279.

29. Lehahn Y, d'Ovidio F, Levy M, Heifetz E (2007) Stirring of the northeast Atlantic spring bloom: A Lagrangian analysis based on multisatellite data. J Geophys Res 112: C08005. doi:10.1029/2006JC003927.

30. Rossi V, Lopez C, Sudre J, Hernandez-Garcia E, Garcon V (2008) Comparative study of mixing and biological activity of the Benguela and Canary upwelling systems. Geophys Res Lett 35: L11602. doi:10.1029/2008GL033610.

31. Rossi V, Ser-Giacomi E, López C, Hernández-García E (2014) Hydrodynamic regions and oceanic connectivity from a transport network help designing marine reserves. Geophys Res Lett 41: 2883–2891. doi:10.1002/2014GL059540.

32. Reygondeau G, Olivier Irisson J, Guieu C, Gasparini S, Ayata S, et al. (April 2013) Toward a dynamic biogeochemical division of the Mediterranean Sea in a context of global climate change. EGU General Assembly Conference Abstracts 15: 10011. Available: http://scholar.google.fr/scholar?q=Toward+a+dynamic+biogeochemical+division+of+the+Mediterranean+Sea+in+a+context+of+global+climate+change&btnG=&hl=en&as_sdt=0%2C5. Accessed 2014 October 7.

33. Longhurst AR (2010) Ecological geography of the sea. San Diego: Academic Press. 560 p.

34. Sydeman WJ, Thompson SA, Garcia-Reyes M, Kahru M, Peterson WT, et al. (2014) Multivariate ocean-climate indicators (MOCI) for the central California Current: Environment change, 1990-2010. Prog Oceanogr 120: 352–369.

35. Canny J (1986). A computational approach to edge detection. IEEE Trans Pattern Anal Mach Intell 6: 67–698.

36. Nieto K, Demarcq H, McClatchie S (2012) Mesoscale frontal structures in the Canary Upwelling System: new front and filament detections algorithms applied to spatial and temporal patterns. Remote Sens Environ 123: 339–346.

37. d'Ovidio F, Fernández V, Hernández-García E, López C (2004) Mixing structures in the Mediterranean Sea from finite-size Lyapunov exponents. Geophys Res Lett 31: L17203. doi:10.1029/2004GL020328.

38. Tew-Kai E, Rossi V, Sudre J, Weimerskirch H, Lopez C, et al. (2010) Top marine predators track Lagrangian coherent structures. Proc Natl Acad Sci U S A 106: 8245–8250. doi:10.1073/pnas.0811034106.

39. Okubo A (1970) Horizontal dispersion of floatable particles in the vicinity of velocity singularities such as convergences. Deep Sea Res Part 1 Oceanogr Res Pap 17: 445–454. doi:10.1016/0011-7471(70)90059-8.

40. Weiss J (1991) The dynamics of enstrophy transfer in two-dimensional hydrodynamics. Physica D 48: 273–294. doi:10.1016/0167-2789(91)90088-Q.

41. Henson SA, Thomas AC (2008). A census of oceanic anticyclonic eddies in the Gulf of Alaska. Deep Sea Res Part 1 Oceanogr Res Pap 55: 163–176. doi:10.1016/j.dsr.2007.11.005.

42. Hartigan JA, Wong MA (1979) Algorithm AS 136: A k-means clustering algorithm. J R Stat Soc Ser C Appl Stat 28: 100–108.

43. Oliver MJ, Glenn S, Kohut JT, Irwin AJ, Schofield OM, et al. (2004) Bioinformatic approaches for objective detection of water masses on continental shelves. J Geogphys Res 109: C07S04. doi:10.1029/2003JC002072.

44. Beaugrand G, Ibanez F (2004) Monitoring marine plankton ecosystems. II: Long-term changes in North Sea calanoid copepods in relation to hydro-climatic variability. Mar Ecol Prog Ser 284: 35–47.

45. Josey SA, Somot S, Tsimplis M (2011) Impacts of atmospheric modes of variability on Mediterranean Sea surface heat exchange. J Geophys Res 116: C02032. doi:10.1029/2010JC006685.

46. Papadopoulos VP, Josey SA, Bartzokas A, Somot S, Ruiz S, et al. (2012) Large-scale atmospheric circulation favoring deep- and intermediate-water formation in the Mediterranean Sea. J Clim 25: 6079–6091. doi:10.1175/JCLI-D-11-00657.1.

47. Fernández V, Dietrich DE, Haney RL, and Tintoré J (2005) Mesoscale, seasonal and interannual variability in the Mediterranean Sea using a numerical ocean model. Prog Oceanogr 66: 321–340.

48. Larnicol G, Ayoub N, Le Traon PY (2002) Major changes in Mediterranean Sea level variability from 7 years of TOPEX/Poseidon and ERS-1/2 data. J Mar Syst 33: 63–89.

49. Malanotte-Rizzoli P, Manca BB, d'Alcalà MR, Theocharis A, Bergamasco A, et al. (1997) A synthesis of the Ionian Sea hydrography, circulation and water mass pathways during POEM-Phase I. Prog Oceanogr 39: 153–204.

50. Pinardi N, Masetti E (2000) Variability of the large scale general circulation of the Mediterranean Sea from observations and modelling: a review. Palaeogeogr Palaeoclimatol Palaeoecol 158: 153–173.

51. Longhurst AR, Sathyendranath S, Platt T, Caverhill C (1995) An estimate of global primary production in the ocean from satellite radiometer data. J Plankton Res 17: 1245–1271.

52. Chase JM, Leibold MA (2002) Spatial scale dictates the productivity-biodiversity relationship. Nature 416: 427–430.

53. Zavatarelli M, Mellor GL (1995) A numerical study of the Mediterranean Sea circulation. J Phys Oceanogr 25: 1384–1414.

54. Hernández-Carrasco I, López C, Hernández-García E, Turiel A (2012) Seasonal and regional characterization of horizontal mixing in the global ocean. J Geophys Res 117: C10007. doi: 10.1029/2012JC008222.

55. d'Ovidio F, Isern-Fontanet J, López C, Hernández-García E, García-Ladona E (2009) Comparison between Eulerian diagnostics and finite-size Lyapunov exponents computed from altimetry in the Algerian basin. Deep Sea Res Part 1 Oceanogr Res Pap 56: 15–31.

56. Cayula JF, Cornillon P (1992) Edge detection algorithm for SST images. J Atmos Oceanic Tech 9: 67–80.

57. Hyrenbach KD, Forney KA, Dayton PK (2000) Marine protected areas and ocean basin management. Aquatic Conserv 10: 437–458.

16

First Autonomous Bio-Optical Profiling Float in the Gulf of Mexico Reveals Dynamic Biogeochemistry in Deep Waters

Rebecca E. Green[1]*, Amy S. Bower[2], Alexis Lugo-Fernández[1]

1 Environmental Studies Section, Bureau of Ocean Energy Management, New Orleans, Lousiana, United States of America, **2** Physical Oceanography Department, Woods Hole Oceanographic Institution, Woods Hole, Massachusetts, United States of America

Abstract

Profiling floats equipped with bio-optical sensors well complement ship-based and satellite ocean color measurements by providing highly-resolved time-series data on the vertical structure of biogeochemical processes in oceanic waters. This is the first study to employ an autonomous profiling (APEX) float in the Gulf of Mexico for measuring spatiotemporal variability in bio-optics and hydrography. During the 17-month deployment (July 2011 to December 2012), the float mission collected profiles of temperature, salinity, chlorophyll fluorescence, particulate backscattering (b_{bp}), and colored dissolved organic matter (CDOM) fluorescence from the ocean surface to a depth of 1,500 m. Biogeochemical variability was characterized by distinct depth trends and local "hot spots", including impacts from mesoscale processes associated with each of the water masses sampled, from ambient deep waters over the Florida Plain, into the Loop Current, up the Florida Canyon, and eventually into the Florida Straits. A deep chlorophyll maximum (DCM) occurred between 30 and 120 m, with the DCM depth significantly related to the unique density layer $\rho = 1023.6$ ($R^2 = 0.62$). Particulate backscattering, b_{bp}, demonstrated multiple peaks throughout the water column, including from phytoplankton, deep scattering layers, and resuspension. The bio-optical relationship developed between b_{bp} and chlorophyll ($R^2 = 0.49$) was compared to a global relationship and could significantly improve regional ocean-color algorithms. Photooxidation and autochthonous production contributed to CDOM distributions in the upper water column, whereas in deep water, CDOM behaved as a semi-conservative tracer of water masses, demonstrating a tight relationship with density ($R^2 = 0.87$). In the wake of the Deepwater Horizon oil spill, this research lends support to the use of autonomous drifting profilers as a powerful tool for consideration in the design of an expanded and integrated observing network for the Gulf of Mexico.

Editor: Wei-Chun Chin, University of California, Merced, United States of America

Funding: This project was funded by Bureau of Ocean Energy Management (BOEM) Contract M10PC00112 to Leidos, Inc. with subcontract to ASB (www.boem.gov). BOEM employees, REG and ALF, were involved in the study design, data analysis, decision to publish, and preparation of the manuscript.

Competing Interests: Leidos, Inc was the lead contractor and Program Manager for the research. There are no patents, products in development or marketed products to declare.

* Email: rebecca.green@boem.gov

Introduction

Long-term monitoring of carbon cycling in the oceans is required to understand oceanic ecosystem response to natural and anthropogenic perturbations, including distinguishing trends due to storm events, climate cycles, oil spills, and global warming. Ocean primary production, as largely contributed to by phytoplanktonic carbon fixation, accounts for approximately half of the global estimated net primary production, an amount roughly equivalent to that on land [1]. This primary production represents the base of the marine food web, supporting nearly all oceanic life and significantly affecting global biogeochemical cycles, including atmospheric CO_2 uptake. Thus, significant changes in phytoplankton biomass as linked to perturbations, such as climate forcing [2–3], can have major implications for marine ecosystem functioning all the way up the food chain. Photosynthetic phytoplankton are also largely responsible for the production of oceanic dissolved organic matter (DOM), which serves as substrate for heterotrophic microbial populations and provides nutrients for autotrophs. Marine DOM represents the largest oceanic pool of reduced carbon, estimated to hold greater than 200 times the carbon inventory of marine biomass [4]. Given the importance of both the marine particulate and dissolved organic matter pools, improved methods are required for jointly assessing and monitoring long-term changes due to perturbations, especially given linkages between the two components.

Autonomous profiling floats represent an emerging capability for monitoring biogeochemical properties of the world's oceans at unprecedented scales. Technological advances in float platforms and sensor technologies allow deployments of longer duration (≥ 5 years), to greater depths (up to 2,000 m), and with higher sampling frequencies as smaller, lower-power sensors are developed [5]. Optical instrumentation recently developed specifically for float applications allows measurement of a suite of biogeochemical parameters, including concentrations of chlorophyll, particulate matter, colored dissolved organic matter (CDOM), dissolved oxygen, and nutrients. Recent miniaturization of sensors now allows for joint measurement of multiple parameters from floats, providing the ability to collect important baseline measurements and repeat monitoring for assessing long-term trends, such as climate-related impacts on ocean productivity, carbon cycling, oxygenation, and acidification [3]. Thus far, optically-equipped floats have been successfully used to provide broad spatial (horizontal and vertical scales) and highly time-resolved measurements of particle types and fluxes, including in the Pacific, Atlantic, and Mediterranean oceans [6–8], providing information on optical variability at previously unobserved scales. However, there are fewer examples of the simultaneous measurement of both particulate and dissolved organic matter cycling from floats, given only recent advances in the technology [9].

Deep waters of the Gulf of Mexico (GOM) represent an important frontier for better characterizing biogeochemical processes using autonomous platform technologies. Deep GOM waters provide valuable ecosystem services, including essential habitats for large pelagic species, deep sea corals, and marine mammals [10]. However, these waters are also heavily utilized by various industries, including commercial fisheries, shipping, and oil and gas production, with potentially harmful effects on the environment, such as the Deepwater Horizon oil spill. Previous work has suggested that a variety of environmental forcing factors can influence biogeochemical cycling in these deep waters, including seasonal mixing, Loop Current (LC) and eddy interactions, distant transport of riverine waters, and upwelling along the shelf edge [11–13]. However, thus far, an understanding of biogeochemical processes in deep GOM waters has been mostly limited to traditional shipboard sampling techniques and remote sensing studies of surface waters. More highly resolved sampling, in both time and 3-D space, is required to tease apart the various processes driving optical variability due to phytoplankton, particulates, and CDOM. To our knowledge, this is the first publication to describe the high-resolution measurements obtained from a bio-optical profiling float in deep waters of the Gulf of Mexico, including collection of both particulate and dissolved organic matter properties.

Methods

The measurements presented here were collected as part of the Bureau of Ocean Energy Management (BOEM)-funded "Lagrangian Study of the Deep Circulation in the Gulf of Mexico", which is measuring currents at depth in both U.S. and Mexican waters. In totality, the study has deployed ~120 acoustically-tracked RAFOS floats [14] at depths of 1500–2500 m to map the deep circulation and its variability, as well as 8 autonomous profiling APEX floats [15], the majority of which (at the time of writing this paper) are still collecting measurements. However, one of the APEX floats has now finished its mission, after 17 months of deployment, and is the topic of this paper (Fig. 1A; Dataset S1). The deployment of the profiling float for this study did not require permits for the following reasons: 1) it was deployed in federal waters of the US or inside the Exclusive Economic Zone and not in State waters, and 2) it was deployed under a study for the BOEM of the US Dept. of the Interior, under the authority of the Outer Continental Shelf Lands Act. This Act requires the Agency to conduct studies to evaluate the potential impacts of the oil and gas industry on the environment.

The APEX float was equipped to provide profiles of both physical and bio-optical measurements. The profiling float was built by Teledyne-Webb Research, Inc. (with float dimensions of 16.5 cm diameter by 127 cm long; Fig. S1) and was interfaced with a pumped conductivity-temperature-depth (CTD) instrument (SBE41-CP, SeaBird), bio-optical sensors (ECO FLbbCD-AP2, WET Labs, Inc.), and two-way Iridium communications, which allowed for both real-time data transmission and sampling plan adjustments. The instrument was controlled to float at a specified depth and to profile at set time intervals throughout the water column. While the profiler was at its specified park depth, it acted as a passive, quasi-Lagrangian current follower. The vertical resolution of float bio-optical sampling was set to provide increasing resolution towards the surface, as follows: 5 m from 0–200 m water depth, 10 m from 200–500 m water depth, 25 m from 500–1,000 m water depth, and 50 m below 1,000 m water depth.

The float profiled in the southeastern Gulf of Mexico, traveling from deep waters of the Florida Plain, along the West Florida Escarpment, and into the Florida Straits (Fig. 1A). It collected a total of 61 water column profiles (equaling 5,514 discrete measurements) of bio-optical and physical properties during a 17-month period. The float was deployed from the R/V Pelican on July 19th, 2011 over the Florida Plain, where the water depth was ~3,200 m, and transmitted high-quality data through December 18th, 2012 when it left the GOM through the Florida Straits and was not retrieved. During its first 140 days of deployment (through Dec. 6th, 2011), the float collected measurements down to 1,500 m in deep GOM waters highly impacted by the LC. After traveling significantly to the southeast of its initial deployment, the float began moving into shallower waters of the West Florida Escarpment and up the Florida Canyon. As the float moved into shallower waters, it likely rested on the bottom between profiles. Then, as the float rose to the surface and descended again, it was advected by the currents, thus landing in a slightly different spot on the bottom. While profiling was initially set to every 14 days (7/19-8/17/2011), it was quickly decreased to every 5 days to upload data more often and clear out the memory backlog (8/17-4/9/2012). Finally, sampling was increased again to every 14 days in the Florida Straits to maximize float time sitting on the bottom, in order to delay its leaving the GOM (4/9-12/18/2012).

The bio-optical sensor suite on the float measured proxies of phytoplankton abundance (chlorophyll fluorescence), total particle concentration (optical backscattering), and dissolved organic matter (CDOM fluorescence). Calibration of the sensors (serial # FLBBCDAP2-2140) was performed by the manufacturer (WET Labs, Inc.) prior to shipping for installation on the floats. Dark counts were determined by the manufacturer using the signal output of the sensor in clean water with black tape over the detector. Sensor scaling factors were determined for each sensor using appropriate standards (i.e., a mono-culture of phytoplankton for chlorophyll fluorescence, microspherical beads for backscattering, and a quinine sulfate dihydrate solution for CDOM fluorescence). A separate field characterization was not performed on the sensors and thus, it is possible that some variation from the dark counts and scale factors determined by the manufacturer may have occurred due to factors in the field. Data from each of the sensors, output in counts, was converted to engineering units using the laboratory calibrations, resulting in chlorophyll concentration (Chl; $\mu g\,l^{-1}$), the volume scattering function at a centroid angle of 140° and a wavelength of 700 nm ($\beta(140°, 700\ nm)$; $m^{-1}\,sr^{-1}$), and CDOM concentration (ppb). Volume scattering data contained significant spikes (perhaps associated with particulate aggregates), and the data was despiked by applying a 3-point running minimum filter followed by a 3-point running maximum filter to separate spikes from the baseline, similar to Briggs et al. [16]. The volume scattering function of seawater, $\beta_{sw}(140°,$

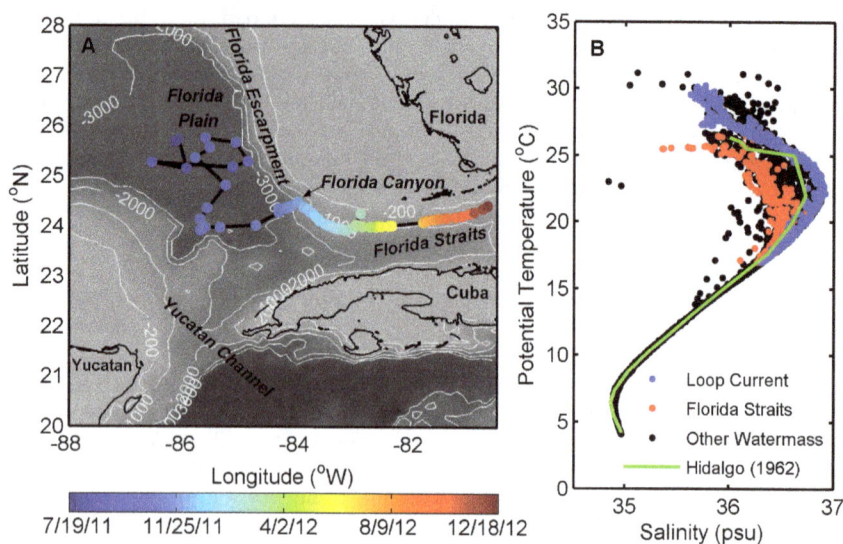

Figure 1. Map of float surface position in the Gulf of Mexico and T-S diagram for the deployment. (A) Float surface position, starting with its deployment on July 19th, 2011 and showing each surfacing (circles), until its last useful profile on December 18th, 2012. Note that profiles are not evenly spaced in time (see Methods). (B) T-S diagram for the float deployment demonstrating characteristic shape for the GOM, including profiles associated with the LC, Florida Straits, and all other water masses sampled. The T-S relationship compares well with an example profile from the historic Hidalgo (1962) cruise.

700 nm) was calculated following Zhang et al. [17] and subtracted from $\beta(140°, 700 \text{ nm})$ to yield the scattering due to particles, $\beta_p(140°, 700 \text{ nm})$, which was converted to integrated particulate backscattering, $b_{bp}(700 \text{ nm})$, according to:

$$b_{bp}(700) = 2\pi\chi\beta_p(140°, 700\text{nm})$$

where $\chi = 1.132$ for this optical configuration (James Sullivan [WET Labs, Inc.], personal communication). While the instruments were not re-calibrated during the deployment, sensor drift was considered to be relatively small given the observed temporal stability of sensor output in the field. For example, at a reference depth of 400 m where low variability was generally observed in optical properties, only small differences were observed between the first and second half of the deployment (differences for Chl, bbp, and CDOM of 0.006 µg l^{-1}, 6×10^{-6} m^{-1}, and 0.01 ppb, respectively).

Analyses in silico were performed to describe observed variability in the bio-optical datasets, including comparison to the physical datasets collected. In addition to temperature (T) and salinity (S) measured by the float, an ancillary dataset of sea surface height anomalies (SSHA) was also employed. For the period corresponding to float deployment, SSHA fields were determined from remotely-sensed altimetric data obtained from the Colorado Center for Astrodynamics Research (CCAR, courtesy of Robert Leben). The criterion used for LC waters was defined as SSHA ≥ 17 cm [18]. All analyses of physical and bio-optical datasets, including creation of figures, were performed using the MATLAB software package (The MathWorks).

Results and Discussion

Float physical and remote sensing data indicated the various unique water masses sampled, including Loop Current and Florida Straits waters. Over all float measurements, temperatures and salinities ranged between 4.2–31.2°C and 34.8–36.9 psu, respectively (Fig. 1B). Float temperature and salinity data

compared well to historical data from the Hidalgo (1962) cruise in the GOM [19], showing the same distinctive T-S relationship. Profiles were often associated with the LC in the upper layer, especially over the Florida Plain, demonstrating unique T-S characteristics, which were distinct from other profiles. In particular, the LC was associated with the maximum temperatures sampled in surface waters, and generally defined the outer envelope of the T-S diagram (Fig. 1B). In contrast, waters in the Florida Straits typically did not retain the unique LC signature and rather defined the inner envelope of the T-S diagram, likely due to mixing with ambient waters (Fig. 1B). In the realm below ~17°C, deeper waters had highly uniform physical properties, fitting a tight T-S relationship. The interaction of the float with the LC was also apparent through comparison with altimetry data, with several crossings of the LC boundary observed, especially during the first part of the deployment in deep waters over the Florida Plain (Fig. 2). Values of SSHA, matched up to float surfacings, ranged from a minimum of −14 cm on September 3rd, 2011 to a maximum of 58 cm on October 9th, 2011.

During the sampling period, variable chlorophyll concentrations contributed evidence supporting a heterogenous picture of deep Gulf of Mexico oligotrophic waters, as punctuated by spatial hot spots and temporal peaks in biomass. The majority of variability in Chl occurred in the upper water column (above 200 m; Fig. S2A) where concentrations ranged from 0.01 to 2.38 µg l^{-1}. Average values observed near-surface (0.14±0.09 µg l^{-1}; Fig. 3A) were similar to those previously reported in offshore GOM waters [13,20]. However, the highly-resolved float measurements demonstrated a greater degree of variability in Chl, with concentrations in the Deep Chlorophyll Maximum (DCM) on average 10 times higher than at the surface, and as much as 30–40 times higher in some locations. The DCM ranged in depth from 30 to 120 m (Fig. 3A), with the average depth greater in deep GOM waters (91±18 m) versus in the Florida Straits (65±18 m). Notably, the depth of the DCM approximately doubled as the float twice moved from outside to inside the LC during the first five months of the deployment (Figs. 2, 3A), indicative of a deeper

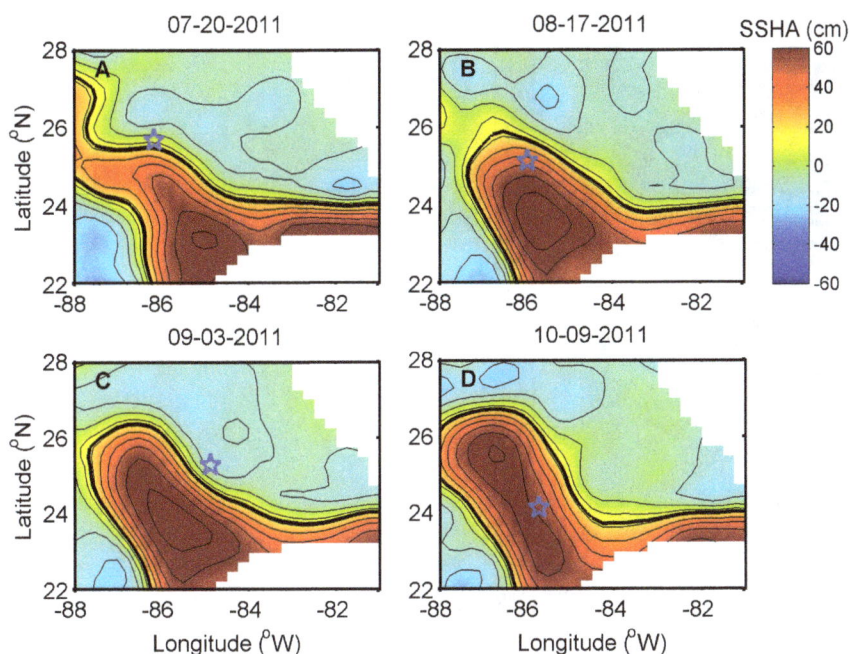

Figure 2. Example comparisons of SSHA fields to float surfacing locations on four dates. (A) July 20th, (B) August 17th, (C) September 3rd, and (D) October 9th, 2011. The float location (blue star) is identified relative to the Loop Current boundary, as determined by the 17-cm isopleth (thick black line).

nitracline in the LC compared to ambient waters [13]. Pycnocline shoaling and resulting subsurface upwelling events offshore of the Southwest Florida Shelf punctuated surface Chl with relatively high values, such as in the deep GOM during 8/29-9/3/2011 and in the Florida Straits during 12/17/2011–2/12/2012 when surface Chl reached ~0.3–0.4 µg l^{-1} (Fig. 3A). Across the entire deployment, DCM depth was highly correlated to the depth of the density layer $\rho = 1023.6$ (Fig. 4a; $R^2 = 0.62$, $p<0.001$), which corresponded to a mean temperature of 25.4°C and salinity of 36.4 psu and roughly to the depth of the pycnocline (Fig. S4).

As an indicator of the total particulate pool, backscattering demonstrated significant complexity throughout the water column, with contributions from diverse particle types. Natural particle assemblages contain a range of living and non-living particles, which can all contribute to optical backscattering in the ocean, depending on their composition and size distribution, including phytoplankton, heterotrophic organisms (mostly bacteria), viruses, detritus, and minerals [21–22]. While backscattering often peaked with chlorophyll (Figs. 4b, S3), it also demonstrated high values at other depths in the water column, indicative of the unique dynamics of the total particulate pool (Fig. 3B). Across the entire float deployment, peak particle concentrations typically occurred at the following depths: (1) coincident with chlorophyll peaks in the upper layer (Fig. 5A), (2) just below the DCM (Fig. 5A), (3) in a surface layer (Fig. 5B), (4) at intermediate depths (200 to 1000 m; Fig. 5C), and (4) near-bottom (Fig. 5D). Elevated b_{bp} values in the upper water column (above 200 m) often occurred at the same locations as high values of chlorophyll (Figs. 5A, S3A–B), as evidenced by the significant relationship between b_{bp} and Chl (Fig. 4b; $R^2 = 0.49$, $p<0.001$). The smaller observed slope between b_{bp} and Chl for most of the dataset (Chl>0.03 mg m^{-3}) compared to other oceanic regimes (Fig. 4B, [23]), suggests a lower backscattering efficiency for phytoplankton and associated particles in the GOM due to differences in particle size distribution and/or composition. As indicated by ocean color imagery,

anomalously high b_{bp} on July 30th, 2012 (Fig. S3b, 4b) appears to have been linked to a plume of terrigenous origin advected offshore into the float's path, introducing a water mass with a significantly different particulate and dissolved composition. Future studies analyzing individual particle characteristics, in addition to bulk optical properties [22], would contribute better understanding of the roles that distinct particle types and characteristics play in determining the optical field in deep GOM waters.

In the oligotrophic waters of the open Gulf of Mexico, the deep chlorophyll maximum and associated biological community play a significant ecological role in structuring the food web. Evidence presented here demonstrates consistently elevated chlorophyll concentrations at depth in the Southeastern GOM (Fig. 3A), and an associated particle assemblage as evidenced by high backscattering values (Fig. 4B). These hot spots of chlorophyll and related primary production occur where nutrient availability is locally enhanced, such as at the pycnocline depth, and are a significant contributor to water column primary production, recently estimated at a median value 0.28 gC m^{-2} d^{-1} for the open Gulf [24]. On an areally-integrated basis, this median estimate resulted in the open Gulf having a larger regional primary production budget than the shallower Gulf regions (i.e., West Florida Shelf, Louisiana Shelf, Texas Shelf, and Mexican Shelf). In terms of the carbon pump, the net result of physical and biological factors in the open Gulf is such that it is also one of the largest net sinks of CO_2 of all Gulf regions, estimated at -0.48 mol C m^{-2} y^{-1} [25]. The export of organic matter (marine snow) from this biological pump in the open GOM helps support a diverse benthic habitat of bacteria, meiofauna, megafauna, fishes, and deep water corals [26–27]. In addition to phytoplankton, productivity hot spots are also associated with higher stocks of zooplankton and micronekton in the deepwater GOM [20,28]. Grazing of algal cells by zooplankton may be responsible for pheophytin peaks observed in previous studies just below the chlorophyll maxima, and thus,

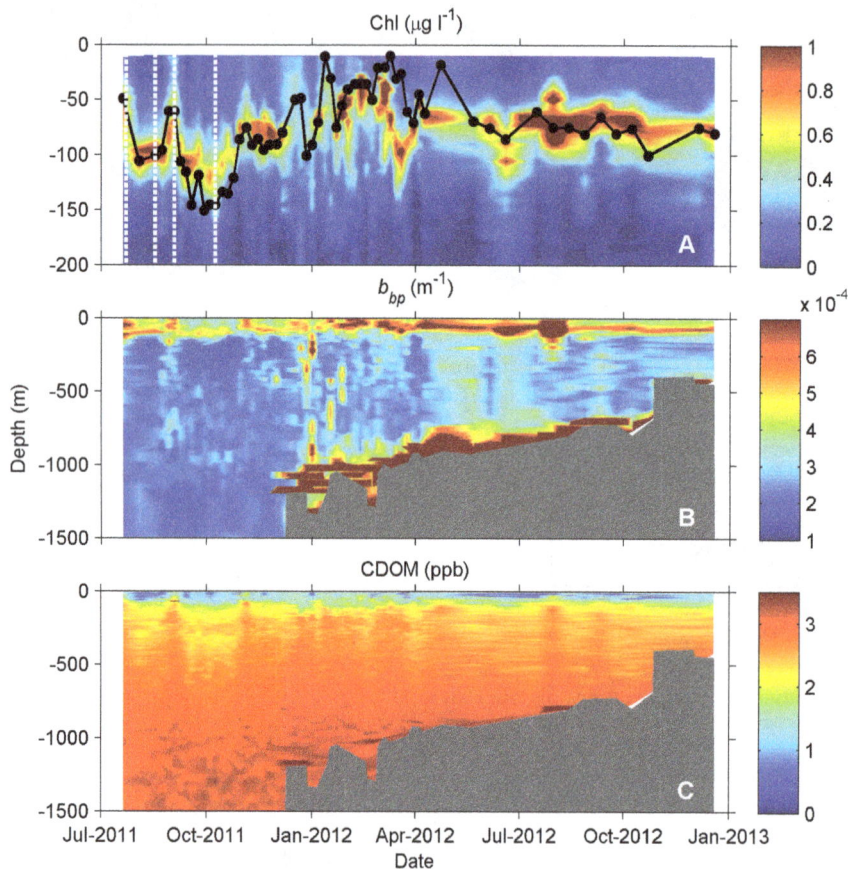

Figure 3. Contour plots demonstrating spatiotemporal variability in bio-optical float profiles. (A) Chl, (B) b_{bp}, and (C) CDOM. Note difference in vertical axes for Chl (upper 200 m) versus b_{bp} and CDOM (entire depth profile) to emphasize depth zones of maximum variability. For reference, the density layer $\rho = 1023.6$ is shown in panel A (black line) with sample times (black dots), and bottom depth is shown in panels B and C (shaded grey). The times corresponding to SSHA imagery in Fig. 2 are indicated in panel A (white dashed lines) to show float location relative to the LC boundary.

may explain the b_{bp} peaks observed in the present study just below the DCM (Fig. 5A). Virus and bacterial concentrations can also be highly correlated with chlorophyll in the oligotrophic southeastern GOM [29]. Together, the association with phytoplankton of these diverse particle types likely contributed to the high backscattering signal observed in and around the DCM in the present work.

In the lower layer of the water column, backscattering often peaked in deep scattering layers (DSL) and demonstrated highest

values near-bottom. The majority of profiles contained backscattering peaks in DSL between 200 to 1,000 m, with layer thickness ranging from 10's to 100's of meters (e.g., Figs. 3B, 5C). These layers may have contributions from a variety of sources, including aggregations of zooplankton and micronekton at depth, as previously observed using acoustics in other parts of the GOM [30], and/or horizontal advection of particles seaward from the continental slope [31]. Zooplankton and related particles are a

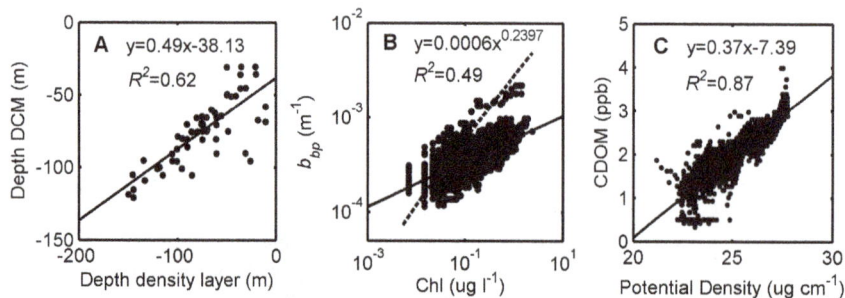

Figure 4. Bio-optical and bio-physical relationships determined based on float profile dataset. (A) depths of DCM vs. density layer $\rho = 1023.6$, (B) Chl vs. b_{bp} in the upper 200 m, and (C) CDOM vs. potential density over the entire water column. In panel B, comparison is shown to results from the algorithm of Morel and Maritorena ([23], dashed line) [2001]. The anomalously high b_{bp} values correspond to the float profile from July 30th, 2012, during which time a plume of terrigenous origin was advected offshore into the float's path.

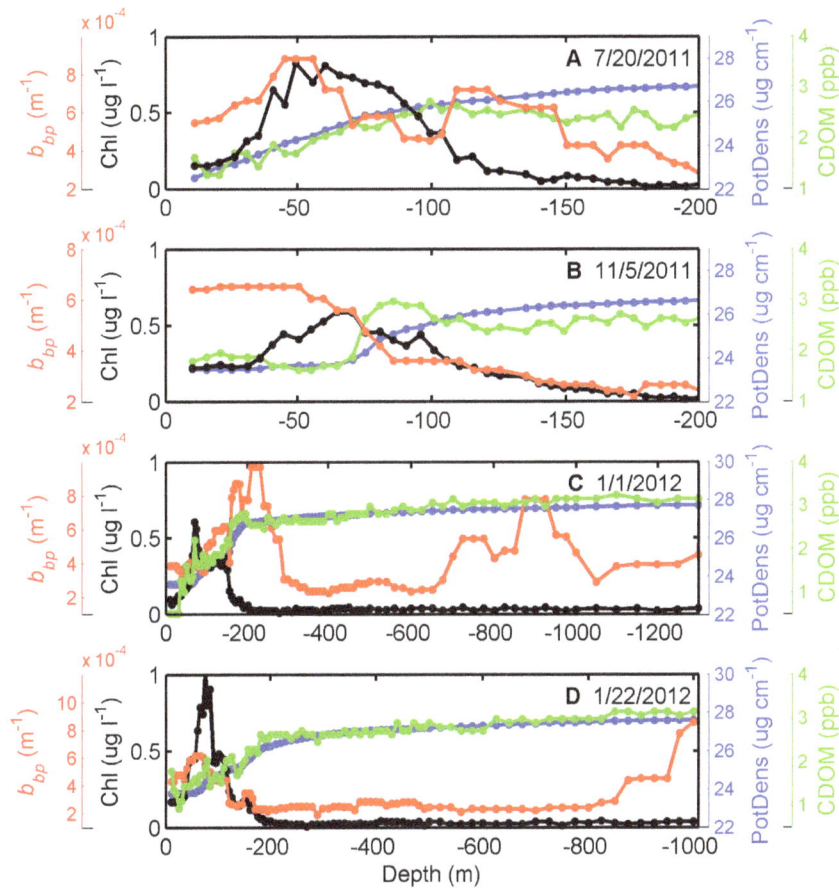

Figure 5. Float profiles exemplifying depth trends and peaks in b_{bp}, Chl, CDOM, and potential density (PotDens). Examples each are provided of b_{bp} peaks at the following depths: (A) coincident with and just below the DCM, (B) in a surface layer, (C) at intermediate depths (700 to 900 m), and (D) near-bottom. In panels A and B, only the upper 200 m of the profiles are shown to emphasize the region of maximum variability in b_{bp}.

likely explanation, especially for profiles in deeper waters away from the slope, and are indicative of a potential prey source for higher trophic levels, including cetaceans, which frequent these waters. Due presumably to contributions from non-living particles, the highest values of b_{bp} occurred near the bottom as the float profiled into shallower waters and close to the maximum water depth (Fig. 3B, 5D), with values as high as 0.01–0.05 m^{-1}. These high near-bottom b_{bp} values are indicative of either natural resuspension in a bottom nepheloid layer or resuspension by the float itself landing on the bottom and disturbing the top sediment layer, while it sat at the bottom in shallower depths between profiles. However, the latter possibility of resuspension by the float landing would likely have been localized in space and would not explain the high backscattering observed 100's of meters above the bottom (e.g., Fig. 5D). If natural resuspension is the cause, then these measurements indicate currents strong enough at the bottom, at depths of 700–1,200 m, to resuspend particulate matter. Such resuspension events in this region are plausible given the predominance of mud as a bottom type [32] and maximum near-bottom current speeds of 40–60 cm s^{-1} (unpublished data from BOEM "Loop Current Dynamics Study"). However, the true cause of high bottom backscattering would need to be further investigated in the future to remove the potential for sampling artifacts. Further bottom boundary layer experiments could also help elucidate environmentally-relevant mass (sediment) flows and net fluxes of resuspended materials. Based on previous research, it

is most likely that the dense water of these particle-laden lower layers in the Florida Straits cannot pass through the shallower sections further downstream [33].

In the upper water column and near-bottom, vertical variability in CDOM profiles demonstrated the various biological and physical sources and sinks that can impact this optically-active dissolved organic matter pool, including autochthonous production, photobleaching, and resuspension. The fast turnover, most bioavailable forms of dissolved organic carbon occur in the surface ocean, in contrast to the longer-lived and more recalcitrant materials which circulate in deep oceanic waters [34]. In the open ocean, possible sources of CDOM production include excretion by organisms, viral lysis, and remineralization of sinking particulate matter, which variously contribute to both the deep CDOM reservoir and mixed layer CDOM; the major sink in the latter is photobleaching as controlled by irradiance and mixed layer depth [35–36]. Across all profiles in our dataset, CDOM was lowest in surface waters ranging from 0.3 to 1.9 ppb (Fig. 3c, S3c), with photobleaching as the major sink in the upper mixed layer, especially during the non-winter months when the water column was more stratified. Similarly, low CDOM has previously been reported in surface Sargasso Sea waters, where stratification and high solar radiation levels lead to bleaching countering local production of CDOM [37]. The high vertical resolution in our profiles did provide evidence of localized contributions from autochthonous production, with CDOM peaks in the upper layer corresponding to peaks in particles. For example, more than half of the profiles clearly demonstrated CDOM peaks co-occurring with elevated chlorophyll concentrations (e.g., Fig. 5B–D), and occasionally coinciding with b_{bp} peaks as well. A steep increase in CDOM of 2–6x was observed between the surface and the pycnocline (Fig. S3c), with such increases also observed in the Sargasso Sea [37]. In our dataset, highest CDOM values were observed near-bottom, reaching a maximum of 4.0 ppb (Fig. 3C), presumably corresponding to resuspension events, though sampling artifacts again can not be ruled out. In New England shelf waters, Boss et al. [38] collected data supporting that bottom sediments can act as a source of dissolved organic carbon during sediment resuspension events. However, this is the first time measurements have suggested this phenomenon in deep GOM waters.

At intermediate depths and greater, changes in CDOM were largely consistent with physical mixing and water mass distributions, suggesting its utility as a semi-conservative tracer at depth in the Gulf of Mexico. CDOM concentrations in deep GOM waters roughly followed the same depth patterns previously observed for nutrients, such as nitrate and phosphate [39], in each of the primary deep water masses, which include: 18°C Sargasso Sea water (depths 200–400 m), Tropical Atlantic Central water (TACW, depths 400–700 m), Antarctic Intermediate Water (AAIW, depths 700–1,000 m), and Upper North Atlantic Deepwater (UNADW, depths ~1,000 m and greater) [19]. Below 200 m, CDOM continued to increase with depth to 1,000 m, though with a much smaller rate of change than in the upper water column; below 1,000 m, CDOM was approximately constant with depth (Fig. 6). Average CDOM concentrations equaled the following values in each of the water masses: 2.5 ppb in 18°C Sargasso Sea water, 2.7 ppb in TACW, 3.0 ppb in AAIW, and 3.1 ppb in UNADW. Below 200 m, CDOM concentrations were strongly and positively related to potential density ($R^2 = 0.87$, $p < 0.01$; Fig. 4c) and temperature ($R^2 = 0.81$; not shown), indicating physical mixing as an important determinant of variability and the role of CDOM as a semi-conservative oceanographic tracer in GOM deep waters. Past studies in the North Atlantic have supported the potential of CDOM as a tracer of ocean circulation processes for subducted water masses [35], and our present results lend evidence for a similar role in the Gulf of Mexico.

During and following the Deepwater Horizon oil spill, a ship-based dataset of CDOM fluorescence was collected in the northcentral Gulf of Mexico in order to track the presence of the subsurface hydrocarbon plume. This dataset spanned the time period from just after the oil spill started until several months after the well was capped (May to October, 2010, [40]). While those CDOM profiles were generally in a similar range of values and demonstrated a similar depth increase to our dataset, many of the profiles showed large spikes in CDOM at depths of ~800–1,200 m corresponding to the presence of the subsurface hydrocarbon plume [41]. However, as expected given the location of the present float dataset (≥400 km to southwest of spill site) and length of time since the oil spill (≥1 year), the deepwater hydrocarbon fluorescence anomaly evidenced in the oil spill dataset was not present in this float data. During future oil spill events, bio-optically equipped profiling floats could prove a useful tool for improved detection of subsurface hydrocarbon plumes, in addition to traditional ship-based measurements.

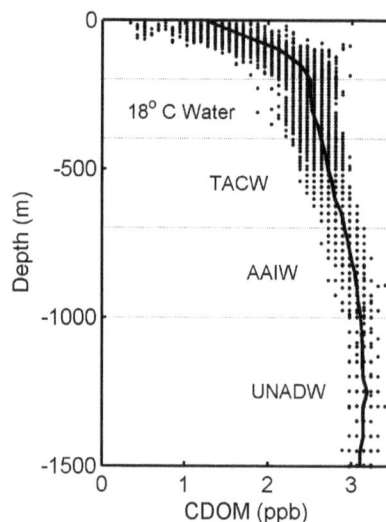

Figure 6. CDOM from all float profiles overlaid with the major Gulf of Mexico water masses below 200 m. The primary deep water masses include: 18°C Sargasso Sea water (200–400 m), Tropical Atlantic Central water (TACW, 400–700 m), Antarctic Intermediate Water (AAIW, 700–1,000 m), and Upper North Atlantic Deepwater (UNADW, ~1,000 m and greater) [19]. Mean CDOM values for all profiles is indicated (solid black line). Note that all CDOM >3.5 ppb were removed from this figure and associated mean, to remove anomalies due to high near-bottom values.

Conclusions

We observed highly dynamic biogeochemistry in both the particulate and dissolved matter pools in deep waters of the southeastern Gulf of Mexico, using an APEX profiling float equipped with bio-optical sensors. Understanding such variability in the open ocean is important because the particulate and dissolved pools play a key role in determining underwater light availability and the resulting impact on biogeochemical cycling. As well, in deeper layers of the water column, bio-optical variability provides insight into the various oceanographic and biological processes at play. However, there are few previous examples of deepwater bio-optical studies in the Gulf of Mexico, in comparison to numerous studies in shallow and shelf regions [42–44]. The present study demonstrated complex variability in the particulate matter pool in deep GOM waters, as measured by chlorophyll fluorescence and optical backscattering (b_b (700 nm)), with peaks observed at various depths throughout the water column. This dataset provided evidence for a dynamic DCM in the GOM impacted by mesoscale processes, as well as evidence for the formation of deep scattering layers between 200–1,000 m, likely of biological origin, and the potential importance of sediment resuspension at depths >500 m. As well, backscattering was significantly related to chlorophyll concentration in the upper water column (Fig. 4B), a parameterization which could improve ocean color, satellite-based retrievals of phytoplankton biomass, as it has in other oceanic regimes [45].

Additionally, the present study provided evidence of the important role that water column density structure, as impacted by water mass variability and vertical mixing, plays in structuring both particulate and dissolved concentrations in the deep GOM in addition to other processes (e.g., photo-oxidation, autochthonous production, grazing, etc.). While previous studies have suggested such a role in the GOM [28], the large number of observations afforded by the present float deployment allowed actual parameterization of the relationships of both chlorophyll and CDOM versus water column density (Fig. 4), including in the wintertime when ship-based measurements are typically rare. Such relationships could significantly improve the current emerging generation of Gulf-wide coupled, biogeochemical models, which aim to capture the spatiotemporal variability in particulate and dissolved matter pools [46–47]. For example, the present float measurements help fill an important deepwater gap in model validation data, with the predominance of data currently in shallow waters, such as the Louisiana-Texas shelf [47].

The emerging role of autonomous underwater vehicles (AUVs) promises to be a critical asset in future ocean observing systems, as they provide economical, long-term deployments and measurements at unprecedented resolution. As well, such autonomous platforms allow for sampling during high-wind periods, when traditional oceanographic methods are impracticable. The present study demonstrates the utility and feasibility of optically-equipped profiling floats for providing new understanding of biogeochemical processes in deep GOM waters, at a critical time for the future of ocean observing in this region. Following the Deepwater Horizon oil spill, in 2012 the U.S. Congress passed the RESTORE Act, which was created to invest oil spill funds into recovering GOM ecosystems that were affected by the disaster. Marine ecosystem monitoring is amongst the activities that can be funded by this legislation, with the oil spill having acutely demonstrated the need for improved oceanographic observing systems [48]. Looking towards the future, our research lends support to the use of autonomous drifting profilers as a powerful tool for consideration in the design of such an integrated observing network for the Gulf of Mexico.

Supporting Information

Figure S1 Picture of APEX float being deployed in the Gulf of Mexico. The antenna and pumped CTD are located at the top of the float, whereas the optical sensors are located near the bottom of the instrument (Photo Credit: CANEK group, CICESE).

Figure S2 Contour plots of bio-optical profiles over the entire depth range the float transited. (A) Chl, (B) b_{bp}, and

(C) CDOM. Bottom depth is shown for reference (shaded grey). Note that profiles are not evenly spaced in time (see Methods).

Figure S3 Contour plots of bio-optical profiles for the upper 200 m to emphasize upper-water column dynamics. (A) Chl, (B) b_{bp}, and (C) CDOM. The reference density layer $\rho = 1023.6$ is shown (black line). Note that profiles are not evenly spaced in time (see Methods). The times corresponding to SSHA imagery in Fig. 1 are indicated in panel A (white dashed lines) to show where the float was located relative to the LC boundary.

Figure S4 Comparison of two physical-mixing indicators: the depth of the density layer $\rho = 1023.6$ and pycnocline depth. The pycnocline depth was calculated for each profile based on the maximum gradient in density in the upper 300 m of the water column.

Dataset S1 Float data used in this analysis. Text file contains the following columns: year, month, day, depth (m), T (°C), S (psu), Chl (μg l^{-1}), b_{bp} (m^{-1}), and CDOM (ppb).

Acknowledgments

The authors appreciate the support of the US Department of Interior, Bureau of Ocean Energy Management, Gulf of Mexico OCS Region during the preparation of this manuscript. The opinions expressed by the authors are their own and do not reflect the opinion or policy of the US government. The authors gratefully acknowledge Heather Furey and Terry McKee of WHOI for the setup and maintenance of the data download system for the APEX floats described in this paper. Thank you also to the captain and crew of the R/V Pelican for assisting with the field deployment. Constructive comments from 3 anonymous reviewers helped to improve this manuscript.

Author Contributions

Conceived and designed the experiments: REG ASB ALF. Performed the experiments: ASB. Analyzed the data: REG ASB ALF. Contributed reagents/materials/analysis tools: REG ASB ALF. Contributed to the writing of the manuscript: REG ASB ALF.

References

1. Field CB, Behrenfeld MJ, Randerson JT, Falkowski P (1998) Primary production of the biosphere: Integrating terrestrial and oceanic components. Science 281: 237–240.
2. Boyce DG, Lewis MR, Worm B (2010) Global phytoplankton decline over the past century. Nature 466: 591–596.
3. Gruber N (2011) Warming up, turning sour, losing breath: ocean biogeochemistry under global change. Phil Trans R Soc A 369: 1980–1996.
4. Hansell DA, Carlson CA, Repeta DJ, Schlitzer R (2009) Dissolved organic matter in the ocean–A controversy stimulates new insights. Oceanography 22: 202–211.
5. Barnard AH, Mitchell TO (2013) Biogeochemical monitoring of the oceans using autonomous profiling floats. Ocean News Technol 19: 16–17.
6. Boss E, Swift D, Taylor L, Brickley P, Zaneveld R, et al. (2008) Observations of pigment and particle distributions in the western North Atlantic from an autonomous float and ocean color satellite. Limnol Oceanogr 53: 2112–2122.
7. Xing X, Morel A, Claustre H, D'Ortenzio F, Poteau A (2011) Combined processing and mutual interpretation of radiometry and fluorimetry from autonomous profiling Bio-Argo floats: Chlorophyll a retrieval. J Geophys Res 116, doi:10.1029/2010JC006899.
8. Estapa ML, Buesseler K, Boss E, Gerbi G (2013) Autonomous, high-resolution observations of particle flux in the oligotrophic ocean. Biogeosciences 10: 1229–1265.
9. Xing X, Morel A, Claustre H, D'Ortenzio F, Poteau A (2012) Combined processing and mutual interpretation of radiometry and fluorimetry from autonomous profiling Bio-Argo floats: 2. Colored dissolved organic matter absorption retrieval. J Geophys Res 117, doi:10.1029/2011JC007632.
10. National Research Council (2013) An ecosystem services approach to assessing the impacts of the Deepwater Horizon oil spill in the Gulf of Mexico. National Academies Press, ISBN 978-0-309-28845-3. 350.
11. Müller-Karger FE, Walsh JJ, Evans RH, Meyers MB (1991) On the seasonal phytoplankton concentration and sea surface temperature cycles of the Gulf of Mexico as determined by satellites. J Geophys Res 96: 12645–12665.
12. Biggs DC, Müller-Karger FE (1994) Ship and satellite observations of chlorophyll stocks in interacting cyclone-anticyclone eddy pairs in the western Gulf of Mexico. J Geophys Res 99: 7371–7384.
13. Qian Y, Jochens AE, Kennicutt II MC, Biggs DC (2003) Spatial and temporal variability of phytoplankton biomass and community structure over the continental margin of the northeast Gulf of Mexico based on pigment analyses. Cont Shelf Res 23: 1–17.
14. Rossby T, Dorson D, Fontaine J (1986) The RAFOS System. J Atmos Oceanic Technol 3: 672–679.
15. Davis RE, Sherman JT, Dufour J (2001) Profiling ALACEs and Other Advances in Autonomous Subsurface Floats. J Atmos Ocean Technol 18: 982–993.
16. Briggs N, Perry MJ, Centinić I, Lee C, D'Asaro E, Gray AM, Rehm E (2011) High-resolution observations of aggregate flux during a sub-polar North Atlantic spring bloom. Deep-Sea Res I 58: 1031–1039.
17. Zhang X, Hu L, He M-X (2009) Scattering by pure seawater: Effect of salinity. Optics Express 17: 5698–5710.
18. Leben RR (2005) Altimeter-derived Loop Current metrics. In: Sturges W, Lugo-Fernandez A, editors. Circulation in the Gulf of Mexico: Observations and Models. Washington, DC: American Geophys Union. 181–202.
19. Nowlin WD Jr., McLellan HJ (1967) A Characterization of the Gulf of Mexico Waters in Winter. J Mar Res 25: 29–59.

20. Biggs DC, Ressler PH (2001) Distribution and abundance of phytoplankton, zooplankton, ichthyoplankton, and micronekton in the deepwater Gulf of Mexico. Gulf Mex Sci 1: 7–29.

21. Stramski D, Bricaud A, Morel A (2001) Modeling the inherent optical properties of the ocean based on the detailed composition of the planktonic community. Appl Opt 40: 2929–2945.

22. Green RE, Sosik HM, Olson RJ (2003) Contributions of phytoplankton and other particles to inherent optical properties in New England continental shelf waters. Limnol Oceanogr 48: 2377–2391.

23. Morel A, Maritorena S (2001) Bio-optical properties of oceanic waters: A reappraisal. J Geophys Res 106: 7163–7180.

24. Lohrenz S, Chakraborty S, Huettel M, Herrera Silveira J, Gunderson K, et al. (2014) Primary production. In: Benway HM, Coble PG, editors. Report of the U.S. Gulf of Mexico Carbon Cycle Synthesis Workshop, March 27–28, 2013. Ocean Carbon and Biogeochemistry Program and North American Carbon Program, 28–38. Available: http://www.us-ocb.org/publications/GMx_report_FINAL.pdf. Accessed 2014 April 28.

25. Robbins LL, Wanninkhof R, Barbero L, Hu X, Mitra S, et al. (2014) Air-sea exchange. In: Benway HM, Coble PG, editors. Report of the U.S. Gulf of Mexico Carbon Cycle Synthesis Workshop, March 27–28, 2013. Ocean Carbon and Biogeochemistry Program and North American Carbon Program, 17–23. Available: http://www.us-ocb.org/publications/GMx_report_FINAL.pdf. Accessed 2014 April 28.

26. Rowe GT, Wei C, Nunnally C, Haedrich R, Montagna P, et al. (2008) Comparative biomass structure and estimated carbon flow in food webs in the deep Gulf of Mexico. Deep-Sea Res II 55: 2699–2711.

27. Cordes EE, McGinley MP, Podowski EL, Becker EL, Lessard-Pilon S, et al. (2008) Coral communities of the deep Gulf of Mexico. Deep-Sea Res I 55: 777–787.

28. Hobson LA, Lorenzen CJ (1972) Relationships of chlorophyll maxima to density structure in the Atlantic Ocean and Gulf of Mexico. Deep-Sea Res 19: 297–306.

29. Boehme J, Frischer ME, Jiang SC, Kellogg CA, Pichard S, et al. (1993) Viruses, bacterioplankton, and phytoplankton in the southeastern Gulf of Mexico: distribution and contribution to oceanic DNA pools. Mar Ecol Prog Ser 97: 1–10.

30. Kaltenberg AM, Biggs DC, DiMarco SF (2007) Deep scattering layers of the Northern Gulf of Mexico observed with a shipboard 38-kHz acoustic Doppler current profiler. Gulf Mex Sci 25: 97–102.

31. Gardner WD, Walsh ID (1990) Distribution of macroaggregates and fine-grained particles across a continental margin and their potential role in fluxes. Deep-Sea Res 37: 401–411.

32. National Oceanographic and Atmospheric Administration (2013) Gulf of Mexico Data Atlas: Bottom sediments – dominant bottom types and habitats. Available: http://www.ncddc.noaa.gov/website/DataAtlas/atlas.htm. Accessed 2014 March 19.

33. Montgomery RB (1941) Transport of the Florida Current off Habana. J Mar Res 4: 198–220.

34. Hansell DA (2013) Recalcitrant dissolved organic carbon fractions. Annu Rev Mar Sci 5: 421–45.

35. Nelson NB, Siegel DA (2002) Chromophoric DOM in the open ocean. In: Hansell DA, Carlson CA, editors. Biogeochemistry of Marine Dissolved Organic Matter. San Diego: Academic Press. 547–578.

36. Swan CM, Siegel DA, Nelson NB, Carlson CA, Nasir E (2009) Biogeochemical and hydrographic controls on chromophoric dissolved organic matter distribution in the Pacific Ocean. Deep-Sea Res I 56: 2175–2192.

37. Nelson NB, Siegel DA, Michaels AF (1998) Seasonal dynamics of colored dissolved material in the Sargasso Sea. Deep-Sea Res I 45: 931–957.

38. Boss E, Pegau WS, Zaneveld JRV, Barnard AH (2001) Spatial and temporal variability of absorption by dissolved material at a continental shelf. J Geophys Res 106: 9499–9507.

39. El-Sayed S (1972) Primary productivity and standing crop of phytoplankton. In: Bushnell VS, editor. Serial Atlas of the Marine Environment, Folio 22: Chemistry, Primary Productivity, and Benthic Marine Algae of the Gulf of Mexico. New York: American Geographical Society, 8–13.

40. Operational Science Advisory Team (2010) Summary report for sub-sea and sub-surface oil and dispersant detection: Sampling and monitoring, Deepwater Horizon Unified Area Command, December 17, 2010. Available: http://www.restorethegulf.gov/sites/default/files/documents/pdf/OSAT_Report_FINAL_17DEC.pdf. Accessed 2014 March 19.

41. National Oceanographic and Atmospheric Administration (2011) Joint Analysis Group, Deepwater Horizon oil spill: Review of preliminary data to examine subsurface oil in the vicinity of MC252#1, May 19 to June 19, 2010. NOAA Technical Report NOS OR&R 25. Available: http://www.ncddc.noaa.gov/activities/healthy-oceans/jag/reports/. Accessed 2014 March 19.

42. Conmy RN, Coble PG, Chen RF, Gardner GB (2004) Optical properties of colored dissolved organic matter in the Northern Gulf of Mexico. Mar Chem 89: 127–144.

43. D'Sa EJ, Miller RL, McKee BA (2007) Suspended particulate matter dynamics in coastal waters from ocean color: Application to the northern Gulf of Mexico. Geophys Res Lett 34, doi:10.1029/2007GL031192.

44. Green RE, Gould Jr RW (2008) A predictive model for satellite-derived phytoplankton absorption over the Louisiana shelf hypoxic zone: Effects of nutrients and physical forcing. J Geophys Res 113, doi:10.1029/2007JC004594.

45. Huot Y, Morel A, Twardowski MS, Stramski D, Reynolds RA (2008) Particle optical backscattering along a chlorophyll gradient in the upper layer of the eastern South Pacific Ocean. Biogeosciences 5: 495–507.

46. DeRada S, Arnone RA, Anderson S (2009) Bio-physical ocean modeling in the Gulf of Mexico. In: Oceans 2009, MTS/IEEE Biloxi–Marine Technology for Our Future: Global and Local Challenges. ISBN: 978-1-4244-4960-6, 1–7.

47. Xue Z, He R, Fennel K, Cai W-J, Lohrenz S, Hopkinson C (2013) Modeling ocean circulation and biogeochemical variability in the Gulf of Mexico. Biogeosciences 10: 1–16.

48. Jochens AE, Watson SM (2013) The Gulf of Mexico Coastal Ocean Observing System: An integrated approach to building an operational regional observing system. Mar Tech Soc J 47: 118–133.

17

Lateral Diffusion of Nutrients by Mammalian Herbivores in Terrestrial Ecosystems

Adam Wolf[1]*, Christopher E. Doughty[2], Yadvinder Malhi[2]

1 Department of Ecology and Evolutionary Biology, Princeton University, Princeton, New Jersey, United States of America, **2** Environmental Change Institute, School of Geography and the Environment, University of Oxford, Oxford, United Kingdom

Abstract

Animals translocate nutrients by consuming nutrients at one point and excreting them or dying at another location. Such lateral fluxes may be an important mechanism of nutrient supply in many ecosystems, but lack quantification and a systematic theoretical framework for their evaluation. This paper presents a mathematical framework for quantifying such fluxes in the context of mammalian herbivores. We develop an expression for lateral diffusion of a nutrient, where the diffusivity is a biologically determined parameter depending on the characteristics of mammals occupying the domain, including size-dependent phenomena such as day range, metabolic demand, food passage time, and population size. Three findings stand out: (a) Scaling law-derived estimates of diffusion parameters are comparable to estimates calculated from estimates of each coefficient gathered from primary literature. (b) The diffusion term due to transport of nutrients in dung is orders of magnitude large than the coefficient representing nutrients in bodymass. (c) The scaling coefficients show that large herbivores make a disproportionate contribution to lateral nutrient transfer. We apply the diffusion equation to a case study of Kruger National Park to estimate the conditions under which mammal-driven nutrient transport is comparable in magnitude to other (abiotic) nutrient fluxes (inputs and losses). Finally, a global analysis of mammalian herbivore transport is presented, using a comprehensive database of contemporary animal distributions. We show that continents vary greatly in terms of the importance of animal-driven nutrient fluxes, and also that perturbations to nutrient cycles are potentially quite large if threatened large herbivores are driven to extinction.

Editor: Mary O'Connor, University of British Columbia, Canada

Funding: AW was supported by the Carbon Mitigation Initiative of Princeton University. CD was supported by the Gordon and Betty Moore Foundation, and YM was supported by the Jackson Foundation. The funders had no role in study design, data collection and analysis, decision to publish, or preparation of the manuscript.

Competing Interests: The authors have declared that no competing interests exist.

* E-mail: adamwolf@princeton.edu

Introduction

Nutrient availability is of primary importance in controlling the primary productivity of the biosphere. The nature of nutrient limitation is mediated between exogenous inputs and various processes taking place *in situ* that control conversion of unavailable nutrients into bioavailable forms; the accumulation of nutrients cycling between different pools; and the rate of losses from these pools [1]. Because a fraction of nutrients are inevitably leaked in any cycle, in the long-term the mean nutrient content of an ecosystem is determined by the balance between the gains and losses of nutrients from the ecosystem [2]. To the extent that exogenous nutrients are important in the nutrient budget of an ecosystem, these are often thought to arrive by abiotic processes, such as dust deposition, erosion, and runoff. These processes can be embodied in coupled ordinary differential equations [e.g. 3, 4].

This paper is an attempt to formally investigate a complementary, biotic, process that can transport nutrients into and across ecosystems: the *lateral* translocation of nutrients by mammalian herbivores, in dung or flesh. Specifically, we investigate horizontal translocation of nutrients as a diffusion process, in which the horizontal flux is proportional to a diffusion coefficient acting on a nutrient gradient. The main topics of this paper are (a) the derivation of a quantitative theoretical framework to understand

lateral diffusion of nutrients by herbivores; (b) the empirical calculation of a diffusion coefficient from a compilation of field studies; (c) the analysis of a reaction-diffusion equation describing the time rate of change of phosphorus availability in a location as a function of horizontal diffusion, first order losses, and external inputs and (d) a global analysis of the magnitude of mammalian herbivore-mediated diffusion. Our goal is to understand the circumstances under which herbivore-mediated processes are dominant processes in ecosystem nutrient budgets, with special attention to the impact of global defaunation on ecosystem function. In this paper, "animal" will refer to mammalian herbivores unless otherwise specified.

There is a large body of work applying advection-diffusion equations to characterize animal movement [5]. However, there is considerably less application of such models to understanding the budgets of materials associated with animal movement, particularly nutrients ingested as biomass and excreted as urine, dung, and eventually falling as the body mass of the dead animal itself.

By contrast, there is a separate body of work focusing on animals and their impact on nutrient accumulation and the rate of nutrient cycling in ecosystems, generally on sites where animals are concentrated. The first deep investigation of this field, G.E. Hutchinson's *Biogeochemistry of Vertebrate Excretion* [6], focused exclusively on guano deposits, that is nutrients from excreta that accumulate when large organisms feed over a "wide trophophoric field and return to a limited site for rest or reproduction." Hutchinson's work in many ways touches on themes that are appropriate in the present article, namely global-scale patterns of physical and biological geography; short (intra annual) timescales of behavior nested within long (Quaternary) timescales of biogeochemistry; and the behavior, diet, and population biology of the species under consideration.

As in Hutchinson [6], research on the role of large mammalian grazers in biogeochemistry has a tendency to focus on the rate of nutrient cycling and consequent productivity in regions of animal concentration [7,8], rather than spatial linkages between nutrient source and sink regions. There are notable exceptions to this pattern, where herbivores provide nutrients to nutrient-limited regions [9–12].

What is the relevance for studying the role of animals in biogeochemical cycles over such long timescales? Typically, analyses indicate that herbivore-mediated nutrient fluxes are small compared with other terms in the nutrient budget [13,14]. However, there is an increasing recognition that probably all ecosystems in the 'Anthropocene' are at disequilibrium following a human perturbation from some prior state [15]. While many such perturbations are obvious, such as land use change, others are subtle or indirect, such as species invasions, eutrophication, CO_2 increase, atmospheric warming, and the like. Among these perturbations, we are interested in exploring the consequences on ecosystem function of the ongoing global defaunation event [16], which may perturb nutrient cycles far into the future in ways that are not fully understood.

The goal of this paper is to develop a theoretical framework for understanding the effects of mammal removal on lateral rates of nutrient transport. The framework is kept general, and could in principle be adapted to different nutrients (including micronutrients such as sodium or calcium) by adding details specific to the element. However, we will focus on phosphorus as the target nutrient in this work, because the timescales associated with its gain and loss terms are long [3,4,17].

Methods

Diffusion of Nutrients Arising from Animal Movement

The exchange of material between two locales is generally treated by one or both of two main processes: advection and diffusion. In advection, the flux of material in the x direction is equal to the concentration of the material n (mass/volume) times a velocity u (length/time), that is $J_x = nu$. In diffusion, the movement from high to low density flux is negatively proportional to the local concentration gradient, $-dn/dx$, with the constant of proportionality termed the "diffusivity" D (length^2/time): $J_x = -D(dn/dx)$. In general, a diffusivity can be derived from a random walk to characterize the aggregate statistics of a population of individually moving agents [5,18,19]. The dynamic equation of a probability density function governed by a random walk with length scale Δx and time scale Δt is given as [Methods S1]:

$$\frac{\partial p}{\partial t} = D \frac{\partial^2 p}{\partial x^2}, \text{where} \lim_{\substack{\Delta x \to 0, \\ \Delta y \to 0}} \frac{(\Delta x)^2}{2\Delta t} = D \qquad (1)$$

Equation [1] describes the probability density of the position of a particle, which could be an animal, a nut or disease it carries, or a particle of food in its gut. The assumptions employed in the random walk model dictate that this model does not treat long-

distance migration. Additionally, we have ignored advection, i.e. bias in movement toward a particular direction, which could be used to consider a tendency to return towards a central place. We propose that there are a variety of conditions under which these assumptions may be met, on larger or smaller scales, such that the lateral transfer of nutrients by herbivores can be reasonably approximated as a diffusion-like process. At millennial scales, landscapes evolve, and migration routes, foraging hotspots, and wallows may be expected to shift location. Disturbance also serves to disrupt habitat and change vegetation type. Interannual variation in climate alters the productivity of the landscape, which drives changes in animal foraging intensity [20]. Boundaries between animal groups will change over time as internal demographics change [21]. Behavioral differences between species create differential patterns of movement. Finally, there may exist a "vasculature" in which large animals transport nutrients large distances, and progressively smaller animals diffuse nutrients more finely into the ecosystem matrix [22]. All of these phenomena suggest that the assumption of randomness may be acceptable to a first approximation, and that it is appropriate to represent the transfer of nutrients as a diffusion-like process.

It would be natural in modeling animal movement to consider Δx to be the daily displacement of a single animal (DD; km/day), where Δt is a day. The present challenge is to consider how to extend this work to predict both the distance and time traversed for a particle of plant material consumed by an animal and excreted some distance apart after some passage of time, as well as the distance and time traversed by the animal itself between consumption of plant material comprising the animal's biomass, and its own eventual mortality. In addition, we need to extend the analysis from a single particle of plant material to the aggregate of all such transported particles, as well as consider its nutrient content.

For ingestion and excretion, the appropriate length scale in the diffusivity is the daily displacement multiplied by the average gut passage time (PT; days), and similarly the time scale for this transport would be PT. Hence, based on Equation [1], whereas the diffusivity for animal position is $D_{animal} \sim DD^2/2$ (km^2/day), the diffusivity for transport of its excreta is $D_{excreta} \sim (DD*PT)^2/(2*PT)$, where the numerator is also in km^2 and the denominator is in days.

The diffusivity for nutrients incorporated into animal bodymass, and especially bones, D_{body}, is associated with different time and length scales than for defecation. The mean residence time of a mineral in the body, e.g. phosphorus in bones, will be the mean time between apatite formation and death. As the time spent as a mature adult becomes long in relation to the time spent growing, this time scale comes to approximate its mean lifetime L (days) [Methods S4; Figures S1 & S2]. The length scale is linked to its home range (HR; km^2). If the HR is interpreted to be the area that contains 95% of the probability density of an animal over its lifetime, then the root mean squared displacement would be $RL = HR^{0.5}/2\pi$. An estimate of the diffusivity for bodymass is then $D_{body} = RL^2/2L = HR/(8\pi^2 L)$.

A similar equation as [1] was developed to estimate the diffusivity of a nutrient transported and redeposited by animals in excreta and biomass [Methods S1]. The development recognized that the mass flux of transported nutrient is determined by the population density of animals (PD; #/km^2) that consume dry matter (DM) to fulfill their metabolic requirements (MR; kgDM/animal/day). The product of PD and MR equates to a population consumption rate of DM (denoted Q), such that $Q\Delta t$ is the mass of DM consumed in Δt (kg DM/km^2). The consumption of the nutrient itself is then determined by $Q(x,t)$ times the nutrient content of the consumed biomass ([P](x,t); kg P/kg DM), where Q[P] has units kg P/km^2. Some fraction ε of consumed nutrient is incorporated into bodymass, while the remainder (1-ε) is excreted. Finally, a normalization is introduced, the abundance of edible biomass (αB; kg edible DM/km^2), to represent the state variable as a mass per area. The resultant equation [Methods S2] is the sum of two diffusion contributions, one capturing the transport of nutrients in excreta, the other the transport in bodymass:

$$\frac{\partial P}{\partial t} = \Phi_{excreta}\frac{\partial^2 P}{\partial x^2} + \Phi_{body}\frac{\partial^2 P}{\partial x^2} \quad (2)$$

where:

$$\Phi_{excreta} = (1-\varepsilon)\frac{Q}{\alpha B}D = (1-\varepsilon)\frac{PD}{\alpha B}*MR*\frac{(DD*PT)^2}{2PT} \quad (3)$$

$$\Phi_{body} = \varepsilon\frac{Q}{\alpha B}D = \varepsilon\frac{PD}{\alpha B}*MR*\frac{HR}{8\pi^2 L} \quad (4)$$

Results and Discussion

Diffusion Coefficients Vary as a Function of Body Size

We note that all of the terms in Φ, save the edible plant biomass αB, are known to vary systematically with herbivore body size M (kg bodymass), including some terms not considered in detail here, such as the energy content of consumed dry matter. There is a rich literature in allometric and metabolic scaling that attempts to explain these patterns, but we note here only that they exist and that we can employ them to approximate the magnitude of Φ for animals of different sizes, including animals for which we have little or no behavioral or physiological data. Whether or not every

power law can be explained by fundamental theory, such power laws are useful empirical descriptors of how a particular phenomenon varies across orders of magnitude of biomass (Table 1).

Data were collated from a variety of sources, and reconciled to a common taxonomic authority, Mammal Species of the World, 3rd Edition [MSW3, 23] (http://www.bucknell.edu/msw3/export. asp). Data were restricted to terrestrial mammals at the species level, totaling 5278 unique taxa. Statistics were only calculated for herbivores, although other taxa with available data (insectivores, carnivores, and those having unknown diet) are plotted for reference. Data collected for this study include longevity, fecundity and metabolic rate from the AnAge database [24]; population density [25]; day range [26]; and home range [27], all of which include M as a predictor variable, as well as a dataset of M per se [28], which was used preferentially if available for a taxon. Passage time was not estimated from primary data, and instead the equation from Demment and Van Soest [29] was employed.

We estimated Φ as a function of M by two routes: first, we calculated the allometries for each term as a function of M (using ordinary least squares) and combined the resulting coefficients to yield an allometric equation for Φ that results from scaling arguments. Second, we multiplied values for the terms in [2] and [3] available in the primary literature to estimate Φ directly for individual species, and fit the allometric equation using the data themselves. Because the primary data comprising Φ include independent primary data, as well as allometric estimates of PR that are exact functions of M, caution is urged in interpreting the goodness of fit of Φ with M.

The correlations of the behavioral and physiological phenomena with M are generally strong, with $r^2 \geq 52\%$ (Table 1, Figure 1). The weakest correlations were found for longevity, which is in part attributable to the diverse taxonomy of the dataset: when only non-primates are considered, this correlation increases to 80% [Figure S2]. However, this term is relatively unimportant because this term only appears in Φ_{body}, which was found to be 10,000x smaller than $\Phi_{excreta}$ (Table 1, Figure 2). Hence, the bodymass term can be safely neglected in nutrient flux calculations. Among terms that contribute to Φ_{exreta}, the weakest correlation with M was for daily displacement. This dataset is fairly current, but nevertheless has the smallest sample size of all the presented data, suggesting that among these factors the ecology of daily movement is the least understood. The allometry of $QD_{excreta}$ calculated using scaling arguments is nearly the same as that calculated from the primary data, suggesting there are not strong correlations between the terms that are not already accounted for by M (Table 1; Figure 2).

Both diffusivities $QD_{excreta}$ and QD_{body} are strong functions of M (Figure 2), highlighting the importance of larger bodied species in transporting nutrients. Both diffusivities increase with body mass, and for the dominant term $QD_{excreta}$, the scaling coefficient is >1, which shows that larger herbivores are increasingly important. Examining Equation [3], we see that large animals are important for diffusion firstly because of their large day ranges ($DD^2 \sim M^{0.736}$) and secondly because of long gut passage times ($PT \sim M^{0.26}$). The influence of higher biomass consumption rates ($MR \sim M^{0.716}$) is almost exactly offset by the lower population density ($PD \sim M^{-0.724}$), leading to little mass dependence of biomass consumption per unit are ($MR*PD \sim M^{-0.08}$). This last feature reflects the "law of energy equivalence" [30], which indicates that the population-level biomass consumption should be equal across a range of M. Hence the essential role of large herbivores is embodied in the D term in Equation [3], namely daily displacement and gut residence time.

Table 1. Allometric fits for behavioral and physiological phenomena used in the calculation of diffusivity†.

Dependent Variable	Units	Equation	n	r^2
Population Density, PD	#/km^2	87.6*M$^{-0.724}$	366	0.71
Metabolic Demand, MR	kgDM/#/day	0.021*M$^{0.716}$	131	0.96
Maximum Longevity, L_{max}	days	4816*M$^{0.164}$	294	0.52
Mean Longevity, L	days	1305*M$^{0.173}$	170	0.57
Day Range, DD	km	0.453*M^{0368}	113	0.52
Home Range, HR	km^2	0.0416*M$^{1.09}$	171	0.76
Range Length, ($\sqrt{HR}$)	km	0.204*M$^{0.546}$	171	0.76
Passage time, PT	days	0.29*M$^{0.26}$	–	–
Fecal Diffusivity‡, Φ_{exreta}	(kgDM/km^2)*(km^2/day)	0.053*M$^{1.011}$	–	–
Fecal Diffusivity†, Φ_{exreta}	(kgDM/km^2)*(km^2/day)	0.050*M$^{1.166}$	15	0.67
Body Diffusivity‡, Φ_{body}	(kgDM/km^2)*(km^2/day)	8.62*10^{-7}*M$^{0.917}$	–	–
Body Diffusivity†, Φ_{body}	(kgDM/km^2)*(km^2/day)	4.84*10^{-7}*M$^{0.897}$	40	0.68

†statistical fit to primary data;
‡estimate computed from scaling coefficients.

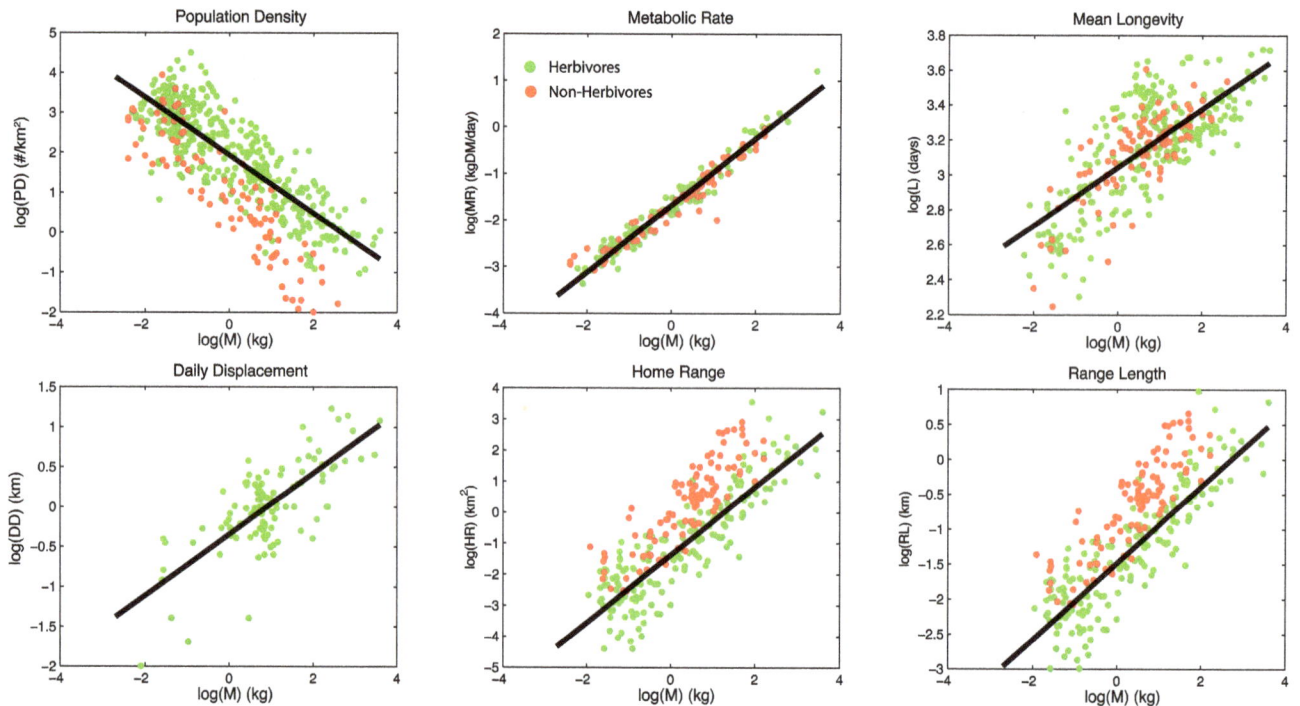

Figure 1. Allometric relations between bodymass M and population density, metabolic rate, mean longevity, daily displacement, home range, and range length.

Transport of Phosphorus in Kruger National Park

We next demonstrate and explore this framework with a specific case study, to test the validity of generalized allometric scaling in a specific local context. A natural locale to explore herbivore impacts on nutrient cycles is Kruger National Park (KNP), in South Africa, because of its large and well-studied animal population [31,32] existing on a landscape with a sharp substrate gradient that impacts the phosphorus concentration and productivity of the vegetation thereon. Herbivores play a prominent role in nutrient biogeochemistry in KNP, but nevertheless a depiction of herbivores in the nutrient cycle neglects the potential for translocation by these vectors [33]. The underlying geology is that of a nutrient poor granitic landscape in the west adjacent to a nutrient rich basalt landscape in the east (Figure 3), creating sharp contrasts in nutrient concentrations in soils [34], and forages quantity and quality [35]. The geometry of this linear substrate boundary makes the park amenable to an analysis in one dimension for clarity *[Methods S3]*. The distribution of animals on the landscape is complex, having a strong component of interannual variability, as well as dietary needs and behavior preferences of wildlife [36].

In estimating the value of Φ, consider the fauna of the park. KNP has 29 mammalian herbivores greater than 10 kg (using masses from [28]). Using the masses and population densities reported in Damuth [25], the predicted biomass density per area is 18747 kg/km^2; however the actual biomass density of the park is estimated as 3931 kg/km^2 [37], approximately 25% of the prediction. This is in part an overestimate of the densities (e.g. elephant population density is said by Damuth to be 1/km^2, whereas in KNP the density is ~0.5/km^2 [38]; moreover many species are endangered (roan tsesseble *Damaliscus lunatus*, [32]) or recovering from past defaunation (rhino, [31]). In addition, not all species occupy all areas of the park, nor overlap completely in their

range [36]. For this analysis, we will analyze Φ as a potential value, as well as a current value which is ¼ of this potential.

The edible biomass αB is approximately 2.5 Mg/ha in KNP [35]. This figure represents only the pasture biomass; variation in this value in space and time, and additional tree foliage are not included in this figure. Furthermore the park has both grazers and browsers that select from either (or both) of these foodstuffs. However, for the purposes of simplifying this analysis, we consider that the biomass can be lumped together with the grazers and browsers consuming it. The potential consumption of each the park's herbivores is estimated as the product of population density and metabolic demand (Figure 4a). The relation between bodymass and potential consumption is relatively flat (Figure 4a), and there are species <1 kg (e.g. the brown rat *Rattus rattus*) that have comparable rates of consumption as species >1000 kg. The mean rate of herbivory we estimate at the population level is approximately 1000 kgDM/km^2 per taxon, regardless of size. Consequently, smaller herbivores play as important role in biomass consumption as large species – the principle of energetic equivalence [30] (Figure 4b). Collectively, we estimate that the herbivores potentially consume ~10% of the park's leaf biomass annually, and up to 15% if smaller herbivores are included. However, given that animal populations in KNP are less than estimated using the allometric equation, their biomass consumption is likely also less.

The consequent diffusivity Φ of this population of herbivores (excretion term only) is summarized in Figure 4c–d. Although Φ calculated from the primary data (when possible) is in general lower, there are some species that have unusually high Φ (*Loxodonta africana*, *Equus burchellii*, *Connochaetes taurinus*, *Damaliscus lunatus*, *Aepyceros melampus* from largest to smallest). Consequently, the Φ calculated using allometry is nearly the same as that calculated using primary data, approximately 7 km^2/year. The coefficient changes little if small taxa are excluded, and even those species up

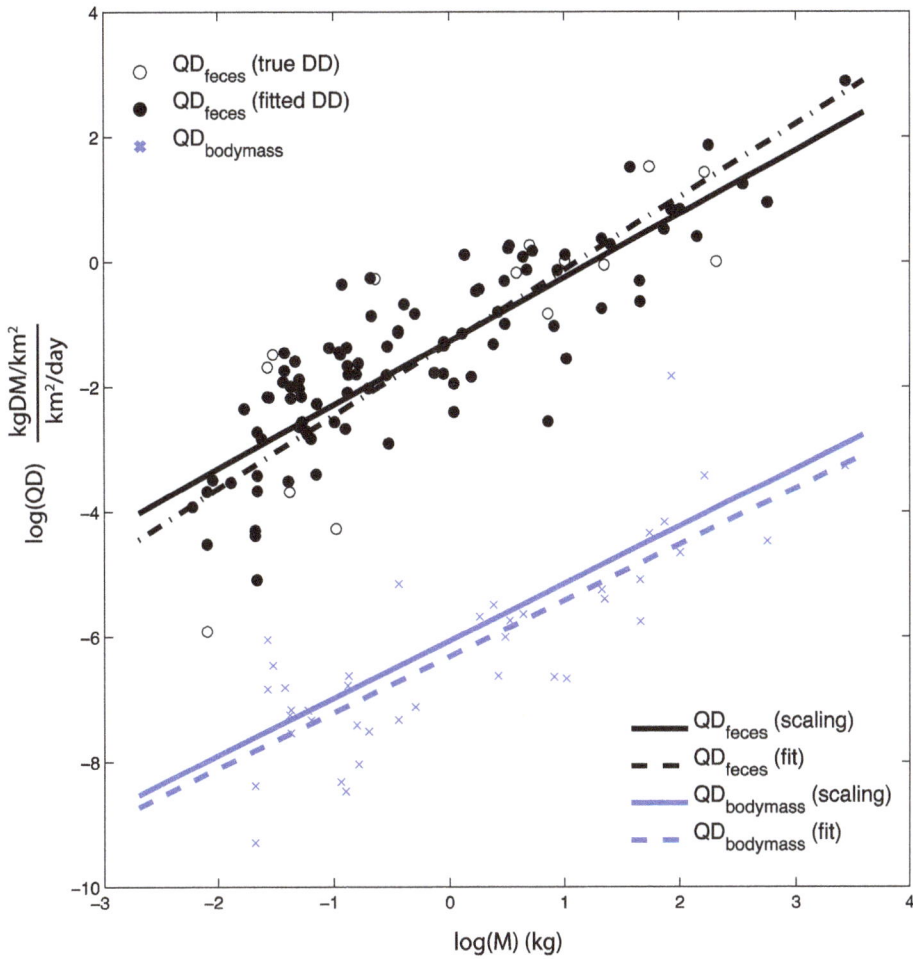

Figure 2. Allometric relations between bodymass M and animal-mediated nutrient diffusivity, Equations [3] and [4] in the main text. Solid lines are estimates calculated using scaling arguments, dashed lines as a fit to primary data. Circles show diffusivity by way of excreta, crosses show diffusivity by way of bodymass.

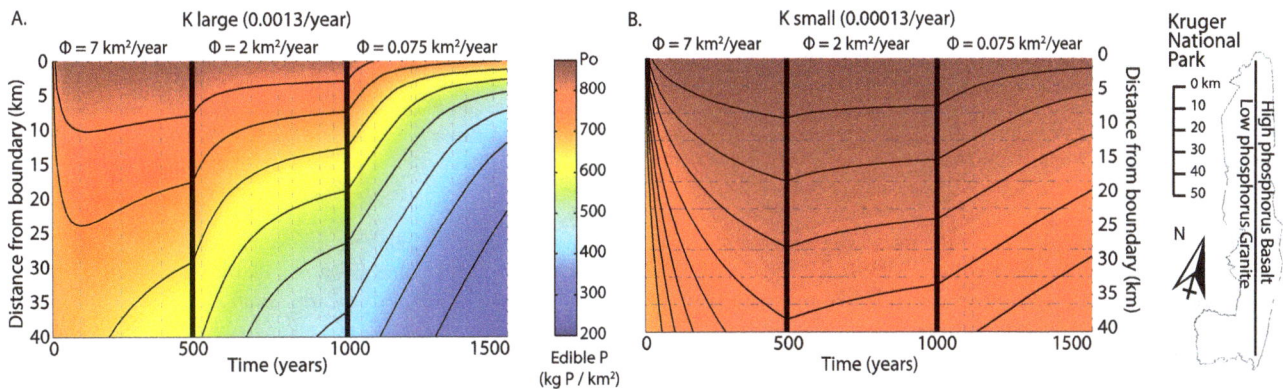

Figure 3. Diffusion into granitic region of KNP. Upper panel shows geometry of the simulated transect, with an inset to show the initial and boundary conditions of a transect across the substrate gradient in the absence of herbivore diffusion. Lower panels show phosphorus stocks in edible vegetation under a succession of herbivore removals, varying from Φ varies from 7 km²/year (estimate prior maximum) to 2 (present-day estimate) to 0.075 (estimate in the absence of herbivores >100 kg). A. P dynamics under an upper estimate of $K = 0.0013$/year; B. P dynamics under a lower estimate of $K = 0.00013$/year. Additional parameter values set to $Po = 875$ kg P km^{-2}, $G = 0.5$ kg P km^{-2} year^{-1}.

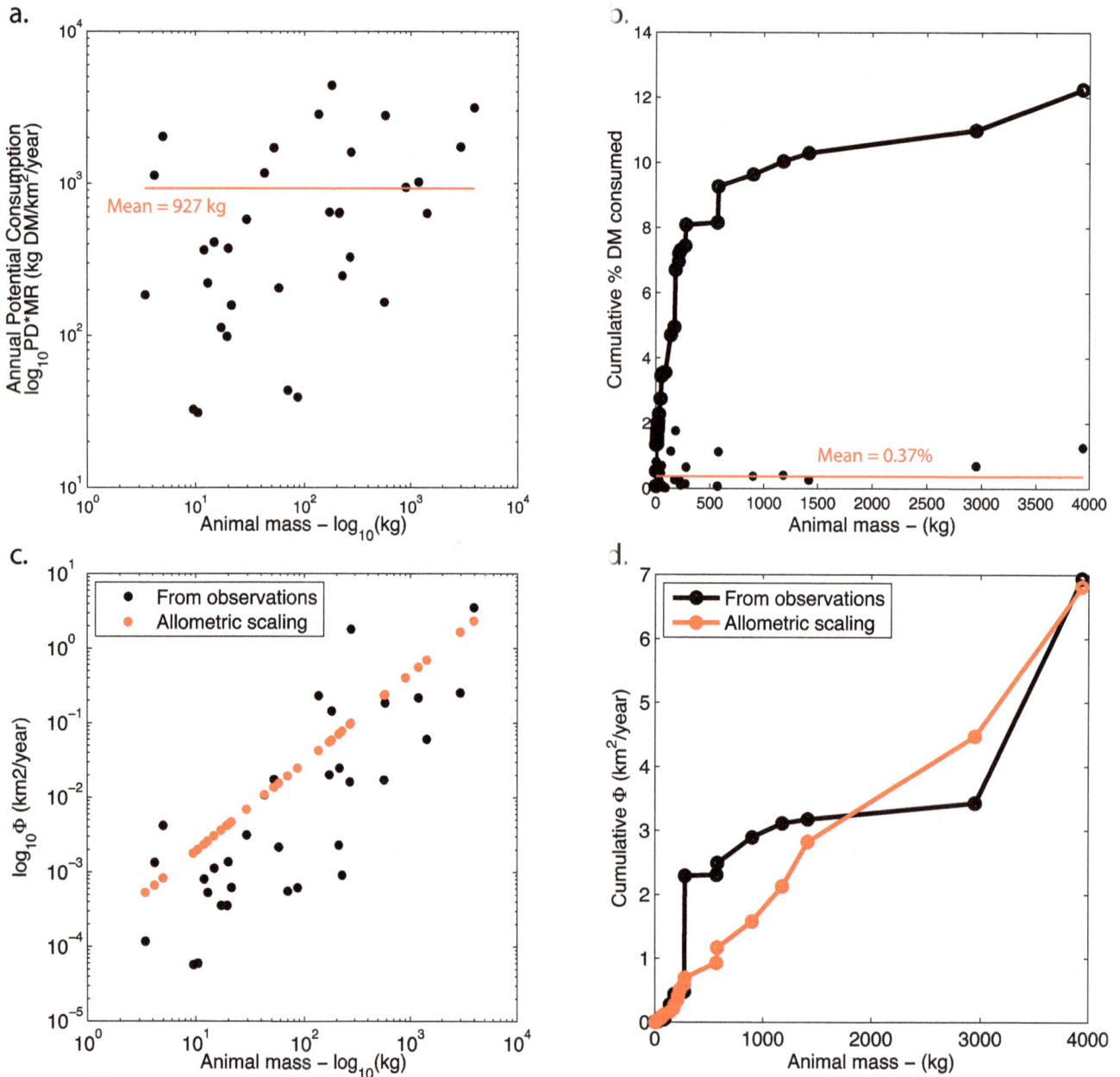

Figure 4. Estimates of herbivory and nutrient diffusivity in Kruger National Park by mammalian herbivores >1 kg in KNP. (a) Potential rates of consumption based on population density and metabolic demand. The mean rate of herbivory per taxon is 927 kg/km^2 or 0.37% of the biomass standing crop. (b) cumulative rate of herbivory across body mass (c) nutrient diffusivity Φ, using observations of component terms (where possible – black points) and allometric scaling (8.672*M$^{1.191}$; red points). (d) cumulative nutrient diffusivity Φ across body mass, based on direct observations of component terms, and allometric scaling of component terms.

to 250 kg account for just 25% of the total, which highlights the importance of the largest megaherbivores [39].

To understand the consequences of this herbivore mediated diffusion, consider a simplified budget of available phosphorus (P) governed by first order losses (K), such as runoff, and zero-order gains (G), such as dust deposition and weathering:

$$\frac{dP}{dt} = -KP + G \qquad (5)$$

This equation is analogous to typical treatments of nutrient cycles using ODEs [4] and has an equilibrium value of G/K. The presence of herbivores adds an additional diffusive term governed by Φ and the spatial gradient in P, forming a reaction-diffusion equation:

$$\frac{\partial P}{\partial t} = \Phi \frac{\partial^2 P}{\partial x^2} - KP + G \qquad (6)$$

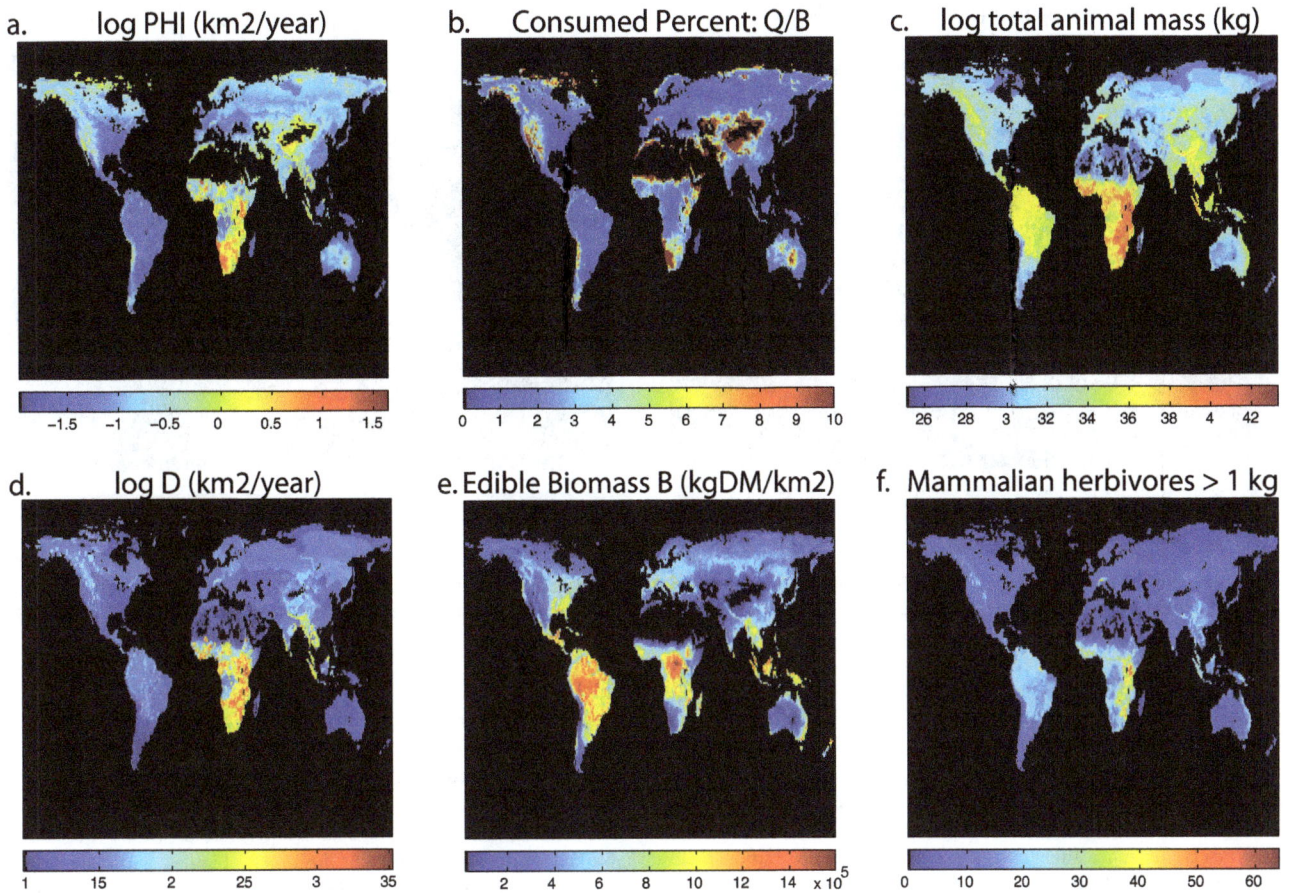

Figure 5. Global distribution of terms in herbivore diffusion of nutrients. (a) nutrient diffusivity $\Phi = DQ/\alpha B$, (b) change in Φ if all threatened species are lost.

Because there is no horizontal transport mechanism in [5], the basalts and granites represent two isolated regions, each govered by their own initial conditions $P_o(x,0)$ and parameters K and G. However, with the presence of herbivores, there exists the possibility for P to be transported from high P to low P regions, so long as $\Phi > 0$. However, the degree to which diffusion rivals other gains and losses in the budget depends on the relative magnitudes of Φ, K, G, and the boundary condition P_o.

The effect of herbivore diffusion from the P-rich basaltic region to the P-poor granitic region is illustrated in Figure 3 using a range of Φ for parameters G, K and P_o approximating that of KNP *[Methods S5]*. The numerical experiment shown simulates the P in vegetation in the granitic region following a succession of herbivore removals in 500 year intervals representing past and future defaunation. The initial condition for the domain is set to 80% the steady state value from [6], i.e. P_o of the basaltic region, 875 kg P km^{-2}. The herbivores in the system beginning at time = 0 with the "potential" diffusivity $\Phi = 7$ km^2/year, followed in 500 years by an herbivore removal to represent the current "actual" diffusivity $\Phi = 2$ km^2/year, and finally at 1000 years the diffusivity is shown with no large herbivores (>100 kg), $\Phi = 0.075$ km^2/year. The analysis was run under two estimates of K, that is the larger estimate of K = 0.0013/year calculated explicitly from the mechanistic model of Buendia et al. [4] and the smaller estimate of K = 0.00013/year calculated implicitly from estimates of G [40,41] and available P [42], under the assumption

that the observed P is a steady state value including no animal inputs of P from animals, i.e. $P_{ss} = G/K$.

A number of features are notable from this analysis. The first observation is that the edible P under low losses (Figure 3, right panel) is improbably large, approximately double the observed value of edible P = 375 kg P km^{-2}, at all levels of animal diffusion. That is, if we believe that the herbivore diffusion as outlined in this paper exists, even if only for small mammals, then the observed amount of P in edible vegetation would be expected to be considerably greater, given the rate of diffusion for even the lowest Φ within the context of the long 500,000 year timeframe of pedogenesis on these soils [43]. Because such a large value of edible P ($\sim$750 kg P km^{-2}) on the granitic soils is not observed, it would appear that the larger, explicitly calculated rate of loss is more plausible, and that the estimate of a low K as K = G/P is flawed by the assumption that $\Phi = 0$. In other words, we argue that the system is better characterized by a higher loss rate that is compensated for by animal inputs from the basaltic substrate (Figure 3, left panel).

When K is large, the presence or absence of herbivores has strong impacts on the spatial gradients of P. In the total absence of herbivores, there is of course a sharp drop in edible P at the boundary. However, with only small herbivores ($\Phi = 0.075$ km^2/year), diffusion is capable of maintaining a nutrient enrichment zone above G/K up to 5–10 km away from the boundary. In the current regime with large herbivores maintained at reduced population densities ($\Phi = 2$ km^2/year), this zone of enrichment

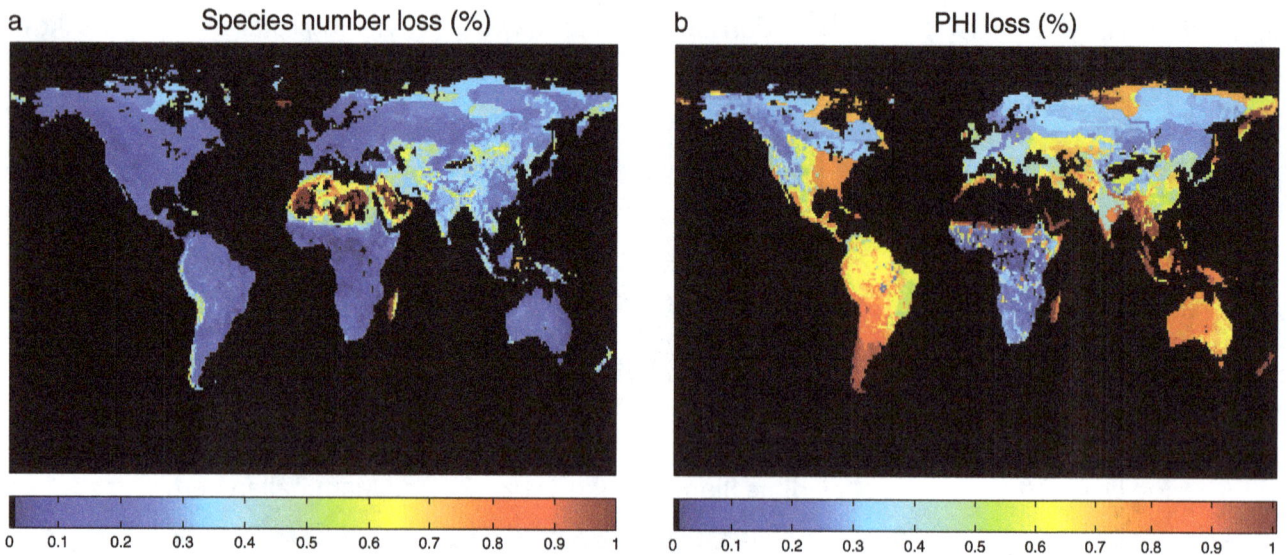

Figure 6. Global distribution of terms in herbivore diffusion of nutrients. (a) movement diffusivity D, (b) percent consumed biomass Q/αB, (c) total animal biomass (ie Σ mass * population), (d) nutrient diffusivity Φ=DQ/αB, (e) edible biomass αB, (f) number of mammalian herbivores >1 kg.

above G/K is extended to 20–30 km. For large herbivore densities ($\Phi = 7$ km^2/year), the effects of diffusion are felt throughout the granitic domain. It is clear in comparing the simulation with large K with small K that the larger the losses, the more important herbivores are in easing spatial gradients in nutrients, and conversely the more their absence is felt if they are removed. In wetter regions with higher rates of P loss, then this would imply that herbivores could play a more important role in those ecosystems in distributing nutrients.

Global Implications

If herbivore mediated diffusion can have large effects on small scales, what is the global distribution of this phenomenon? We used the IUCN spatial database on mammal species and their ranges [44] to develop a gridded, global estimate of Φ. Although such a global gridded product should be treated with caution when applied to any specific local context, it can nevertheless provide valuable insight into broad global patterns in the capacity of animals to shift nutrients laterally within a locality. With few exceptions, each IUCN taxon was resolved to the MSW3 mammalian species list [23] and assigned a body mass from a bodymass database [28], likewise keyed to MSW3. Of the 5278 terrestrial mammals in MSW3, 2429 of these had information on body mass, largely from Smith et al. [28], although some others originated with the datasets outlined earlier. Species for which no bodymass data were available were interpolated phylogenetically, i.e. assigned to the mean value for the genus or family if necessary. Edible biomass (i.e. annual foliage production) at 1° resolution was estimated using the CASA carbon cycle model [45], summing both tree and grass/forb foliage.

It is apparent that there is great variation among the continents in the potential for animals to transport nutrients (Figure 5a). We note that Φ is the product of two distinct terms, namely the D term that reflects the ability of animals to transport material long distances (Figure 5d), and an herbivory term that reflects the consumption (Q) of available edible biomass (αB) (Figure 5b). Of these two terms, we noted earlier that Q varied little among species varying in M, and here Q is set to 750 kg DM/km^2, which approximates the mean across the data presented in Figure 1.

Therefore Q is more or less a function of the number of taxa present, here restricted to those >1 kg (Figure 5f). The D term, by contrast, is a strong allometric function of body size (0.0598* $M^{0.9962}$). Predictably the Q/αB term is highest where αB is lowest, and in fact deserts (the lowest 10 percent of values here) are masked out (Figure 5b). Φ however (Figure 5a) is strongly determined by D (Figure 5d), which is greatest in Africa, particularly south and east Africa, as well as Southeast Asia and the Tibetan plateau. Africa, and to a lesser extent tropical Asia, remain the megafauna-rich continents, yet in the late Pleistocene similar high abundances of megafauna would have been found in most other continents.

The global asymmetries in Φ are striking: the Kruger example we presented earlier is at the higher end of Φ globally, with many areas reflecting a level of Φ that is most analogous to the ecosystem after all herbivores have been removed. It is not surprising that most biogeochemical research has tended to ignore this term as nearly irrelevant, for in Europe, eastern North America, and most of South America this diffusion term is 1/20th or 1/100th of values typical in Africa.

Naturally, the global analysis presented here omits many of the details that are known to be at play in herbivory at this scale. For one, the analysis is restricted to mammalian herbivores, which is restrictive given the importance of other clades in transporting nutrients [46,47]. Second, we ignored relationships between herbivory and forage productivity and quality [48], instead coming up with an independent estimate that relied on IUCN species range maps and body size as predictors of biomass consumption. To the extent that species richness corresponds to productivity, our estimates are in agreement; however, this is often not the case, in particular comparing productive tropical regions such as the Amazon and Congo basins, which greatly differ in their abundance of mammals. Third, there is considerable local heterogeneity in nutrients that this global analysis ignores. This local heterogeneity in nutrients is the "potential gradient" that diffusivity acts on to create a flux, and without knowledge of this heterogeneity we can make no estimate as to the magnitude of nutrient fluxes that are borne by mammalian herbivores.

There are other aspects of local heterogeneity that deserve more careful attention as well, in particular those that impact the parameters in the model, such as population density (PD) or daily displacement (DD). We have ignored any dependence of these parameters on the underlying nutrient quality, for example the potential that high P might support higher PD or lower P might drive larger DD.

To the extent these to phenomena work in opposite directions, they might cancel each other, but nevertheless they present real challenges to the model we use and should be critically evaluated in the future. Finally, we have completely ignored other trophic levels in this analysis, particularly higher-level consumers (including humans), which would also act to limit PD.

Although these limitations are potentially important, and will shade or modify the effort to apply this work to any one place, we believe the general finding still holds. That is, that in the presence of local heterogeneity, areas with higher Φ will show a greater capacity for lateral nutrient fluxes, and that these fluxes are potentially of comparable magnitude to other major fluxes in the system.

Conclusion

There is a rich story of the imprint of species extinctions on the global distribution of Φ (Figure 5a). It is worth considering that locales that are now considered oligotrophic, such as tropical regions like the Amazon basin, Congo basin, and southeast Asia may once have had a substantial supply of P by animal vectors despite having little renewal of surface soils by Pleistocene glaciation. In fact, as humans have gradually supplanted non-human herbivores as the major consumers of primary productivity [49], the character of P redistribution has likely also undergone a shift: whereas natural Φ probably acts like a vascular system, creating entropy by dispersing nutrients to the matrix, humans bring nutrients from the matrix and concentrate them in animal operations, much like the subjects of G.E. Hutchinson's monograph.

In summary, we have presented a mathematical framework to quantify the diffusion of nutrients by herbivores, demonstrated its applicability in the specific local context of Kruger National Park, and used these insights to mao the approximate global patterns of lateral nutrient diffusion. We propose that lateral nutrient diffusion is a previously unrecognized ecosystem service, provided by roaming large herbivores, which fuels productivity by taking nutrients from places of excess and depositing them in places of deficit. How is this ecosystem service threatened globally? A first order estimate can be obtained by exploring the consequences of extinction or movement restriction of all species identified as threatened in the IUCN redlist [44]. The fraction of species that are not extinct but currently threatened are illustrated in Figure 6a. This map highlights threats in areas that have have low intrinsic productivity (Figure 5e) and few herbivores (Figure 5f), but generally the fraction of species threatened ranges from 10–30%. By contrast, we can see in Figure 6b that extinctions to these threatened species portend large changes to Φ. This contrast indicates that the species losses are especially concentrated among taxa with high capacity to transport nutrients, i.e. large mammalian herbivores. Species extinctions historically have been felt in larger taxa [50], and in many parts of the world there do not remain many large herbivores (Figure 5f). Nevertheless, threats are felt among the remaining species, such that Φ is in many locales threatened to drop by 75–100% (Figure 6b). In addition, even if megafauna continue to persist, their population densities are greatly reduced and their ability to roam (and hence Φ) is highly constrained by habitat fragmentation and restriction to reserves. Hence the lateral flow of nutrients in wild animals is likely to be declining rapidly. It is interesting to speculate (but beyond the scope of the current study) if in many regions this loss may be compensated for by wide-ranging domesticated fauna, especially cattle and buffalos, which may play a similar but more circumscribed role in lateral nutrient diffusion.

The primary conclusion of this paper is to highlight the potential importance of lateral nutrient diffusion of nutrients by vertebrate herbivores. The framework we have developed is necessarily approximate when applied to local situations, and needs to be tested with focused empirical studies in specific ecosystems.

Supporting Information

Methods S1 Calculating diffusivity from a random walk.

Methods S2 Diffusion of nutrients transported by animals.

Methods S3 Solution to 1-D PDE for diffusion away from a source region.

Methods S4 Mean age of death in a population (includes Figures S1 and S2).

Methods S5 Parameterization of reaction-diffusion model for Kruger National Park.

Acknowledgments

We thank Simon Levin, Jim Murray, Shaun Levick, Charles Yackulic and Adam Pellegrini for their thoughts and comments on ideas presented in this manuscript. Izak Smit and Rina Grant-Biggs of Kruger National Parks provided data and guidance on the model application to KNP, as did Oliver Chadwick.

Author Contributions

Conceived and designed the experiments: AW CD YM. Performed the experiments: AW CD. Analyzed the data: AW CD. Contributed reagents/materials/analysis tools: AW CD. Wrote the paper: AW.

References

1. Vitousek PM (2004) *Nutrient cycling and limitation : Hawai'i as a model system.* Princeton: Princeton University Press. 223 p.
2. Hedin LO, Armesto JJ, Johnson AH (1995) Patterns of Nutrient Loss from Unpolluted, Old-Growth Temperate Forests - Evaluation of Biogeochemical Theory. Ecology 76: 493–509.
3. Porder S, Vitousek PM, Chadwick OA, Chamberlain CP, Hilley GE (2007) Uplift, erosion, and phosphorus limitation in terrestrial ecosystems. Ecosystems 10: 158–170.
4. Buendia C, Kleidon A, Porporato A (2010) The role of tectonic uplift, climate, and vegetation in the long-term terrestrial phosphorous cycle. Biogeosciences 7: 2025–2038.
5. Okubo A, Levin SA (2001) *Diffusion and ecological problems : modern perspectives* 2nd Ed. New York: Springer. 467 p.
6. Hutchinson GE (1950) The biogeochemistry of vertebrate excretion. Survey of contemporary knowledge of biogeochemistry, vol 3. New York: American Museum of Natural History. 554 p.

7. Wardle DA, Bardgett RD, Klironomos JN, Setälä H, van der Putten WH, et al. (2004) Ecological linkages between aboveground and belowground biota. Science 304: 1629–1633.

8. McNaughton SJ, Banyikwa FF, McNaughton MM (1997) Promotion of the cycling of diet-enhancing nutrients by African grazers. Science 278: 1798–1800.

9. Hilderbrand GV, Hanley TA, Robbins CT, Schwartz CC (1999) Role of brown bears (Ursus arctos) in the flow of marine nitrogen into a terrestrial ecosystem. Oecologia 121: 546–550.

10. Stevenson PR, Guzman-Caro DC (2010) Nutrient Transport Within and Between Habitats Through Seed Dispersal Processes by Woolly Monkeys in North-Western Amazonia. Am J Primatol 72: 992–1003.

11. Abbas F, Merlet J, Morellet N, Verheyden H, Hewison AJM, et al. (2012) Roe deer may markedly alter forest nitrogen and phosphorus budgets across Europe. Oikos: 121: 1271: 1278.

12. Frank DA, Inouye RS, Huntly N, Minshall GW, Anderson JE (1994) The Biogeochemistry of a North-Temperate Grassland with Native Ungulates - Nitrogen Dynamics in Yellowstone-National-Park. Biogeochemistry 26: 163–188.

13. Woodmansee RG (1978) Additions and Losses of Nitrogen in Grassland Ecosystems. Bioscience 28: 448–453.

14. Pletscher DH, Bormann FH, Miller RS (1989) Importance of Deer Compared to Other Vertebrates in Nutrient Cycling and Energy-Flow in a Northern Hardwood Ecosystem. Am Midl Nat 121: 302–311.

15. Malhi Y (2012) The productivity, metabolism and carbon cycle of tropical forest vegetation. J Ecol 100: 65–75.

16. Barnosky AD, Matzke N, Tomiya S, Wogan GOU, Swartz B, et al. (2011) Has the Earth's sixth mass extinction already arrived? Nature 471: 51–57.

17. Walker TW, Syers JK (1976) Fate of Phosphorus during Pedogenesis. Geoderma 15: 1–19.

18. Ovaskainen O, Crone EE (2009) Modeling animal movement with diffusion. In: Cantrell S, Cosner C, Ruan S Editors. Spatial Ecology. Boca Raton: Chapman & Hall. 63–83.

19. Skellam JG (1951) Random Dispersal in Theoretical Populations. Biometrika 38: 196–218.

20. Bailey DW, Gross JE, Laca EA, Rittenhouse LR, Coughenour MB, et al. (1996) Mechanisms that result in large herbivore grazing distribution patterns. J Range Manage 49: 386–400.

21. White KAJ, Lewis MA, Murray JD (1996) A model for wolf-pack territory formation and maintenance. J Theor Biol 178: 29–43.

22. May RM (1978) Evolution of Ecological Systems. Sci Am 239: 161: 175.

23. Wilson DE, Reeder DM (2005) Mammal species of the world : a taxonomic and geographic reference 3rd Ed. Baltimore: Johns Hopkins University Press. 2142 p.

24. de Magalhaes JP, Costa J (2009) A database of vertebrate longevity records and their relation to other life-history traits. J Evolution Biol 22: 1770–1774.

25. Damuth J (1987) Interspecific Allometry of Population-Density in Mammals and Other Animals - the Independence of Body-Mass and Population Energy-Use. Biol J Linn Soc 31: 193–246.

26. Carbone C, Cowlishaw G, Isaac NJB, Rowcliffe JM (2005) How far do animals go? Determinants of day range in mammals. Am Nat 165: 290–297.

27. Kelt DA, Van Vuren DH (2001) The ecology and macroecology of mammalian home range area. Am Nat 157: 637–645.

28. Smith FA, Lyons SK, Ernest SKM, Jones KE, Kauffman DM, et al. (2003) Body mass of late quaternary mammals. Ecology 84: 3403–3403.

29. Demment MW, Van Soest PJ (1985) A Nutritional Explanation for Body-Size Patterns of Ruminant and Nonruminant Herbivores. Am Nat 125: 641–672.

30. Damuth J (2007) A macroevolutionary explanation for energy equivalence in the scaling of body size and population density. Am Nat 169: 621–631.

31. Pienaar UDV (1963) The large mammals of the Kruger National Park - Their distribution and present day status. Koedoe 6: 1–37.

32. Seydack AH, Grant CC, Smit IP, Vermeulen WJ, Baard J, et al. (2012) Large herbivore population performance and climate in a South African semi-arid savanna. Koedoe 54(1).

33. Scholes MC, Scholes RJ, Otter LB, Woghiren AJ (2003) Biogeochemistry: The Cycling of Elements. In Du Toit JT, Biggs H, Rogers KH Editors. The Kruger Experience. Washington: Island Press. 130–148.

34. Venter FJ (1986) Soil patterns associated with the major geological units of the Kruger National Park. Koedoe 29: 125–138.

35. Grant CC, Peel MJS, van Ryssen JBJ (2000) Nitrogen and phosphorus concentration in faeces: an indicator of range quality as a practical adjunct to existing range evaluation methods. African Journal of Range & Forage Science 17: 81–92.

36. Smit IPJ (2011) Resources driving landscape-scale distribution patterns of grazers in an African savanna. Ecography 34: 67–74.

37. Palmer A, Peel MJS, Kerley G (2006) Arid and semiarid rangeland production systems of Southern Africa: Wildlife. Secheresse 17: 362–370.

38. Smit IPJ, Ferreira SM (2010) Management intervention affects river-bound spatial dynamics of elephants. Biol Conserv 143: 2172–2181.

39. Owen-Smith RN (1988) Megaherbivores : the influence of very large body size on ecology. New York: Cambridge University Press. 369 p.

40. Mahowald N, et al. (2008) Global distribution of atmospheric phosphorus sources, concentrations and deposition rates, and anthropogenic impacts. Global Biogeochem Cy 22: GB4026.

41. Okin GS, Mahowald N, Chadwick OA, Artaxo P (2004) Impact of desert dust on the biogeochemistry of phosphorus in terrestrial ecosystems. Global Biogeochem Cy 18: GB2005.

42. Hartshorn AS, Coetsee C, Chadwick OA (2009) Pyromineralization of soil phosphorus in a South African savanna. Chem Geol 267: 24–31.

43. Khomo L (2008) Weathering and soil properties on old granitic catenas along climo-topographic gradients in Kruger National Park. PhD Thesis. Johannesburg, South Africa: University of Witwatersrand.

44. IUCN (2010) IUCN Red List of Threatened Species. Version 2010.4.

45. Field CB, Behrenfeld MJ, Randerson JT, Falkowski P (1998) Primary production of the biosphere: Integrating terrestrial and oceanic components. Science 281: 237–240.

46. Kitchell JF, et al. (1999) Nutrient cycling at the landscape scale: The role of diel foraging migrations by geese at the Bosque del Apache National Wildlife Refuge, New Mexico. Limnol Oceanogr 44: 828–836.

47. Young HS, McCauley DJ, Dunbar RB, Dirzo R (2010) Plants cause ecosystem nutrient depletion via the interruption of bird-derived spatial subsidies. P Natl Acad Sci USA 107: 2072–2077.

48. Cebrian J, Lartigue J (2004) Patterns of herbivory and decomposition in aquatic and terrestrial ecosystems. Ecol Monogr 74: 237–259.

49. Doughty CE, Field CB (2010) Agricultural net primary production in relation to that liberated by the extinction of Pleistocene mega-herbivores: an estimate of agricultural carrying capacity? Environ Res Lett 5: 044001.

50. Burney DA, Flannery TF (2005) Fifty millennia of catastrophic extinctions after human contact. Trends Ecol Evol 20: 395–401.

18

Development of a 3D Coupled Physical-Biogeochemical Model for the Marseille Coastal Area (NW Mediterranean Sea): What Complexity Is Required in the Coastal Zone?

Marion Fraysse[1,2,3]*, Christel Pinazo[2,3], Vincent Martin Faure[2,3], Rosalie Fuchs[2,3], Paolo Lazzari[4], Patrick Raimbault[2,3], Ivane Pairaud[1]

1 IFREMER, Laboratoire Environnement Ressources Provence Azur Corse, La Seyne sur Mer, France, **2** Aix Marseille Université, CNRS/INSU, IRD, Mediterranean Institute of Oceanography (MIO), UM 110, Marseille, France, **3** Université de Toulon, CNRS/INSU, IRD, Mediterranean Institute of Oceanography (MIO), UM 110, La Garde, France, **4** Dept. of Oceanography, Istituto Nazionale di Oceanografia e di Geofisica Sperimentale (OGS), Trieste, Italy

Abstract

Terrestrial inputs (natural and anthropogenic) from rivers, the atmosphere and physical processes strongly impact the functioning of coastal pelagic ecosystems. The objective of this study was to develop a tool for the examination of these impacts on the Marseille coastal area, which experiences inputs from the Rhone River and high rates of atmospheric deposition. Therefore, a new 3D coupled physical/biogeochemical model was developed. Two versions of the biogeochemical model were tested, one model considering only the carbon (C) and nitrogen (N) cycles and a second model that also considers the phosphorus (P) cycle. Realistic simulations were performed for a period of 5 years (2007–2011). The model accuracy assessment showed that both versions of the model were able of capturing the seasonal changes and spatial characteristics of the ecosystem. The model also reproduced upwelling events and the intrusion of Rhone River water into the Bay of Marseille well. Those processes appeared to greatly impact this coastal oligotrophic area because they induced strong increases in chlorophyll-a concentrations in the surface layer. The model with the C, N and P cycles better reproduced the chlorophyll-a concentrations at the surface than did the model without the P cycle, especially for the Rhone River water. Nevertheless, the chlorophyll-a concentrations at depth were better represented by the model without the P cycle. Therefore, the complexity of the biogeochemical model introduced errors into the model results, but it also improved model results during specific events. Finally, this study suggested that in coastal oligotrophic areas, improvements in the description and quantification of the hydrodynamics and the terrestrial inputs should be preferred over increasing the complexity of the biogeochemical model.

Editor: Moncho Gomez-Gesteira, University of Vigo, Spain

Funding: This work was supported by the PACA region (http://www.regionpaca.fr/), IFREMER grant (www.ifremer.fr/), GIRAC (http://www.polemerpaca.com/Environnement-et-amenagement-du-littoral/Gestion-de-l-eau-en-zone-cotiere/GIRAC) and PNEC-EC2CO MASSILIA (http://www.insu.cnrs.fr/actions-sur-projets/ec2co) projects, sustained by the water agency (AERMC,www.eaurmc.fr/) and the PERSEUS European FP7 project (http://www.perseus-fp7.eu/). The funders had no role in study design, data collection and analysis, decision to publish, or preparation of the manuscript.

Competing Interests: The authors have declared that no competing interests exist.

* E-mail: marion.fraysse@univ-amu.fr

Introduction

Coastal regions, located at the interface between oceanic and terrestrial systems, play a crucial role in earth system functioning [1]. Greater than 60% of the world's population lives less than 60 km from the sea, increasing the human pressure on these systems [2]. Understanding the fate of anthropogenic inputs from major cities and their impacts on the adjacent marine ecosystems is essential for the protection and management of coastal waters. Although it is clear that coastal systems are locally strongly impacted by human activities, it remains difficult to distinguish between climatic and anthropogenic forcing [1–3]. Modeling approaches can assist with this difficulty and are useful tools for studying such a complex coastal environment.

Marseille is the second largest city in France, and the metropolitan area of Marseille extends beyond the city limits with a population of 1 038 940 and a density 1 718 people per km^2 [4]. The density of contaminant-generating industries in the city of

Marseille and the quantity of sewage are highly representative of large, modern Mediterranean cities. Hydrodynamic and biogeochemical processes affect the transport and form of chemical compounds; for example, contaminant speciation often depends on suspended matter and on organic compounds (Particulate Organic Carbon (POC) and Dissolved Organic Carbon (DOC)) [5]. Marseille was thus chosen for the development of a numerical tool (a chain of models) to assess the raw inputs (from city to sea) and exports (from mid-sea to open sea) of chemical contaminants. This tool was developed based on the coupling of a hydrodynamic model [6], a sedimentary model, a biogeochemical model and a model of chemical contamination. This paper presents the coupled hydrodynamic-biogeochemical compartment.

The Marseille coastal area is located in the eastern part of the Gulf of Lions (GoL) in the western Mediterranean Sea (Figure 1). The GoL is one of the most productive areas of the Mediterranean Sea [5], even if the Mediterranean Sea remains oligotrophic. The biogeochemical functioning of the GoL is strongly impacted by inputs from the Rhone River as it is the most significant source of freshwater and nutrients in the Mediterranean Sea; these inputs have a direct influence on the primary production. On an annual basis, approximately 50% of the primary production in the GoL can be attributed to terrigenous inputs [7,8]. The hydrodynamics of the GoL are complex and highly variable [9] because there is strong temporal and spatial variability in the forcing occurring at the eastern part of the GoL [10].

The biogeochemical functioning of the Bay of Marseille (BoM) is complex and highly driven by hydrodynamics. The hydrodynamics of the Marseille coastal area were studied in details using a modeling approach [6]. The primary forcing components are the two dominant winds (north-northwesterly winds, which favor upwelling, and southeasterly winds, which favor downwelling) and the oligotrophic Northern Current (NC), which flows along the continental slope toward the west [11] and occasionally intrudes on the shelf [6,12–14]. North-northwesterly wind gusts induce upwelling zones off the "Cote Bleue" and off "Calanques" [6,11]. During an upwelling event, cold, rich waters are advected upwards [15], which can induce (in favorable cases) an increase in primary production. A case study undertaken in the BoM [16] showed that primary production tripled at a coastal station influenced by upwelling compared with a reference offshore station. Observation [17] and modeling studies of the Region Of Freshwater Influence (ROFI) [18–20] show the predominant westwards direction of the Rhone River plume. A less common orientation of the Rhone River plume, toward the East up to 40 km from the Rhone River mouth and offshore in the BoM, was recently observed [21]. The presence of water from the Rhone River in the BoM was established in a modeling study by Pairaud et al. [6]. The effects of the eastward intrusion events of the Rhone River plume were

observed on biological production, local phytoplankton blooming and chromophoric dissolved organic matter (CDOM) production was measured [22]. The BoM is also impacted by urban rivers, a Wastewater Treatment Plant (WWTP) and atmospheric deposition, but their impacts on the biogeochemical functioning of the Bay and nutrient limitation remain to be studied.

The knowledge available on nutrient limitation is at the scale of the Mediterranean Sea and its Western Basin. Thus, it is essential to understand how photosynthetic production is or is not limited by nitrogen (N) and phosphorus (P) because the cycles of key nutrient elements, such as N and P, have been massively altered by anthropogenic activities [23]. In the Mediterranean Sea, N or P concentrations are generally considered to be limiting factors for algal production [24]. Van Wambeke et al. [25] confirm that P limitation of bacterioplankton is a generalized phenomenon in the Mediterranean Sea. Indeed, with a $NO_3:PO_4$ ratio of 65–80 [26,27], the Rhone River contributes to the relative P deficit of the Mediterranean Sea [5]. Atmospheric inputs also provide N in excess relative to P (mean DIN/DIP ratio of 60 in the Western Mediterranean Basin) and contribute to the P limitation [28]. In addition, in the northwestern Mediterranean Sea, P limitation for both phytoplankton and bacteria was suggested by Thingstad et al. [29]. Therefore, a modeling approach could aid in the understanding of the limitation functioning of the Marseille coastal area.

3D physical and biogeochemical modeling approaches were used in the GoL to study upwelling [30], nitrate fluxes between the margin and the open sea [31], the functioning of the planktonic ecosystem during spring and its impact on particulate organic carbon deposition [32] and the biogeochemical functioning of eddies [33]. Recently, Fontana et al. [34] performed assimilation

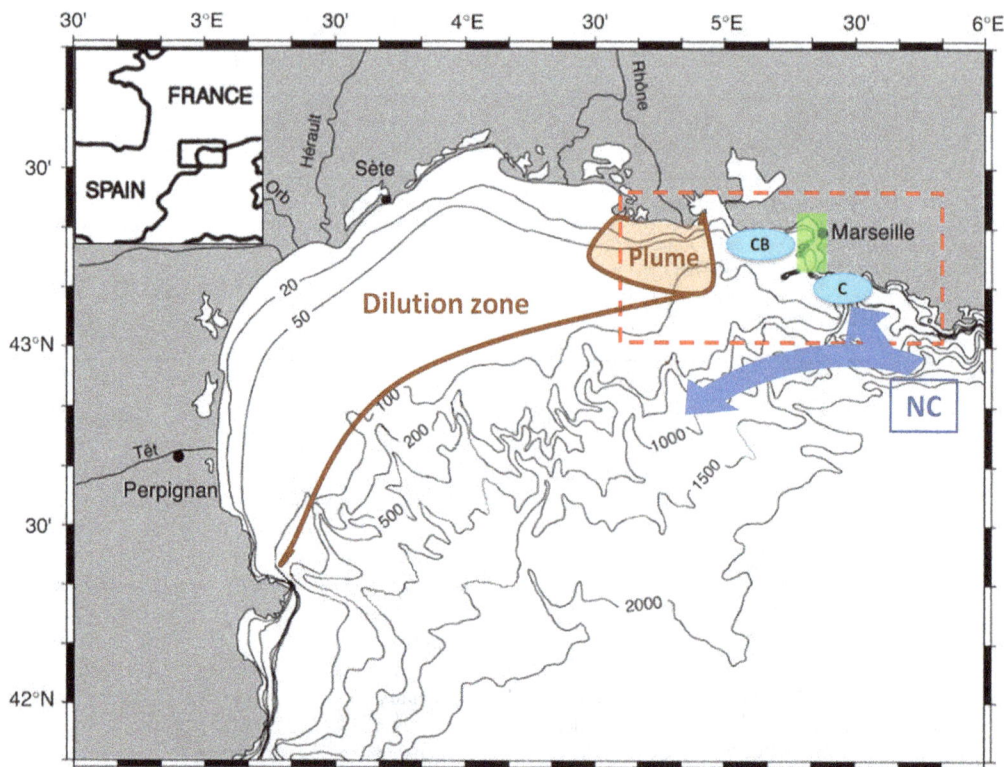

Figure 1. Map of the Gulf of Lion and the model domain. The RHOMA model domain (red dotted lines), the Bay of Marseille (green rectangle), the "Cote Bleue" and "Calanques" upwelling spots (blue circles), the Rhone River plume and intrusion zone (in brown) and the Northern current (NC) are represented.

of chlorophyll-a remote sensing data in a 3D coupled physical-biogeochemical model for the area near the mouth of the Rhone River. This study represents the first 3D, high-resolution coupled physical-biogeochemical modeling study of the BoM. Coastal zones are highly variable both hydrodynamically and biogeochemically; therefore, the modeling approach cannot be the same as used in the open sea. Franks [35] strongly suggested that the model should be designed around the scientific question being addressed. Indeed, the choice of the complexity of the biogeochemical model usually depends on the study to be conducted [36]. In our case, a model light in computational cost has many advantages for operational applications and for coupling it with other models (sediment and contamination models). Moreover, the complexity of 3D coupled models can make them particularly difficult to evaluate. Arhonditsis and Brett [37] performed a systematic analysis of 153 biological models that incorporated plankton. Only 47% of the models assessed had any performance-related accuracy assessment, and only 30% determined some measure of goodness of fit with respect to observed values. In this paper, we attempted to develop the simplest model to address our questions and then to assess its ability to reproduce both temporal and spatial variability and the primary processes of this coastal ecosystem.

The objective of this study was to develop and to assess a physical-biogeochemical coupled model to characterize the biogeochemical functioning of the Marseille coastal zone. An initial version of the coupled model was developed, and then the biogeochemical model was improved by adding the P cycle. Then, we assessed both versions of the model and whether the addition of the P cycle yields an improvement in the model simulation. In the final section, we discuss the complexity required in coastal ecosystem modeling and the contribution of model results to the understanding of the Marseille coastal area.

Materials and Methods

1. Models

1.1. The hydrodynamic model: MARS3D RHOMA. The hydrodynamic model used for this study was the free surface, three-dimensional MARS3D model (3D hydrodynamic Model for Applications at regional Scale, IFREMER). The high resolution MARS3D-RHOMA configuration was applied and validated to the forecast of the oceanic circulation off Marseille [6], with a horizontal resolution of 200 m and 30 sigma vertical levels. The time step was fixed at 30 s. Atmospheric forcing, hydrodynamic open boundary conditions and the numerical schemes of the model were described by Pairaud et al. [6]. The model grid resolution used in this work is a downgrade of the version described in Pairaud et al. [6], with a 400-m horizontal resolution. Similar validation was performed on the 400-m version, and the processes were well reproduced.

1.2. Coupled model. The coupled model domain covered an area of 100 km×48 km between the mouth of the Rhone River and Cape Sicié. The study area was discretized horizontally using a uniform mesh of 252×120 cells at 400-m resolution. The vertical direction was divided into 30 sigma levels (refining the resolution close to the surface and bottom).

The hydrodynamic model and the biogeochemical model were coupled online (Equation 1). The biogeochemical model calculated all variables tendency (Source-Minus-Sinks, SMS) for each grid point every 20 minutes, whereas the physical model performed the advection-diffusion of the biogeochemical concentrations (C) with a 30-second time-step. The analytical formulations of the advection and diffusion terms are the same as those for the

temperature and salinity in the physical model.

$$
\frac{\partial C}{\partial t} + u\frac{\partial C}{\partial x} + v\frac{\partial C}{\partial y} + w\frac{\partial C}{\partial z} =
$$
$$
\frac{\partial}{\partial x}\left(K_x\frac{\partial C}{\partial x}\right) + \frac{\partial}{\partial y}\left(K_y\frac{\partial C}{\partial y}\right) + \frac{\partial}{\partial z}\left(K_z\frac{\partial C}{\partial z}\right) + SMS
$$

(1)

The coupling was introduced as the extinction light coefficient calculated by the biogeochemical model used in the function of the chlorophyll-a concentration, which induced a retro-action on the temperature calculation of the physical extinction light coefficient; thus, it was truly a coupled online model. Total radiance was read from the meteorological file (MM5) by the hydrodynamic model.

1.3. The biogeochemical model: ECO3M MASSILIA. The biogeochemical model was implemented using the Eco3M (Ecological Mechanistic and Modular Modeling) modeling platform [38,39]. A new biogeochemical model (ECO3M-MASSILIA) was developed for this study. The model structure used is primarily based on the pelagic plankton ecosystem model published by Faure et al. [40,41] but without the P cycle.

The biogeochemical model was split into 5 compartments (phytoplankton, heterotrophic bacteria, dissolved and particulate organic matter and dissolved inorganic matter). This model allowed for a variable intracellular content of phytoplankton and bacteria. Another particularity was that the zooplankton were not represented as a state variable of the model; rather, their physiological fluxes (e.g., grazing or excretion) were considered as explicit functions. The closure formulation of the model was based on the assumption that all of the matter grazed by the zooplankton and the higher trophic levels was returned to one of the pools of organic or inorganic matter (dissolved and particulate organic matter and dissolved inorganic matter).

In this paper, we present and discuss only the modifications applied to the initial version of the model. First, the ECO3M-MASSILIA-noP was developed with new parameterization adapted for the Mediterranean Sea and the addition of grazing limitation by temperature. Then, the ECO3M-MASSILIA-noP was modified with the addition of the P cycle (ECO3M-MASSILIA-P). Thus, the pelagic ecosystem was summarized in 12 or 17 state variables (Table 1), and the cycles of C and N were described by the ECO3M-MASSILIA-noP, with the addition of P for the ECO3M-MASSILIA-P (Figure 2). We refer the reader to Annex S1 for the detailed equations of the biogeochemical model.

The parameterization of the biogeochemical model was adapted for the Marseille coastal area (Annex S2). Previous studies of the phytoplankton community in the Marseille – Rhone area concluded that diatoms dominated the phytoplankton community [42,43]. When possible, phytoplankton parameters were chosen to represent diatoms. For the purpose of parameter refinement, we implemented zero-dimensional models, which allowed us to perform many simulations with different parameters sets.

A coastal study in the northwestern Mediterranean Sea demonstrated that the grazing rate had strong seasonality: the grazing is lower in winter and higher in summer [44]. Another study reported that the development of the zooplankton was impacted by the water temperature [45]. This seasonality was introduced into the model as a function linking the grazing rate to the temperature (T) (Equation 2). The limitation function f(T) [46] was applied to the grazing rate.

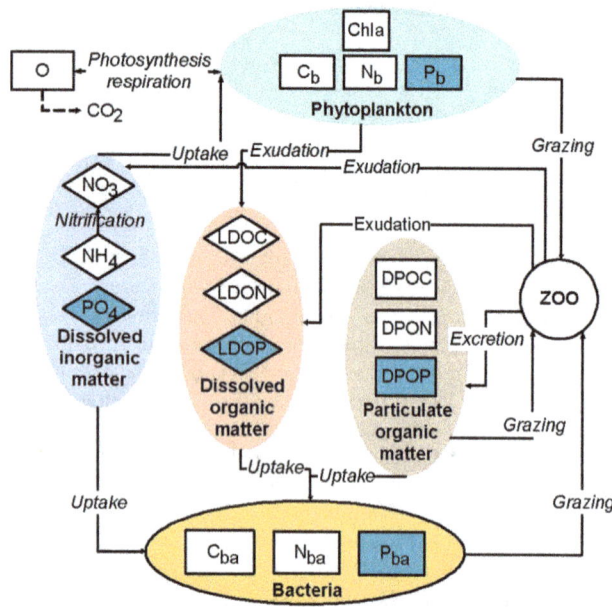

Figure 2. The pelagic biogeochemical model. The ECO3M-MASSILIA-noP (only white-colored variables) and ECO3M-MASSILIA-P (white- and purple-colored variables) versions of the biogeochemical model are represented.

Table 1. Biogeochemical model variables (ECO3M-MASSILIA-P version).

	Variable	Definition
1	C_B	Phytoplankton Carbon
2	N_B	Phytoplankton Nitrogen
3	Chl.a	Phytoplankton Chlorophyll-a
4	CBA	Bacterial Carbon
5	NBA	Bacterial Nitrogen
6	DPOC	Detrital Particulate Organic Carbon
7	DPON	Detrital Particulate Organic Nitrogen
8	LDOC	Labile Dissolved Organic Carbon
9	LDON	Labile Dissolved Organic Nitrogen
10	NO3	Nitrates
11	NH4	Ammonium
12	O	Oxygen
13	PB	Phytoplankton Phosphorus
14	PBA	Bacterial Phosphorus
15	DPOP	Detrital Particulate Organic Phosphorus
16	LDOP	Labile Dissolved Organic Phosphorus
17	PO4	Phosphorus

Only variables from 1 to 12 for ECO3M-MASSILIA-noP version. Units in $\mu mol.L^{-1}$ except Chlorophyll-a which is in $\mu g.L^{-1}$.

$$f(T) = 0.0498 \cdot \exp(0.13 \cdot T) \qquad (2)$$

The literature on the limitation by P [25,29] and the potential deficiency in P of diverse inputs into the study area [26–28] highlight the importance of considering this element. For the addition of the P cycle in ECO3M-MASSILIA-P, five supplementary states variables were added (Figure 2) : dissolved inorganic phosphorus (PO_4), labile dissolved organic phosphorus (LDOP), detrital particulate organic phosphorus (DPOP), phosphorus phytoplankton biomass (Pb) and phosphorus bacterial biomass (Pba). For coherence, the representation of the P cycle mimicked that of the N cycle. The structural modifications are described hereafter for the P equations that differed from those of the N. The full model equations are detailed in Annex S1.

Concerning the phytoplankton compartment, the phytoplankton biomass in phosphorus (Pb) was introduced as a state variable. Pb depends on the PO_4 uptake and zooplankton grazing (Equation 3).

$$\frac{\partial Pb}{\partial t} = +uptake - grazing \qquad (3)$$

The equations of the phytoplankton biomass in carbon and chlorophyll were modified. The carbon biomass equation was modified through the term for the limitation of the phytoplankton maximal growth rate (P^C_m) by nutrient (Q^*) (Equation 4). The co-limitation of phytoplankton by nutrients was introduced into the equation for phytoplankton carbon biomass as an independent nutrient co-limitation of type I [47]. The co-limitation between N and P was calculated using the Liebig's Law of the minimum [48,49]. The phytoplankton limitation by P was calculated in the same manner as by N: with the intracellular quota (Q) and the

Geider formulation [50]. Consequently, phytoplankton production is controlled by the most limiting internal ratio (Q^*).

$$Q* = \min\left(\left(\frac{Q^N_C - \min Q^N_C}{\max Q^N_C - \min Q^N_C}\right), \left(\frac{Q^P_C - \min Q^P_C}{\max Q^P_C - \min Q^P_C}\right)\right) \qquad (4)$$

The phytoplankton chlorophyll-a concentration was a diagnostic variable related to the phytoplankton biomass by carbon (C_B) and the variable ratio Chla:C (Q^{Chla}_C) (Equation 5). Equation 6 describes the Q^{Chla}_C of phytoplankton depending on the phytoplankton N-to-C ratio and the limitation by the most limiting internal ratio (Q^*). As Q^* decreases toward zero, Q^{Chla}_C also decreases, as does the limitation on Chla. In Smith and Tett [51] and Faure et al. [52], the variable Q^* only represented the internal N:C ratio. To also consider the limitation of chlorophyll-a production when phytoplankton was limited by P, we modified the equation for Q^* (Equation 4).

$$[Chla] = Q^{Chla}_C \cdot C_B \qquad (5)$$

With $Q^{Chla}_C = Q^N_C \cdot (\max Q^{Chla}_N \cdot Q* + \min Q^{Chla}_N (1 - Q*))$ $\qquad$ (6)

As for phytoplankton, heterotrophic bacteria were described in terms of carbon, nitrogen and phosphorus. To include the P cycle in the microbial loop, the bacterial biomass in phosphorus (Pba) was added as a new state variable. Pba varied with the uptake and grazing of organic matter (DPOP and LDOP) and inorganic matter (PO_4) (Equation 7).

$$\frac{\partial Pba}{\partial t} = + uptake\,(DPOP, LDOP, PO_4) - grazing \qquad (7)$$

We simulated the temporal evolution of heterotrophic bacteria according to the cell quota theory. This concept [53] considers that growth rate (μ^{BA}_{max}) is a function of limiting nutrients within the cell. A new system of co-limitation by C, N and P was needed to control the bacterial production (BP) (Equation 8). We choose to represent the co-limitation (f) of the bacterial production by the Liebig's law (Equation 9), similarly to the representation of phytoplankton production. However, even if Liebig's law was used for the co-limitation of the phytoplankton, the limitation of the maximum growth rate (μ^{BA}_{max}) of bacteria was calculated differently. Q^C_{BA}, Q^N_{BA} and Q^P_{BA} represents the carbon, nitrogen and phosphorus cell quota of bacteria (see Annex S1 for more details). As the ratio (min Q^X_{BA}:Q^X_{BA} increased, the limitation of the bacteria by the element X also increased. The most restrictive element corresponds to the higher ratio (min Q^X_{BA}:Q^X_{BA}.

$$BP = \mu^{BA}_{max}.(1-f).NBA \qquad (8)$$

$$\text{With } f = \max\left(\frac{\min Q^C_{BA}}{Q^C_{BA}}, \frac{\min Q^N_{BA}}{Q^N_{BA}}, \frac{\min Q^P_{BA}}{Q^P_{BA}}\right) \qquad (9)$$

1.4. Modeling strategy. To determine the spin-up period of the model, tests were conducted for the year 2007 by replacing winter initial conditions with summer initial conditions. We considered the maximum spin-up period to be completed when all state variables were equal between the two runs. The period lasted 90 days; therefore, the model results can be studied from April 2007 onwards (Annex S3).

The hydrodynamic open boundary conditions and initial conditions of the MARS3D-RHOMA configuration are described in Pairaud et al. [6]. The biogeochemical open boundary conditions and initial conditions were built from a coupled model applied to a larger area (GoL) at a horizontal resolution of 1.2 km and 30 vertical sigma levels. The hydrodynamic model (MARS 3D-GOL configuration) was based on the MARS3D-MENOR configuration validated by Nicolle et al. [54]. The MARS 3D-GOL configuration was coupled online with the biogeochemical ECO3M-MASSILIA-P [55]. The biogeochemical open boundary conditions of this larger model were provided by the OPATM-BFM pre- and operational runs 2007–2011 performed in the framework of the Mersea and MyOcean projects [56].

Atmospheric dry deposition (dry and wet deposits) was sampled weekly by the national MOOSE program (Mediterranean Oceanic Observing System on Environment) and the "Service d'Observation" of the Mediterranean Institute of Oceanography (S.O. MIO). Inputs from rain (wet deposition) were measured for each rain event. The device (collecteur MTX-Italia) was installed on an island located in the bay off Marseille. Atmospheric samples were analyzed for soluble components (nitrate (NO$_3$), ammonium (NH$_4$), phosphate (PO$_4$), dissolved organic carbon (DOC), dissolved organic nitrogen (DON) and dissolved organic phosphorus (DOP)) and insoluble particulate matter (particulate organic carbon (POC), particulate organic nitrogen (PON) and particulate phosphorus (POP)). Wet atmospheric concentrations were applied to the rainfall represented by the atmospheric model MM5. Dry deposition was applied as a mean flux between each pair of sample dates.

Between 2007 and 2011, daily averaged Rhone River discharges at the Beaucaire station were available and provided by the "Compagnie Nationale du Rhone". The "Grand Rhone" located in the modeling domain represents only 90% of total Rhone River discharge [57,58]. Marseille has 4 main Urban Rivers (Aygalade, Belvedère-Figuière, Huveaune-Jarret and Bonneveine) and a Wastewater Treatment Plant (WWTP), which flows in at Cortiou (Annex S4). The Marseille Urban Rivers discharges were available with a time step of 6 minutes, and the daily discharge from the Marseille WWTP was available (Data provided by the DEA-MPM (Direction de l'Eau et de l'Assainissement-Marseille Provence Métropole)). The Berre Lagoon, which is a shallow semi-confined ecosystem, is connected to the Mediterranean Sea via the Caronte channel. The latter was represented in the coupled model as a river with a constant discharge of 20 m^3.s^{-1}. River and WWTP concentrations.

Biogeochemical concentrations in the Rhone River (NO$_3$, NH$_4$, PO$_4$, POC, PON, POP, DOC, DON and DOP) were measured daily and provided by the national MOOSE program and the SO MIO. As our model considered only the labile fraction of organic matter, a percentage representing the labile fraction of organic matter was applied to the daily Rhone River concentrations. However, studies of the lability of Rhone River organic matter were rare. The labile fraction of particulate organic carbon (DPOC) was considered to be 18% of POC [59]. Déliat [60] estimated that approximately 20% of the Rhone River DOC was biodegradable; thus, we computed LDOC as equal to 20% of DOC. The labile fraction of DON (LDON) was also estimated to be 20% of DON [60]. Experiments in Loch Creran (Scotland) showed that bioavailable DOP (BDOP) accounted for 88±8% of DOP (average±SD) [61]. The same percentages were applied to the particulate organic matter concentrations to obtain DPON and DPOP.

The concentrations in the urban rivers were not always available from in-situ data for all of the variables and hence were derived from empirical relationships (Table 2). The Rhone River concentrations (described above) were applied to the concentrations of nutrients in other Marseille Urban Rivers. Organic matter concentrations in the Marseille Urban Rivers and from the WWTP were deduced from suspended particulate matter (SPM) concentrations [62] with constant ratios. The same hypotheses on the lability of the Rhone River organic matter were applied to the urban river concentrations. The concentrations of NH$_4$ and NO$_3$ from the Marseille WWTP were available daily (Data provided by the DEA-MPM (Direction de l'Eau et de l'Assainissement-Marseille Provence Métropole)). The PO$_4$ concentration was fixed to 13.4 µmol (Faure, comm. Pers.). Marine phytoplankton and bacteria species were considered to be absent from all of the rivers and WWTP inputs. The Caronte concentrations were fixed at constant values (Table 2) obtained by averaging the measured concentrations [63].

2. Observational data sets

2.1. In-situ data. To validate the coupled model, we used a long time series of hydro-biogeochemical data collected twice monthly at the Somlit station (43°14.30′N; 5°17.30′E) located in the BoM (Annex S4). High vertical resolution profiles of temperature, salinity and oxygen were obtained between 0 and 55 m using a conductivity temperature-depth-oxygen profiler (CTDO, Seabird 19+). Water samples were collected at 3 depths with hydrological Niskin bottles for the determination of inorganic and organic nutrients concentrations. Samples for nitrate, nitrite

Table 2. Concentrations of River inputs ($\mu mol.L^{-1}$).

River	NO$_3$	NH$_4$	PO$_4$	LDOC	LDON	LDOP	LPOC	LPON	LPOP
Rhone	data	data	data	0.2*DOC	0.2*DON	0.88*DOP	0.18*POC	0.2*POC	0.88*POP
Caronte	0.75	2.72	0.4	45	3.73	0.11	1.8	1.36	0.078
WWTP	data	data	13.4	135	17	3	8403	864	372
Bonneveinee	RR	RR	RR	52.4	6.62	1.17	667	68.60	29.6
Huveaune	RR	RR	RR	38.5	4.86	0.86	869	89.4	38.57
Emiss 1	RR	RR	RR	553	69.9	12.37	2430	250	107.8
Emiss 2	RR	RR	RR	14.4	1.82	0.32	f(RD)	f(RD)	f(RD)
Aygalade	RR	RR	RR	52.4	6.62	1.17	f(RD)	f(RD)	f(RD)
Belvedere	RR	RR	RR	52.4	6.62	1.17	f(RD)	f(RD)	f(RD)

RR: Rhone River daily measured concentrations; f(RD) : concentrations linked to urban rivers discharges.

and phosphate were collected into 60 ml polyethylene nutrients flasks and were frozen at $-20°C$ until analysis at laboratory according to Aminot and Kerouel [64]. Samples for acid silicic were collected in 60 ml polyethylene flasks and stored at $5°C$ until analysis at laboratory according to [64]. Samples for ammonium determination were collected in triplicate in 60 ml polycarbonate tubes. The reagent was immediately added to the tubes and ammonium level was determined by fluorometry according to Holmes et al. [65]. Particulate organic carbon (POC) and particulate organic nitrogen (PON) in suspended matter collected on Whatman GF/F glass micro-fibre filters pre-combusted for 4 h at $450°C$, were determined by using the high combustion method ($1000°C$) on a CN Integra mass spectrometer [66]. Chlorophyll concentrations were estimated by fluorometry [67] on suspended matter collected on Whatman GF/F filter. The data were provided by the SOMLIT network (Service d'Observation en Milieu Littoral, http://somlit.epoc.u-bordeaux1.fr).

2.2. Satellite remote sensing data. We also compared the model-predicted fields of surface chlorophyll-a concentrations with the corresponding remote sensing-derived concentrations. The MODIS (Moderate Resolution Imaging Spectroradiometer) and MERIS (MEdium Resolution Imaging Spectrometer) ocean color sensors have a spatial resolution of approximately 1 km, which is more course than the spatial resolution of our model (400 m). For the comparisons, we used the remotely sensed chlorophyll-a concentration processed with the algorithm OC5 [68,69] from the IFREMER (Institut Français de Recherche pour l'Exploitation de la Mer) database. The OC5 method [68] is empirical and derived from the OC4 algorithm of NASA (or OC3M-547 for MODIS and OC4E for MERIS). This method gives results similar to OC4 in open waters but provides more realistic values over the continental shelf [70]. The quality of the remote sensing data was evaluated in Annex S5. The remote sensing data contained observational errors, which was also the case for the in-situ data. Nevertheless, the large number of existing data, their availability throughout the year and their spatial coverage make remote sensing data essential to the consideration of spatial patterns and gradients. Thus, the comparison with the model results had to consider the observational error.

3. Statistical comparison with observations

To assess the accuracy of the 3D coupled online model, we used statistics highlighting correspondences between the model results (M) and in-situ and remote sensing observations (O). The study of model performance often requires multiple stages of analysis, the evaluation of the ability of the model to reproduce instantaneous station values (and trends) and the ability of the model to recreate spatial characteristics (and trends) of processes [71]. Stow et al. [72] explained that model start to have skill when the observational and predictive uncertainty halos overlap; in the ideal case, the halos overlap completely. Thus, accuracy assessment requires a set of quantitative metrics and procedures for comparing model output with observational data in an appropriate manner for the particular application. Here, we used different statistical indicators [72,73] :

– The percentage model bias (PB, model error normalized by the observations) measures whether the model underestimated or overestimated the observations [74]. The closer to zero the value is, the better the model.

– The cost function (CF) gives a non-dimensional value that is indicative of the "goodness of fit" between two sets of values [73]. Radach and Moll [75] proposed an interpretation of the values of the cost functions adapted from the OSPAR Commission [76] rating. Model results are classified as very good (cost function between 0 and 1), good (1 and 2), reasonable (2 and 3) and poor (higher than 3) [77].

– The correlation coefficient (R) is a measure of the strength and direction of the linear relationship between two sets of values [74].

– The average absolute error (AAE) quantifies the magnitude rather than the direction of each discrepancy (Equation 10). The closer to zero the value is, the better the model.

$$AAE = \frac{\sum_{i=1}^{n} |O_i - M_i|}{n} \qquad (10)$$

– The Root Mean Square Deviation (RMSD) is also a measure of the difference between values predicted by the model and the values actually observed from the environment [74]. The RMSD has the same units as the quantity being evaluated.

To compare the variables, we used signed, normalized, unbiased RMSD (snuRMSD) to quantify the magnitude direction of each discrepancy and the normalized bias to quantify the direction of each discrepancy [78]. Those indicators are used to construct target diagrams [78].

Model developments and parameterization were performed for the years 2007 and 2008, and then the observations available for the years 2009, 2010 and 2011 were used only to assess the accuracy of both versions of the model.

Results

We assessed the ability of the coupled model to reproduce the main characteristics of the Marseille coastal area for the two versions of the biogeochemical model, with and without the P cycle.

An evaluation of a 3D ecosystem model starts with the hydrodynamics. Pairaud et al. [6] performed the hydrodynamics study for the years 2007 and 2008. They computed model/ observation statistics and showed the ability of the model to capture monthly to seasonal variability in the thermal structure and to reproduce the observed features over the shelf, such as the warming or cooling of the sea due to upwelling and downwelling events and the punctual extension of the Rhone River plume.

Realistic 3D simulations were performed using the coupled model for 2007 through 2011. First, we present the model-observation comparisons, where we examined the temporal dynamics in the surface and bottom layers at the Somlit station and computed statistics. Then, we present spatial maps to assess the representation of the spatial gradient and their variability during the year. Finally, we focused on short time scale shelf processes and their impacts on the biogeochemical functioning of the pelagic coastal ecosystem.

1. Observations/model comparisons at the Somlit station

1.1. Temporal dynamics. To estimate the reliability of the model in qualitative terms, we described the agreement between simulated and observed monthly means, in-situ measurements of temperature, chlorophyll-a, NO_3, particulate organic matter and PO_4 (Figure 3–4). At the Somlit coastal station (Annex S4), both model versions (with and without the P cycle) captured the annual cycle, with strong seasonal oscillations in temperature, NO_3 and chlorophyll-a. The error bars showed the variability at a smaller time scale than the month of the simulation. If the winter temperatures were estimated well, the maximum summer temperatures seemed to be significantly underestimated. Nitrate concentrations and seasonal evolution were well quantified by the model. Higher chlorophyll-a variability was observed over shorter time scales in spring and summer than in winter and autumn (Figure 3). Both versions of the model reproduced the winter period well when the vertical mixing induced high NO_3 concentrations and low temperature in the surface layer. The spring was characterized by high chlorophyll-a concentrations (spring bloom) associated with a decrease in NO_3. Peaks in the chlorophyll-a concentrations during this season were slightly overestimated by both model versions (Figure 3). During summer, nutrient depletion was reproduced well, while the chlorophyll-a concentration was overestimated. Some isolated values of nitrate and chlorophyll-a were captured by the model.

Both versions of the model demonstrated their abilities to capture inter-annual variability, with a higher spring peak in chlorophyll-a (Figure 3), POC and PON (Figure 4) in 2010 than in 2011. In 2009, the chlorophyll-a concentration increased in November and December, which was not the case for 2007, and the model reproduced this inter-annual variability well.

The model suggests a clear and regular seasonal variation for phosphate, similar to those for nitrate, which is not the pattern observed in the field by bi-weekly sampling. The concentrations in PO_4 remained very low (lower than 0.2 $\mu mol.L^{-1}$) throughout the

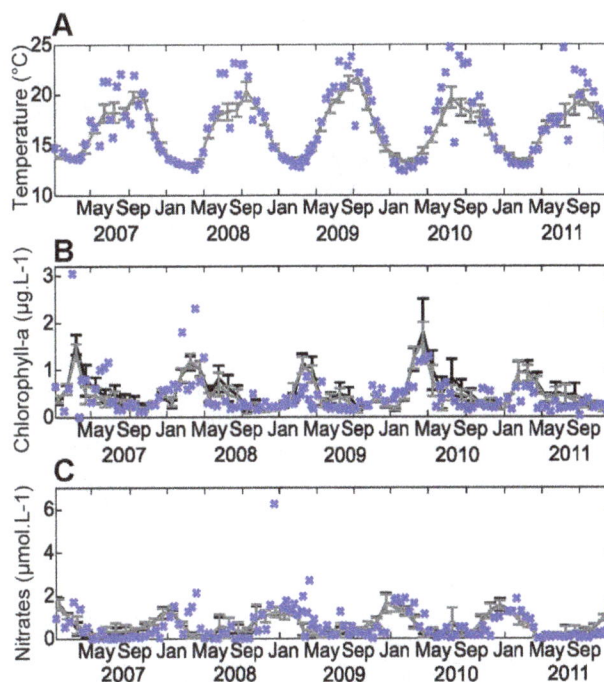

Figure 3. Surface model/data time series comparisons for temperature, chlorophyll-a and NO₃. Time series of the monthly mean temperature (°C)(A), chlorophyll-a concentration (μg.L-1) (B) and NO₃ concentration (μmol.L-1) (C) from the model with the P cycle (grey) and the model without P (black), with one standard deviation (error bar), compared with the in-situ data (blue) at the Somlit station.

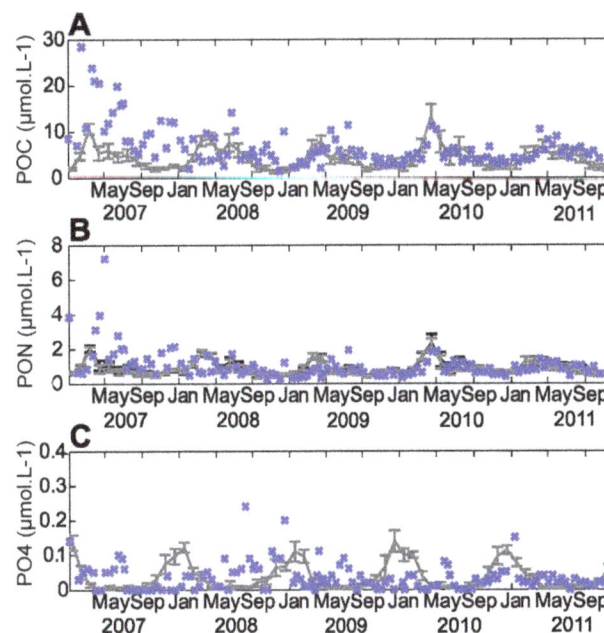

Figure 4. Surface model/data time series comparisons for POC, PON and PO₄. Time series of the simulated monthly mean POC (μmol.L-1) (A), PON (μmol.L-1) (B) and PO₄ (μmol.L-1) (C) concentrations from the model with the P cycle (grey) and the model without P (black), with one standard deviation (error bar), compared with the in-situ data (blue) at the Somlit station.

year, which made it difficult to distinguish a seasonal signal in the in-situ PO_4 data. The two model versions had rather similar results over the five years. However, there were a few discrepancies between them in chlorophyll-a, POC and PON, particularly in summer and spring when the model version without the P cycle overestimated the chlorophyll-a concentration more than the model version with the P cycle (Figure 3, Figure 4). Thus, the addition of the P cycle improved the results during spring and summer.

1.2. Statistical model/observations comparisons. This qualitative comparison was completed through the calculation of statistical indicators between the Somlit time series and the results of the simulations (only for the years 2009 to 2011). The surface and bottom results comparisons are presented in Table 3 and Table 4, respectively. Although the Somlit station is a coastal station near Marseille, the low mean concentrations of NO_3, chlorophyll-a and PO_4 (both in model and in observations) indicated that oligotrophic conditions were predominant. The standard deviation of the in-situ data was frequently equal to or very close to the mean concentration at the surface for NO_3 (0.68±0.64 µmol.L-1), PO_4 (0.03±0.03 µmol.L-1) and chlorophyll-a (0.4±0.32 µg.L-1), which highlights that this oligotrophic ecosystem had a strong variability, possibly due to pulsed nutrients. This variability made the model/observations comparisons difficult, so that different statistical indicators were used.

First, we computed the cost function. According to the OSPAR classification [76], our results were very good or good except for bottom chlorophyll-a, POC, PON and PO_4. Then, we computed the average absolute error (AAE, Equation 10) and the RMSD, which measure the average magnitude of the errors in a dataset, without considering their direction. AAE and RMSD were expressed in the same units as the variable. In the surface layer, the addition of the P cycle induced a decrease or stagnation in the AAE for all of the variables, except for the POC, for which the increase in the AAE was not significant. The model version with the P cycle performed better at representing the chlorophyll-a concentration, with an AAE of 0.33 µmol.L^{-1} versus 0.38 µmol.L^{-1} for the model version without the P cycle. In the bottom layer, the AAEs were higher than in the surface layer, except for the NH_4, salinity and O_2. The addition of the P cycle induced slight variations or stagnations in the AAE for all the variables. The model with the P cycle performed slightly worse for

the bottom chlorophyll-a concentrations, with an AAE of 0.52 versus 0.49 µmol.L^{-1} for the model without the P cycle.

Target diagrams [78] provide summary and visual information regarding the bias and the error between a model and observations. In order to compare variables with each other normalized statistics were used as described in Jolliff et al. [78]. In the Target diagram presented (Figure 5), the Y-axis corresponds to the normalized bias (Bias*) and the X-axis corresponds to the normalized unbiased RMSD (uRMSD*). Variables located in the upper part of the diagram (Y>0) were overestimated by the model (chlorophyll-a, PON and PO_4). For the X-axis, the model standard deviation is larger than the standard deviation for the observations when variables are located in the right part of the target (X>0). The distance between any point and the origin is then the value of the total RMSD (see Jolliff et al. [78] for more details). As shown by the time series and previous comparisons, the discrepancies between the two versions of the biogeochemical model are small both at the surface (Figure 5 A–B) and at the bottom (Figure 5 C–D). The main differences were in the chlorophyll-a concentrations, for which the model version with the P cycle was more efficient in the surface layer with a lower normalized bias and a smaller uRMSD* than the model version without the P cycle. Both model versions overestimated the amplitude of the variations in the state variables except for TEMP, NO_3 and NH_4. The target diagrams also demonstrate that models were less efficient in representing the bottom layer, with an increased bias for PON and POC.

To summarize, the timing and magnitude of the surface chlorophyll-a and NO_3 concentrations were generally well matched. The shortcomings of the simulation were an overestimation of the chlorophyll-a concentrations in spring and a slight overestimation in summer. The model also had more difficulty in simulating the bottom than the surface at the Somlit station. However, using the in-situ data to evaluate the ability of the model as discussed above refers to data from a single station sampled every fortnight and with undoubted local bias and particularity. We thus used remote sensing data for the validation of the representation of spatial processes.

2. Evaluation of spatial processes representation

Ocean color observations derived from remote sensing data were used to assess the abilities of the model to capture spatial gradients and seasonality over the whole study area.

Table 3. Surface model and data comparison at the Somlit station from 2009 to 2011.

	TEMP	SAL	CHL		NO₃		NH₄		O₂		POC		PON		PO4
			P	No P	P	No P	P	No P	P	No P	P	No P	P	No P	P
n	78	78	78	78	73	73	77	77	74	74	73	73	73	73	64
Mean (in situ)	17.15	37.97	0.40	0.40	0.68	0.68	0.31	0.31	241.12	241.12	5.13	5.13	0.79	0.79	0.03
Mean (model)	16.59	37.99	0.52	0.58	0.58	0.56	0.13	0.13	231.55	233.46	3.59	4.29	0.88	0.93	0.04
Std (in situ)	3.63	0.17	0.32	0.32	0.64	0.64	0.68	0.68	14.19	14.19	2.09	2.09	0.35	0.35	0.03
Std (model)	2.68	0.36	0.52	0.64	0.57	0.59	0.20	0.26	18.66	20.89	2.85	3.71	0.59	0.67	0.05
CF	0.27	1.24	1.01	1.16	0.80	0.81	0.40	0.43	1.30	1.37	1.20	1.19	1.17	1.19	1.39
Bias (%)	3.21	−0.05	−28.75	−44.41	14.17	17.05	59.95	59.65	3.97	3.18	30.03	16.30	−10.96	−16.84	−15.35
AAE	0.99	0.21	0.33	0.38	0.51	0.51	0.27	0.29	18.39	19.47	2.49	2.48	0.41	0.41	0.04
RMSD	1.37	0.33	0.52	0.63	0.7	0.7	0.73	0.75	22.56	23.94	3.17	3.38	0.55	0.6	0.05
R	0.96	0.44	0.34	0.34	0.34	0.36	0.00	−0.01	0.24	0.20	0.40	0.47	0.41	0.46	0.10

n: the number of in-situ data available for comparison, std : the standard deviation.

Table 4. Bottom model and data comparison at the Somlit station from 2009 to 2011.

	TEMP	SAL	CHL		NO$_3$		NH$_4$		O$_2$		POC		PON		PO4
			P	No P	P	No P	P	No P	P	No P	P	No P	P	No P	P
n	74	74	74	74	71	71	73	73	67	67	69	69	69	69	63
Mean (in situ)	14.56	38.04	0.36	0.36	1.03	1.03	0.14	0.14	243.53	243.53	4.16	4.16	0.66	0.66	0.03
Mean (model)	14.78	38.16	0.81	0.77	0.5	0.52	0.11	0.11	233.38	233.5	6.03	6.3	1.5	1.47	0.09
Std (in situ)	1.75	0.09	0.22	0.22	0.89	0.89	0.11	0.11	12.38	12.38	1.6	1.6	0.28	0.28	0.03
Std (model)	1.25	0.11	0.56	0.52	0.58	0.58	0.15	0.15	17.41	17.92	3.3	3.55	0.66	0.65	0.06
CF	0.42	1.75	2.39	2.23	0.88	0.87	1.08	1.08	1.27	1.29	1.89	2.03	3.21	3.09	2.32
Bias (%)	−1.55	−0.32	−122.52	−111.72	51.96	50.23	20.15	16.66	4.17	4.12	−44.98	−51.49	−128.24	−123.06	−164.91
AAE	0.74	0.15	0.52	0.49	0.78	0.77	0.12	0.12	15.68	15.94	3.02	3.25	0.89	0.85	0.07
RMSD	0.95	0.17	0.71	0.65	1.09	1.08	0.18	0.18	19.49	19.76	3.84	4.15	1.06	1.03	0.08
R	0.86	0.29	0.25	0.26	0.21	0.22	0.06	0.05	0.41	0.41	0.19	0.2	0.24	0.26	0.04

2.1. Climatology of surface chlorophyll-a patterns. The maps of time-averaged surface chlorophyll-a concentrations permit the evaluation of the ability of the model to reproduce horizontal patterns of chlorophyll-a. The outputs from both versions of the model were compared with observed maps of surface chlorophyll-a concentrations from MODIS and MERIS during the 4 seasons of the year (Figure 6). The Rhone River plume area was characterized by high chlorophyll-a concentrations, in contrast with the eastern portion of the study area, where concentrations in chlorophyll-a remained very low. The models captured the west-east gradient in surface chlorophyll-a concentration. The BoM seems to be a transition zone between the rich water of the Rhone River plume and the oligotrophic water located eastward. The models reproduced well the spatial extension of the Rhone River intrusion plume, except very close to the mouth, where the concentrations were underestimated by the model. This underestimation was caused by the inability to distinguish terrestrial and fluvial chlorophyll-a from marine chlorophyll-a in the remote sensing data. The marine chlorophyll-a is the only one included in the model.

The chlorophyll-a remote sensing data were highly variable in both the spatial distribution and the concentrations; this variability was well captured by both versions of the model. In winter, both version of the model reproduced well the chlorophyll-a concentrations in the BoM, but the concentrations associated with the Rhone River plume were underestimated. During spring and summer, the model fields deviate from the observations in overestimating the remotely sensed chlorophyll-a concentrations in the Rhone River plume. Nevertheless, the spring remains the season during which the surface layer was the most productive and the chlorophyll-a concentration was at its maximum in the BoM. The summer period had lower concentrations than in the spring in the surface layer, particularly in the BoM. The fall was characterized by the lowest chlorophyll-a concentrations of the year in the entire study area; the signal of the Rhone River was weak in comparison with the three other seasons.

The spatial comparisons exhibited little discrepancy between the two versions of the model. During spring and summer, the chlorophyll-a signal associated with the Rhone River plume was lower and more similar to the observations in the model with the P cycle both spatially and in intensity. Therefore, the addition of the P cycle causes an improvement in the representation of the effects of the Rhone River on chlorophyll-a concentrations.

3. Influence of short time scale shelf processes forcing on co-limitation functioning

In the evaluation of the representation of short time scale shelf processes in the coupled models, we focused on the year 2008, for which hydrodynamic shelf processes off Marseille were analyzed in detail by Pairaud et al. [6]. The Marseille coastal area was under strong forcing influences (upwelling events, intrusion of Rhone River plume in the BoM), which induced significant daily variability in the biogeochemical variables. Figure 7 presents the observations and daily model results for 2008 used to evaluate the ability of the model to reproduce processes over short time scales. In early April, a peak in chlorophyll-a was present in the model results and in the remote sensing and in-situ data, and we noticed strong discrepancies between the in-situ data and the remote sensing data. The model values were between those of the remote sensing and in-situ data, leading to the assumption that chlorophyll-a concentrations could be underestimated from remote sensing data in spring. In summer, the BoM experienced enrichment events, when modeled chlorophyll-a concentrations increased from 0.5 to 1.5 μmol.L^{-1}. Those events matched the observations very well, except for the event of late August 2008. The chlorophyll-a concentrations remained low during fall with slight variations in intensity for both the models and observations. Weak discrepancies were observed at short time scales between the two versions of the model, except in summer, when enrichment events were clearly less intense in the model with the P cycle. To explain these discrepancies, we focused on phytoplankton limitation by P and N.

Figure 8 presents phytoplankton limitation by nutrients calculated using the cell quota formulation in the surface and bottom layers at the Somlit station. As the limitation increases in value toward 1, the nutrient limitation on phytoplankton increases. Temperature was also presented on the same graph (Figure 8) to evaluate the links between nutrient co-limitation and temperature. The results of the model showed that phytoplankton limitation by nutrients was greater in the surface layer (Figure 8A) than in the bottom layer (Figure 8B), as was the variability in limitation. As shown in Figure 3, high concentrations of nutrients were available from November to March due to winter vertical mixing, so that phytoplankton limitation by nutrients remained very low in the surface and bottom layers. An increase in limitation occurred from March to the beginning of May, corresponding to the spring bloom. The end of the spring bloom was marked by the

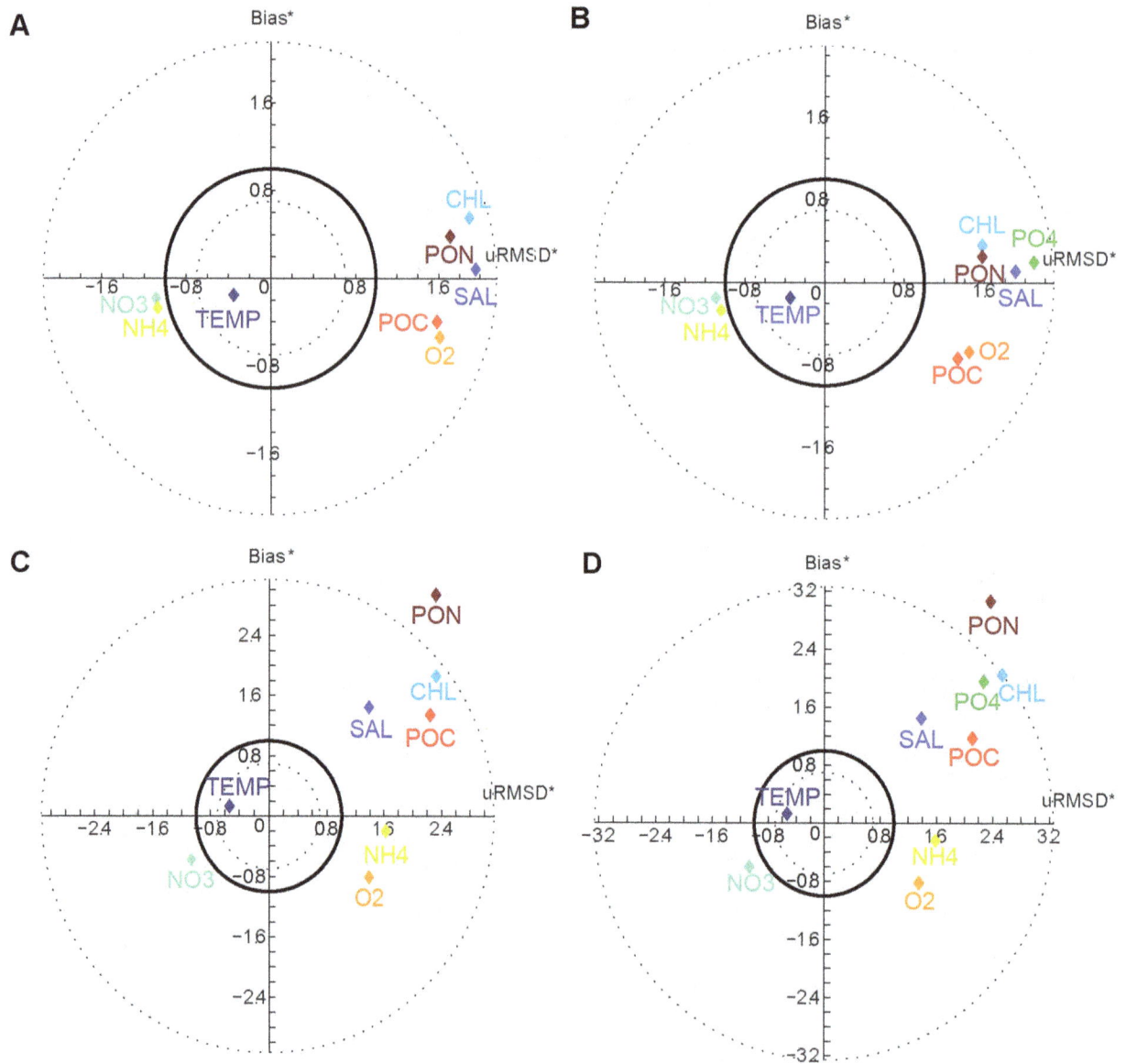

Figure 5. Target diagrams of the model/data comparison. Target diagrams of a set of variables for the model with the P cycle (A–C) and without the P cycle (B–D). The model/data comparison is made using in-situ data for the years 2009 to 2011 at the surface (A–B) and at 55 m (bottom C–D) at the Somlit station. The normalized bias (dimensionless) was computed versus the sign unbiased normalized RMSD (dimensionless).

exhaustion of nutrient supplies in the surface layer, the stabilization of limitation at high levels and the stabilization of the chlorophyll-a concentration at approximately 0.3 µg.L^{-1} at the surface (Figure 7). Short, strong decreases in limitation happened during summer at the surface. Then, the fall was associated with a decrease in limitation by nutrients. Therefore, phytoplankton limitation by nutrients had an apparent seasonal signal, and increases in the chlorophyll-a concentration in the surface layer were linked with short decreases in phytoplankton nutrient limitation. However, phytoplankton nutrient limitation was not directly correlated with chlorophyll-a because temperature and light limitation were also considered to affect chlorophyll-a seasonality.

The model results also showed that phytoplankton were more limited by P (grey points) in the surface layer and more by N (green points) in the bottom layer. However, switches between P and N limitation occurred several times during the year, highlighting the

strong variability of ecosystem functioning most likely due to physical shelf processes.

3.1. Influence of upwelling. The Marseille coastal area is strongly impacted by upwelling events associated with strong upwards vertical velocity (maximum of 5 cm.s^{-1} during an upwelling event in November 2008), which led to an important decrease in surface temperature [6]. Significant upwelling events occurred during the summer of 2008, and these strong temperature decreases were always associated with N limitation (Figure 8A). This result was consistent with the modeled N limitation of phytoplankton development in bottom layer (Figure 8B).

The maps in Figure 9 present a comparison of surface temperature, chlorophyll-a concentration and phytoplankton nutrient limitation during an upwelling event in July 2008. During this event, the area impacted by the ascent of deep water was characterized by low temperature (T<20°C) (Figure 9A) and

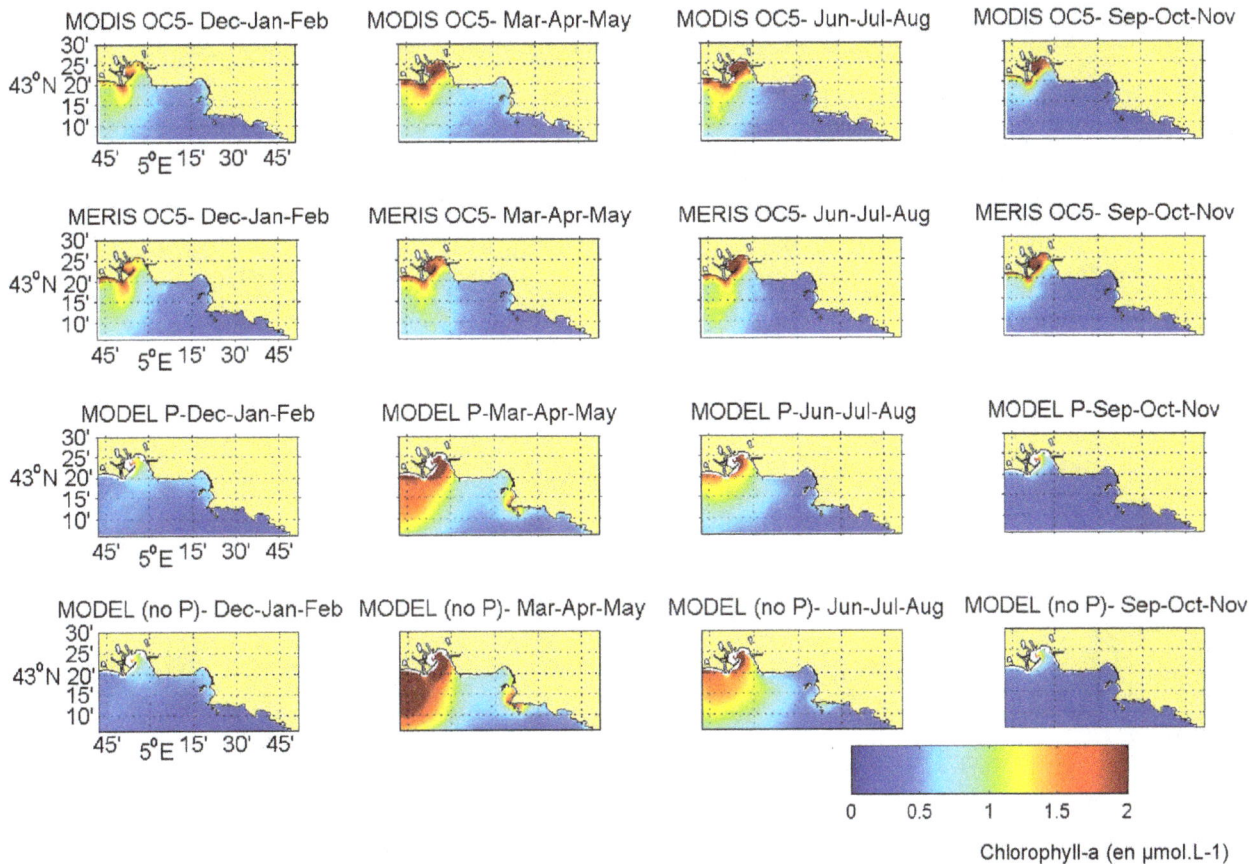

Figure 6. Mean surface chlorophyll concentration for winter, spring, summer and fall. Mean surface chlorophyll concentration for winter (December–January–February), spring (March–April–May), summer (June–July–August) and fall (September–October–November) for the years 2007 to 2011 from the remote sensing data (MODIS and MERIS) and the simulation (with and without the P cycle). The model results presented were averaged across the 10 first meters of the water column.

limitation by N rather than P (Figure 9B). The lower temperature was localized to the two primary upwelling points ("Cote Bleue" and "Calanques"), but the majority of the Marseille coastal area was impacted by upwelling as described by Pairaud et al. [6]. The remote sensing data and modeled chlorophyll-a concentrations were in good agreement (Figure 9 C–D); the concentrations were

high in the two primary upwelling points in response to the nutrient enrichment induced by the upwelling. A weak increase in chlorophyll-a was observed far from the two main upwelling spots but still located in the temperature-impacted upwelling area. This could be explained by the rapid uptake of nutrients near the upwelling points. The nutrients were consumed by phytoplankton

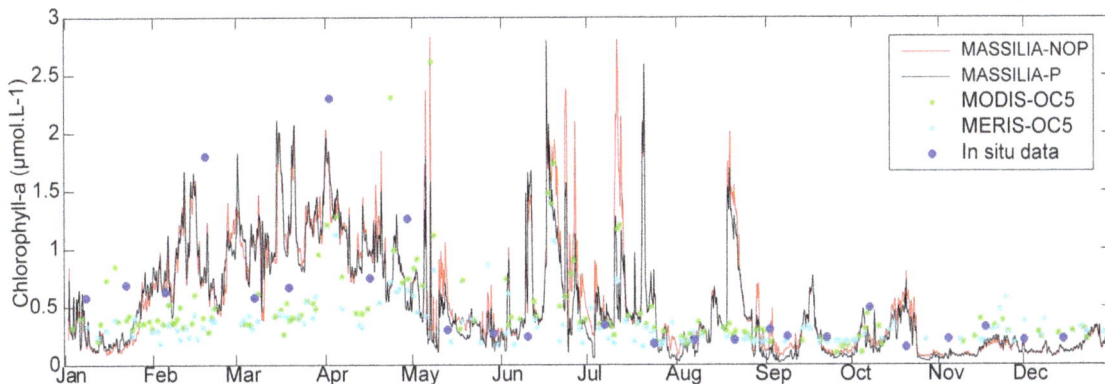

Figure 7. Surface chlorophyll concentrations for the year 2008. Evolution of the surface chlorophyll concentrations from the remote sensing data of MODIS (green point) and MERIS (blue point), in-situ data (blue point) and the simulation with the P cycle (black line) and without the P cycle (blue line) at the Somlit station.

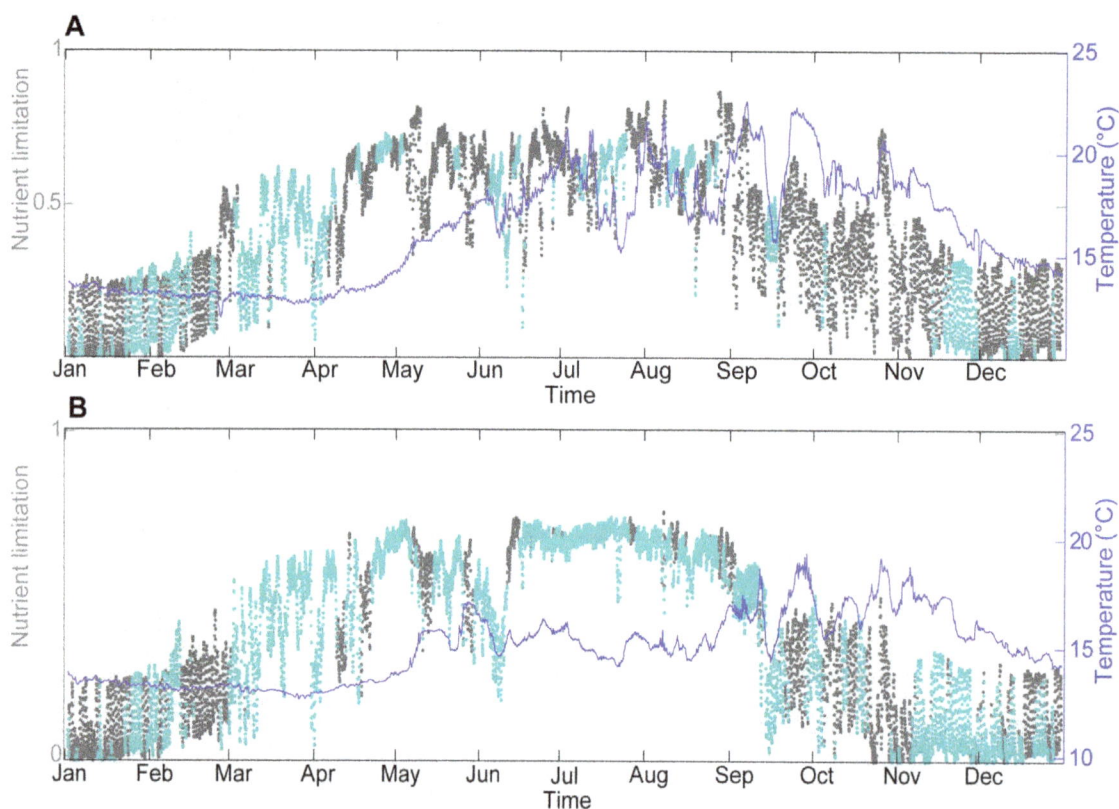

Figure 8. Evolution of temperature and phytoplankton nutrient limitation for the year 2008. Phytoplankton limitation by P (grey point) and N (blue point) and model temperature (blue line) for the year 2008 at the Somlit station at the surface (A) and at 60 m (B).

before being transported away from the upwelling points due to the strong nutrient limitation of phytoplankton in the summer and in the surface layer. These results explained why the surface temperature was strongly impacted and the surface chlorophyll-a concentration more weakly impacted at the Somlit station during upwelling events (Figure 7–8A).

3.2. Influence of the intrusion of water from the Rhone River in the BoM. The intrusion event of water from the Rhone River into the BoM that was studied by Pairaud et al. [6] occurred in the second part of June 2008 and was detected in the model results at the Somlit station as a decrease in salinity (Figure 10A). The study of the summer decreases in phytoplankton nutrient limitation revealed that they were linked to low salinity events. Rich Rhone River water induced an increase in nutrients and thus a decrease in phytoplankton nutrient limitation (Figure 10A). Both versions of the model were in agreement with the observed chlorophyll-a concentrations (Figure 10B), highlighting the strong biological response to the Rhone River water intrusion events. The intensity of the increases in chlorophyll-a were lower in the version of the model with the P cycle during most of the low salinity events. Therefore, according to the results of the model, phytoplankton was limited by P during most of the low salinity events. This finding was in agreement with spatial comparisons of chlorophyll-a concentrations in the Rhone River plume (Figure 6), which were less extended and better represented in the model version with the P cycle. Therefore, the P cycle slightly improved the results, such as during the low salinity event in mid-July where a net improvement was noticed.

Discussion

Due to a worldwide movement toward ecosystem-based management [79], the demand for quantitative tools to support ecosystem-based management initiatives increases [79,80]. There is now a proliferation of end-to-end ecosystem models which attempt to represent the entire ecological system and the associated abiotic environment [81]. This kind of models has been developed in various areas [82,83] and some are listed in the MEECE European project (http://www.meece.eu). However this kind of models has lots of parameters, needs a lot of input data, human and material resources to be developed, which make them difficult to transpose in numerous areas.

In this study, we preferred a different modeling approach and we developed a model adapted to targeted questions: study of the impact of the physical processes on low trophic levels and study of the inputs of chemical contaminants in the trophic chain. So, we chose to develop a physical biogeochemical coupled model with a high spatial and temporal resolution and with an on-line coupling technique in order to describe finely the link between physics and biogeochemistry. This coupled model focused on the low trophic level in order to study the key processes which permit the input of contaminant in the trophic chain. The biogeochemical model with a low number of parameters relatively to end-to-end models is easier to adapt to other oligotrophic coastal area influenced by large river inputs like the Gulf of Gabes in Tunisia and the Nile delta in eastern Mediterranean Sea. Indeed, the biogeochemical model ECO3M-MASSILIA-P was modified from an initial model [40,41,84].

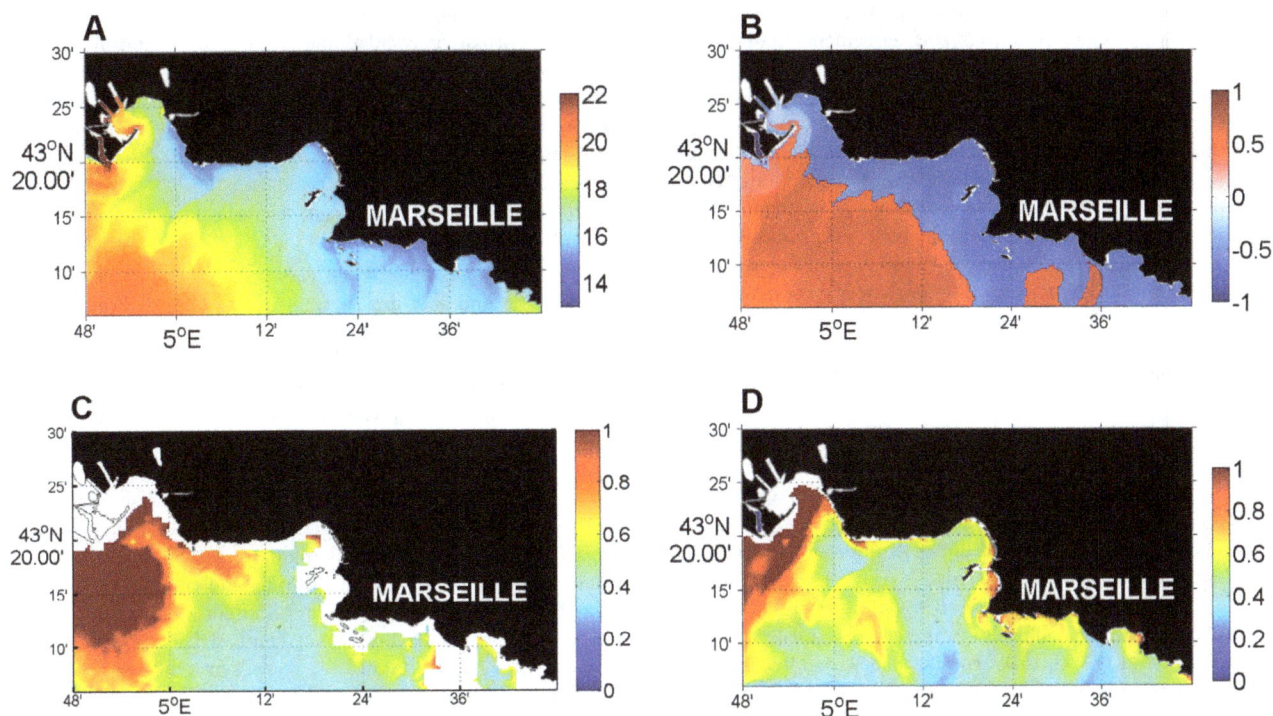

Figure 9. Upwelling event of 15/07/2008. Simulated temperature (A), simulated phytoplankton limitation by P (red) and N (blue) (B), chlorophyll-a concentrations from MODIS and simulated chlorophyll-a concentration (average across the first 10 meters of depth) compared for 15/07/2008.

During the biogeochemical model development, the following question of the complexity level of the biogeochemical model was asked: What was the level of detail necessary and sufficient to describe this coastal ecosystem? The Mediterranean Sea is generally known to be P limited; thus, we decided to add the P cycle to the model. In choosing the most reliable model, particular attention was paid to the accuracy of the model and the assessment of the shortcomings of the two new versions of the biogeochemical model.

1. Model accuracy assessment

The models simulated 5 years, from 2007 to 2011. The model development and parameterizations were performed for 2007 and 2008, whereas model accuracy assessment of both final model versions was performed for 2009 to 2011, which allowed a robust evaluation of the model results.

The results showed that both versions of the model had significant ability to simulate the biogeochemical functioning of the Marseille coastal area. Indeed, the models reproduced with relatively good accuracy the seasonal and the inter-annual dynamics of chlorophyll-a and nutrient concentrations. Nevertheless, the chlorophyll-a concentration was slightly overestimated by the model during the spring bloom and in summer. The models reproduced the spatial pattern of this area well, with a good representation of the west-east gradient in chlorophyll-a concentration associated with the Rhone River plume extensions. The general ability assessment of model results also showed that there was strong variability in biogeochemical concentrations at short time scales (from a few hours to several days), mostly in spring and summer. The model was able to reproduce different types of shelf events, such as upwelling or the intrusion of Rhone River water into the BoM, and their associated biogeochemical response. The use of target diagrams allowed a visual comparison of statistical indicators. The results appeared more accurate at the surface than at the bottom, and the discrepancies between the two versions of the biogeochemical model were rare.

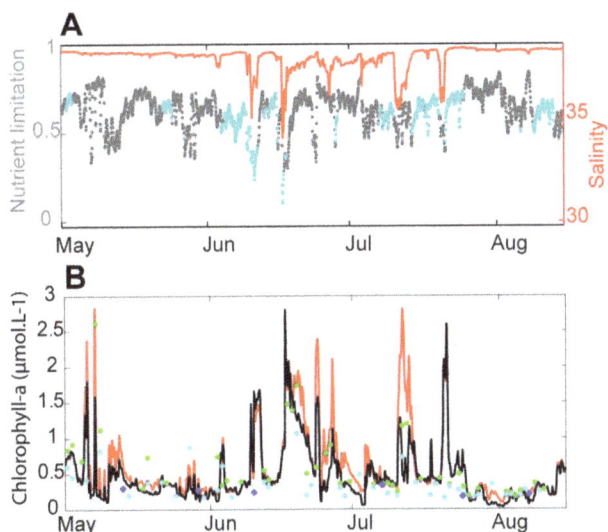

Figure 10. Impact of the intrusion of Rhone River water on phytoplankton nutrient limitation and chlorophyll-a concentrations. Evolution from May to mid-August 2008 at the Somlit station of the simulated phytoplankton limitation by P (grey point) and N (blue point) and the simulated salinity (red line) (A), the surface chlorophyll-a concentrations from remote sensing observations of MODIS (green point) and MERIS (blue point), the in-situ data (bleu point) and the simulation with the P cycle (black line) and without the P cycle (blue line) (B).

Nevertheless, the implementation of the P cycle in the biogeochemical model improved the agreement between the model predictions and the surface chlorophyll-a observations for a given event. In addition, the model with the P cycle performed well at reproducing the plume and processes associated with the Rhone River. However, in the bottom layer, its use increased the error in predicting the chlorophyll-a concentration due to the overestimation of the PO_4 concentration. We attributed the improved fit in the surface chlorophyll-a concentrations to the ability of the model to reproduce the P limitation in the spring and summer periods. In winter, phytoplankton production was principally limited by light because the strong vertical mixing brought nutrients in excess to the surface layer. During this period, the P cycle did not impact the surface chlorophyll-a concentration.

2. Model complexity

The general assessment of the results of both versions of the model showed that the accuracies of the models with and without the P cycle were similar. The addition of the P cycle to the biogeochemical model improved the description of the ecosystem functioning, but it also injected errors.

The errors associated with the P cycle came from different sources. The primary difficulty was that the measured PO_4 concentrations in the Marseille coastal area were often lower or very close to the detection limit. At the Somlit station in the surface layer, the concentration of PO_4 ranged from 0 to 0.3 $\mu mol.L^{-1}$, with a mean value of 0.03 $\mu mol.L^{-1}$ for the years 2009 to 2011, which indicated that concentrations of PO_4 were generally lower than or close to their detection limits, as the precision of the PO_4 concentrations was ± 0.02–0.03 $\mu mol.L^{-1}$. Therefore, comparisons between the model and the observations were not possible. Then, as previously stated by several authors [85–87], there is a critical need for new experimental data examining N and P dynamics under different and extreme input ratios and growth rates [48]. Thus, the need for improved biogeochemical understanding of N/P co-limitation increased the difficulty of modeling the P cycle. Furthermore, Poggiale et al. [88] demonstrated that models were sensitive to uptake formulations; even if two formulas provided similar values, large numerical differences in the stability criteria may occur. Thus, errors were certainly introduced by the choice of the parameter values and the mathematical functions representing the P cycle.

Even if it appears plausible from a biological a priori standpoint that more complex models should mirror reality better, other recent studies have suggested that it is not always the case. A meta-analysis of mechanistic aquatic biogeochemical models also found that increased model complexity did not improve the fit [37]. Kriest et al. [89] suggested that increasing complexity of unturned, unoptimized models that were simulated with parameters commonly used in large-scale model studies did not necessarily improve performance. Los et al. [90] also indicated that adding more complexity did not necessarily improve the quality of the model results in terms of their ability to reproduce measurements and hence their applicability as prognostic tools. Instead, they argued that there should be a balance between ecological and physical resolution in relation to the specific question to be addressed. Crout et al. [91] evaluated models using model reduction for three models. In all cases, they identified reduced models that outperformed the more complicated ones, "suggesting some over-parameterization has occurred during model development" [91].

In the study of the Marseille coastal area, we followed an inverse approach, with the simplest model (model without P cycle) at the beginning and consecutively increasing model complexity with the

P cycle. The addition of the P cycle increased the computation time by approximately 30% and did not automatically improve the fit with the observations. Therefore, the model version (with or without the P cycle) should be chosen depending on the question to be addressed and by considering the general accuracy assessment achieved in this work.

3. The Marseille Coastal area: a challenging zone for a modeling approach

The region of interest is very complex (biologically and dynamically) and was thus a challenge to model. Principally, 4 difficulties were encountered: (i) there was a strong west-east gradient in oligotrophy, namely the eastern portion of the studied area was mainly oligotrophic, while the western portion was more productive; (ii) few observations were available for comparisons with the model results; (iii) the studied region was a coastal area highly influenced by large river inputs (the Rhone River and the Marseille urban rivers); and (iv) the hydrodynamic shelf processes were extremely variable at short time scales, and they strongly impacted most of the coastal domain.

The BoM was generally oligotrophic and characterized by low concentrations of chlorophyll-a and nutrients, with mean concentrations of chlorophyll-a, PO_4 and DIN of 0.4 $\mu g.L^{-1}$, 0.03 $\mu mol.L^{-1}$ and 1 $\mu mol.L^{-1}$, respectively. In comparison, in the surface layer of the Baltic Sea, the mean concentrations of chlorophyll-a, PO_4 and DIN were approximately 3.5 $\mu g.L^{-1}$, 0.33 $\mu mol.L^{-1}$ and 3.6 $\mu mol.L^{-1}$ [92]. As in the Baltic Sea, a modeling study in the Channel and Southern Bight of the North Sea showed mean concentrations of chlorophyll-a of 5.35 $\mu g.L^{-1}$, mean PO_4 of 0.82 $\mu mol.L^{-1}$ and mean DIN of 17.57 $\mu mol.L^{-1}$ [77]. Therefore, in the BoM, the concentrations were ten times less than the concentration range in the Channel and Southern Bight of the North Sea and in the Baltic Sea. The BoM contrasted with the Rhone River plume, which was characterized by high concentrations of nutrients and chlorophyll-a. Thus, the large range of trophic conditions in this small model domain put the model to the test.

The scarcity of observations to compare with the model results was a problem during the model development phase. The Somlit station was sampled only twice monthly, and as shown by the model results, many processes occurred at shorter time scales. One risk to avoid was to tune the model to obtain good results in comparison with the in-situ data at the Somlit station. We also evaluated the spatial pattern of the processes reproduced by the model and compared them with remote sensing data, which provided spatial patterns and gave daily information (during non-cloudy conditions). Nevertheless, the errors associated with the remote sensing data were high.

Coastal areas are highly influenced by cross-shore and earth/sea exchange. The Marseille coastal area has two large open boundary conditions (OBC). Thus, a good representation of the OBC was essential to accurately describe the cross-shore exchanges. Coastal modeling also requires long-term monitoring of inputs to represent the discharge and concentrations associated with each river well. Although the Rhone River was well-studied and the daily discharge and concentrations of the primary state variables were available, it was necessary to construct from the literature or from the few available observations the concentrations for the missing variables for the Marseille urban rivers and the Rhone River. Progress could be made in the description of the concentrations of the rivers by monitoring each river to better describe the organic matter and nutrient inputs from the rivers.

In addition, the low impact of the addition of the P cycle to the model suggested that physics drives the majority of the coupled model results. Hydrodynamic shelf processes appeared extremely variable at short time scales, and they strongly impacted most of the coastal domain. In coastal areas, hydrodynamics were of primary importance for coupled modeling. An overestimation of the plume extension of the Rhone River by the hydrodynamic model induced a strong error in the chlorophyll-a concentration at the Somlit station, which demonstrated that the choice of the physical model is very important for biogeochemistry, as stated in many studies [93,94].

4. Contribution of the model results to the understanding of the Marseille coastal area

As previously discussed, although this coastal ecosystem was complex and thus a challenge to model, the coupled model performed well at reproducing the primary characteristics and processes. Therefore, the coupled model provided interesting information on the functioning of this coastal ecosystem.

First, the temporal evolution of surface chlorophyll-a clearly distinguished the four main seasons. Spring was characterized by high concentrations of chlorophyll-a (spring bloom) and was the most productive season. In contrast, the fall appeared to be the least productive period. This finding is not in accordance with general knowledge of the Northwestern Mediterranean Sea as a small autumnal bloom is supposed to occur during this period. Our coupled model reproduced the autumnal bloom only for 2009.

From a spatial point of view, the Marseille coastal area was characterized by a strong west-east oligotrophy gradient. Outside the ROFI, the coastal domain near Marseille generally remained oligotrophic, whereas at the mouth of the Rhone River, the phytoplankton biomass was always high [95]. Nevertheless, the Rhone River was limited by P because the NO_3:PO_4 ratio of the Rhone River inputs was approximately 65–80 [26,27]. This P limitation was also detectable in the model results because the chlorophyll-a concentrations in the Rhone River plume were higher in the model without the P cycle than in the model with the P cycle, better mimicking the observations.

The urban rivers and the Marseille WWTP delivered high concentrations of nutrients and organic matter, but the eutrophication risk remained extremely low in the BoM because the Urban River inputs and sewage from the WWTP were rapidly diluted by strong hydrodynamic events [6] due to the short residence time. Nevertheless, the Gulf of Fos was characterized by shallow depths and higher chlorophyll-a concentrations (up to 10 $\mu mol.L^{-1}$ [34]), which indicated that the eutrophication risk was higher. Previous studies [96] demonstrated that the greatest risk of algal blooms occurs during periods of calm weather because the Gulf of Fos is flushed intermittently but strongly by wind-driven lateral circulation [34].

The coupled model permitted the evaluation of the eutrophication risk in the BoM, but the model was also very useful in contextualizing the in-situ data. Indeed, the Somlit station is impacted by many shelf events (e.g., upwelling and the intrusion of Rhone River water). For example, an increase in nutrient and chlorophyll-a concentrations associated with a decrease in salinity could be caused by the intrusion of Rhone River water, heavy rainfall events or an extension of the urban river plumes.

The coupled model presented a significant advantage over satellite images and in-situ measurements by allowing the study of the processes at short time scales and in three dimensions. It was proven to be a pertinent tool in the study of the biological response associated with physical processes. The study of upwelling events in summer 2008 showed that a large part of the Marseille coastal area was impacted by upwelling events. However, the chlorophyll-

a response to the nutrient enrichment caused by upwelling was mostly noticeable at the two main upwelling points ("Cote Bleue" and "Calanques"). Another interesting event was the intrusion of Rhone River water into the BoM. The influence of the Rhone River water in the BoM has been observed hydrodynamically [6,21], but the only impacts on the ecosystem that had previously been investigated were in the CDOM concentrations [22]. The coupled model revealed that the intrusion of Rhone River water into the BoM induced an obvious increase in chlorophyll-a concentrations.

Finally, further studies are necessary to better explore and characterize the biogeochemical functioning of this coastal ecosystem. The coupled model allowed the quantification of the impact of atmospheric inputs (organic matter and nutrients) on the biogeochemical functioning of this coastal ecosystem. Preliminary results (not shown) suggested that atmospheric deposits might be negligible, except during episodes of heavy rainfall. The initial results concerning the intrusion of Rhone River water also raised new questions: What caused the eastward transport of Rhone River water? What quantities of nutrients are transported into the BoM during Rhone River water intrusion events? Thus, the coupled model could be a useful tool in quantifying the relative impact of the Rhone River and the Urban Rivers on the biogeochemical functioning of the BoM using a mass balance approach.

Conclusion

In this paper, we presented the development and the evaluation of a 3D coupled physical-biogeochemical model for the Marseille coastal area. The high-resolution coupled model demonstrated its ability to recreate realistic situations. It underlined that the biogeochemical functioning of the BoM was very complex because the physical processes and river inputs were important drivers of the biogeochemical model results. Indeed, the ecosystem of the BoM quickly switched between oligotrophic conditions and important enrichment events due to external forcing.

The P cycle was added to the biogeochemical model because this element was well known to limit biological production in the Mediterranean Sea. The addition of P improved the description of ecosystem functioning and in some cases improved the model results, but it also introduced errors into the model. In using the model results, awareness of its shortcomings and positive features is important. Therefore, the general accuracy assessment in this paper and the questions addressed should be considered in choosing the version of the biogeochemical model (with and without the P cycle).

The coupled model appeared to be a useful tool for the analysis of processes and the estimation of budgets in a very dynamic environment where it is difficult to extrapolate from discrete measurements. Thus, despite its imperfections, large quantities of information are available in the results of the coupled model. The model data could help researchers to design field campaigns by better anticipating the biogeochemical front and gradient. In addition, the model results allow better evaluation of the impact of urban inputs on the coastal area and could thus help policy managers propose solutions.

In conclusion, this study suggested that improvements in the description of hydrodynamics and terrestrial inputs should be preferred over increasing the complexity of the biogeochemical model in this coastal oligotrophic area.

Supporting Information

Annex S1   Equations of the biogeochemical model of the Marseille coastal area (ECO3M-MASSILIA-P).

Annex S2   Parameters.

Annex S3   Spin up of the MARS3D-RHOMA/ECO3M-MASSILIA coupled model for the year 2008 at the Somlit station at the surface. "Model 1" corresponding to the summer initial conditions and "Model 2" corresponding to the winter initial conditions.

Annex S4   Maps of the Marseille city inputs.

Annex S5   Quality of the remote sensing data.

Acknowledgments

The authors acknowledge the staff of the Somlit national network for the littoral observations, (INSU-CNRS) for providing the data time series, IFREMER for providing the ocean color remote sensing data computed with the OC5 algorithm and the Compagnie Nationale du Rhone for the data on Rhone River discharges and Meteo France. The Rhone concentration data were provided by the national MOOSE program (Mediterranean Oceanic Observing System on Environment) and the Service d'Observation of the Mediterranean Institute of Oceanography (MIO). The authors gratefully acknowledge N. Garcia, V. Lagadec and M. Fornier for analytical and field assistance. This study was part of the "MERMEX WP3-C3A" and international "IMBER" projects. We are grateful to DEA-MPM (Direction des Eaux et Assainissement- Marseille Provence Metropole) and SERAM (Société d'Exploitation du Réseau d'Assainissement de Marseille) for providing the Marseille urban river and WWTP discharge. Many thanks also to Benedicte Thouvenin, Pierre Garreau, Francis Gohin, Romaric Verney, Cassandre Jany and Cedric Garnier for fruitful discussion. We thank American Journal Experts for English corrections and the two anonymous reviewers for their helpful comments to improve this paper.

Author Contributions

Analyzed the data: MF CP IP PR RF VF PL. Wrote the paper: MF CP IP RF VF PR PL. Conceived and designed modelling configurations: MF CP IP RF VF PL PR.

References

1. Le Tissier MD., Buddemeier R, Parslow J, Swaney DP, Crossland CJ, et al. (2006) The role of the coastal ocean in the disturbed and undisturbed nutrient and carbon cycles A management perspective. LOICZ, Geesthacht, Germany.
2. Goberville E, Beaugrand G, Sautour B, Tréguer P, Somlit T (2010) Climate-driven changes in coastal marine systems of western Europe. Marine Ecology Progress Series 408: 129–147.
3. Behrenfeld MJ, O'Malley RT, Siegel DA, McClain CR, Sarmiento JL, et al. (2006) Climate-driven trends in contemporary ocean productivity. Nature 444: 752–755.
4. INSEE (2009) Résultats du recensement de la population. Available: http://www.recensement.insee.fr/chiffresCles.action?zoneSearchField=MARSEILLE&codeZone=241300391-GFP&idTheme=3. Accessed 17 April 2013.
5. Durrieu de Madron X, Guieu C, Sempéré R, Conan P, Cossa D, et al. (2011) Marine ecosystems' responses to climatic and anthropogenic forcings in the Mediterranean. Progress in Oceanography 91: 97–166.
6. Pairaud IL, Gatti J, Bensoussan N, Verney R, Garreau P (2011) Hydrology and circulation in a coastal area off Marseille: Validation of a nested 3D model with observations. Journal of Marine Systems 88: 20–33.
7. Coste B (1974) Role des apports nutritifs mineraux rhodaniens sur la production organique des eaux du Golfe du Lion. Tethys 6: 727–740.
8. Morel A, Bricaud A, André JM, Pelaez-Hudlet J (1990) Spatial- temporal evolution of the Rhone plume as seen by CZCS imagery. Consequences upon the primary production in the Gulf of Lions. Water Pollution Research Reports In: Martin: 45–62.
9. Petrenko A, Dufau C, Estournel C (2008) Barotropic eastward currents in the western Gulf of Lion, north-western Mediterranean Sea, during stratified conditions. Journal of Marine Systems 74: 406–428.
10. Allou A, Forget P, Devenon J-L (2010) Submesoscale vortex structures at the entrance of the Gulf of Lions in the Northwestern Mediterranean Sea. Continental Shelf Research 30: 724–732.
11. Millot C (1990) The Gulf of Lions' hydrodynamics. Continental Shelf Research 10: 885–894.
12. Millot C, Wald L (1980) The effect of mistral wind on the ligurian current near Provenc. Oceanologica Acta 3: 399–402.
13. Petrenko A (2003) Variability of circulation features in the Gulf of Lion NW Mediterranean Sea. Importance of inertial currents. Oceanologica Acta 26: 323–338.
14. Gatti J (2008) Intrusions du courant nord méditerranéen sur la partie est du plateau continental du golfe du Lion University Aix-Marseille II.
15. El Sayed MA, Aminot A, Kerouel R (1994) Nutrients and trace metals in the northwestern Mediterranean under coastal upwelling conditions. Continental Shelf Research 14: 507–530.
16. Minas HL (1968) A propos d'une remontée d'eaux "profondes" dans les parages du golfe de Marseille (oct. 1964), conséquences biologiques. Cahiers Océano-graphiques 20: 647–674.
17. Broche P, Devenon J-L, Forget P, de Maistre J-C, Naudin J-J, et al. (1998) Experimental study of the Rhone plume. Part I: physics and dynamics. Oceanologica Acta 21: 725–738.
18. Estournel C, Broche P, Marsaleix P, Devenon JL, Auclair F, et al. (2001) The Rhone River plume in unsteady conditions: numerical and experimental results. Estuarine, Coastal and Shelf Science 53: 25–38.
19. Arnoux-Chiavassa S, Rey V, Fraunié P (2003) Modeling 3D Rhône river plume using a higher order advection scheme. Oceanologica Acta 26: 299–309.
20. Reffray G, Fraunie P, Marsaleix P (2004) Secondary flows induced by wind forcing in the Rhone region of freshwater influence. Ocean Dynamics 54: 179–196.
21. Gatti J, Petrenko A, Devenon J-L, Leredde Y, Ulses C (2006) The Rhone river dilution zone present in the northeastern shelf of the Gulf of Lion in December 2003. Continental Shelf Research 26: 1794–1805.
22. Para J, Coble PG, Charrière B, Tedetti M, Fontana C, et al. (2010) Fluorescence and absorption properties of chromophoric dissolved organic matter (CDOM) in coastal surface waters of the northwestern Mediterranean Sea, influence of the Rhône River. Biogeosciences 7: 4083–4103.
23. Elser JJ, Bracken MES, Cleland EE, Gruner DS, Harpole WS, et al. (2007) Global analysis of nitrogen and phosphorus limitation of primary producers in freshwater, marine and terrestrial ecosystems. Ecology letters 10: 1135–1142.
24. Moutin T, Raimbault P, Golterman HL, Coste B (1998) The input of nutrients by the Rhône river into the Mediterranean Sea: recent observations and comparison with earlier data. Hydrobiologia 373–374: 237–246.
25. Van Wambeke F, Christaki U, Giannakourou A, Moutin T, Souvemerzoglou K (2002) Longitudinal and vertical trends of bacterial limitation by phosphorus and carbon in the Mediterranean Sea. Microbial ecology 43: 119–133.
26. Ludwig W, Dumont E, Meybeck M, Heussner S (2009) River discharges of water and nutrients to the Mediterranean and Black Sea: Major drivers for ecosystem changes during past and future decades? Progress in Oceanography 80: 199–217.
27. Ludwig W, Bouwman AF, Dumont E, Lespinas F (2010) Water and nutrient fluxes from major Mediterranean and Black Sea rivers: Past and future trends and their implications for the basin-scale budgets. Global Biogeochemical Cycles 24: GB0A13.
28. Markaki Z, Loÿe-Pilot MD, Violaki K, Benyahya L, Mihalopoulos N (2010) Variability of atmospheric deposition of dissolved nitrogen and phosphorus in the Mediterranean and possible link to the anomalous seawater N/P ratio. Marine Chemistry 120: 187–194.
29. Thingstad TF, Zweifel UL, Rassoulzadegan F (1998) P limitation of heterotrophic bacteria and phytoplankton. Limnol Oceanogr 43: 88–94.
30. Pinazo C, Marsaleix P, Millet B, Estournel C, Véhil R (1996) Spatial and temporal variability of phytoplankton biomass in upwelling areas of the northwestern mediterranean: a coupled physical and biogeochemical modelling approach. Journal of Marine Systems 7: 161–191.
31. Tusseau-Vuillemin M-H, Mortier L, Herbaut C (1998) Modeling nitrate fluxes in an open coastal environment (Gulf of Lions): Transport versus biogeochemical processes. Journal of Geophysical Research 103: 7693.
32. Auger PA, Diaz F, Ulses C, Estournel C, Neveux J, et al. (2011) Functioning of the planktonic ecosystem on the Gulf of Lions shelf (NW Mediterranean) during spring and its impact on the carbon deposition: a field data and 3-D modelling combined approach. Biogeosciences 8: 3231–3261.

33. Campbell R, Diaz F, Hu Z, Doglioli A, Petrenko A, et al. (2013) Nutrients and plankton spatial distributions induced by a coastal eddy in the Gulf of Lion. Insights from a numerical model. Progress in Oceanography 109: 47–69.

34. Fontana C, Grenz C, Pinazo C, Marsaleix P, Diaz F (2009) Assimilation of SeaWiFS chlorophyll data into a 3D-coupled physical–biogeochemical model applied to a freshwater-influenced coastal zone. Continental Shelf Research 29: 1397–1409.

35. Franks PJS (2009) Planktonic ecosystem models: perplexing parameterizations and a failure to fail. Journal of Plankton Research 31: 1299–1306.

36. Ourmières Y, Brasseur P, Lévy M, Brankart J-M, Verron J (2009) On the key role of nutrient data to constrain a coupled physical–biogeochemical assimilative model of the North Atlantic Ocean. Journal of Marine Systems 75: 100–115.

37. Arhonditsis G, Brett M (2004) Evaluation of the current state of mechanistic aquatic biogeochemical modeling. Marine Ecology Progress Series 271: 13–26.

38. Baklouti M, Diaz F, Pinazo C, Faure V, Quéguiner B (2006) Investigation of mechanistic formulations depicting phytoplankton dynamics for models of marine pelagic ecosystems and description of a new model. Progress in Oceanography 71: 1–33.

39. Baklouti M, Faure V, Pawlowski L, Sciandra A (2006) Investigation and sensitivity analysis of a mechanistic phytoplankton model implemented in a new modular numerical tool (Eco3M) dedicated to biogeochemical modelling. Progress in Oceanography 71: 34–58.

40. Faure V, Pinazo C, Torreton J-P, Jacquet S (2010) Modelling the spatial and temporal variability of the SW lagoon of New Caledonia I: a new biogeochemical model based on microbial loop recycling. Marine pollution bulletin 61: 465–479.

41. Faure V, Pinazo C, Torreton J-P, Douillet P (2010) Modelling the spatial and temporal variability of the SW lagoon of New Caledonia II: realistic 3D simulations compared with in situ data. Marine pollution bulletin 61: 480–502.

42. Harmelin-Vivien M, Loizeau V, Mellon C, Beker B, Arlhac D, et al. (2008) Comparison of C and N stable isotope ratios between surface particulate organic matter and microphytoplankton in the Gulf of Lions (NW Mediterranean). Continental Shelf Research 28: 1911–1919.

43. Beker B, Romano J-C, Arlhac D (1999) Le suivi de la variabilité des peuplements phytoplanctoniques en milieu marin littoral: le golfe de Marseille en 1996–1997. Oceanis 25: 395–415.

44. Gutiérrez-Rodríguez A, Latasa M, Scharek R, Massana R, Vila G, et al. (2011) Growth and grazing rate dynamics of major phytoplankton groups in an oligotrophic coastal site. Estuarine, Coastal and Shelf Science 95: 77–87.

45. Halsband-Lenk C, Hirche H-J, Carlotti F (2002) Temperature impact on reproduction and development of congener copepod populations. Journal of Experimental Marine Biology and Ecology 271: 121–153.

46. Rose JM, Caron DA (2007) Evidence Does low temperature constrain the growth rates of heterotrophic protists? and implications for algal blooms in cold waters. Limnology and Oceanography 52: 886–895.

47. Saito MA, Goepfert TJ, Ritt JT (2008) Some thoughts on the concept of colimitation: Three definitions and the importance of bioavailability. Limnology and Oceanography 53: 276–290.

48. Bougaran G, Bernard O, Sciandra A (2010) Modeling continuous cultures of microalgae colimited by nitrogen and phosphorus. Journal of theoretical biology 265: 443–454.

49. Rhee GY (1978) Effects of N:P atomic ratios and nitrate limita- tion on algal growth, cell compositions, and nitrate uptake. Limnology and Oceanography 23: 10–25.

50. Geider RJ, MacIntyre HL, Kana TM (1998) A Dynamic Regulatory Model of Phytoplanktonic Acclimation to Light, Nutrients, and Temperature. Limnology and Oceanography 43: pp. 679–694.

51. Smith CL, Tett P (2000) A depth-resolving numerical model of physically forced microbiology at the European shelf edge. Journal of Marine Systems 26: 1–36.

52. Faure V, Pinazo C, Torréton J-P, Douillet P (2006) Relevance of various formulations of phytoplankton chlorophyll a:carbon ratio in a 3D marine ecosystem model. Comptes rendus biologies 329: 813–822.

53. Droop MR (1968) Vitamin B12 and marine ecology. IV. The kinetics of uptake growth and inhibition in Monochrysis lutheri. Journal of Marine Biology Association 48: 689–733.

54. Nicolle A, Garreau P, Liorzou B (2009) Modelling for anchovy recruitment studies in the Gulf of Lions (Western Mediterranean Sea). Ocean Dynamics 59: 953–968.

55. Pinazo C, Fraysse M, Fuchs R, Thouvenin B, Pairaud I, et al. (2012) Modelling the Gulf of Lions coastal area with a 3D coupled physical and biogeochemical approach. 50th ECSA Conference. Venice, Italy.

56. Lazzari P, Solidoro C, Ibello V, Salon S, Teruzzi A, et al. (2012) Seasonal and inter-annual variability of plankton chlorophyll and primary production in the Mediterranean Sea: a modelling approach. Biogeosciences 9: 217–233.

57. Ibanez C, Pont D, Prat N (1997) Characterization of the Ebre and Rhone Estuaries: A Basis for Defining and Classifying Salt-Wedge Estuaries. Limnology and Oceanography 42: pp. 89–101.

58. Pont D, Simonnet J-P, Walter A V (2002) Medium-term Changes in Suspended Sediment Delivery to the Ocean: Consequences of Catchment Heterogeneity and River Management (Rhône River, France). Estuarine, Coastal and Shelf Science 54: 1–18.

59. Sempere R, Charriere B, Van Wambeke F, Cauwet G (2000) Carbon inputs of the Rhne River to the Mediterranean Sea: Biogeochemical implications. Global Biogeochemical Cycles 14: 669–681.

60. Déliat G (2001) Matière organique dissoute des zones côtières: sources, distribution et biodégradabilité Université Paris VI.

61. Lønborg C, Davidson K, Álvarez–Salgado XA, Miller AEJ (2009) Bioavailability and bacterial degradation rates of dissolved organic matter in a temperate coastal area during an annual cycle. Marine Chemistry 113: 219–226.

62. Jany C, Thouvenin B (2012) Récapitulatif des hypothèses utilisées pour les rejets de MES dans les simulations pour METROC, Rapport Ifremer, RST.ODE/LER/PAC/12-19.

63. Gouze E (2008) Bilan de matière de l'étang de Berre : influence des apports des tributaires et des processus de régénération dans le maintien de l'eutrophisation Université d'Aix marselle II.

64. Aminot A KR (2004) Hydrologie des écosystèmes marins. Paramètres et analyses. Quae.

65. Holmes MR, Aminot A, Kerouel R, Hooker BA, Peterson JB (1999) A simple and precise method for measuring ammonium in marine and freshwater ecosystems. Can J Fish Aquat Sci 56: 1801–1808.

66. Raimbault P, Garcia N, Cerutti F (2008) Distribution of inorganic and organic nutrients in the South Pacific Ocean – evidence for long-term accumulation of organic matter in nitrogen-depleted waters. Biogeosciences 5: 281–298.

67. Raimbault P, Neveux J LF (2004) Dosage rapide de la chlorophylle a et des phaeopigments a par fluorimétrie après extraction au méthanol. Comparaison avec la méthode classique d'extraction à l'acétone. Oceanis 30: 189–205.

68. Gohin F, Druon JN, Lampert L (2002) A five channel chlorophyll concentration algorithm applied to SeaWiFS data processed by SeaDAS in coastal waters. International Journal of Remote Sensing 23: 1639–1661.

69. Gohin F, Loyer S, Lunven M, Labry C, Froidefond J-M, et al. (2005) Satellite-derived parameters for biological modelling in coastal waters: Illustration over the eastern continental shelf of the Bay of Biscay. Remote Sensing of Environment 95: 29–46.

70. Gohin F (2011) Annual cycles of chlorophyll-a, non-algal suspended particulate matter, and turbidity observed from space and in-situ in coastal waters. Ocean Science 7: 705–732.

71. Shutler JD, Smyth TJ, Saux-Picart S, Wakelin SL, Hyder P, et al. (2011) Evaluating the ability of a hydrodynamic ecosystem model to capture inter- and intra-annual spatial characteristics of chlorophyll-a in the north east Atlantic. Journal of Marine Systems 88: 169–182.

72. Stow C, Jolliff J, McGillicuddy DJ, Doney SC, Allen JI, et al. (2009) Skill assessment for coupled biological/physical models of marine systems. Journal of Marine Systems 76: 4–15.

73. Allen JI, Holt JT, Blackford J, Proctor R (2007) Error quantification of a high-resolution coupled hydrodynamic-ecosystem coastal-ocean model: Part 2. Chlorophyll-a, nutrients and SPM. Journal of Marine Systems 68: 381–404.

74. Allen JI, Somerfield PJ, Gilbert FJ (2007) Quantifying uncertainty in high-resolution coupled hydrodynamic-ecosystem models. Journal of Marine Systems 64: 3–14.

75. Radach G, Moll A (2006) Review of three-dimensional ecological modelling related to the North Sea shelf system—Part 2: model validation and data needs. Oceanography and Marine Biology; An Annual Review 44: 1–60.

76. OSPAR Commission (1998) Report of the ASMO Modelling Workshop on Eutrophication Issues (5–8 November 1996, The Hague, The Netherlands).

77. Lacroix G, Ruddick K, Park Y, Gypens N, Lancelot C (2007) Validation of the 3D biogeochemical model MIRO&CO with field nutrient and phytoplankton data and MERIS-derived surface chlorophyll a images. Journal of Marine Systems 64: 66–88.

78. Jolliff JK, Kindle JC, Shulman I, Penta B, Friedrichs MAM, et al. (2009) Summary diagrams for coupled hydrodynamic-ecosystem model skill assessment. Journal of Marine Systems 76: 64–82.

79. Rose Ka, Roth BM, Smith EP (2009) Skill assessment of spatial maps for oceanographic modeling. Journal of Marine Systems 76: 34–48.

80. Pikitch EK, Santora C, Babcock EA, Bakun A, Bonfil R, et al. (2004) Ecosystem-Based Fishery Management. science 305: 346–347.

81. Fulton Ea (2010) Approaches to end-to-end ecosystem models. Journal of Marine Systems 81: 171–183.

82. Travers M, Shin Y-J, Jennings S, Machu E, Huggett Ja, et al. (2009) Two-way coupling versus one-way forcing of plankton and fish models to predict ecosystem changes in the Benguela. Ecological Modelling 220: 3089–3099.

83. Libralato S, Solidoro C (2009) Bridging biogeochemical and food web models for an End-to-End representation of marine ecosystem dynamics: The Venice lagoon case study. Ecological Modelling 220: 2960–2971.

84. Fuchs R, Dupouy C, Douillet P, Caillaud M, Mangin A, et al. (2012) Modelling the impact of a La Niña event on a South West Pacific Lagoon. Marine Pollution Bulletin 64: 1596–1613.

85. Flynn KJ (2008) The importance of the form of the quota curve and control of non-limiting nutrient transport in phytoplankton models. Journal of Plankton Research 30: 423–438.

86. Klausmeier Ca, Litchman E, Levin Sa (2007) A model of flexible uptake of two essential resources. Journal of theoretical biology 246: 278–289.

87. Leonardos N, Geider RJ (2005) Elemental and Biochemical Composition of Rhinomonas Reticulata (Cryptophyta) in Relation To Light and Nitrate-To-Phosphate Supply Ratios1. Journal of Phycology 41: 567–576.

88. Poggiale J-C, Baklouti M, Queguiner B, Kooijman SALM (2010) How far details are important in ecosystem modelling: the case of multi-limiting nutrients in phytoplankton-zooplankton interactions. Philosophical transactions of the Royal Society of London Series B, Biological sciences 365: 3495–3507.

89. Kriest I, Khatiwala S, Oschlies A (2010) Towards an assessment of simple global marine biogeochemical models of different complexity. Progress in Oceanography 86: 337–360.

90. Los FJ, Villars MT, Van der Tol MWM (2008) A 3-dimensional primary production model (BLOOM/GEM) and its applications to the (southern) North Sea (coupled physical–chemical–ecological model). Journal of Marine Systems 74: 259–294.

91. Crout NMJ, Tarsitano D, Wood a T (2009) Is my model too complex? Evaluating model formulation using model reduction. Environmental Modelling & Software 24: 1–7.

92. Wan Z, She J, Maar M, Jonasson L, Baasch-Larsen J (2012) Assessment of a physical-biogeochemical coupled model system for operational service in the Baltic Sea. Ocean Science 8: 683–701.

93. Skogen MD, Moll A (2005) Importance of ocean circulation in ecological modeling: An example from the North Sea. Journal of Marine Systems 57: 289–300.

94. Edwards KP, Barciela R, Butenschön M (2012) Validation of the NEMO-ERSEM operational ecosystem model for the North West European Continental Shelf. Ocean Science 8: 983–1000.

95. Younes W, Bensoussan N, Romano J-C, Arlhac D, Lafont M-G (2003) Seasonal and interannual variations (1996–2000) of the coastal waters east of the Rhone river mouth as indicated by the SORCOM series. Oceanologica Acta 26: 311–321.

96. Tett P, Gilpin L, Svendsen H, Erlandsson CP, Larsson U, et al. (2003) Eutrophication and some European waters of restricted exchange. Continental Shelf Research 23: 1635–1671.

19

Isotopes and Trace Elements as Natal Origin Markers of *Helicoverpa armigera* – An Experimental Model for Biosecurity Pests

Peter W. Holder[1]*, Karen Armstrong[1], Robert Van Hale[2], Marc-Alban Millet[3¤a], Russell Frew[2¤b],
Timothy J. Clough[4], Joel A. Baker[3¤c]

1 Bio-Protection Research Centre, Lincoln University, Canterbury, New Zealand, **2** Department of Chemistry, Otago University, Dunedin, New Zealand, **3** School of Geography Environment and Earth Sciences, Victoria University of Wellington, Wellington, New Zealand, **4** Department of Soil and Physical Sciences, Lincoln University, Canterbury, New Zealand

Abstract

Protecting a nation's primary production sector and natural estate is heavily dependent on the ability to determine the risk presented by incursions of exotic insect species. Identifying the geographic origin of such biosecurity breaches can be crucial in determining this risk and directing the appropriate operational responses and eradication campaigns, as well as ascertaining incursion pathways. Reading natural abundance biogeochemical markers using mass spectrometry is a powerful tool for tracing ecological pathways as well as provenance determination of commercial products and items of forensic interest. However, application of these methods to trace insects has been underutilised to date and our understanding in this field is still in a phase of basic development. In addition, biogeochemical markers have never been considered in the atypical situation of a biosecurity incursion, where sample sizes are often small, and of unknown geographic origin and plant host. These constraints effectively confound the interpretation of the one or two isotope geo-location markers systems that are currently used, which are therefore unlikely to achieve the level of provenance resolution required in biosecurity interceptions. Here, a novel approach is taken to evaluate the potential for provenance resolution of insect samples through multiple biogeochemical markers. The international pest, *Helicoverpa armigera*, has been used as a model species to assess the validity of using naturally occurring δ^2H, $^{87}Sr/^{86}Sr$, $^{207}Pb/^{206}Pb$ and $^{208}Pb/^{206}Pb$ isotope ratios and trace element concentration signatures from single moth specimens for regional assignment to natal origin. None of the biogeochemical markers selected were individually able to separate moths from the different experimental regions (150–3000 km apart). Conversely, using multivariate analysis, the region of origin was correctly identified for approximately 75% of individual *H. armigera* samples. The geographic resolution demonstrated with this approach has considerable potential for biosecurity as well as other disciplines including forensics, ecology and pest management.

Editor: Daniel Doucet, Natural Resources Canada, Canada

Funding: This work was funded by a New Zealand Tertiary Education Commission PhD scholarship and a National Isotope Map of New Zealand Precipitation Cross Department Research Project grant (10749/2007). The funders had no role in study design, data collection and analysis, decision to publish, or preparation of the manuscript.

Competing Interests: The authors have declared that no competing interests exist.

* E-mail: peter.holder@lincoln.ac.nz

¤a Current address: Department of Earth Sciences, Durham University, England, United Kingdom
¤b Current address: Food and Environmental Protection Laboratory, IAEA/FAO, Vienna, Austria
¤c Current address: School of Environment, University of Auckland, Auckland, New Zealand

Introduction

Biosecurity encompasses the provision of services that minimise the impact of exotic pest species on a nation's economy, environment and public health. In agriculturally based economies, such as that of New Zealand, biosecurity systems protect industries worth billions of dollars against constant risk of exotic pest introduction [1], which have large direct and indirect financial costs [2]. As biosecurity risks escalate with the increased international mobility of people and trade products [3], these systems need to become more efficient. This includes an emerging requirement to ascertain the natal geographic origin of intercepted exotic pests, as this is commonly unknown for organisms that are detected in surveillance networks. Such a capability could be used to differentiate between non-established individuals and members of established (locally breeding) populations. This information would direct appropriate response actions in post-border investigations and eradication campaigns, as an unestablished exotic pest requires a much lower scale response than an established population. Similarly, knowing immediate prior origins can help verify a region's pest free status for specific high impact pests, and so maintain trade access [4] by confirming intercepted individuals as vagrant rather than locally established. Point-of-origin data could also be used to identify biosecurity risk pathways and so inform biosecurity policy for pre-border protection.

Although tracing the geographical origins and dispersal of insects are important components within many aspects of entomological science, there are currently no suitable methods available that can determine the immediate origin of biosecurity interceptions. Tracing the dispersal of insects by classical methods, such as mark and recapture, is clearly unavailable for biosecurity investigation, where it is necessary to interpret naturally occurring, unlabelled specimens – as is also the case for many other ecological and pest management studies [5]. Likewise, genetic methods that use the similarity of heritable DNA markers to infer invasion histories or original sources of an introduction are inappropriate for resolving such recent and dynamic relationships [6]. DNA markers can help to assign an individual to a likely population, and therefore by inference the geographic place at which that genetic population is known to occur [7,8]; however, they cannot discriminate a new invader from a less recent one given the intergenerational time necessary for DNA mutations to be acquired. Consequently the DNA signature of an insect that had just arrived (F_0, i.e. of exotic origin and non-established) would look the same as one that could putatively have arrived from the same place one or more generations prior ($>F1$, i.e. of local origin); therefore an intercept could not be distinguished as having just arrived or not.

On the other hand, stable isotope ratio and trace element concentration signatures ('biogeochemical markers') can be direct indicators of provenance. These markers are not heritable, but are intrinsically incorporated into the tissues of all members of a population via their food and water sources as the organisms develop [9]. Hence, the markers that vary spatially due to differences in geology [10], elevation and climate [11], such as $^{87}Sr/^{86}Sr$ and δ^2H, may provide the desired understanding of the immediate origin of intercepted samples and distinguish an insect as either F_0 or $\geq F_1$ with respect to establishment status. Various natural abundance biogeochemical markers have been successfully applied to track a wide range of dispersing organisms and items of commercial or forensic interest [12]. However, to date, such markers have been underutilised for provenance determination in entomology, and our understanding in this field is still in a phase of basic development. Early investigations considered concentrations of the small series of common elements able to be analysed with the spectrometry techniques available at the time, (e.g., P, S, Cl, K, Ca, Fe, Cu, Zn) [13–16]. However, these elements are biologically active [17] and thus their concentrations are subject to variation linked to physiological differences between individual insects. Consequntly, these markers were confounded by polyphagy, adult feeding and gender differences affecting elemental expression, which masked the point-of-origin signals [18,19]. More recently, stable isotopes have been considered and spatial separation of insect populations across continental δ^2H and $\delta^{13}C$ contours has been demonstrated [20,21]. However, the scale of resolution from these light elements can be too coarse for confident provenance determination [22,23], with often insufficient difference between study areas and/or the within-region environmentally driven variation in signal being greater than the between-region differences [24]. This is of particular consequence in forensic or biosecurity applications, where the typically small sample sizes impede statistically confident provenance assignment [25]. The specific impetus for the current study was the inability to determine the origin of two important biosecurity pests collected post-border in Auckland, New Zealand in 2005 and 2006 – painted apple moth (*Teia anartoides*, Lymantriidae) and fall web worm (*Hyphantria cunea*, Arctiidae). Based on the successful elucidation of monarch butterfly migration routes [26], provenance assignment for these Auckland incursions using δ^2H and $\delta^{13}C$ was attempted [27]. However, interpretation of the results was inconclusive, as the accuracy and limitations of this methodology were unknown in a biosecurity context. In contrast

to the Hobson et al. [26] study, that used a single host-plant system within a pre-defined time and space, the Auckland specimens belong to polyphagous species and were accidentally introduced; as is typical with biosecurity interceptions. Therefore, these insects were from an unknown and unpredictable host, place and point in time, which impeded isoscape-to-insect corrections.

A number of reviews have proposed that provenance discrimination may be enhanced by multivariate analyses of several markers [12,28–31], although few studies have empirically tested this, e.g., [32–34]. Consequently, the research hypothesis for this study was that the level of spatial resolution and confidence in provenance assignment for biosecurity samples could be improved by combining the continental scale, temperature-linked distribution patterns of δ^2H, with the finer spatial scale of geological markers such as the isotopes of Sr and Pb and trace element concentrations. In testing this hypothesis, we also assess both the practical feasibility of such a method, and whether the regional spatial resolution achieved is sufficient for biosecurity applications.

Methods

Model insect and host plant system

Helicoverpa armigera (Hübner 1805) [Lepidoptera: Noctuidae: Heliothinae] (tomato fruit worm) was used as an experimental model of an invasive pest. The fundamental biological parameters of this species are well understood, it is readily field collectable and its pan-global distribution facilitated geographically extensive sample collection in locations appropriate to the research objectives. Further, *H. armigera* is a major pest of food, fibre, oil and ornamental crop plants [35]. There is an ongoing interest in elucidating this species migratory patterns and population dynamics, with view to improving the effectiveness of pest management strategies against it [36,37]. *Zea mays* ('corn') was selected as the most suitable model plant for the inter-regional comparison. It is grown extensively in the areas of research interest and is a productive *H. armigera* host, on which this insect has comparatively low levels of parasitism. Further, *Zea mays* does not support the morphologically similar species *Helicoverpa punctigera* (Wallengren), facilitating the field collection of the correct species.

Study design and sample collection

The bio-geographical regions of Mid-Canterbury (MC), Bay of Plenty (BP), and Auckland (AK) in New Zealand, and the corn growing areas around Toowoomba (Queensland – QLD) and Wagga Wagga (New South Wales – NSW) in Australia were used for comparison (Figure 1). These regions were selected because they represent geological and climatic contrasts and similarities, the model insect-host system occurs in them all, and they are important areas with regard to New Zealand biosecurity [38,39].

Sample collection was carried out over January – May (southern hemisphere late summer) in two consecutive years, 2008 and 2009, in order to also examine inter-year variation. The collection dates were adjusted between the years so as to occur at the same development phase of both the corn crop (beginning of Kernel Dent Stage) and *H. armigera* phenology (pre-diapause late instar larvae and pupae) in the two summers. *Helicoverpa armigera* were collected from a minimum of 12 separate sites (paddocks) at each of the five different regions, however, some sites did not yield any adult moth samples, and others provided several (Table 1).

To ensure the specimens were from known locations, and to avoid the potential influence of multiple host plant sources, late instar *H. armigera* larvae were collected from corn cobs for subsequent rearing, and/or pupae were excavated from under the host plants at each site. The larvae were reared on their original

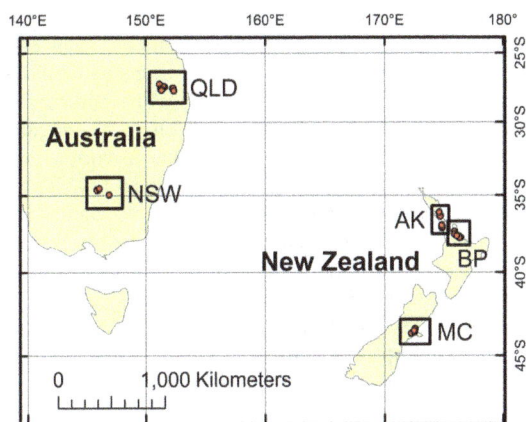

Figure 1. Australasian regions used to test biogeochemical markers for provenance assignment of *H. armigera*. These regions represent biogeochemical contrasts and similarities, and they are important areas with regard to biosecurity for both Australia and New Zealand. MC = Mid Canterbury, BP = Bay of Plenty, AK = Auckland, New Zealand; NSW = Wagga Wagga, QLD = Toowoomba, Australia.

cob, until pupation. These and the excavated pupae were held and emerged under a constant 25°C, 16:8 h light: dark regime. Emerged moths were held without food or water for four days, to avoid the influence of adult feeding and to allow the wings to complete sclerotization, then euthanized and stored frozen (−20°C), dry, for later identification and analysis.

Insect identification

The identification of the collected moths was confirmed as *H. armigera* by screening to genus using fore-wing patterns and to species or species group using hind-wing markings [40]. For specimens where species determination was not possible using exterior morphological examination the identification was confirmed using characteristics of the genitalia [41] and DNA barcoding [42] (GenBank accession numbers KF661352 – KF661389).

Ethics statement

No animal care approval was required for the collection and handling of *H. armigera*. The specimens were collected on commercial properties with the permission of the land owners. Live samples from Australia were transferred to New Zealand quarantine facilities under a 'Permit to Import Live Animals' from Biosecurity New Zealand (Ministry of Primary Industries) (Permit numbers 2008033670, 2009036197).

Sample preparation and chemical analyses

Each moth was partitioned to provide samples for the various analyses. A set of wings was dissected for δ^2H analysis and the remainder of the moth bodies were used for Sr and Pb isotope and trace element concentration analyses.

Samples used for δ^2H measurement were washed three times with a solution of 2:1 chloroform: methanol to remove oils and then air dried for 12 h. Six, ≈200 µg pieces (three replicate pairs) were dissected from the distal costal section of the wing and loosely crimped into 3×5 mm silver elemental analyzer cups (OEA Laboratories, UK). Samples were then equilibrated in a pair of static, sealable chambers with one of two water vapours (−258.0‰ or +60.0‰ VSMOW) at 110 °C for 1 hour with vacuum drying at 110°C before and after equilibration, modified from [43]. δ^2H measurements were conducted using a vacuum purged Costech Zero Blank autosampler on a Thermo TC/EA coupled to a Thermo Delta V IRMS in continuous flow mode, at Otago University, New Zealand. The raw δ^2H values were corrected to the nine IAEA-CH-7 reference standards (δ^2H_{VSMOW} −100.3‰) measured at intervals during each batch. Paired results from the equilibrations with the two waters were used to calculate the non-exchangeable hydrogen isotope ratio using equation 3 of Schimmelmann et al. [44]. KHS (−54.1±0.6‰) [45] was used as the quality assurance standard. Average precision of measurement over the three months that the analyses were carried out was ±0.8 ‰.

In preparation for the solution chemistry used for trace element and Sr–Pb isotope analyses, individual moths were 'washed' by passing two 30 second 250 kPa+ streams of high purity N_2 over them in a filtered chamber, as described by Font et al.[46]. All subsequent specimen handling, chemistry and drying was conducted under ultra-clean conditions, within PicoTrace Class 10 laminar flow workstations. Samples were digested using three Seastar 15 M HNO_3+30% H_2O_2 closed digestion – evaporation cycles in Savillex Teflon beakers at 120°C; then cooled and taken up into solution in 1 M HNO_3. A weighed aliquot, comprising approximately 20% of this solution, was then subject to trace element analysis, using an Agilent 7500cs ICPMS *via* a Cetac ASX-520 autosampler and a 100 µl/min Microflow nebuliser spray chamber (Victoria University of Wellington Geochemistry Laboratory, New Zealand) (Tables S1 & S2, Figure S1). Element concentrations were determined by bracketing each set of five samples with a multi-element calibration standard solution made up from mono-elemental standard solutions (BDH Laboratory Supplies, England). The remaining portion of the solutions were dried down for Sr and Pb separation column procedures, as described by Pin & Bassin [47] and Baker et al. [48] respectively, using 1 ml pipette tips fitted with pre-cleaned 30 µm pore-size polypropylene frits and pre-cleaned Sr Spec (Eichrom Technologies, IL. USA) and AG1-X8 (Bio-Rad Laboratories, CA. USA) resins. Sr isotope ratios were measured on a Thermo-Finnegan Triton TIMS at the Laboratoire Magmas et Volcans, Clermont-Ferrand, France. The Sr samples were taken up in 1 M

Table 1. Number of *H. armigera* adult moths reared (*n*) and number of sites represented.

		MC	BP	AK	NSW	QLD
2008	*n*	17	8	24	22	26
	sites	9	6	11	7	9
2009	*n*	44	38	61	58	36
	sites	8	13	12	11	9

H_3PO_4 mixed with tantalum salt as an activator, loaded onto single Re filaments that were previously outgassed at 4.0 ampere (A) for 30 min and then dried down slowly at 1 A. The filaments were then heated up to a temperature of 1400 to1500°C, until a high enough ion beam was reached. Measurements were made in multidynamic mode with two cycles, ion beams being shifted one collector down during the 2nd cycle and samples run until the signal started to drop off in order to maximise internal error. Instrument mass bias was corrected for using a $^{86}Sr/^{88}Sr$ ratio of 0.1194 and an exponential mass fractionation law. The accuracy of the $^{87}Sr/^{86}Sr$ data was assessed by repeated analyses of $\approx$30 ng Sr from BHVO-2, which reflected the amount of Sr available for analysis from each moth. This gave an average value of 0.703508±0.000035 (2SD, $n = 4$). The average internal precision of all moth $^{87}Sr/^{86}Sr$ analyses was 0.000143 (2SE). Pb isotope ratios were determined with a Nu Instruments MC ICPMS at Victoria University of Wellington. The Pb samples, dissolved in 0.5 wt% Seastar HNO_3, were introduced to the MS via a DSN-100 desolvating nebulizer (Table S3). Data was acquired using two blocks of 25 integrations of 5 seconds each. NBS 981 calibration standards bracketed each three samples. Repeated analysis of $\approx$4 ng Pb (to match the average moth sample Pb abundance) JB-2 rock standard gave an average of 2.08718±0.00012 (2SD) for $^{208}Pb/^{206}Pb$ and 0.848593±0.000056 for $^{207}Pb/^{206}Pb$ ($n = 7$). The average internal precisions for the actual moth sample Pb isotopes analyses were ±0.00098 2SE for $^{208}Pb/^{206}Pb$ and ±0.00049 for $^{207}Pb/^{206}Pb$. Total procedural Pb blanks in this study yielded <15 pg Pb, which represents <0.55% of the average moth sample Pb abundance and required an insignificant blank correction, given the internal precision of the analyses.

All insects were subjected to H isotope analyses. However, logistical constraints necessitated that just six moth samples per region per year were processed for the other biogeochemical markers. Further, due to analytical error, trace element results for the 2008 season was acquired for only four moths from MC, AK, NSW and QLD and two for BP. The markers obtained from the 2008 samples were δ^2H, $^{207}Pb/^{206}Pb$ and $^{208}Pb/^{206}Pb$ and elemental concentrations for Li, Al, Sc, Cr, Mn, Ni, Zn, Ga, As, Rb, Sr, Cd, Cs, Ba, W and Pb. To improve the discrimination between the regions, the selection of variables was refined for the 2009 material, and Li, Al, Ca, Sc, Ti, Cr, Co, Ni, Cu, Zn, As, Rb, Sr, Cd, Cs, Ba, La, Ce, W and Pb concentrations were obtained, as well as δ^2H, $^{87}Sr/^{86}Sr$ and $^{207}Pb/^{206}Pb$ and $^{208}Pb/^{206}Pb$.

Statistical analyses

The multivariate datasets from the moth bodies were assessed for regional discrimination for both the 2008 and 2009 data sets. It was necessary to use non-parametric methods, as experimental constraints resulted in fewer samples than the number of variables. Furthermore, with parametric methods, statistical assumptions regarding normally distributed data in multivariate space would have been potentially violated [49] and outlying data points may have led to over-emphasised groups [50]. As such, the datasets were assessed for overall regional difference using PERMA-NOVA+ (version 1.0.3) (PRIMER-E version 6.1.13) permutation-al multivariate analysis of variance main test (i.e., overall Pseudo-F); and differences between the individual regions were evaluated using post-hoc PERMANOVA pair-wise tests. The data were log (x+1) transformed, normalised and the analyses carried out using a Euclidean distance resemblance matrix. Both tests used 9999 permutations. The moth multivariate datasets were then assessed for regional grouping and discrimination using a canonical analysis of principal coordinates (PERMANOVA+ CAP analysis).

A dimension reduction process was assessed for the potential to achieve a combination of isotope and trace element values that maximised the separation between the regions by removing non-informative variables. This was accomplished by first ranking the variables according to their relative contribution to the original CAP regional grouping (assessed using the linear correlations between the variables and the CAP ordination axes) for CAP axes 1–3. The least informative variables were eliminated by nominally selecting and discounting those that had a correlation coefficient less than half the largest correlation coefficient on all three CAP axes [51]. The CAP analysis was then re-run with both years' datasets, without the least informative variables. Regional assignment of the moth samples was then tested by 'Leave-one-out Allocation of Observations to Groups' cross-validation and re-run pair-wise PERMANOVA tests.

The level of spatial resolution of the multivariate analyses was compared to that of individual variables, after we had identified the most informative individual variables as those having the highest correlaion with the multivariate CAP ordination axes. The regional discrimination potential of these individual variables was assessed using the same cross-validation processes as described above, and further analysed using univariate ANOVA and pair-wise Fishers unrestricted LSD tests ($\alpha = 0.05$) (GenStat 14.1). Moth δ^2H sample sizes were uneven at each site and region, and therefore unbalanced ANOVA were employed for this assessment. Regression analyses were also conducted on the un-grouped (i.e., not-mean values) data to test goodness of fit versus latitude. Retrospective power analyses for the δ^2H, $^{87}Sr/^{86}Sr$, $^{207}Pb/^{206}Pb$ (univariate) datasets was also conducted. This was carried out by calculating the differences between the means of the regions and then determining the minimum sample size required for each comparison to be 5% significantly different, using a two-sided, two-sample t-test, with a power of 90% (GenStat 14.1). (The power of a statistical test is defined as the probability that the test will correctly reject the null hypothesis when the null hypothesis is false – i.e. the probability of not committing a Type II error). A standard deviation pooled over all regions was used.

Results and Discussion

The sample preparation and analytical methodology presented here has enabled the analyses of multiple biogeochemical markers from single insect samples, despite the low concentrations of many of the elements. The regional discrimination potential of the multivariate data was examined initially. This assessment concomitantly identified the most informative individual variables and thus allowed the subsequent comparison of the regional differentiation potential of multivariate vs univariate analyses, which are considered below.

A multivariate test of provenance differentiation

PERMANOVA analyses identified an overall regional difference for both 2008 and 2009 moth multivariate biogeochemical marker datasets (2008 Pseudo-$F_{4, 13} = 2.4051$, p(perm) = 0.0033; 2009 Pseudo-$F_{4, 25} = 2.79$, p(perm) = 0.0001). The geographical resolution achieved between individual study areas is illustrated in Figure 2 (animated in Movie S1 & S2) and the associated pair-wise tests (Table S4). In the 2008 dataset, BP moths were not significantly different from moths from any of the other regions, possibly due to the small number of samples from BP in that year ($n = 2$) affecting the comparison with the other regions. The NSW and QLD 2008 moths were also not significantly different from each other. However, the AK and MC moths were distinguishable both from the Australian moths and each other. In contrast, for

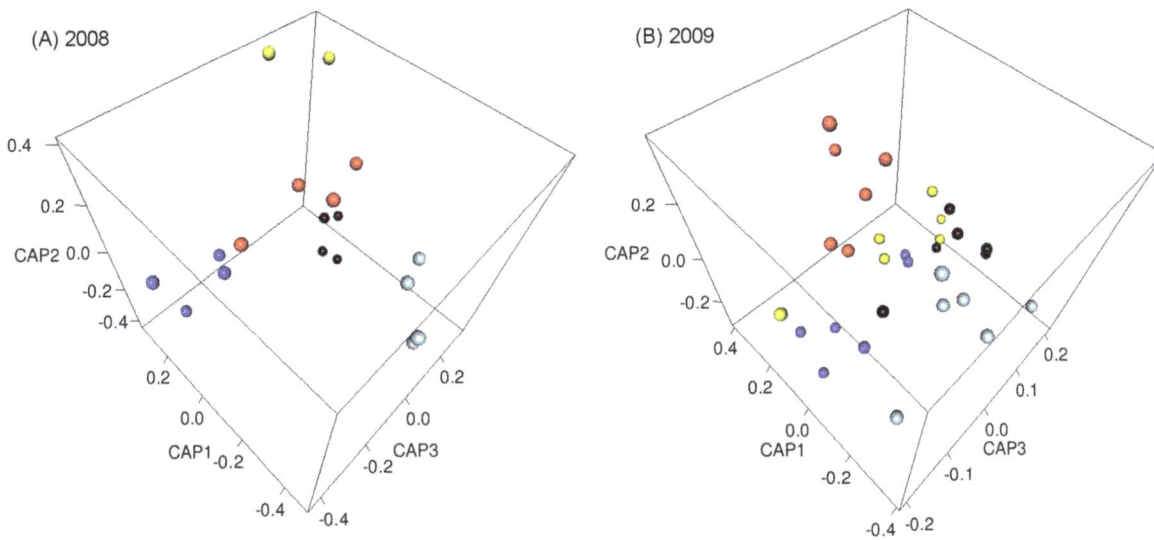

Figure 2. The geographical resolution achieved between the experimental regions using multiple biogeochemical markers. Canonical analysis of principle co-ordinates plots of δ^2H, trace element concentration (ng/g) and $^{206}Pb/^{208}Pb$, $^{207}Pb/^{208}Pb$ data from *H. armigera* adult specimens, reared from Australian and New Zealand sites: (A) March – May 2008 and (B) Jan – March 2009. The 2009 moth data was optimised to remove non-informative variables and also includes $^{87}Sr/^{86}Sr$ values. Black = MC, yellow = BP, dark blue = AK, light blue = NSW, red = QLD.

the 2009 moth dataset, the pair-wise regional comparisons were all significantly different, except for the BP versus MC, and BP versus AK comparisons.

The more powerful geographical separation in the 2009 dataset was primarily due to the addition of $^{87}Sr/^{86}Sr$ data. This marker provided robust separation of the moths from the two Australian regions, as well as a lesser but still significant difference between the Australian and New Zealand regions (Figure 3). In addition, the 2009 dataset incorporated a greater number of informative trace elements, including Co, Ce and La, all of which contributed to the improved regional separation (Table S5).

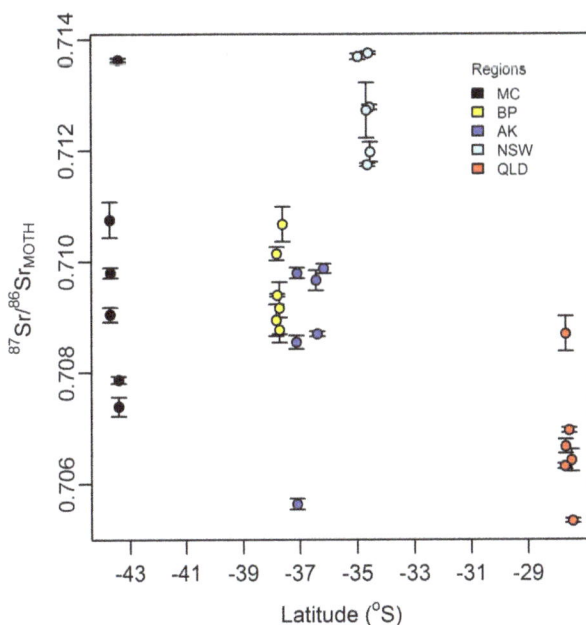

Figure 3. *H. armigera* $^{87}Sr/^{86}Sr$ **distribution, relative to degress latitude south.** Error bars = analytical 2SD.

The CAP regional assignment cross-validation tests gave misclassification errors of 22.2% with the 2008 dataset and of 26.7% with the larger 2009 dataset (Table 2). Leaving out the least informative markers (i.e., dimension reduction) was beneficial with the 2009 regional comparison, with the misclassification error being reduced from 36.7%. In contrast, attempts at 'optimisation' with the 2008 data in this way increased misclassification error. This indicates that successful provenance determination requires a balanced appraisal of all available markers. It is necessary to consider the potential for the signal to be confounded by biological processes, the degree of overlap between the potential source regions, and the variation within the regions [12,52].

A significant finding regarding provenance assignment for biosecurity is that all the moths from the New Zealand regions are distinguished from the Australian moths using the 2009 dataset. As with the regional pair-wise tests above, the superior inter-country allocation achieved with the latter dataset is attributed primarily to the inclusion of $^{87}Sr/^{86}Sr$ as a variable. This result, along with the 73.3% cross-validation success rate, suggests that determining whether a suspect sample has originated from its collection point, or not – i.e., in a biosecurity scenario – is more likely to be successful than not. However, 100% accurate re-allocation was achieved in only one in five regions with the 2009 dataset and two out of five regions with the smaller 2008 dataset. Misclassification between the regions is attributed to the similarities in the mean values and over-lapping ranges of several of the variables. Hence, single insect samples ($n = 1$) may be difficult to reliably assign to place of origin in such circumstances, although discrimination between regions is expected to be more reliable when the sample-sizes are larger.

The most informative variables in the 2008 dataset were: $^{207}Pb/^{206}Pb$, $^{208}Pb/^{206}Pb$, δ^2H, the elemental ratios Pb/Sr, Rb/Sr, Ba/Sr, and the concentrations of Rb and Sr, and Li, Cr, Ga, Ba and Pb (Table S5). In the 2009 dataset, the most informative variables were: $^{87}Sr/^{86}Sr$, δ^2H, concentrations of Pb, As, Sr, Ba, Cs, all the elemental ratios considered, Pb isotopes, and the variables with significant correlation to the 3rd CAP ordination axes, Ti, Co, Ni, La and Ce. To understand the impact that these

Table 2. Validation tests of regional assignment for individual *H. armigera* samples.

		Classified						
		MC	BP	AK	NSW	QLD	Total	%correct
2008								
Original Group	MC	4	0	0	0	0	4	100
	BP	1	0	0	0	1	2	0
	AK	0	0	3	0	1	4	75
	NSW	0	0	0	3	1	4	75
	QLD	0	0	0	0	4	4	100
2009								
Original Group	MC	4	1	1	0	0	6	66.667
	BP	1	4	1	0	0	6	66.667
	AK	2	1	3	0	0	6	50
	NSW	0	0	0	6	0	6	100
	QLD	0	1	0	0	5	6	83.333

Tests of CAP generated groupings using 'Leave-one-out Allocation of Observations to Groups' method. 2008 scored 14/18 correct, misclassification error = 22.2%; 2009 scored 22/30 correct, misclassification error = 26.7%.

might have on the ability to assign origins, and to test the hypothesis that provenance discrimination using multivariate analyses is superior to univariate analysis, these most informative markers individually are considered below.

Provenance differentiation using δ^2H_M

The plot of *H. armigera* wing δ^2H values (δ^2H_{MOTH}) against latitude confirms a latitudinal continental scale cline in both 2008 and 2009 (Figure 4). The 'δ^2H_M per degree latitude' regression is 1.6 and 1.5‰ per degree in the 2008 and 2009 datasets respectively, which is slightly less than the ≈2‰ per degree described by Hobson & Wassenaar et al. [26] for monarch butterflies over eastern North America. Further, the regional δ^2H_M means are significantly different in both years (2008 $F_{4, 92} = 33.67$; p<0.001; 2009 $F_{4, 210}$ 56.93, p<0.001). However, the δ^2H_M versus latitude R^2 indicates that at only 46% of the variation was due to latitude for the 2008 dataset, and 35% for the 2009 data. This suggests that biological and/or localized environmental variation within regions is of equal or greater influence than latitude.

Pair-wise comparisons of the δ^2H_M means reveal that, on a population level, the moths from the most southern region, MC, were able to be distinguished from the moths from the more northerly regions, being significantly "lighter" (having lower δ^2H values) than all the other regions in both years ($\alpha = 5\%$). Beyond this however, δ^2H_M values of the other regions were too similar (Table S6) and/or have too much overlap to be reliably distinguished. The often large sample sizes required to achieve significant differences between the regional δ^2H_M means (calculated retrospectively, Table S7) reiterates the broad scale of spatial resolution and inconsistent individual sample provenance assignment achieved by δ^2H. Where the δ^2H_M means are distinctly different, there is strong potential for δ^2H to discriminate moths from different regions. Hence MC can be distinguished from all the other regions by sample sizes of 12 or fewer moths and some comparisons required *n* of only 3 or 4. Conversely, where the δ^2H_M means are close and/or variation is high, the required sample sizes are impractically large, and more than typically collected in biosecurity incursions (commonly 2–6 insects). Further, small sample sizes have high misallocation errors (= low power). For example, with *n* = 2 moths, MC, the most distinct region, contrasted to the other regions gave power values ranging from 0.13–0.41 (calculated using a GenStat 14.1, 2-sided, 2-sample t-Test, significance 0.05).

The provenance discrimination achieved with the δ^2H_M univariate analyses for individual samples is compared to that achieved with the multivariate analysis in Table 3. Overall, the total misclassification error for the δ^2H_M univariate analyses was around 55%, which is approximately twice that of the multivariate analysis.

This limited geographical resolution is attributed to the large degree of both intra-region and intra-site variation in the δ^2H moth values. The intra-region δ^2H_M variation spanned 24.8–44.6‰ (for both years), with the average variation being 38‰ in 2008 and 38.9‰ in the 2009 dataset. This is higher than the differences between all the region's means. The intra-site variation comprises the largest component of the within-region variability, being 29.0‰ in the 2008 dataset and 27.8‰ in 2009. This degree of variation is greater than the ≈26‰ (interpolated) intra-region heterogeneity of monarch butterflies [26] and the intra-site variation of ≈28‰ reported in *Inachis io* (Lepidoptera: Nymphalidae) [23]. However, the results herein have similar variability to the intra-site variation that has been observed in *Arhopalus ferus* beetles (Cerambycidae) (up to 31.1‰) [53] and an unidentified

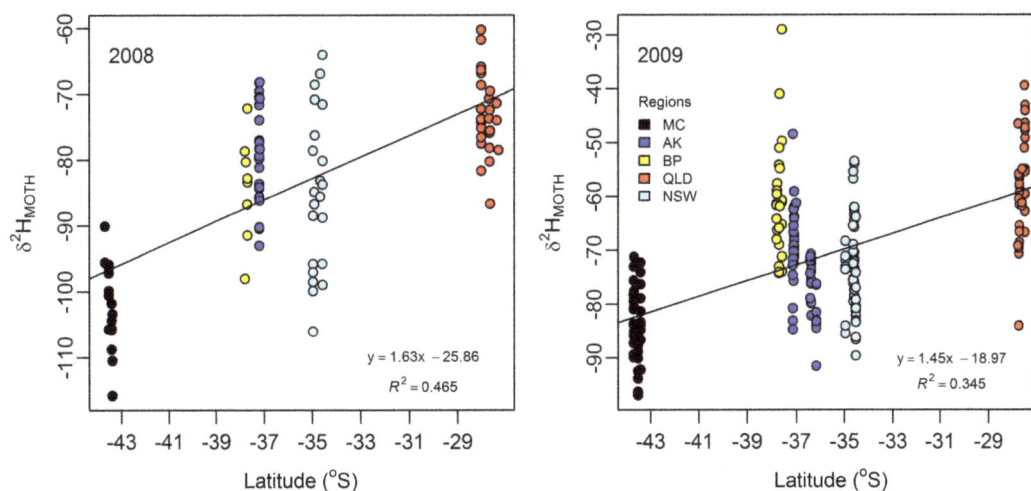

Figure 4. Relationship between *H. armigera* wing δ^2H values and latitude.

insect species (possibly beetle, 40‰) [54]. These results confirm the that quantifying within-population δ^2H heterogeneity is as necessary for insects as it is for birds [55,56]. Such within-population heterogeneity needs to be taken into account when using insect δ^2H information in geographical assignment and is required to propagate error in predictive geographical assignment modelling [57]. It also needs to be taken into account when using insect δ^2H information in paleoclimate reconstruction, cf. [58].

Furthermore, the relative differences between the regions were inconsistent for the two years, with the δ^2H_M values for the 2009 dataset being significantly "heavier" (having higher δ^2H values) than the 2008 dataset ($F_{1,4} = 27.87$, $p = 0.006$). Although it is important to appreciate that the collections were made at different weeks in each year, the inter-annual variation in δ^2H_M observed indicates that applications using insect δ^2H need to correct or specifically calibrate the data for each period of interest [59].

While this spatial and temporal heterogeneity of δ^2H expression makes it difficult to rely upon this as a single marker, it clearly still provides a level of spatial discrimination that can be informative.

Provenance differentiation using $^{87}Sr/^{86}Sr_M$

The *H. armigera* $^{87}Sr/^{86}Sr$ values ($^{87}Sr/^{86}Sr_M$) from the five regions (Figure 3) were significantly different overall ($F_{4, 25} = 14.04$, p<0.001). The NSW moths had the highest $^{87}Sr/^{86}Sr$ (mean $^{87}Sr/^{86}Sr = 0.71278$), and QLD the lowest (mean $^{87}Sr/^{86}Sr = 0.70673$) (a significant difference ($\alpha = 5\%$), Table S8). The New Zealand moth $^{87}Sr/^{86}Sr$ values were intermediate to the Australian regions, with all the New Zealand regional means having values of approximately 0.709. Pair-wise comparisons confirmed that the New Zealand moth specimens are significantly different from both NSW and QLD moths ($\alpha = 5\%$). However, the New Zealand regions were not significantly different from each other, with median $^{87}Sr/^{86}Sr$ values being separated by only 0.0003.

The capacity of $^{87}Sr/^{86}Sr$ data to separate vulnerable biosecurity regions in New Zealand from the relevant risk regions in Australia indicates that, on a population level, Sr isotopes are a potentially powerful tool for provenance determination of intercepted specimens. The minimum sample size required to achieve significant differences between the Australian and New Zealand regions with $^{87}Sr/^{86}Sr$ alone was 12 or fewer insects (Table S9). However, the power associated with sample sizes $n = 2$

Table 3. A comparison of the regional discrimation achieved by multivariate and univariate analyses.

	Original Group	Multivariate assignment	Univariate assignment			
			δ^2H	δ^2H	$^{87}Sr/^{86}Sr$	$^{207}Pb/^{206}Pb$
n used for test		30	30	214	30	30
% correctly allocated to original group	MC	66.7%	16.7%	62.8%	33.3%	50%
	BP	66.7%	33.3%	32%	33.3%	0%
	AK	50%	66.7%	39%	33.3%	16.7%
	NSW	100%	50%	29.6%	100%	16.7%
	QLD	83.3%	66.7%	66.7%	83.3%	16.7%
Total misclassification error		26.7%	53.3%	55.1%	43.3%	80.0%

The assignement of individual *H. armigera* samples to their original region; generated by CAP 'Leave-one-out Allocation of Observations to Groups' method. Uses the 2009 data only; for δ^2H_{MOTH} both the sub-sample used in the multivariate analysis and full δ^2H_M dataset are given.

(a realistic interception sample size) for AK, the highest biosecurity risk centre in New Zealand, versus NSW and QLD is only 0.46 and 0.14 respectively. Further, the New Zealand moths were not able to be assigned to region using $^{87}Sr/^{86}Sr$ without impractically large sample sizes. Correspondingly, the total error when using $^{87}Sr/^{86}Sr_M$ in the univariate reassignment test was 43.3% of individual moths misclassified, as compared to 26.7% misclassification error in the multivariate test (Table 3). Therefore, the regional discrimination potential of strontium isotopes as a single variable cannot be assumed, even for places that are geologically distinct and geographically widely separated.

A prominent characteristic of the $^{87}Sr/^{86}Sr_M$ data is the within region heterogeneity. MC had the most diverse range, possibly reflecting the geological heterogeneity of the alluvial flood plain. Values varied from 0.7136, which is similar to both Canterbury's rhyolite volcanic [60] and metasiltstone metamorphic rocks [61], to 0.7074 which is consistent with the values reported for Miocene volcanic rocks on the adjacent Banks Peninsular [62]. The AK $^{87}Sr/^{86}Sr_M$ heterogeneity is also consistent with the geological diversity of the region; a single "low" value (0.7056) lying within the range of values reported for nearby greywacke [63] and the other five values from divergent parts of the Auckland isthmus clustered around 0.7091, possibly reflecting metapelite metamorphic rocks [63] and/or input from marine aerosols [64] (both around 0.709). With regard to the QLD, no geographically close rock or soil $^{87}Sr/^{86}Sr$ values have been found in the literature, although the $^{87}Sr/^{86}Sr_M$ from this region may reflect local trachyte (approximately 0.706) or rhyolite rocks (0.7077) [65]. BP and NSW had the lowest $^{87}Sr/^{86}Sr$ dispersion, with ranges of 0.0019 and 0.002 respectively. The degree of within-population $^{87}Sr/^{86}Sr$ variation found in the *H. armigera* populations is consistent with that reported in other terrestrial ecology references. For example, within population $^{87}Sr/^{86}Sr$ ranges up to 0.0018 have been observed in black-throated blue warblers (but n only 2) [66], 0.0025 in snail (Pulmonata, family not given) populations [67] and 1SD values up to 0.00113 reported in tree swallow [68]. In species of Geometridae and Notodontidae (Lepidoptera, species not given) caterpillars at single forest sites, Blum et al. [69] found a $^{87}Sr/^{86}Sr$ range of 0.00252, and Blum et al. [70] 0.00307, which are both similar to the within-site $^{87}Sr/^{86}Sr$ variance shown here for *H. armigera* (0.00039–0.00203).

As with δ^2H, therefore, insect $^{87}Sr/^{86}Sr$ is also very heterogeneous and has limited utility as a single marker for provenancing. However, the geologically linked expression observed in $^{87}Sr/^{86}Sr_M$, along with its contribution to the regional differentiation achieved in the multivariate test above, indicates that a combination of geological and climate markers can provide confident regional provenance assignment.

Provenance differentiation using Pb isotope ratios

In the 2008 data, there was significant overall difference between the regional $^{207}Pb/^{206}Pb_M$ means ($F_{4, 23} = 9.94$, p = 0.000), but not for $^{208}Pb/^{206}Pb_M$ ($F_{4, 23} = 1.80$, p = 0.163), although a pairwise comparison of the 2008 means revealed that NSW $^{208}Pb/^{206}Pb_M$ was significantly different to all other regions ($\alpha = 5\%$) (Figure 5). Five out of the seven 2008 NSW moths had Pb isotope ratios very significantly shifted from the expected NSW mixing line ($^{207}Pb/^{206}Pb$ approximately 0.895, $^{208}Pb/^{206}Pb$ 2.148), to an 'exotic value group' cluster with the median values of $^{207}Pb/^{206}Pb$ 0.757 and $^{208}Pb/^{206}Pb$ 2.195. No site bias was detected, with the exotic value group being from sites evenly spread over the entire NSW collection region (over a distance of approximately 100km i.e., Ganmain to Coleambally, NSW) and one site yielded both exotic value and non- exotic value samples.

To verify that these exotic values were not the result of systematic error, another pair of 2008 NSW moths were subject to separate analytical preparation run and mass spectrometry. These had similar exotic and non-exotic values, which confirm the validity of the earlier analyses. It appears that the affected moths have acquired Pb from sources in addition to the host plant, as their Pb isotope ratios are comprehensively different to that of the associated soils and host plants (average $^{207}Pb/^{206}Pb_{SOIL}$ 0.835, $^{208}Pb/^{206}Pb_{SOIL}$ 2.073; $^{207}Pb/^{206}Pb_{PLANT}$ 0.895, $^{208}Pb/^{206}Pb_{PLANT}$ 2.148). The additional source path may be respiratory inhalation, with the exotic Pb source being aerosols or dust particulates. Invertebrate acquisition of Pb by inhalation and accumulation of low concentration Pb contamination has previously been shown in snails (*Cepaea nemoralis*) [71]. The origin of the exotic signal in the present study is theorised to be particulate dust from within few a hundred kilometres west of the collection area. The *H. armigera* exotic value group described here had $^{207}Pb/^{206}Pb_M$ values similar to the range known for soils at Lake Frome, central South Australia ($^{207}Pb/^{206}Pb$, 0.7720, $^{208}Pb/^{206}Pb$, 2.066) [72] and near Adelaide, South Australia [73]. These locations align with the general pattern of dust storms in this region of Australia moving in a southeast direction [74]. In contrast, there was no significant difference between the regional Pb isotope ratio means in the 2009 data ($^{207}Pb/^{206}Pb_M$ $F_{4, 25} = 0.54$, p = 0.709; $^{208}Pb/^{206}Pb_M$ $F_{4, 25} = 0.83$, p = 0.520).

The moth Pb isotope values for 2008 and 2009 were not statistically different, despite the group of exotic values in the 2008 NSW moth dataset ($^{207}Pb/^{206}Pb$ $F_{1, 4}$ 0.0, p = 0.949; $^{208}Pb/^{206}Pb$ $F_{1, 4}$ 2.79, p = 0.17). This is likely to be a consequence of both years' data being widely dispersed, with the 2009 values clustered centrally within the more scattered 2008 data range (Figure 5).

The CAP regional grouping procedure showed that lead isotopes can provide information regarding geographical origin (Table S5). However, lead isotopes appear to be less informative than δ^2H and $^{87}Sr/^{86}Sr$ (Tables S7 & S9 versus Table S10), and had 80% univariate reassignment miscalculation error, which is more than three times that of the multivariate analysis. On-the-other-hand, this work has shown lead isotope data can be obtained from single insect samples, and that the sensitive fine scale resolution available from lead isotope analyses holds considerable promise for tracing ecological linkages and pollution sources which are hitherto not able to be elucidated in entomological science.

Provenance differentiation using trace element concentrations

The essential elements, which are those linked to common metabolic processes [17], were not geo-location informative, with the elements of atomic number $\leq$ Arsenic being generally less informative than the elements $\geq$ atomic number of Rb. Trace element variables that gave the best regional separation across both years are Sr, Cs, Ba and Pb, as well as the Pb/Sr elemental ratio (Table S5). Except for Ba, all of these were univariately significantly different between the regions (Figure 6). However, the values and the relative contributions of the elemental concentrations were not consistent between years. Further, none of the trace elements alone reliably discriminated moths from all of the different geographic regions, as the statistical differences were between only two or three of the five regions. For example, the BP and AK moths had the highest mean Rb and Cs concentrations in both years, and the MC moths the highest Cd values, yet the other regions were not significantly or consistently different (please note however, the results for 2008 BP may not be representative of the entire region, given the n = 2 sample size). The lack of a single geographical trace element marker is consistent with other

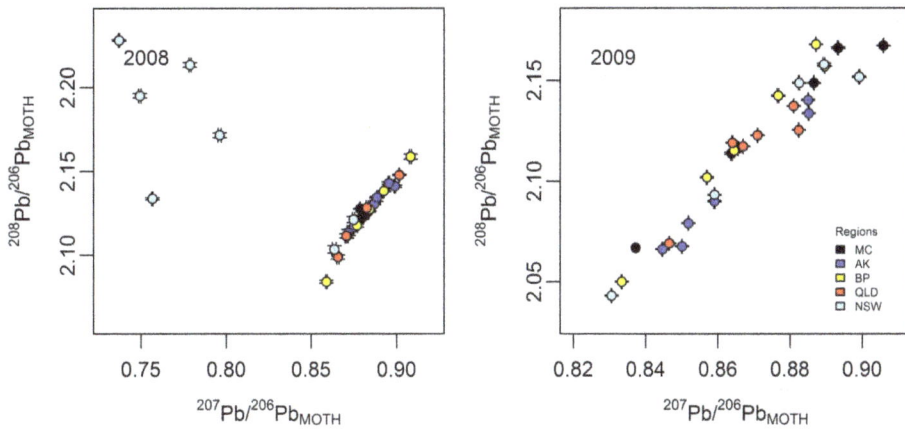

Figure 5. Pb isotope scatter plots from *H. armigera*. Error bars = analytical 2SD. Note: the axes for the 2008 dataset is larger scale than 2009.

ecological provenance determination studies, despite the significant differences in regional mean values, e.g., [75]. Nevertheless, elemental concentrations clearly contribute to geographical resolution.

In the 2008 moth trace element data set, the Australian regions had significantly higher Sr concentrations than the AK and MC moths. In contrast, Sr concentrations in the 2009 moth dataset were not significantly different overall ($F_{4,25} = 1.69$, $p = 0.183$), although BP and QLD were significantly different from each other

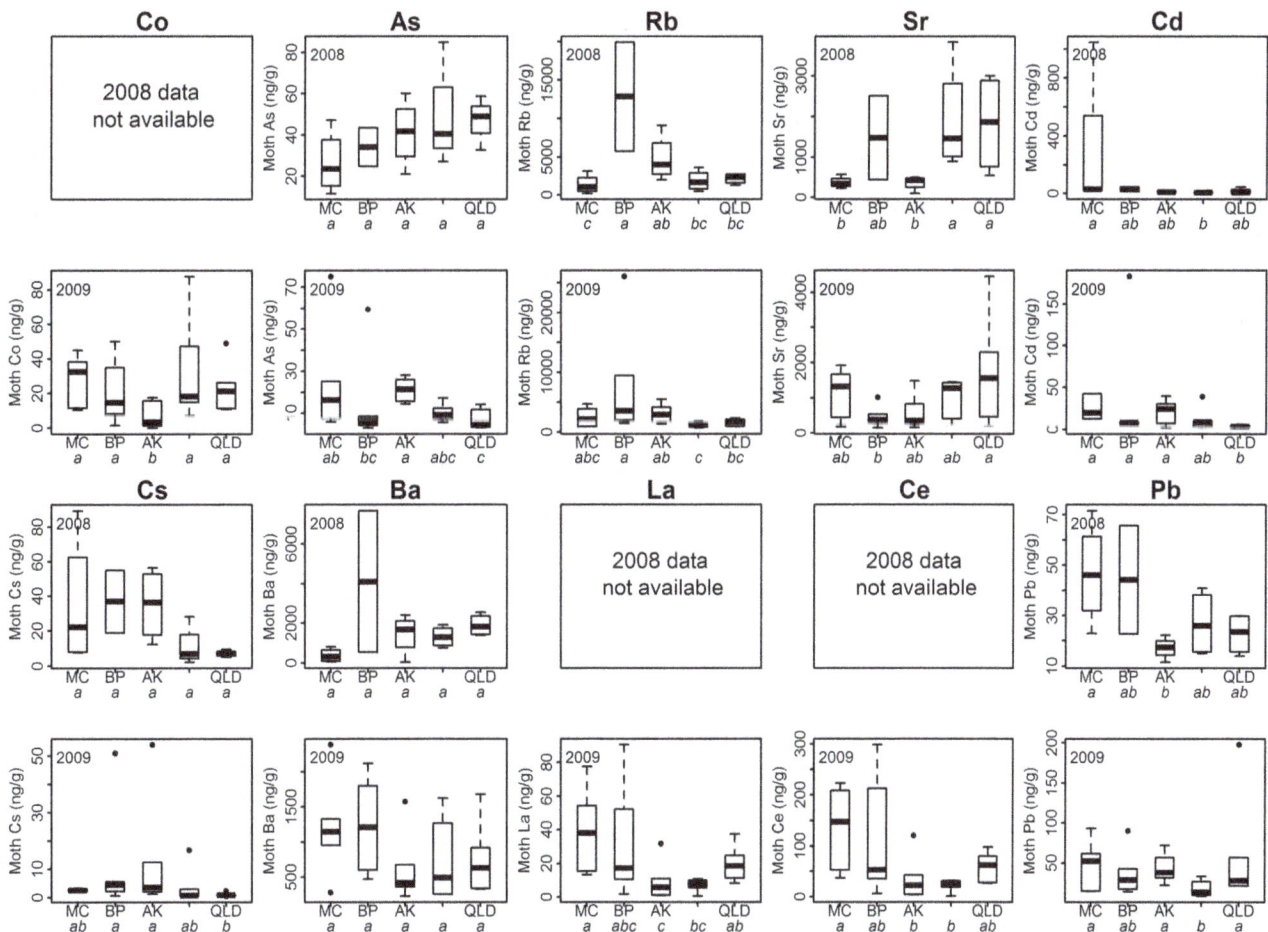

Figure 6. Trace element concentration data for *H. armigera*. Only the most informative elements are shown. Data is displayed as median, quartiles and the minimum/maximum value within 1.5 inter-quartile range; values outside 1.5 IQR are designated by a circle. Regions assigned a different lower case italic letters are significantly different (Fishers unrestricted LSD = 5%). 'Data not available' = information for that element was not recorded or lost due to analytical error.

in a pair wise test ($\alpha = 0.05$). The geographical resolution potential of Sr identified here, agrees with the pistachio provenance study of Anderson & Smith [33], with Sr giving the largest source region discrimination potential of all the elements they analysed.

New Zealand moth samples had higher average Cd levels than Australian samples, consistent with studies regarding the elevated levels of Cd in New Zealand agricultural soils [76]. However, despite some regional means being significantly different, moth Cd concentration was not a strong driver in the regional separation CAP analysis. This is due to the large degree of intra-region variation in moth Cd concentration values in all the regions, which results in poor allocation power on an individual moth basis.

Moth average Cs concentrations were also consistently higher in the New Zealand compared to Australian samples, although the statistical distribution of the Cs values may limit the potential of Cs as a biosecurity marker – when sample sizes are typically <6 insects. Most moths had Cs values <10 ng/g, although the larger mean values in both years were skewed by 2–5 moths with Cs values of >50 ng/g. However, all the Cs values >30 ng/g occurred in New Zealand samples and the highest values were most common in BP and AK moths. Thus Cs may be a useful geographical marker for New Zealand with larger sample sizes.

These trace element results are consistent with the avian studies of Norris et al. [77] and Szep et al. [32]. They reported a similar suite of elements (Mg, Cd, Sr, Ba, Rb, Cd, Pb) to be the most informative, and similar degrees of intra-regional heterogeneity – resulting from between site differences (cf. within site variation). This intra-regional variation facilitates better near-distance discrimination than light element stable isotopes, which typically separate populations on continental scales. However, the findings of Torres-Dowdall et al. [78] urge a cautionary interpretation of trace element data. They reported poor re-allocation accuracy for red knot shorebirds (*Calidris canutus*), due to both the lack of trace element marker resolution and because several elemental concentrations, including Sr and Pb, changed as the adult birds aged. The chemical profiles of feathers are believed to be affected by direct absorption from contaminants [79], preening behaviour and chemical leaching [80,81]. Therefore, although the biochemical processes and age-related changes will be different between birds and insects, as elemental profiles have been shown to also change during the moths' adult stadia [82], trace element profiles from whole moths may not be a reliable indicator of point-of-origin.

Conclusion

This study is the first evaluation of multiple isotope and trace element markers as a means of insect provenance assignment, as well as the first use of Sr isotopes for this purpose in entomological science. It is also believed to be the first study that has considered Pb isotopic information from insects.

The provenance assignment achieved demonstrates that, with the small samples sizes typical of biosecurity interceptions, none of the biogeochemical markers assessed can individually separate insects reared in different regions of biosecurity importance in New Zealand and Australia. In contrast, a multivariate combination of δ^2H, $^{87}Sr/^{86}Sr$, $^{208}Pb/^{206}Pb$, $^{208}Pb/^{207}Pb$ and selected element concentrations was able to distinguish the region of origin of *H. armigera* for 73.3% of individual moths. This supports the hypothesis that provenance discrimination achievable from multivariate analyses is superior to that of univariate analysis, e.g., [12]. In addition, the value of using multiple independent variables has been highlighted. Specifically, δ^2H, is a proxy for climate and therefore approximations of latitude, whereas the Sr and Pb isotopic ratios of the moths appear to be primarily that of

the source point soils and underlying geology and are independent of climate.

However, it is well recognised that all natural abundance markers have their weaknesses [28]. As such, the advances described above need to be considered in light of various biotic and abiotic limiting factors that are yet to be specifically defined. Identifying and accounting for these limitations is recommended as future research priorities. However, that should not detract from the ongoing use and further development of biogeochemical markers in entomological applications, which could be improved by considering some overarching issues revealed in this study.

Firstly, because of within-region heterogeneity in marker expression there is a strong relationship between confidence of provenance assignment, sample size and the degree of isotopic difference in the potential sources [57]. Similar multifarious marker expression has been observed elsewhere for single or paired isotope systems [67], and needs to be also taken into account in multivariate tracing.

Secondly, the relative discriminating power of the individual variables was inconsistent between the two years that were sampled. In particular, the insect δ^2H data needs to be calibrated by reference to precipitation δ^2H data for each period of interest. However, if emphasis is given to those variables that gave significant regional discrimination in both years, as well as those less likely to be affected by inconsistent biological and environmental parameters, which is assumed to include $^{87}Sr/^{86}Sr$ [10], temporal discrepancies can likely be minimised.

Lastly, our understanding of how soil and precipitation biogeochemical signals are expressed in insects is limited to the few studies that have actually quantified this relationship. Specifically, such information is available for *H. armigera* [82] and for δ^2H only, the hoverfly *Episyrphus balteatus* [Diptera] [83], monarch butterflies [26] and several dragon fly species [21]. Therefore, provenance assignment of other insect species currently requires reference populations of the same species from the candidate areas, e.g., [53]. Quantifying these 'transmission factors' (e.g., ^2H fractionation) for a wider range of plant-insect systems will facilitate wider entomological application of this technology in areas such as ecology, forensics and pest management, as well as paleo-climatic reconstruction.

Supporting Information

Figure S1 An assessment of the linearity of ICPMS measurement using a dilution series. Ratios of selected elements' concentrations in diluted solutions (1:2–1:4) of an in-house moth body standard over the long term averages of the non-diluted PH-armig moth standard (1:1). The average distortion on the analytical values, comparing the non-diluted moth standard averages to the most heavily diluted (1:4) was 3.5%. This indicates that there were minimal matrix effects suffered in the ICP-MS analysis.

Table S1 ICP-MS instrument settings, conditions and method used for trace element analysis of insect samples.

Table S2 ICPMS trace element measurement precision. The averages of dilute (10%) calibration standard, in-house moth body standard and NBS 1575 Pine needle external standard from each analytical run. All concentrations are ng/g calculated using sample and dilution weights. %CV = coefficient of variation. The average recovery of elements for NBS 1575 is versus the following published values: A = Certificate of Analysis (Reed, 1993); B = (Freitas et al., 2008); C = (Saitoh et al., 2002); D = (Asfaw & Wibetoe, 2006); E = (Taylor et al., 2007).

Table S3 Typical instrument operating conditions of the Victoria University Nu MC-ICP-MS and the DSN-100 parameters used for Pb isotope analysis.

Table S4 Pair-wise tests of regional differences for *H. armigera* populations, showing significant differences between the collection regions. Generated by PERMANOVA analyses of the multivariate datasets (using Euclidean distance resemblance matrices). † = p<0.10; * = p<0.05; ** = p<0.01.

Table S5 Relative contribution of the individual markers to the CAP regional grouping. Expressed as Pearson's correlation coefficient (i.e., linear measure of assocaition) between the individual markers (ignoring all others) and the CAP ordination axes within the mulitvariate data cloud. 2-sided significance test expressed as † = 10%; * = 5%; ** = 1%.; 2008 data df = 16, 2009 df = 18. 2009 data is an optimized suite.

Table S6 *H. armigera* wing δ^2H summary table. Showing regional δ^2H_M averages ± 1SD; values within a row that are followed by a different letter are significantly different (Fishers unrestricted LSD = 5%).

Table S7 Retrospective power analyses for *H. armigera* δ^2H. To detect significant differences between the regional means (Δ‰), at a two-sided significance level of 0.05 with a power of 0.90 using a two-sample t-test, the calculated sample size (n) would be required for each sample. A standard deviation pooled across all regions was used in each power analysis.

Table S8 *H. armigera* $^{87}Sr/^{86}Sr$ summary table. Showing regional $^{87}Sr/^{86}Sr$ averages ± 1SD; values within a row that are followed by a different letter are significantly different (Fishers unrestricted LSD = 5%). n = 6 for each region.

Table S9 Retrospective power analysis for the *H. armigera* $^{87}Sr/^{86}Sr$ data. To detect significant differences between the regional means (Δ), at a two-sided significance level of 0.05 with a power of 0.90 using a two-sample t-test, replication of the calculated n for each sample is required.

Table S10 Retrospective power analysis for the *H. armigera* $^{207}Pb/^{206}Pb$ data. To detect significant differences between the regional means (Δ), at a two-sided significance level of 0.05 with a power of 0.90 using a two-sample t-test, replication of the calculated n for each sample is required.

Movie S1 Animation of the geographical resolution achieved between the experimental regions using multiple biogeochemical markers, 2008. Canonical analysis of principle co-ordinates plots of δ^2H, trace element concentration (ng/g) and $^{206}Pb/^{208}Pb$, $^{207}Pb/^{208}Pb$ data from *H. armigera* adult specimens, reared from Australian and New Zealand sites, March – May 2008. Black = MC, yellow = BP, dark blue = AK, light blue = NSW, red = QLD.

Movie S2 Animation of the geographical resolution achieved between the experimental regions using multiple biogeochemical markers, 2009. Canonical analysis of principle co-ordinates plots of δ^2H, trace element concentration (ng/g) and $^{206}Pb/^{208}Pb$, $^{207}Pb/^{208}Pb$ data from *H. armigera* adult specimens, reared from Australian and New Zealand sites, Jan – March 2009. The 2009 moth data was optimised to remove non-informative variables and also includes $^{87}Sr/^{86}Sr$ values. Black = MC, yellow = BP, dark blue = AK, light blue = NSW, red = QLD.

Acknowledgments

The production of this work was supported by the Better Border Biosecurity consortium (?http://b3nz.org/). The contribution of Len Wassenaar (Environment Canada) to the project conception and on-going support is gratefully acknowledged, as is the sampling design of David Baird (VSN NZ Ltd) and statistical support of Marti Anderson (Massey University), James Ross and Dave Saville (both Lincoln University). Advice was generously provided by Laura Font (University of Amsterdam), Scott Hardwick (Landcare Research) and Dave Murray (Queensland Department of Primary Industries). We thank Régis Doucelance (Clermont-Ferrand Magmas et Volcans laboratory) for the TIMS analyses, and acknowledge the technical support of Anastasija Chomic and Sam Brown (both Lincoln University), Dianne Clarke, Emad Ehtesham, and Dave Barr (all Otago University).

Author Contributions

Conceived and designed the experiments: PWH KA RF TJC JAB. Performed the experiments: PWH RVH MAM. Analyzed the data: PWH RVH MAM. Contributed reagents/materials/analysis tools: RF JAB PWH RVH MAM. Wrote the paper: PWH KA RVH JAB MAM. Proposed the the sample partition method and designed the analytical cascade: JAB. Built and operated the steam equilibration method: RVH. Developed and operated the CPS ICPMS method: MAM.

References

1. Goldson SL, Rowarth JS, Caradus JR (2005) The impact of invasive invertebrate pests in pastoral agriculture: a review. New Zealand Journal of Agricultural Research 48: 401–415.
2. Pimentel D (2011) Biological Invasions: Economic and Environmental Costs of Alien Plant, Animal, and Microbe Species, 2nd Edition. Florida: CRC Press 463 p.
3. ISSG (2008) IUCN/SSC Invasive Species Specialist Group. ISSG. Available: http://www.issg.org/
4. FAO (2002) Guidelines for Phytosanitary Certificates. International Standards for Phytosanitary Measures: FAO. Available: http://www.fao.org/DOCREP/004/Y3241E/Y3241E00.HTM
5. Lavandero B, Wratten S, Hagler J, Jervis M (2004) The need for effective marking and tracking techniques for monitoring the movements of insect predators and parasitoids. International Journal of Pest Management 50: 147–151.
6. Fitzpatrick BM, Fordyce JA, Niemiller ML, Reynolds RG (2012) What can DNA tell us about biological invasions? Biological Invasions 14: 245–253.
7. Barr NB (2009) Pathway analysis of *Ceratitis capitata* (Diptera: Tephritidae) using mitochondrial DNA. J Econ Entomol 102: 401–411.
8. Manel S, Schwartz MK, Luikart G, Taberlet P (2003) Landscape genetics: combining landscape ecology and population genetics. Trends in Ecology & Evolution 18: 189–197.
9. Hobson KA, Wassenaar LI (2008) Tracking Animal Migration with Stable Isotopes; Hobson KA, Wassenaar LI, editors. Amsterdam: Academic Press. 160 p.
10. Capo RC, Stewart BW, Chadwick OA (1998) Strontium isotopes as tracers of ecosystem processes: theory and methods. Geoderma 82: 197–225.
11. Bowen GJ (2010) Isoscapes: Spatial Pattern in Isotopic Biogeochemistry. Annual Review of Earth and Planetary Sciences 38: 161–187.

12. Oulhote Y, Le Bot B, Deguen S, Glorennec P (2011) Using and interpreting isotope data for source identification. TrAC Trends in Analytical Chemistry 30: 302–312.

13. McLean JA, Bennett RB (1978) Characterization of two *Gnathotrichus sulcatus* populations by X-ray energy spectrometry. Environmental Entomology 7: 93–96.

14. McLean JA, Shepherd RF, Bennett RB (1979) Chemoprinting by X-ray energy spectrometry: We are where we eat. In: Rabb RL, Kennedy GG, editors. Movement of highly mobile insects: concepts and meth[o]dology in research: proceedings of a conference "Movement of selected species of Lepidoptera in the southeastern United States," April 9–11, 1979. Raleigh, N.C.: Dept. of Entomology, North Carolina State University. pp. 369–379.

15. Turner RH, Bowden J (1983) X-ray-microanalysis applied to the study of insect migration with special reference to rice bug, *Nilaparvata lugens*. Scanning Electron Microscopy: 873–878.

16. Bowden J, Brown G, Stride T (1979) The application of X-ray spectrometry to analysis of elemental composition (chemoprinting) in the study of migration of *Noctua pronuba* L. Ecological Entomology 4: 199–204.

17. Mertz W (1981) The essential trace elements. Science 213: 1332–1338.

18. Bowden J, Digby PGN, Sherlock PL (1984) Studies of elemental composition as a biological marker in insects .1. The influence of soil type and host-plant on elemental composition of *Noctua-pronuba* (L) (Lepidoptera, Noctuidae). Bulletin of Entomological Research 74: 207–225.

19. Dempster JP, Lakhani KH, Coward PA (1986) The use of chemical-composition as a population marker in insects - a study of the Brimstone Butterfly. Ecological Entomology 11: 51–65.

20. Wassenaar LI, Hobson KA (1998) Natal origins of migratory monarch butterflies at wintering colonies in Mexico: New isotopic evidence. Proceedings of the National Academy of Sciences of the United States of America 95: 15436–15439.

21. Hobson KA, Soto DX, Paulson DR, Wassenaar LI, Matthews JH (2012) A dragonfly (δ^2H) isoscape for North America: a new tool for determining natal origins of migratory aquatic emergent insects. Methods in Ecology and Evolution 3: 766–772.

22. Abney MR, Sorenson CE, Gould F, Bradley JR (2008) Limitations of stable carbon isotope analysis for determining natal host origins of tobacco budworm, *Heliothis virescens*. Entomologia Experimentalis et Applicata 126: 46–52.

23. Brattström O, Wassenaar LI, Hobson KA, Åkesson S (2008) Placing butterflies on the map – testing regional geographical resolution of three stable isotopes in Sweden using the monophagus peacock *Inachis io*. Ecography 31: 490–498.

24. Spence KO, Rosenheim JA (2005) Isotopic enrichment in herbivorous insects: a comparative field-based study of variation. Oecologia 146: 89–97.

25. Lancaster J, Waldron S (2001) Stable isotope values of lotic invertebrates: Sources of variation, experimental design, and statistical interpretation. Limnology and Oceanography 46: 723–730.

26. Hobson KA, Wassenaar LI, Taylor OR (1999) Stable isotopes (δD and δ^{13}C) are geographic indicators of natal origins of monarch butterflies in eastern North America. Oecologia 120: 397–404.

27. Husheer T, Frew R (2005) Stable Isotope investigation of painted apple moth and fall webworm. IsoTrace New Zealand Limited report to Biosecurity New Zealand: 13 p.

28. Hobson KA (2005) Using stable isotopes to trace long-distance dispersal in birds and other taxa. Diversity and Distributions 11: 157–164.

29. Kelly SD, Heaton K, Hoogewerff J (2005) Tracing the geographical origin of food: the application of multi-element and multi-isotope analysis. Trends in Food Science and Technology 16: 555–567.

30. Rossmann A (2001) Determination of stable isotope ratios in food analysis. Food Reviews International 17: 347–381.

31. Rubenstein DR, Hobson KA (2004) From birds to butterflies: animal movement patterns and stable isotopes. Trends in Ecology & Evolution 19: 256–263.

32. Szep T, Moller AP, Vallner J, Kovacs B, Norman D (2003) Use of trace elements in feathers of sand martin *Riparia riparia* for identifying moulting areas. Journal of Avian Biology 34: 307–320.

33. Anderson KA, Smith BW (2005) Use of chemical profiling to differentiate geographic growing origin of raw pistachios. Journal of Agricultural and Food Chemistry 53: 410–418.

34. Kelly SD, Baxter M, Chapman S, Rhodes C, Dennis J, et al. (2002) The application of isotopic and elemental analysis to determine the geographical origin of premium long grain rice. European Food Research and Technology 214: 72–78.

35. Fitt GP (1989) The Ecology of *Heliothis* Species in Relation to Agroecosystems. Annual Review of Entomology 34: 17–53.

36. Feng H, Gould F, Huang Y, Jiang Y, Wu K (2010) Modeling the population dynamics of cotton bollworm *Helicoverpa armigera* (Hubner) (Lepidoptera: Noctuidae) over a wide area in northern China. Ecological Modelling 221: 1819–1830.

37. Brevault T, Achaleke J, Sougnabe SP, Vaissayre M (2008) Tracking pyrethroid resistance in the polyphagous bollworm, *Helicoverpa armigera* (Lepidoptera: Noctuidae), in the shifting landscape of a cotton-growing area. Bulletin of Entomological Research 98: 565–573.

38. Biosecurity-New-Zealand (2006) Pathway Analysis Report 2005-06. Ministry of Agriculture and Forestry. BMG 05-06/12. 51 p.

39. Biosecurity-New-Zealand (2011) Pathway Analysis Report 2010-11. Ministry of Agriculture and Forestry. BMG 10-11/12.

40. Common IFB (1953) The Australian Species of *Heliothis* (Lepidoptera: Noctuidae) and their Pest Status. Australian Journal of Zoology 1: 319–344.

41. Pogue MG (2004) A new synonym of *Helicoverpa zea* (Boddie) and differentiation of adult males of *H. zea* and *H. armigera* (Hübner) (Lepidoptera: Noctuidae: Heliothinae). Annals of the Entomological Society of America 97: 1222–1226.

42. Armstrong KF, Ball SL (2005) DNA barcodes for biosecurity: invasive species identification. Philosophical Transactions of the Royal Society B-Biological Sciences 360: 1813–1823.

43. Sauer PE, Schimmelmann A, Sessions AL, Topalov K (2009) Simplified batch equilibration for D/H determination of non-exchangeable hydrogen in solid organic material. Rapid Communications in Mass Spectrometry 23: 949–956.

44. Schimmelmann A, Lewan MD, Wintsch RP (1999) D/H isotope ratios of kerogen, bitumen, oil, and water in hydrous pyrolysis of source rocks containing kerogen types I, II, IIS, and III. Geochimica et Cosmochimica Acta 63: 3751–3766.

45. Wassenaar LI, Hobson KA (2010) Two new keratin standards (δ^2H, δ^{18}O) for daily laboratory use in wildlife and forensic isotopic studies. The 7th International Conference on Applications of Stable Isotope Techniques to Ecological Studies. University of Alaska, Fairbanks, Alaska, USA.

46. Font L, Nowell GM, Graham Pearson D, Ottley CJ, Willis SG (2007) Sr isotope analysis of bird feathers by TIMS: a tool to trace bird migration paths and breeding sites. Journal of Analytical Atomic Spectrometry 22: 513–522.

47. Pin C, Bassin C (1992) Evaluation of a strontium-specific extraction chromatographic method for isotopic analysis in geological materials. Analytica Chimica Acta 269: 249–255.

48. Baker J, Peate D, Waight T, Meyzen C (2004) Pb isotopic analysis of standards and samples using a ^{207}Pb-^{204}Pb double spike and thallium to correct for mass bias with a double-focusing MC-ICP-MS. Chemical Geology 211: 275–303.

49. Baxter MJ (2008) Mathematics, statistics and archaeometry: The past 50 years or so*. Archaeometry 50: 968–982.

50. StatSoft_Inc (2011) Electronic Statistics Textbook. In: Hill T, Lewicki P, editors. STATISTICS: Methods and Applications. Tulsa,OK, USA: StatSoft.

51. Timm NH (2002) Applied Multivariate analysis. New York: Springer-Verlag.

52. Montgomery J, Evans JA, Cooper RE (2007) Resolving archaeological populations with Sr-isotope mixing models. Applied Geochemistry 22: 1502–1514.

53. Holder PW (2013) *Arhopalus ferus* (Cerambycidae) hydrogen isotope data as a place of origin marker. Available: http://dx.doi.org/10.6084/m9.figshare.813315

54. Schimmelmann A, Miller RF, Leavitt SW (1993) Hydrogen isotopic exchange and stable isotope ratios in cellulose, wood, chitin, and amino compounds. In: Swart PK, Lohmann KC, McKenzie J, Savin S, editors. Climate Change in Continental Isotopic Records. Washington, DC: American Geophysical Union. pp. 367–374.

55. Wassenaar LI, Hobson KA (2006) Stable-hydrogen isotope heterogeneity in keratinous materials: mass spectrometry and migratory wildlife tissue subsampling strategies. Rapid Communications in Mass Spectrometry 20: 2505–2510.

56. Langin K, Reudink M, Marra P, Norris D, Kyser T, et al. (2007) Hydrogen isotopic variation in migratory bird tissues of known origin: implications for geographic assignment. Oecologia 152: 449–457.

57. Wunder MB, Norris DR (2008) Improved estimates of certainty in stable-isotope-based methods for tracking migratory animals. Ecological Applications 18: 549–559.

58. Gröcke DR, Hardenbroek M, Sauer PE, Elias SA (2011) Hydrogen Isotopes in Beetle Chitin. In: Gupta NS, editor. Chitin: Springer Netherlands. pp. 105–116.

59. Farmer A, Cade B, Torres-Dowdall J (2008) Fundamental limits to the accuracy of deuterium isotopes for identifying the spatial origin of migratory animals. Oecologia 158: 183–192.

60. Barley ME (1987) Origin and evolution of mid-cretaceous, garnet-bearing, intermediate and silicic volcanics from Canterbury, New Zealand. Journal of Volcanology and Geothermal Research 32: 247–267.

61. Adams CJ, Maas R (2004) Rb-Sr age and strontium isotopic characterisation of the Torlesse Supergroup in Canterbury, New Zealand, and implications for the status of the Rakaia Terrane. New Zealand Journal of Geology and Geophysics 47: 201–217.

62. Timm C, Hoernle K, Van Den Bogaard P, Bindeman I, Weaver S (2009) Geochemical Evolution of Intraplate Volcanism at Banks Peninsula, New Zealand: Interaction Between Asthenospheric and Lithospheric Melts. Journal of Petrology 50: 989–1023.

63. Adams CJ, Maas R (2004) Age/isotopic characterisation of the Waipapa Group in Northland and Auckland, New Zealand, and implications for the status of the Waipapa Terrane. New Zealand Journal of Geology and Geophysics 47: 173–187.

64. Faure G, Mensing TM (2005) Isotopes: Principles and Applications. Hoboken, N.J.: Wiley. 897 p.

65. GeoRoc (2012) Geochemistry of Rocks of the Oceans and Continents. In: Sarbas B, editor. Mainz Germany. Available: http://georoc.mpch-mainz.gwdg.de/georoc/

66. Chamberlain CP, Blum JD, Holmes RT, Feng X, Sherry TW, et al. (1997) The use of isotope tracers for identifying populations of migratory birds. Oecologia 109: 132–141.

67. Frei KM, Frei R (2011) The geographic distribution of strontium isotopes in Danish surface waters - A base for provenance studies in archaeology, hydrology and agriculture. Applied Geochemistry 26: 325–340.

68. Sellick MJ, Kyser TK, Wunder MB, Chipley D, Norris DR (2009) Geographic Variation of Strontium and Hydrogen Isotopes in Avian Tissue: Implications for Tracking Migration and Dispersal. PLoS ONE 4: e4735.

69. Blum JD, Taliaferro EH, Weisse MT, Holmes RT (2000) Changes in Sr/Ca, Ba/Ca and ^{87}Sr/^{86}Sr ratios between trophic levels in two forest ecosystems in the northeastern U.S.A. Biogeochemistry 49: 87–101.

70. Blum JD, Taliaferro EH, Holmes RT (2001) Determining the sources of calcium for migratory songbirds using stable strontium isotopes. Oecologia 126: 569–574.

71. Notten MJM, Walraven N, Beets CJ, Vroon P, Rozema J, et al. (2008) Investigating the origin of Pb pollution in a terrestrial soil-plant-snail food chain by means of Pb isotope ratios. Applied Geochemistry 23: 1581–1593.

72. Kamber BS, Marx SK, McGowan HA (2010) Comment on "Lead isotopic evidence for an Australian source of Aeolian dust to Antarctica at times over the last 170,000 years" by P. De Deckker, M. Norman, I.D. Goodwin, A. Wain and F.X. Gingele Palaeogeography, Palaeoclimatology, Palaeoecology 285 (2010)205-223. Palaeogeography Palaeoclimatology Palaeoecology 298: 432–436.

73. Gulson BL, Tiller KG, Mizon KJ, Merry RH (1981) Use of lead isotopes in soils to identify the source of lead contamination near Adelaide, South Australia. Environmental Science & Technology 15: 691–696.

74. Revel-Rolland M, De Deckker P, Delmonte B, Hesse PP, Magee JW, et al. (2006) Eastern Australia: A possible source of dust in East Antarctica interglacial ice. Earth and Planetary Science Letters 249: 1–13.

75. Kaimal B, Johnson R, Hannigan R (2009) Distinguishing breeding populations of mallards (Anas platyrhynchos) using trace elements. Journal of Geochemical Exploration 102: 176–180.

76. Longhurst RD, Roberts AHC, Waller JE (2004) Concentrations of arsenic, cadmium, copper, lead, and zinc in New Zealand pastoral topsoils and herbage. New Zealand Journal of Agricultural Research 47: 23–32.

77. Norris DR, Lank D, Pither J, Chipley D, Ydenberg R, et al. (2007) Trace element profiles as unique identifiers of western sandpiper (Calidris mauri) populations. Canadian Journal of Zoology 85: 579–583.

78. Torres-Dowdall J, Farmer AH, Abril M, Bucher EH, Ridley I (2010) Trace Elements Have Limited Utility for Studying Migratory Connectivity in Shorebirds that Winter in Argentina. The Condor 112: 490–498.

79. Goede AA, Debruin M (1986) The use of bird feathers for indicating heavy-metal pollution. Environmental Monitoring and Assessment 7: 249–256.

80. Edwards WR, Smith KE (1984) Exploratory experiments on the stability of mineral profiles of feathers. Journal of Wildlife Management 48: 853–866.

81. Bortolotti GR, Szuba KJ, Naylor BJ, Bendell JF (1988) Stability of mineral profiles of spruce grouse feathers. Journal of Wildlife Management 52: 736–743.

82. Holder PW (2013) Isotopes and trace elements as geographic origin markers for biosecurity pests. A thesis submitted in partial fulfilment of the requirements for the Degree of Doctor of Philosophy: Lincoln University.

83. Ouin A, Menozzi P, Coulon M, Hamilton AJ, Sarthou JP, et al. (2011) Can deuterium stable isotope values be used to assign the geographic origin of an auxiliary hoverfly in south-western France? Rapid Communications in Mass Spectrometry 25: 2793–2798.

20

Nitrogen Biogeochemistry in the Caribbean Sponge, *Xestospongia muta*: A Source or Sink of Dissolved Inorganic Nitrogen?

Cara L. Fiore[1], David M. Baker[2], Michael P. Lesser[1]*

1 University of New Hampshire, Department of Molecular, Cellular and Biomedical Sciences, Durham, New Hampshire, United States of America, **2** Geophysical Laboratory, Carnegie Institute of Washington, Washington, District of Columbia, United States of America

Abstract

Background: Sponges have long been known to be ecologically important members of the benthic fauna on coral reefs. Recently, it has been shown that sponges are also important contributors to the nitrogen biogeochemistry of coral reefs. The studies that have been done show that most sponges are net sources of dissolved inorganic nitrogen (DIN; NH_4^+ and NO_3^-) and that nitrification, mediated by their symbiotic prokaryotes, is the primary process involved in supplying DIN to adjacent reefs.

Methodology/Principal Findings: A natural experiment was conducted with the Caribbean sponge *Xestospongia muta* from three different locations (Florida Keys, USA; Lee Stocking Island, Bahamas and Little Cayman, Cayman Islands). The DIN fluxes of sponges were studied using nutrient analysis, stable isotope ratios, and isotope tracer experiments. Results showed that the fluxes of DIN were variable between locations and that *X. muta* can be either a source or sink of DIN. Stable isotope values of sponge and symbiotic bacterial fractions indicate that the prokaryotic community is capable of taking up both NH_4^+ and NO_3^- while the differences in $\delta^{15}N$ between the sponge and bacterial fractions from the NH_4^+ tracer experiment suggest that there is translocation of labeled N from the symbiotic bacteria to the host.

Conclusions/Significance: Nitrogen cycling in *X. muta* appears to be more complex than previous studies have shown and our results suggest that anaerobic processes such as denitrification or anammox occur in these sponges in addition to aerobic nitrification. Furthermore, the metabolism of this sponge and its prokaryotic symbionts may have a significant impact on the nitrogen biogeochemistry on Caribbean coral reefs by releasing large amounts of DIN, including higher NH_4^+ concentrations that previously reported.

Editor: Melanie R. Mormile, Missouri University of Science and Technology, United States of America

Funding: This project was funded by grants from the National Oceanic and Atmospheric Administration Ocean Exploration Program, Undersea Research Program, National Institute for Undersea Science and Technology as well as the National Science Foundation. The funders had no role in study design, data collection and analysis, decision to publish, or preparation of the manuscript.

Competing Interests: The authors have declared that no competing interests exist.

* E-mail: mpl@unh.edu

Introduction

Sponges are an ecologically dominant component in many marine ecosystems, including coral reefs, where they contribute to the consolidation of reefs, prevent erosion, filter large quantities of seawater and provide habitat and food for many invertebrates and fishes [1–3]. Because of their ability to efficiently filter picoplankton, sponges can also contribute significantly to the coupling of productivity in the overlying water column to the benthos [1,4,5]. More recently, sponges and their prokaryotic symbionts have become an important area of research to quantify the fluxes of DIN by sponges and the biogeochemical cycling of nutrients on coral reefs [4,6–8].

Nitrogen cycling on tropical coral reefs is particularly important, as nitrogen is a limiting nutrient and the success of several coral reef taxa (e.g., corals) is dependent on their symbiotic partners and the efficient re-cycling of nitrogen between host and symbionts [9]. Compared to ambient seawater the excurrent water of actively pumping sponges is often enriched in DIN such as NO_3^- (or $NO_2^- + NO_3^-$) as a result of nitrification [10]. This has been documented for sponges from coral reefs, mangroves, seagrass beds [7,8,11–13], as well as sponges from temperate and cold-water environments [14–16]. In fact, coral reef sponges have been documented to have rates of nitrification that are significantly higher (5.8–16.0 mmol m^{-2} d^{-1} NO_3^-) [7,13] than what has been reported for benthic habitats such as microbial mats (up to 1.4 mmol m^{-2} d^{-1} NO_3^- [17]) or coral reef sediment (1.7 mmol m^{-2} d^{-1} NO_3^- [18]).

All pathways of nitrogen biogeochemistry have been reported to occur in sponges [10] including nitrogen fixation which was initially measured using the acetylene reduction method [19–21]. Stable isotope tracer studies (i.e., $^{15}N_2$) later confirmed the presence of nitrogen fixation, albeit at low rates, in several species of sponges from coral reefs [21,22]. Recently, the first sponge-derived *nif*H gene sequences and transcripts, which encode for the iron protein component of the nitrogenase enzyme responsible for nitrogen fixation, were documented in two sponge species from the Florida Keys [23]. *nif*H genes have also been recovered from *Xestospongia muta* from multiple locations in the Caribbean [Fiore and Lesser unpublished] and hypoxic/anoxic conditions, known to occur in sponges [10], would be required for activity since nitrogenase is inactivated by molecular oxygen and only fixes nitrogen under anaerobic or micro-aerobic conditions [10]. With the presence of anaerobic microhabitats in sponges [10,24], other anaerobic nitrogen transformations including sulfate reduction, denitrification and anaerobic ammonium oxidation (anammox) have been observed and quantified using stable isotopic tracer methods, radiolabeled isotopes and recovery of gene specific sequences for key enzymes [16,24,25]. Interestingly, genomic analysis of the candidate phylum Poribacteria, which is found in sponges from numerous marine habitats [26,27], suggests that Poribacteria may also be capable of denitrification [28].

Nitrification in sponges, which produces the bulk of the DIN released [8,13], has been documented using collection and incubation methods combined with nutrient analyses [12,13,16]. Southwell et al. [7,8] was the first to use an *in situ* method to identify actively nitrifying sponges and estimate the flux of DIN onto the adjacent coral reef, and found no significant difference in flux of DIN between the incubation and in situ methods. The interest in nutrient fluxes mediated by sponges and their symbionts as well as the nutrient biogeochemistry of coral reefs has resulted in a surge of research into the prokaryotic community composition of sponges and the processes they mediate. Recently, this has included the use of high throughput sequencing methods, such as 454 pyrosequencing of the 16S rRNA gene [[27,29,30],Fiore and Lesser unpublished] which have increased our understanding of the prokaryotic composition of many sponge species from different marine habitats. This genetic information has provided useful complementary insight for characterizing prokaryotic mediated nutrient cycling in these sponges.

On Caribbean coral reefs *Xestospongia muta* is an ecologically dominant member of the benthic community, and on Conch Reef, FL (USA) the number of *X. muta* has been shown to be significantly increasing over time [31]. *Xestospongia muta* is also characterized as a high microbial abundance sponge [32] but little was known about the composition of this community other than it contained Cyanobacteria [33] until recent studies documented a diverse prokaryotic community in *Xestospongia muta* and other members of this genus [[34,35],Fiore and Lesser unpublished]. Additionally, *X. muta* outside of the Florida Keys have not been as well studied generally with quantitative 454 pyrosequencing comparing the prokaryotic symbionts of *X. muta* from the Florida Keys, Cayman Islands and Bahamas having been recently completed [Fiore and Lesser unpublished].

The primary goal of this study was to quantify DIN fluxes in *X. muta* from the same three populations where our 16S rRNA 454 pyrosequencing study was done. We ask whether sponges from these same populations in the Caribbean have different fluxes of DIN, and potentially how any differences in DIN fluxes, may be related to the taxonomy of their symbiotic prokaryotes using a comparative approach and a natural experiment [36].

Materials and Methods

Sample Locations

Replicate sponges (n = 6) were sampled at approximately 15 m depth between 9 and 10 AM and again between 4 and 5 PM when indicated from each of three locations: Rock Bottom Reef, Little Cayman, Cayman Islands (LC) (19°42'7.36" N, 80°3'24.94" W), North Perry Reef, Lee Stocking Island, Bahamas (LSI)

(23°47'0.03" N, 76°6'5.14" W), and Conch Reef, Key Largo, FL (FL) (24°57'0.03" N, 80°27'11.16" W). All populations were sampled during the late spring and early summer of 2011 where the maximum irradiance of photosynthetically active radiation (PAR; 400–700 nm) at noon for all three locations is ~500–600 µmol quanta m^{-2} s^{-1} [Lesser unpublished]. Necessary permits were obtained for all three locations: the Marine Conservation Board, Cayman Islands; Department of Marine Resources, Bahamas; NOAA ONMS permit number FKNMS-2011-066 for Conch Reef, Florida Keys.

Nutrient analyses and rates of sponge pumping

Ambient and excurrent water samples for nutrient analysis were collected from individual sponges (n = 6) at each location for nutrient analysis by slowly filling 100 ml syringes and placing all water samples on ice for transport to shore. Ambient water was obtained by filling the syringe adjacent to each sponge (within 20 cm of the sponge body wall) and excurrent water was obtained by placing weighted Tygon ® tubing inside the sponge close to the base of the spongocoel that was attached to a 100 ml syringe and drawing water into the syringe slowly (~1 ml s^{-1}). Syringes were then purged of approximately 10 ml and then 40 ml was saved and frozen for NH_4^+ and $NO_2^- + NO_3^-$ analysis (NO_x^-). For the nutrient analyses the water samples were thawed and filtered (0.22 µm, Whatman, USA) to remove particulate matter then re-frozen and sent to the Nutrient Analytical Facility at Woods Hole Oceanographic Institute (WHOI, Woods Hole, MA, USA) for analysis using a Lachet QuickChem 8000 (flow injection analysis system) according to standard protocols to determine concentration of NH_4^+ and NO_x^-. Instrumental errors associated with the measurements were calculated as relative standard deviation (RSD) and includes: NH_4^+ −0.6% measured RSD, and $NO_2^- + NO_3^-$ −0.59% measured RSD. Sponges were marked near their base with labeled flagging tape attached to nails embedded in the substrate to facilitate repeated measurements on the same sponges.

The volume flow or pumping rates for each individual sponge was determined as previously described [37]. A small amount (~1 ml) of fluorescein dye was injected using a syringe and 16 gauge needle into the sponge just below the base of the spongocoel and the time(s) that the dye front took from its first appearance at the base of the spongocoel to the top of the spongocoel was recorded to obtain the centerline fluid velocity to calculate volume flux or pumping rate. We understand that unlike previous studies on tubular sponges where plug flow can be reasonably assumed (e.g., [37]) the morphology of *X. muta* likely creates more complicated excurrent plumes where the velocity across the osculum is not uniform [38]. This is easily observed using the timing of multiple dye tracks on *X. muta* injected in different locations with dye tracks closer to the sponge wall being slower than the centerline flow [Lesser unpublished]. As a result we recognize that our measurements of volume flow or pumping rates are likely to be an overestimate. That said our estimates of volume flow are in agreement with the results of Southwell et al. [8] using similar techniques for *X. muta*. Additionally, in our hands we have never observed cessation of pumping, or other artifacts, as a result of exposure to fluorescein in both thin walled and thick walled sponges [5,37]. Both spongocoel and total sponge volume were calculated by measuring sponge height, base circumference, osculum diameter, and spongocoel depth and inner diameter with a measuring tape (to ±1.0 mm) and volume calculated as previously described [31]. The mass (kg) of individual sponges was then calculated by multiplying the individual total sponge volume (l), obtained as described above, by the average density of *X. muta* sponges (0.617 g cm^{-3}) which was determined from direct measurements of the displacement volume and mass of pieces ($n = 5$) of sponge (including both mesohyl and pinacoderm). The flux of nutrients was then calculated by multiplying the ΔDIN (the difference in nutrient concentration between the ambient and excurrent water in μmol l^{-1} by the flow rate (cm s^{-1}) and normalized to both sponge volume and mass for comparisons between sites.

Pumping rates and nutrient flux data were tested for assumptions of ANOVA and if the data failed either normality or homoscedasticity a constant integer to all values was added followed by log transformation. The transformed data passed Bartlett's test [39] for homoscedasticity but often failed the Shapiro-Wilks [40] test for normality. Because Bartlett's test is sensitive to deviations in normality [41] we choose to proceed with ANOVA, which is known to be robust to deviations from normality [42], on the transformed data. To determine if time of day was a significant factor in the flux of DIN, a two-way ANOVA with interaction was performed using the statistical program R [43] with time (AM and PM) and location as fixed factors for the flux of NO$_x^-$, NH$_4^+$, total DIN and pumping rate as the response variables. A repeated measures ANOVA was not performed because the general requirement of this approach is three time points. Since the effect of time and interaction of time with location was not significant the flux of NO$_x^-$, NH$_4^+$, total DIN and pumping rate, ΔDIN and pumping rates for each location were calculated by averaging the AM and PM values for each individual sponge. Collapsing the design to a single factor analysis to examine differences between locations was then assessed using a one-way ANOVA with location as a fixed factor [41].

Flow Cytometry

Ambient and excurrent water samples were collected as described above for the nutrient analyses for another set of sponges ($n = 4$) from LSI only. Approximately 3 ml from each collected water sample were fixed in electron microscopy grade paraformaldehyde at a final concentration of 0.5% in filtered (0.22 μm) seawater and frozen at $-50°$C. Frozen water samples were sent to the Bigelow Laboratory for Ocean Sciences J.J. MacIsaac Aquatic Cytometry Facility where they were stored in liquid nitrogen until analysis. Each sample was analyzed for cell abundances using a Becton Dickinson FACScan flow cytometer with a 30 mW, 488 nm laser. Simultaneous measurements of forward light scattering (FSC, relative size), $90°$ light scatter (SSC), chlorophyll fluorescence (>650 nm), and phycoerythrin fluorescence (560–590 nm) were made simultaneously on each sample as previously described [5]. Calculations of cyanobacteria, prochlorophyte, and heterotrophic cell concentration and filtering efficiency were performed as previously described [5]. Technical replicates ($n = 2$) were averaged for each sample and the cell abundance of heterotrophic bacteria was determined using PicoGreen (Molecular Probes), a dsDNA specific dye, which stains all prokaryotes (emission fluorescence 515–525 nm). Subtraction of the chl a containing picoplankton from the total prokaryotes yielded the heterotrophic bacterial component of the community while cyanobacterial and prochlorophyte cells were differentiated by the presence or absence, respectively, of phycoerythrin fluorescence. All filtered cells were converted to carbon and nitrogen equivalents using the following conversions; heterotrophic bacteria: 20 fg C cell^{-1} [44], *Prochlorococcus*: 53 fg C cell^{-1} [45], *Synechococcus*: 470 fg C cell^{-1} [46], heterotrophic bacteria: 3.3 fg N cell^{-1} [47], *Prochlorococcus*: 9.4 fg N cell^{-1} [48], *Synechococcus*: 35 fg N cell^{-1} [48]. Data were log transformed or

arcsin transformed as necessary and an ANOVA followed by Tukey's HSD were performed to test for significant differences in the number of filtered cells between cell types (cyanobacteria, prochlorophytes, heterotrophic bacteria and total cells), filtration efficiency and total particulate carbon (POC) and nitrogen (PON) consumed by sponges.

Stable isotopic analyses and tracer experiments

Sponge samples that were frozen without buffer ($n = 3$ each location) were later lyophilized, ground to a powder with a mortar and pestle, and then acid treated with 1 M HCl to remove carbonate and rinsed with distilled water and allowed to dry. An analysis of samples from FL, separated into the outer pigmented layers of the sponge and the non-pigmented inner tissues (containing the pinacoderm and outer mesohyl respectively), showed no significant differences in stable isotope signatures [Fiore and Lesser, unpublished data] so whole cross-sections of sponge samples, consisting of both pinacoderm and mesohyl, from all locations were analyzed. Samples were then sent to the Marine Biological Laboratory (MBL) for the analysis of particulate C and N as well as the natural abundance of the stable isotopes δ^{15}N and δ^{13}C. Samples were analyzed using a Europa ANCA-SL elemental analyzer-gas chromatograph attached to a continuous-flow Europa 20–20 gas source stable isotope ratio mass spectrometer. The carbon isotope results are reported relative to Vienna Pee Dee Belemnite and nitrogen isotope results are reported relative to atmospheric air and both are expressed using the delta (δ) notation in units per mil (‰). The analytical precision of the instrument is ± 0.1‰, and the mean precision of sample replicates for δ^{13}C was ± 0.4‰ and δ^{15}N was ± 0.2‰. A one-way ANOVA was used to test for significant differences between locations for δ^{13}C, δ^{15}N and C:N ratios followed by the post hoc multiple comparison Tukey's HSD test as needed.

Two stable isotope tracer experiments were conducted during the summer of 2011 at LSI: the first used Na^{15}NO$_3$ (5 mg l^{-1} final concentration) plus H^{13}CO$_3$ (50 mg l^{-1}) and the second used ^{15}NH$_4$ (0.31 mg l^{-1}) as tracers (Sigma-Aldrich, USA). The method was the same for each experiment: 11 individual *X. muta* (average volume 172 ± 77 ml or mass 0.106 ± 0.048 kg; mean $\pm$SD) were collected by cutting through the bottom of the sponge but keeping the tissue from the base of the spongocoel intact, from approximately 12 m at North and South Perry reefs at LSI and held in a large holding tank with flow through seawater for 5 d to recover from being removed from the reef. Care was taken to ensure that sponges were never exposed to air and that light levels were maintained at the same levels as found at ~12 m using neutral density screens over the outdoor flowing seawater tanks. Sponges were checked for pumping activity using fluorescein dye and incubated statically with the tracer compound(s) for 4 h. Subsequently, T$_0$ sponges ($n = 3$) were then removed and stored for analysis.

The remaining sponges were placed in individual aquaria with flow through seawater and sponges were sampled at 3 h ($n = 2$), 6 h ($n = 3$) and at 12 h ($n = 3$) for each fraction. Frozen samples were initially processed by separating the bacteria and sponge fractions following the methods of Freeman and Thacker [49] and Freeman et al. [50] except for two steps: an initial centrifugation was performed at $520 \times$ g for 4 min, and the resulting sponge pellet was rinsed an additional two times. The purity of the sponge and bacterial fractions were assessed using light and epifluorescence light microscopy as described by Freeman and Thacker [49]. The sponge fractions always contained large cells (8–10 μm diameter) consisting of at least 85% per microscopic field and exhibiting low natural fluorescence, whereas the bacterial fractions contained only small cells (<1–2 μm diameter) and high natural fluorescence. While efforts were made to separate and purify the sponge fraction as much as possible from all prokaryotic cells, it is possible that some prokaryotes that were located intracellularly were not detected (non-fluorescent) in the sponge fraction. These methods have been shown to be effective for other sponge species [49] and were optimized for use with *X. muta*. Additionally, a one-way ANOVA of the C:N ratios for the two fractions yielded significant differences ($F_{1,13} = 8.38$, $p = 0.01$ (NH_4^+ tracer experiment); $F_{1,13} = 24.6$, $p < 0.01$ ($NO_3^- + HCO_3^-$ tracer experiment) indicating that good separation of these fractions occurred. It is likely, however, that some contamination occurred and was considered when interpreting the results of these experiments. Samples were then lyophilized and $\sim$1.0 mg was weighed and placed into silver capsules (Costech, CA, USA) and acidified three times with 20 μl of 12 M HCl. Samples were allowed to dry in between acidifications, then oven dried at 50°C for 48 h. Samples were combusted in a Carlo-Erba NC2500 elemental analyzer, and the resulting gas was analyzed in a Thermo Delta V isotope ratio mass spectrometer via a Conflo III open-split interface. The analytical precision of the instrument was ±0.2‰, and the mean standard deviation of sample replicates for $\delta^{13}C$ was ±0.4‰ and for $\delta^{15}N$ it was ±0.8‰ for enriched samples and ±0.1‰ and ±0.1‰ for natural abundance samples, respectively. For the tracer experiments the data were log transformed as necessary to meet the assumptions of parametric statistics and a two-way ANOVA with interaction, with fraction and time as fixed factors, was used to assess treatment effects.

Results

Stable Isotopic Signatures of Sponges

The values of $\delta^{13}C$ from each location were not significantly different from each other (ANOVA, $F_{2,6} = 1.16$, $p = 0.38$). The $\delta^{13}C$ of sponge samples ranged from -19.1 to -18.4‰ (Fig. 1). The $\delta^{15}N$ of sponge samples ranged from 4.0–4.4‰ (Fig. 1), and were not significantly different between locations (ANOVA, $F_{2,6} = 0.52$, $p = 0.62$). The ratios of C:N were significantly different between locations (ANOVA, $F_{2,6} = 22.57$, $p = 0.002$), with post hoc pairwise comparisons showing that FL sponges had significantly higher C:N ratios than LSI (Tukey's HSD, $p<0.05$), and that LC sponges significantly higher than LSI (Tukey's HSD, $p<0.05$). There was no significant difference between LC and FL (Tukey's HSD, $p>0.05$). Despite the significant results for C:N ratios the mean values did not vary greatly, with a range of 4.33–4.83.

Inorganic Nitrogen Fluxes in *Xestospongia muta*

The difference in nutrient concentration between the ambient and excurrent of NH_4^+, $NO_3^- + NO_2^-$ (NO_x^-) and total DIN for *X. muta* varied considerably between individual sponges as expected for sponges over a large size range (Table 1). The ΔNH_4^+ values were not significantly different between locations (ANOVA, $F_{2,15} = 3.19$, $p = 0.07$) as were the ΔNO_x^- values (ANOVA, $F_{2,15} = 2.58$, $p = 0.11$). ΔDIN values, however, were significantly different between sites (ANOVA, $F_{2,15} = 6.82$, $p = 0.008$) with post hoc pairwise comparisons showing that FL sponges were significantly lower than both LSI and LC sponges LSI (Tukey's HSD, $p<0.05$) which were not significantly different than each other (Tukey's HSD, $p>0.05$). No measurements of ambient NO_x^- exceeded 4 μM eliminating the potential for ambient nutrient concentrations to be confounded by oceanographic features such as internal waves [51,52]. Sponge pumping

rates varied with size (Table 1) and did not differ significantly with location (ANOVA, $F_{2,15} = 1.61$, $p = 0.23$).

The volume and mass normalized fluxes of NH_4^+ (Table 1, Fig. 2 a) were not significantly different between locations (ANOVA, $F_{2,15} = 0.45$, $p = 0.65$ (volume); $F_{2,15} = 0.85$, $p = 0.45$ (mass)). The fluxes of NO_x^- normalized to sponge volume (Table 1, Fig. 2 b) were significantly different between locations (ANOVA, $F_{2,15} = 4.89$, $p = 0.02$) with FL sponges significantly lower than LSI and LC not significantly different than either FL or LSI (Fig. 2A) but when normalized to mass did not show a significant effect of location (ANOVA, $F_{2,15} = 3.56$, $p = 0.054$). The flux of total DIN ($NO_x^- + NH_4^+$) normalized to volume and to mass were not significantly different among locations (ANOVA, $F_{2,15} = 1.24$, $p = 0.32$ (volume); $F_{2,15} = 2.09$, $p = 0.16$ (mass)).

Feeding Study

Xestospongia muta (n = 4) from LSI instantaneously filtered an average of 1.5×10^7 cells ml^{-1} and there was a significant effect of cell type (ANOVA, $F_{3,12} = 4.45$, $p = 0.03$) (Fig. 3A). The number of both total cells and heterotrophic bacteria filtered was significantly higher than that of prochlorophytes (Tukey's HSD, $p<0.05$) but not cyanobacteria (Tukey's HSD, $p>0.05$), while the number of cyanobacterial cells filtered was indistinguishable (Tukey's HSD, $p>0.05$) from the prochlorophyte or total cell and heterotrophic cell groupings (Fig. 3A). For the filtration efficiency of each cell type there were no significant differences (ANOVA, $F_{3,12} = 0.45$, $p = 0.72$) (Fig. 3B). The total amount of POC for each retained cell type was greatest for cyanoacteria, but there was no significant difference between cell types (ANOVA, $F_{3,12} = 2.77$, $p = 0.09$). Differences between the amount of PON for each retained cell type was significant (ANOVA, $F_{3,12} = 4.68$, $p = 0.02$) and greatest for total cells and heterotrophic bacteria compared to prochlorophytes (Tukey's HSD, $p = 0.03$) but not cyanobacteria (Tukey's HSD, $p>0.05$), while the PON of cyanobacterial cells was indistinguishable (Tukey's HSD, $p>0.05$) from the prochlorophyte or total cell and heterotrophic cell groupings (Fig. 3C).

Nitrogen tracer experiment: Nitrate and Bicarbonate

While sponge and bacterial fractions became more enriched from 3 to 6 hours there was no significant effect of enrichment of ^{15}N from the NO_3^- tracer in those fractions (ANOVA, $F_{7,14} = 0.84$, $p = 0.57$). There was also no significant difference between sponge and bacterial fractions or over time or the interaction of fraction and time (fraction, $F_{1,3} = 0.89$, $p = 0.36$; time, $F_{1,3} = 1.16$, $p = 0.36$; interaction term, $F_{1,3} = 0.28$, $p = 0.84$) (Fig. 4A). Additionally, the enrichment of ^{13}C from the bicarbonate tracer experiment was significant (ANOVA, $F_{7,14} = 20.9$, $p<0.0001$) with fraction being non-significant ($F_{1,3} = 0.13$, $p = 0.13$) and time being significant ($F_{1,3} = 44.3$, $p<0.001$) with a non-significant interaction term, ($F_{1,3} = 3.2$, $p = 0.056$) (Fig. 4B). As a result post-hoc multiple comparison tests were only performed for time. Both the sponge and bacterial fractions became more enriched in ^{13}C then the sponge fraction over time with all sampling periods being significantly different than T_0 (Tukey's HSD $p<0.05$) and not significantly different (Tukey's HSD $p>0.05$) from each other (Fig. 4B).

Nitrogen tracer experiment: Ammonium

There was significant enrichment of ^{15}N from the NH_4^+ tracer in the experimental sponges (ANOVA, $F_{7,13} = 3.38$, $p = 0.03$) with the effects tests for fraction ($F_{1,3} = 6.4$, $p = 0.03$) and time ($F_{1,3} = 5.0$, $p = 0.02$) being significant and the interaction term non-significant ($F_{1,3} = 1.8$, $p = 0.19$) (Fig. 4C). Post-hoc multiple comparison testing for time revealed a significant (Tukey's HSD

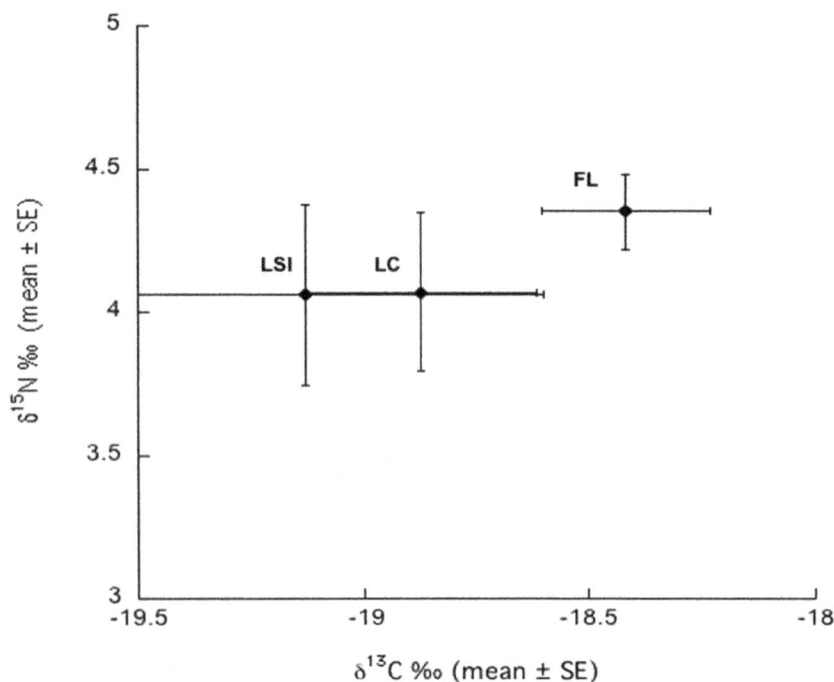

Figure 1. $\delta^{15}N$ and $\delta^{13}C$ values (mean ±SE) for *Xestospongia muta* (n = 3) for each location. FL = Florida Keys LC = Little Cayman, LSI = Lee Stocking Island, Bahamas.

p<0.05) increase in enrichment over time with the bacterial fraction exhibiting greater enrichment (Fig. 4C).

Discussion

This study of the inorganic nitrogen fluxes in *Xestospongia muta* is the first to show differences in net fluxes of DIN both within and between populations of this ecologically dominant sponge on Caribbean coral reefs. *Xestospongia muta* densities are increasing [31], so quantifying both the fluxes of DIN and understanding the underlying processes driving the nitrogen biogeochemistry in this sponge is important for understanding the DIN availability on reefs in the Caribbean.

In a survey of nitrification in sponges on Conch Reef, Florida Southwell et al. [7,8] reported evidence of nitrification in nine out of twelve sponge species, including *X. muta*. The rates of nitrification measured in these sponges varied, but overall they were at least two orders of magnitude higher than other habitats (e.g., benthos, coral rubble). In the current study, we observed similar rates of *X. muta* pumping activity reported for Conch Reef sponges by Southwell et al. [8]. However, unlike Southwell et al. [7,8], *X. muta* from Conch Reef exhibited a negative flux of NO_x^-, indicating that either denitrification or anammox processes were taking place or possibly dissimilatory nitrate reduction. Our results from LSI and LC are consistent with Southwell et al. [7,8] and for LSI the fluxes of NO_x^- are significantly higher than fluxes of NO_x^- from FL compared to Southwell et al. [8].

Xestopongia muta were actively pumping for all measurements taken during this study, which was not significantly different over time of day or between locations. While pumping rates were not significantly different, the observed variability in the unidirectional pumping of sponges has the potential to create microhabitats where both anaerobic nitrogen transformations (e.g., denitrification) and aerobic nitrogen transformations (e.g., nitrification) could occur [10,16,24,53]. Additionally, recent studies of the bacterial

communities of *X. muta* using16S rRNA sequencing [35,Fiore and Lesser unpublished] have reported many bacterial groups that are capable of denitrification and anammox (i.e., Burkholderiales, Pseudoalteromonadaceae, Poribacteria, Planctomycetes).

Interestingly, Southwell et al. [8] found that NO_x^- made up the majority of the DIN pool from *X. muta* and that NO_x^- was almost entirely NO_3^-. In this study we observed, in addition to positive net NO_3^- fluxes, a greater net efflux of NH_4^+ for all samples of *X. muta*. For some *X. muta* populations (i.e., LSI and LC) the flux of NH_4^+ had a significant impact on total DIN fluxes. These differences in the fluxes of NH_4^+, probably generated from the utilization of nitrogen rich POM by the sponge host, are unusual given there is an active nitrifying community [7,8], and a prokaryotic photosynthetic community [22] that could readily utilize NH_4^+ in this sponge [54].

The fluxes of DIN from sponges such as *X. muta* can have a significant impact on the availability and composition of DIN on coral reefs [7,8]. Results of the current study indicate that fluxes of DIN from *X. muta*, the primary contributor to DIN on Caribbean coral reefs [7,8], is more complex than previously thought and is significantly different between locations. Additionally, these fluxes vary over time as a previous study of the same population of *X. muta* from LSI in 2010 showed both positive and negative fluxes of NO_x^- with a net ΔNO_x^- of -0.27 μM±0.06 (mean ±SE) [Fiore and Lesser unpublished].

Natural abundance stable isotope values have been commonly used to trace sources of C and N through the food chain [55]. Nitrogen fixation yields an average $\delta^{15}N$ signature of approximately 0.0‰ [56,57], and trophic enrichment typically results in a +2.2 to +3.5‰ increase per trophic level for $\delta^{15}N$ [58,59]. Therefore, several studies have used a cutoff of ≤2.0‰ to indicate N from a fixed source [23,60,61]. Additionally, carbon fixation by marine phytoplankton typically results in $\delta^{13}C$ values of about -19 to -24‰ [55], with an average of +0.5 to +1.0‰ enrichment

Table 1. Calculated volume, mass, and flux parameters for samples of *Xestospongia muta* at each location.

Location	Spongocoel (L)	Volume (L)	Mass (kg)	Flow rate (L h-1)	ΔDIN NH_4^+ (µmol L^{-1})	ΔDIN NO_x (µmol L^{-1})	Flux NH_4^+ (µmol h^{-1} L^{-1})	Flux NO_x (µmol h^{-1} L^{-1})	Flux DIN (µmol h^{-1} L^{-1})	Flux NH_4^+ (µmol h^{-1} kg^{-1})	Flux NO_x (µmol h^{-1} kg^{-1})	Flux DIN (µmol h^{-1} kg^{-1})
FL	4	43	26.8	4050	-0.15	-0.99	-13	-103	-116	-21	-167	-188
FL	20	101	62.2	18090	0.00	0.65	0	117	117	0	189	189
FL	39	36	22.4	62370	0.35	-0.30	668	-496	172	904	-804	100
FL	33	111	68.2	21938	0.10	-0.09	26	-26	1	43	-42	1
FL	14	76	47.0	15377	0.90	0.26	-188	51	-137	-387	82	-305
FL	7	37	22.9	6480	-0.05	-0.01	-9	-2	-9	-5	-3	-7
LC	12	64	39.3	4656	5.10	2.60	328	116	444	1653	1037	2690
LC	5	12	7.4	2649	1.80	-0.30	2665	-181	2484	1313	-122	1191
LC	4	50	31.1	5423	0.35	0.90	38	92	130	681	1887	2567
LC	6	49	30.3	7109	0.40	0.35	81	71	151	333	1021	1354
LC	4	17	10.5	2145	3.30	0.25	213	41	254	4279	324	4603
LC	14	63	38.9	8463	0.45	0.20	60	31	90	350	272	622
LSI	5	22	13.6	6912	0.36	1.23	-180	438	258	-292	710	418
LSI	10	45	27.6	16640	0.11	0.53	40	66	106	64	107	172
LSI	29	81	50.2	28800	0.50	0.63	213	159	372	344	258	603
LSI	2	18	11.0	11040	-0.12	0.87	-162	1202	1040	-262	1948	1686
LSI	3	23	14.0	2687	-0.19	0.30	12	15	27	19	24	43
LSI	2	14	8.6	11220	2.53	0.67	4313	793	5106	6990	1286	8276

A

B

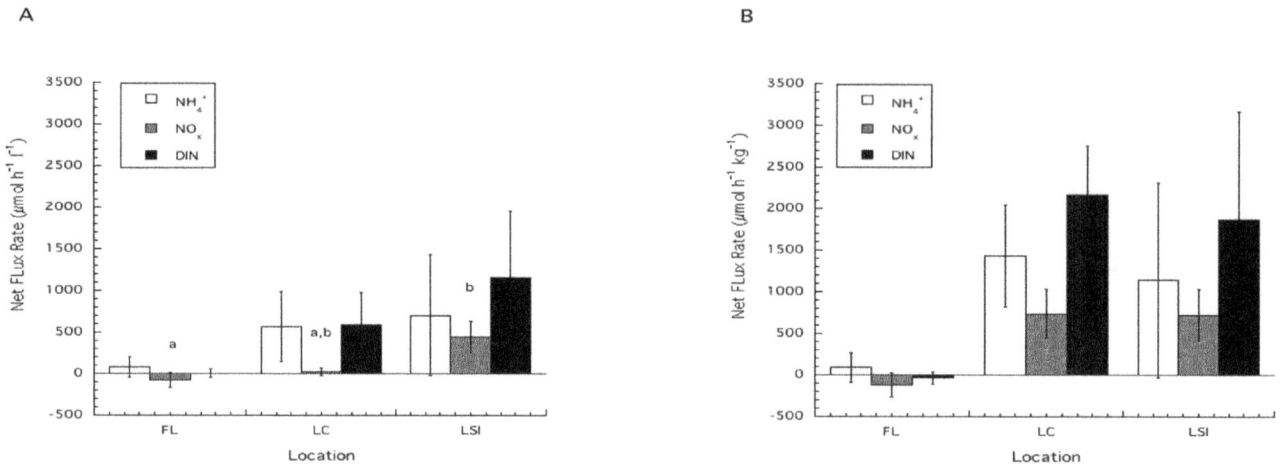

Figure 2. The average flux (mean ±SE) of NH$_4^+$, NO$_x^-$ and DIN. The average flux is shown for each location normalized to sponge volume (A) and mass (B). Treatment groups with similar superscripts are not statistically different from each other.

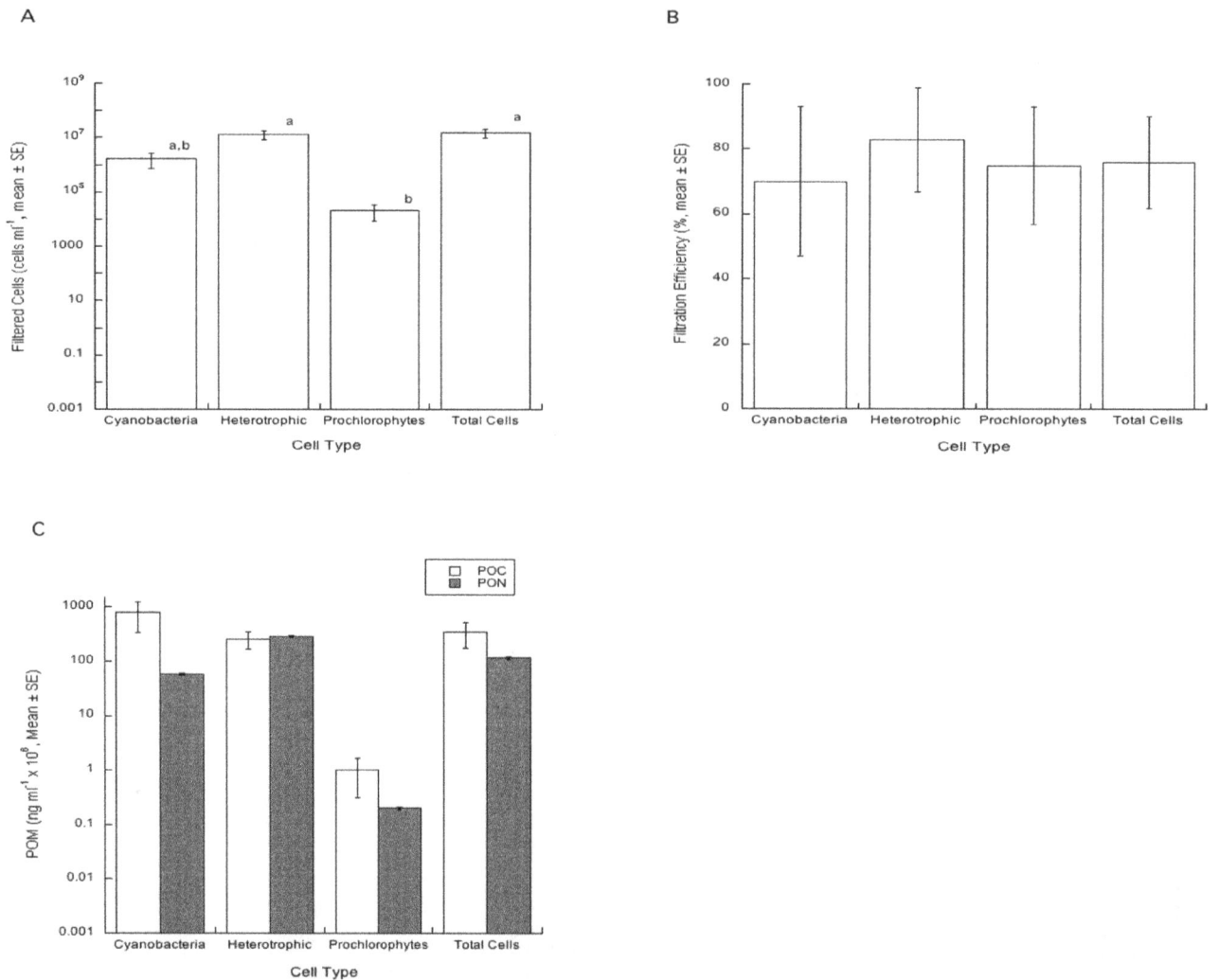

A

B

C

Figure 3. Filtration of bacterioplankton from the water column by *X. muta*. Number of filtered cells (A), filtration efficiency (B), and particulate organic matter as carbon and nitrogen (C) available from filtered cells for *X. muta* from LSI (n = 4). Treatment groups (mean ±SE) with similar superscripts are not statistically different from each other.

A

B

C

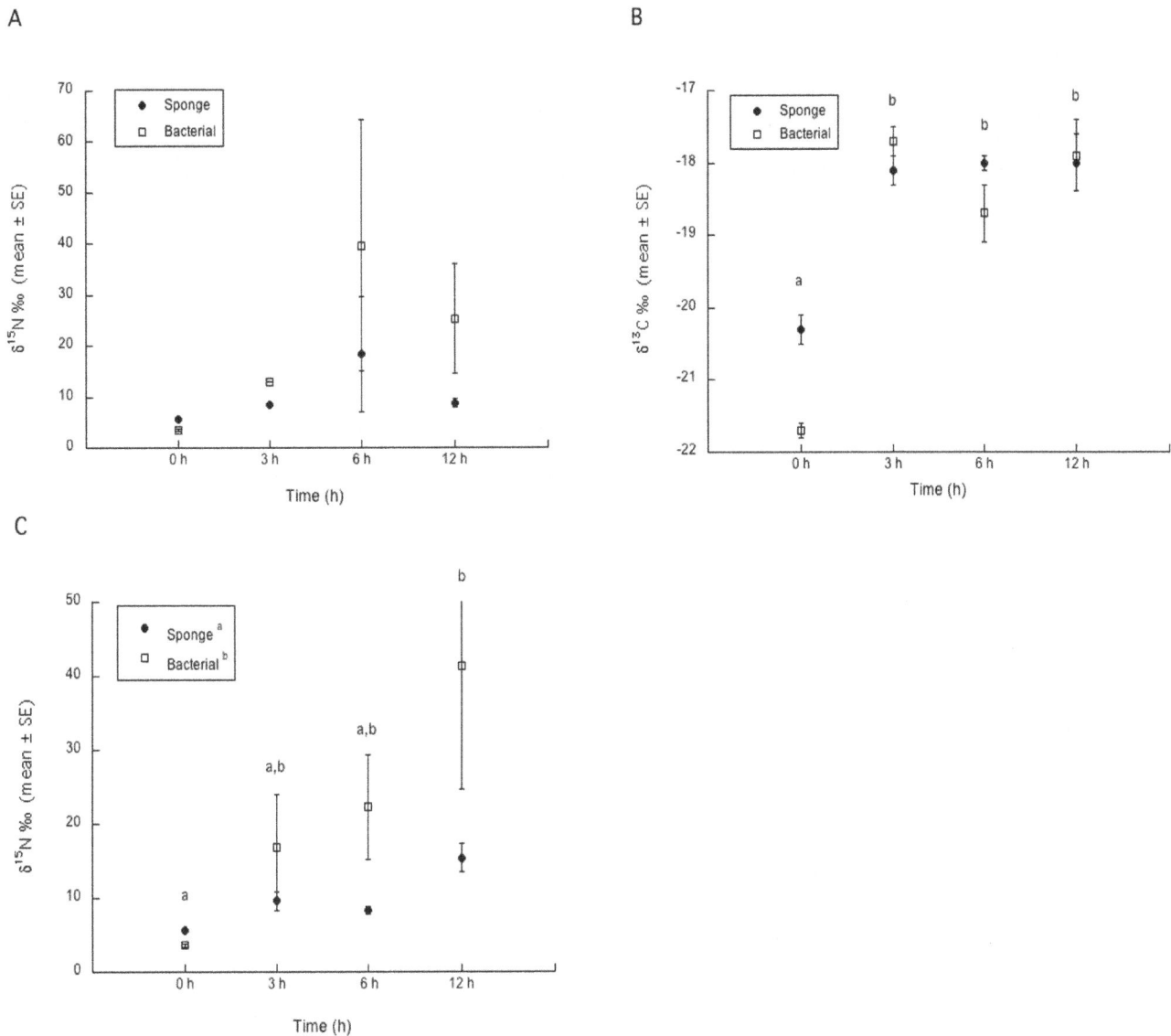

Figure 4. δ^{15}N and δ^{13}C isotopic values from uptake experiments with *X. muta*. δ^{15}N over time (A) for ^{15}N nitrate and δ^{13}C over time (B) for ^{13}C bicarbonate enriched sponge and bacterial fractions, and δ^{15}N over time for ^{15}N ammonium enriched sponge and bacterial fractions (C). Samples collected at 3 h were under low irradiances while samples collected at 6 h had been exposed to sunlight (Under neutral density screens representative of irradiances at the depth of collection) and samples from 12 h were collected at night (for both experiments). Treatment groups (mean ±SE) with similar superscripts are not statistically different from each other.

per trophic level [62]. Based on previous studies that have used stable isotope analysis to investigate the relationship between sponges and their symbionts [6,7,49], a cutoff for δ^{13}C of −18‰ or lower was used as an indication of photoautotrophic carbon fixation for *X. muta*. The bulk stable isotopic values measured for both C and N in *X. muta* tissue, comprising both the host tissue and prokaryotic biomass, were not significantly different between sites and similar to those documented in previous studies on *X. muta* [22,23].

Using the cutoff values described above, there is no stable isotopic evidence that nitrogen fixation was occurring in *X. muta* (Fig. 1) although *nif*H genes have been sequenced from *X. muta* [Fiore and Lesser unpublished]. The δ^{13}C values of the sponge samples show evidence of photoautotrophy, although we cannot say definitively that there is transfer of C from symbionts to the sponge. Freeman and Thacker [49] demonstrated that high

microbial abundance (HMA) sponges, such as *X. muta*, can obtain either C or N, or both, from their symbionts. Any interpretation of tissue stable isotopic signatures must, however, include heterotrophy on picoplankton for sponges. Based on the feeding study of LSI sponges, and despite the high abundance of resident bacteria, sponges are actively and non-selectively filtering most of the bacteria from the ambient water, which would supply significant amounts of POC and PON that could be potentially used by the host which is consistent with studies on other sponge species from the Caribbean [5,37]. Taken together, the results of this study suggest some site related differences with sponges from the more open ocean sites of LSI and LC being more dependent on photoautotrophic sources of C and coastal FL sponges being more dependent on POM for their C requirements (Fig. 1).

The tracer experiments provide additional insight into what may be occurring in terms of the dynamics of C and N uptake by the prokaryotic community. Results of the HCO_3^- tracer experiment indicate that there is uptake by prokaryotes and then immediate equilibrium of C with the sponge tissue. These results show that there is an active autotrophic community present in *X. muta* and provides evidence for transfer of C from symbiont to host. Uptake of HCO_3^- is not limited to autotrophs, and heterotrophic bacteria may have also taken up HCO_3^- for use in aneplurotic pathways [63].

The NH_4^+ tracer experiment indicates that the symbiotic prokaryotic community of *X. muta* actively assimilated the NH_4^+ and the host sponge is the likely source of NH_4^+. The small increase in sponge host $\delta^{15}N$ during the NH_4^+ experiment may also indicate direct uptake of NH_4^+ by the sponge, as has been demonstrated for corals [64]. Alternatively, as in the NO_3^- tracer experiment, the accumulation of tracer in the sponge fraction may also be explained by transfer of N from the bacteria to the host.

NO_3^- can also be utilized by the prokaryotic community, as demonstrated by the increased $\delta^{15}N$ of the bacterial fractions incubated with $^{15}NO_3^-$ (Fig 4 A). Photosynthetically driven NO_3^- uptake has been demonstrated in planktonic communities [65], and may explain the increase in NO_3^- uptake when the sponges were exposed to natural solar radiation. However, it should be noted that heterotrophic bacteria could take up NO_3^- as well [66]. As the symbiotic nitrifiers in the sponge produce NO_3^- it would provide a source of NO_3^- for uptake by the photosynthetic community as well as a substrate for denitrification. Similarly, NH_4^+ is also a likely substrate for aerobic ammonia oxidation by the crenarcheote community, which is supported by the recovery of crenarchaeal *amo*A genes in *X. muta* [34], expression of *amo*A genes [34] and tracer studies [8]. If assimilatory and dissimilatory processes are competing for NO_3^-, NO_2^-, and NH_4^+, which have been documented in other communities [67], then this may have a significant role in nitrogen cycling in the sponge holobiont. Lastly, the anaerobic oxidation of NH_4^+ (anammox) is another process that may utilize both NH_4^+ and the NO_2^- generated from nitrification. The rates, however, of anammox are relatively low in the water column [68], as is the only documented rate for anammox in sponges [16]. It is possible that anammox may occur within anoxic microhabitats of *X. muta*, and support for this is provided by the presence of planctomycete bacteria in *X. muta* [35,Fiore and Lesser unpublished] but the nutrient flux data clearly show that a net efflux of NH_4^+ is still occurring in all sponges suggesting an abundance of this substrate for either nitrification or anammox. Other factors that may be important in explaining the observed variability in net fluxes of DIN include the variability in tissue O_2 concentration as a result of pumping activity [10] and the concentration of other compounds that are known to influence N cycling such as H_2S. Variable O_2 concentrations within the sponge tissue will determine the relative rates of nitrification and denitrification [10,53], while H_2S is known to inhibit nitrification and denitrification [69,70]. H_2S may be present in *X. muta*, as bacteria involved in sulfur cycling have been recovered from this sponge (Chromatiales, Syntrophobacteraceae, Fiore and Lesser unpublished). If nitrification and denitrification are tightly coupled, then variations in H_2S or O_2 concentrations may indeed influence the rates of these processes and the net fluxes of various species of DIN.

If we consider each outcome in terms of net flux of DIN from *X. muta* from the current study, we can model which dissimilatory processes are likely occurring that can then be used to formulate testable hypotheses (Fig. 5) for future studies. The model also allows us ask questions on the broader ecological impacts of sponge-derived DIN; for example, LC sponges, which generally had a net positive flux of both NH_4^+ and NO_x^- (Fig. 5) and often NH_4^+ was a significant component to total DIN (Fig. 2), may differentially influence N cycling on the surrounding coral reef relative to FL or LSI sponges. We do not know the extent to which sponge-derived DIN influences the biogeochemistry and ecology of the surrounding habitat, but studies on multi-species sponge assemblages, and coral reef communities dominated by active suspension feeding sponges, have shown the significant role of active suspension feeding and the coupling of POC and PON from the water column to the benthos [71,72]. The composition of DIN released into the water column by sponges would influence how it might be utilized, and who utilizes it in the surrounding environment, as NH_4^+ is more readily incorporated into biomass than NO_3^- which can then potentially support local increases in planktonic community production [73]. As discussed by Southwell et al. [8], excess inorganic nutrients, such as release of DIN by sponges, may have detrimental effects on coral reef ecosystems by stimulating an increase in the growth of fleshy algae in the absence of herbivores [8]. It is important that further research be done to determine the ecosystem level effects of DIN release by sponges, and particularly from *X. muta* in regards to Caribbean coral reefs as it is believed to be a primary contributor of DIN released by sponges [8].

We have shown that the flux of DIN from three populations of *X. muta* is highly variable which may have a significant impact on the availability of DIN on coral reefs given the high abundance of these sponges. Nitrification had been previously demonstrated to occur in *X. muta*, and we show here that other nitrogen transformations including denitrification and/or anammox may occur in these sponges as well as the importance of active suspension feeding on the nitrogen rich pool of picoplankton. Further work is needed to better characterize the flux of DIN from *X. muta* and other sponges on Caribbean coral reefs including whether sponges can fix N_2. This will require additional investigations on the functional activity of the symbiotic prokaryotic community of sponges using a combination of experimental and molecular approaches (i.e., transcriptomics) that will yield insight into the taxonomy and function of this community, and how this impacts nutrient fluxes and biogeochemical cycling on Caribbean coral reefs.

Acknowledgments

We thank Marc Slattery, Deborah Gochfeld, Erica Hunkin, Cole Easson, Sylvester Lee, Julia Stevens, Christopher Freeman, Julie Olson, Mauritius Bell, and the Aquarius team for help in the field. Marshall Otter at the stable isotope laboratory at the Marine Biological Laboratory performed the stable isotope analyses. Marilyn Fogel and Roxane Bowden at the Carnegie Institution of Washington for additional support and assistance with stable isotope analyses. Paul Henderson at the Nutrient Analytical Facility at Woods Hole Oceanographic Institute conducted the nutrient analyses and the flow cytometry analyses were performed at the J.J. MacIsaac Facility for Aquatic Cytometry at the Bigelow Laboratory for Ocean Sciences.

Author Contributions

Conceived and designed the experiments: CLF MPL. Performed the experiments: CLF MPL. Analyzed the data: CLF DBM MPL. Contributed reagents/materials/analysis tools: CLF DBM MPL. Wrote the paper: CLF DBM MPL.

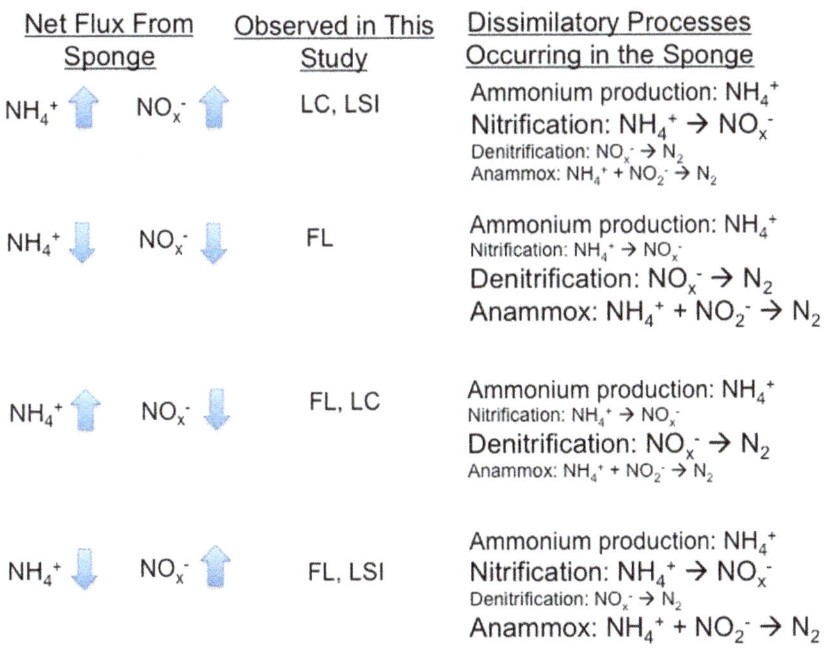

Figure 5. Potential dissimilatory N transformations occurring in _X. muta_ based on the observed net flux of NH_4^+ and NO_x^- from the sponge. Font size for a given process indicates the relative importance of that process. Locations may appear more than once due to differences in individual sponges at each location.

References

1. Reiswig H (1971) In situ pumping activities of tropical Demospongiae. Mar Biol 9: 38–50.

2. Diaz CM and Rützler K (2001) Sponges: an essential component of Caribbean coral reefs. Bull Mar Sci 69: 535–546.

3. Ribeiro SM, Omena EP, Muricy G (2003) Macrofauna associated to _Mycale microsigmatosa_ (Porifera, Demospongiae) in Rio de Janeiro State, SE Brazil. Estuar Coast Shelf Sci 57: 951–959.

4. Ribes M, Coma R, Atkinson MJ, Kinzie RA (2005) Sponges and ascidians control removal of particulate organic nitrogen form coral reef water. Limnol Oceangr 50: 1480–1489.

5. Lesser MP (2006) Benthic-pelagic coupling on coral reefs: Feeding and growth of Caribbean sponges. J Exp Mar Biol Ecol 328: 277–288.

6. Weisz JB, Hentschel U, Lindquist N, Martens CS (2007) Linking abundance and diversity of sponge-associated microbial communities to metabolic differences in host sponges. Mar Biol 152: 475–483.

7. Southwell MW, Popp BN, Martens CS (2008 a) Nitrification controls on fluxes and isotopic composition of nitrate form Florida Keys sponges. Mar Chem 108: 96–108.

8. Southwell MW, Weisz JB, Martens CS, Lindquist N (2008 b) In situ fluxes of dissolved inorganic nitrogen from the sponge community on Conch Reef, Key Largo, Florida. Limnol Oceangr 53: 986–996.

9. Muscatine L, Porter JW (1977) Reef corals: mutualistic symbioses adapted to nutrient-poor environments. Bioscience 27: 454–460.

10. Fiore CL, Jarett JK, Olson ND, Lesser MP (2010) Nitrogen fixation and nitrogen transformations in marine symbioses. Trends Microbiol 18: 455–463.

11. Reiswig H (1981) Partial carbon and energy budgets of the bacteriosponge Verongia fistularis (Porifera: Demospongiae) in Barbados. Mar Ecol 2: 273–293.

12. Corredor JE, Wilkinson CR, Vicente VP, Morell JM, Otero E (1988) Nitrate release by Caribbean reef sponges. Limnol Oceangr 33: 114–129.

13. Diaz M, Ward B (1997) Sponge-mediated nitrification in tropical benthic communities. Mar Ecol Prog Ser 156: 97–107.

14. Eroteida J, Ribes M (2007) Sponges as a source of dissolved inorganic nitrogen: nitrification mediated by temperate sponges. Limnol Oceangr 52: 948–958.

15. Bayer K, Schmitt S, Hentschel U (2008) Physiology, phylogeny and _in situ_ evidence for bacterial and archaeal nitrifiers in the marine sponge Aplysina aerophoba. Environ Microbiol 10: 2942–2955.

16. Hoffmann F, Radax R, Woebken D, Holtappels M, Lavik G, et al. (2009) Complex nitrogen cycling in the sponge Geodia barretti. Environ Microbiol 11: 2228–2243.

17. Bonin PC, Michetoy VD (2006) Nitrogen budget in a microbial mat in the Camargue (southern France). Mar Ecol Prog Ser 322: 75–84.

18. Capone DG, Dunham SE, Horrigan SG, Duguay LE (1992) Microbial nitrogen transformations in unconsolidated coral-reef sediments. Mar Ecol Prog Ser 80: 75–88.

19. Wilkinson C, Fay P (1979) Nitrogen fixation in coral reef sponges with symbiotic cyanobacteria. Nature 279: 527–529.

20. Shieh WY, Lin YM (1994) Association of heterotrophic nitrogen-fixing bacteria with a marine sponge of _Halichondria_ sp. Bull Mar Sci 54: 557–564.

21. Wilkinson CR, Summons RE, Evans E (1999) Nitrogen fixation in symbiotic marine sponges: ecological significance and difficulties in detection. Mem Queensland Mus 44: 667–673.

22. Southwell MW (2007) Sponges impacts on coral reef nitrogen cycling, Key Largo, Florida. Dissertation, University of North Carolina at Chapel Hill.

23. Mohamed NM, Colman AS, Tai Y, Hill RT (2008) Diversity and expression of nitrogen fixation genes in bacterial symbionts of marine sponges. Environ Microbiol 10: 2910–2921.

24. Hoffmann F, Larsen O, Thiel V, Rapp HT, Pape T, et al. (2005a) An anaerobic world in sponges. Geomicrobiol J 22: 1–10.

25. Mohamed NM, Saito K, Tal Y, Hill RT (2009) Diversity of aerobic and anaerobic ammonia-oxidizing bacteria in marine sponges. ISME J 4: 38–48.

26. Taylor MW, Radax R, Stegor D, Wagner M (2007) Sponge-associated microorganisms: evolution, ecology, and biotechnological potential. Microbiol Mol Biol Rev 71: 1–53.

27. Schmitt S, Tsai P, Bell J, Fromont, Ilan JM, et al. (2011) Assessing the complex sponge microbiota: core, variable and species-specific bacterial communities in marine sponges. ISME J 6: 564–576.

28. Siegl A, Kamke J, Hochmuth T, Piel JOR, Richter M, et al. (2010) Single-cell genomics reveals the lifestyle of Poribacteria, a candidate phylum symbiotically associated with marine sponges. ISME J 5: 61–70.

29. Webster NS, Taylor MW, Behnam F, Lücker S, Rattei T, et al. 2010. Deep sequencing reveals exceptional diversity and modes of transmission for bacterial sponge symbionts. Environ Microbiol 12: 2070–2082.

30. Lee OO, Wang Y, Yang J, Lafi FF, Al-Suwailem A, Qian PY (2010) Pyrosequencing reveals highly diverse and species-specific microbial communities in sponges from the Red Sea. ISME J 5: 650–664.

31. McMurray SE, Blum JE, Pawlik JR (2008) Redwood of the reef: growth and age of the giant barrel sponge Xestospongia muta in the Florida Keys. Mar Biol 155: 159–171.

32. Hentschel U, Usher KM, Taylor MW (2006) Marine sponges as microbial fermenters. FEMS Microbiol Ecol 55: 167–177.

33. Steindler L, Huchon D, Avni A, Ilan M (2005) 16S rRNA Phylogeny of Sponge-Associated Cyanobacteria. Appl Environ Microbiol 71: 4127–4131.

34. Lopez-Legentil S, Erwin OM, Pawlik JR, Song B (2010) Effects of sponge bleaching on ammonia-oxidizing *Archaea*: distribution and relative expression of ammonia monooxygenase genes associated with the barrel sponge *Xestospongia muta*. Microbial Ecol 60: 561–571.

35. Montalvo NF, Hill RT (2011) Sponge-Associated Bacteria Are Strictly Maintained in Two Closely Related but Geographically Distant Sponge Hosts. Appl Environ Microbiol 77: 7207–7216.

36. Diamond J (1986) Overview: laboratory experiments, field experiments, and natural experiments. In: Community Ecology. Diamond J,Case TJ, New York: Harper and Row, 3–22.

37. Trussel GC, Lesser MP, Patterson MR, Genovese SJ (2006) Depth-specific differences in growth of the reef sponge Callyspongia vaginalis: role of bottom-up effects. Mar Ecol Prog Ser 323: 149–158.

38. Weisz JB, Lindquist UN, Martens CS (2008) Do associated microbial abundances impact demosponge pumping rates and tissue densities? Oecologia 155: 367–376.

39. Bartlett MS (1937) Properties of sufficiency and statistical tests. Proc Royal Soc London Ser 160: 268–282.

40. Royston P (1995) A remark on Algorithm AS 181: the *W* test for normality. Appl Stat 44: 547–551.

41. Sokal RR, Rohlf JF (1995) Biometry, 3rd ed. W. H. Freeman and Company. 880.

42. Schmider E, Ziegler M, Danay E, Beyer L, Bühner M (2010) Is it really robust? Reinvestigating the robustness of ANOVA against violations of the normal distribution assumption. Methodology: Euro J Res Meth Behav Social Sci 6: 147–151.

43. R Core Team (2012) R: A language and environment for statistical computing. Vienna, Austria, R Foundation for Statistical Computing. ISBN 3-900051-07-0, R project website. Available: http://www.R-project.org/. Accessed 2012 Feb 19.

44. Ducklow HW, Kirchman DL, Quinby HL, Carlson CA, Dam HG (1993) Stocks and dynamics of bacterioplankton carbon during the spring bloom in the eastern North Atlantic Ocean. Deep Sea Res 40: 245–263.

45. Morel A, Ahn YH, Partensky F, Vaulot D, Claustre H (1993) *Prochlorococcus* and *Synechococcus*: a comparative study of their optical properties in relation to their size and pigmentation. J Mar Res 51: 617–649.

46. Campbell L, Nolla HA, Vaulot D (1994) The importance of *Prochlorococcus* to community structure in the central North Pacific Ocean. Limnol Oceanogr 39: 954–960.

47. FaggerBakke KM, Heldal M, Norland S (1996) Content of carbon, nitrogen, oxygen, sulfur and phosphorus in native aquatic and cultured bacteria. Mar Ecol Prog Ser 10: 15–27.

48. Bertilsson S, Berglund O, Karl DM, Chisholm SW (2003) Elemental composition of marine *Prochlorococcus* and *Synechococcus*: implications for the ecological stoichiometry of the sea. Limnol Oceanogr 48:1721–1731.

49. Freeman CJ, Thacker RW (2011) Complex interactions between marine sponges and their symbiotic microbial communities. Limnol Oceangr 56: 1577–1586.

50. Freeman CJ, Thacker RW, Baker DM, Fogel ML (2013) Quality or quantity: is nutrient transfer driven more by symbiont identity and productivity than by symbiont abundance? J ISME doi:10.1038/ismej.2013.7

51. Leichter JJ, Wing SR, Miller SL, Denny MW (1996) Pulsed delivery of subthermocline water to Conch Reef (Florida Keys) by internal tidal bores. Limnol Oceanogr 41: 1490–1501.

52. Leichter JJ, Stewart HL, Miller SL (2003) Episodic nutrient transport to Florida coral reefs. Limnol Oceanogr 48:1394–1407.

53. Schläppy M-L, Schottner SL, Lavik G, Kuypers MMM, de Beer D, et al. (2010a) Evidence of nitrification and denitrification in high and low microbial abundance sponges. Mar Biol 157: 593–602.

54. Muro-Pastor MI, Reyes JC, Florencio FJ (2005) Ammonium assimilation in cyanobacteria. Photosynth Res 83: 135–150.

55. Fry B (2006) Stable Isotope Ecology. New York: Springer. 308.

56. Mariotti A (1983) Atmospheric nitrogen is a reliable standard for natural ^{15}N abundance measurements. Nature 303: 685–687.

57. Peterson BJ, Fry B (1987) Stable isotopes in ecosystem studies. Ann. Rev. Ecol. Syst. 18: 293–320.

58. Vander Zanden JM, Rasmussen JB (2001) Variation in δ^{15}N and δ^{13}C trophic fractionation: Implications for aquatic food web studies. Limnol Oceanogr 46: 2061–2066.

59. McCutchan JH, Lewis WM, Kendall C, McGrath CC (2003) Variation in trophic shift for stable isotope ratios of carbon, nitrogen, and sulfur. Oikos 102: 378–390.

60. Carpenter EJ, Harvey HR, Fry B, Capone DG (1997) Biogeochemical tracers of the marine cyanobacterium *Trichodesmium*. Deep-Sea Res 44: 27–38.

61. Montoya JP, Carpenter EJ, Capone DG (2002) Nitrogen fixation and nitrogen isotope abundances in zooplankton of the oligotrophic North Atlantic. Limnol Oceanogr 47: 1617–1628.

62. Michener RH, Schell DM (1994) Stable isotope ratios as tracers in marine aquatic food webs. In: Lajtha K, Michener RH, Stable isotopes in ecology and environmental science. Blackwell Scientific. 138–157.

63. DeLorenzo S, Brauer SL, Edgmont CA, Herfort L, Tebo BM, et al. (2012) Ubiquitous dissolved inorganic carbon assimilation by marine bacteria in the Pacific Northwest coastal ocean as determined by stable isotope probing. PLOS ONE 7: e46695.

64. Yellowlees D, Rees TAV, Fitt WK (1994) Effect of ammonium-supplemented seawater on glutamine synthetase and glutamate dehydrogenase activities in the host tissue and zooxanthellae of *Pocillopora damicornis* and on ammonium uptake rates of the zooxanthellae. Pac Sci 48: 291–295.

65. Maguer J-F, L'Helguen S, Caradec J, Klein C (2011) Size-dependent uptake of nitrate and ammonium as a function of light in well-mixed temperate coastal waters. Cont Shelf Res 31: 1620–1631.

66. Kirchman DL (1994) The uptake of inorganic nutrients by heterotrophic bacteria. Microb Ecol 28: 255–271.

67. Mackey KRM, Bristow L, Parks DR, Altabet MA, Post AF, et al. (2011) The influence of light on nitrogen cycling and the primary nitrite maximum in a seasonally stratified sea. Prog Oceanogr 91: 545–560.

68. Kuypers MMM, Sliekers AO, Lavik G, Schmid M, Jorgensen BB, et al. (2003) Anaerobic ammonium oxidation by anammox bacteria in the Black Sea. Nature 422: 608–611.

69. Caffey JM, Sloth NP, Kaspar H, Blackburn TH (1993) Effect of organic loading on nitrification and denitrification in marine sediment microcosms. FEMS Microbiol Ecol 12: 159–167.

70. Purubsky WP, Weston NB, Joye SB (2009) Benthic metabolism and the fate of dissolved inorganic nitrogen in intertidal sediments. Estuar Coast Mar Sci 83: 392–402.

71. Ribes M, Coma R, Atkinson MJ, Kinzie III RA (2003) Particle removal by coral reef communities: picoplankton is a major source of nitrogen. Mar Ecol Prog Ser 257: 13–23.

72. Perea-Blásquez A, Davy SK, Bell JJ (2012) Estimates of particulate organic carbon flowing from the pelagic environment to the benthos through sponge assemblages. Hydrobiologia 687: 237–250.

73. O'Neil JM, Capone DG (2008) Nitrogen Cycling in Coral Reef Environments. In: Capone DG, Bronk DA, Mullholland MR, Carpenter EJ, Nitrogen in the Marine Environment. Burlington: Academic Press. 949–989.

Permissions

The contributors of this book come from diverse backgrounds, making this book a truly international effort. This book will bring forth new frontiers with its revolutionizing research information and detailed analysis of the nascent developments around the world.

We would like to thank all the contributing authors for lending their expertise to make the book truly unique. They have played a crucial role in the development of this book. Without their invaluable contributions this book wouldn't have been possible. They have made vital efforts to compile up to date information on the varied aspects of this subject to make this book a valuable addition to the collection of many professionals and students.

This book was conceptualized with the vision of imparting up-to-date information and advanced data in this field. To ensure the same, a matchless editorial board was set up. Every individual on the board went through rigorous rounds of assessment to prove their worth. After which they invested a large part of their time researching and compiling the most relevant data for our readers.

The editorial board has been involved in producing this book since its inception. They have spent rigorous hours researching and exploring the diverse topics which have resulted in the successful publishing of this book. They have passed on their knowledge of decades through this book. To expedite this challenging task, the publisher supported the team at every step. A small team of assistant editors was also appointed to further simplify the editing procedure and attain best results for the readers.

Apart from the editorial board, the designing team has also invested a significant amount of their time in understanding the subject and creating the most relevant covers. They scrutinized every image to scout for the most suitable representation of the subject and create an appropriate cover for the book.

The publishing team has been an ardent support to the editorial, designing and production team. Their endless efforts to recruit the best for this project, has resulted in the accomplishment of this book. They are a veteran in the field of academics and their pool of knowledge is as vast as their experience in printing. Their expertise and guidance has proved useful at every step. Their uncompromising quality standards have made this book an exceptional effort. Their encouragement from time to time has been an inspiration for everyone.

The publisher and the editorial board hope that this book will prove to be a valuable piece of knowledge for researchers, students, practitioners and scholars across the globe.

List of Contributors

Elisabet Ejarque-Gonzalez and Andrea Butturini
Departament d'Ecologia, Facultat de Biologia, Universitat de Barcelona, Barcelona, Catalunya, Spain

Francesca Rossi
Laboratoire Ecologie des Systémes marins côtiers, Université Montpellier 2, Montpellier, France

Britta Gribsholt
Department of Ecosystems, Royal Netherlands Institute for Sea Research, Yerseke, the Netherlands

Frederic Gazeau
Centre National de la Recherche Scientifique-Institut National des Sciences de l'Univers, Laboratoire d' Oceanographie de Villefranche, Villefranche-sur-mer, France
Université Pierre et Marie Curie, Observatoire Océanologique de Villefranche, Villefranche-sur-mer, France

Valentina Di Santo
Department of Biology, Boston University, Boston, Massachusetts, United States of America

Jack J. Middelburg
Department of Ecosystems, Royal Netherlands Institute for Sea Research, Yerseke, the Netherlands
Department of Earth Sciences – Geochemistry, Faculty of Geosciences, Utrecht University, Utrecht, The Netherlands

Violaine Jacq, Céline Ridame, Fanny Kaczmar and Alain Saliot
Université Pierre et Marie Curie, UMR LOCEAN -IPSL/CNRS/IRD/MNHN, Paris, France

Stéphane L'Helguen
Université de Brest, CNRS/IRD, UMR 6539, LEMAR, OSU-IUEM, Plouzané, France

Michael J. Wilkins, Carrie D. Nicora, Lee Ann McCue and Mary S. Lipton
Biological Sciences Division, Pacific Northwest National Laboratory, Richland, Washington, United States of America

Kelly C. Wrighton, Kim M. Handley, Chris S. Miller and Jillian F. Banfield
Department of Earth and Planetary Science, University of California, Berkeley, California, United States of America

Kenneth H. Williams, Alison P. Montgomery and Philip E. Long
Earth Sciences Division, Lawrence Berkeley National Laboratory, Berkeley, California, United States of America

Ludovic Giloteaux and Derek R. Lovley
Department of Microbiology, University of Massachusetts Amherst, Amherst, Massachusetts, United States of America

Mayali, Peter K. Weber, Shalini Mabery and Jennifer Pett-Ridge
Physical and Life Science Directorate, Lawrence Livermore National Laboratory, Livermore California, United States of America

Andrea G. Vincent
Department of Ecology and Environmental Sciences, Umeå University, Umeå, Sweden
Department of Forest Ecology and Management, Swedish University of Agricultural Sciences, Umeå, Sweden

Maja K. Sundqvist and David A. Wardle
Department of Forest Ecology and Management, Swedish University of Agricultural Sciences, Umeå, Sweden

Reiner Giesler
Department of Ecology and Environmental Sciences, Umeå University, Umeå, Sweden
Climate Impacts Research Centre, Department of Ecology and Environmental Sciences, Umeå University, Abisko, Sweden

Letizia Tedesco
Marine Research Centre, Finnish Environment Institute, Helsinki, Finland

Marcello Vichi
Istituto Nazionale di Geofisica e Vulcanologia, Bologna, Italy
Centro Euro-Mediterraneo sui Cambiamenti Climatici, Bologna, Italy

Liming Zhang
College of Resource and Environment, Fujian Agriculture and Forestry University, Fuzhou, China
State Key Laboratory of Soil and Sustainable Agriculture, Institute of Soil Science, Chinese Academy of Sciences, Nanjing, China

Dongsheng Yu, Xuezheng Shi, Shengxiang Xu and Yongcong Zhao
State Key Laboratory of Soil and Sustainable Agriculture, Institute of Soil Science, Chinese Academy of Sciences, Nanjing, China

Shihe Xing
College of Resource and Environment, Fujian Agriculture and Forestry University, Fuzhou, China

James R. Christian
Canadian Centre for Climate Modelling and Analysis, Victoria, B.C., Canada
Fisheries and Oceans Canada, Institute of Ocean Sciences, Sidney, BC, Canada

Mo Chen
State Key Laboratory of Lake Science and Environment, Nanjing Institute of Geography and Limnology, Chinese Academy of Sciences, Nanjing, China
Graduate University of Chinese Academy of Sciences, Beijing, China

Tian-Ran Ye and He-Long Jiang
State Key Laboratory of Lake Science and Environment, Nanjing Institute of Geography and Limnology, Chinese Academy of Sciences, Nanjing, China

Lee R. Krumholz
Department of Microbiology and Plant Biology, University of Oklahoma, Norman, Oklahoma, United States of America

Hang Wang, HongYi Li, Zhi Jian Zhang and Xin Hua Xu
College of Natural Resource and Environmental Sciences, China Academy of West Development, Zhejiang University, Hangzhou, Zhejiang Province, China

Jeffrey D. Muehlbauer
Curriculum for the Environment & Ecology, University of North Carolina at Chapel Hill, Chapel Hill, North Carolina, United States of America

Qiang He
Department of Civil & Environmental Engineering, University of Tennessee, Knoxville, Tennessee, United States of America

Chun Lei Yue
Institute of Ecology, Zhejiang Forestry Academy, Hangzhou, China

DaQian Jiang
Department of Earth and Environmental Engineering, Henry Krumb School of Mines, Columbia University, New York, New York, United States of America

Gennadi Lessin
Plymouth Marine Laboratory, Prospect Place, The Hoe, Plymouth, United Kingdom

Urmas Raudsepp
Marine Systems Institute, Tallinn University of Technology, Tallinn, Estonia

Adolf Stips
European Commission, Joint Research Centre, Institute for Environment and Sustainability, Water Research Unit, Ispra, Italy

Rocio I. Ruiz-Cooley and Matthew D. McCarthy
Ocean Sciences Department, University of California Santa Cruz, Santa Cruz, California, United States of America

Paul L. Koch
Earth and Planetary Sciences Department, University of California Santa Cruz, Santa Cruz, California, United States of America

Paul C. Fiedler
Southwest Fisheries Science Center, National Marine Fisheries Service, National Oceanic and Atmospheric Administration, La Jolla, California, United States of America

Tomoharu Inoue, Shin Nagai, Rikie Suzuki, Hadi Fadaei and Reiichiro Ishii
Department of Environmental Geochemical Cycle Research, Japan Agency for Marine-Earth Science and Technology (JAMSTEC), Yokohama, Japan

Satoshi Yamashita, Kimiko Okabe and Hisatomo Taki
Department of Forest Entomology, Forestry and Forest Products Research Institute (FFPRI), Tsukuba, Ibaraki, Japan

Yoshiaki Honda and Koji Kajiwara
Center of Environmental Remote Sensing, Chiba University, Chiba, Japan

Anne-Elise Nieblas and Sylvain Bonhommeau
Unité Mixte Recherche Ecosystémes Marins Exploités 212, Institut Français de Recherche pour l'Exploitation de la Mer (IFREMER), Séte, France

Kyla Drushka
Applied Physics Laboratory, University of Washington, Seattle, Washington, United States of America

Gabriel Reygondeau
Center for Macroecology, Evolution and Climate, National Institute for Aquatic Resources, Technical University of Denmark (DTU Aqua), Charlottenlund, Copenhagen, Denmark

Vincent Rossi
Instituto de FÍsica Interdisciplinary Sistemas Complejos, Institute for Cross-Disciplinary Physics and Complex Systems, (CSIC-UIB), Campus Universitat de les Illes Balears, Palma de Mallorca, Spain

Hervé Demarcq
Unité Mixte de Recherche Ecosystémes Marins Exploités 212, Institut de Recherche pour le Développement (IRD), Séte, France

Laurent Dubroca
European Commission, Joint Research Center, Institute for Environment & Sustainability, Water Resources, Ispra, Italy

Rebecca E. Green and Alexis Lugo-Fernández
Environmental Studies Section, Bureau of Ocean Energy Management, New Orleans, Lousiana, United States of America

Amy S. Bower
Physical Oceanography Department, Woods Hole Oceanographic Institution, Woods Hole, Massachusetts, United States of America

Adam Wolf
Department of Ecology and Evolutionary Biology, Princeton University, Princeton, New Jersey, United States of America

Christopher E. Doughty and Yadvinder Malhi
Environmental Change Institute, School of Geography and the Environment, University of Oxford, Oxford, United Kingdom

Marion Fraysse
IFREMER, Laboratoire Environnement Ressources Provence Azur Corse, La Seyne sur Mer, France
Aix Marseille Université, CNRS/INSU, IRD, Mediterranean Institute of Oceanography (MIO), UM 110, Marseille, France

Université de Toulon, CNRS/INSU, IRD, Mediterranean Institute of Oceanography (MIO), UM 110, La Garde, France

Christel Pinazo, Vincent Martin Faure, Rosalie Fuchs and Patrick Raimbault
Aix Marseille Université, CNRS/INSU, IRD, Mediterranean Institute of Oceanography (MIO), UM 110, Marseille, France
Université de Toulon, CNRS/INSU, IRD, Mediterranean Institute of Oceanography (MIO), UM 110, La Garde, France

Paolo Lazzari
Dept. of Oceanography, Istituto Nazionale di Oceanografia e di Geofisica Sperimentale (OGS), Trieste, Italy

Ivane Pairaud
IFREMER, Laboratoire Environnement Ressources Provence Azur Corse, La Seyne sur Mer, France

Peter W. Holder and Karen Armstrong
Bio-Protection Research Centre, Lincoln University, Canterbury, New Zealand

Robert Van Hale and Russell Frew
Department of Chemistry, Otago University, Dunedin, New Zealand

Marc-Alban Millet and Joel A. Baker
School of Geography Environment and Earth Sciences, Victoria University of Wellington, Wellington, New Zealand

Timothy J. Clough
Department of Soil and Physical Sciences, Lincoln University, Canterbury, New Zealand

Cara L. Fiore and Michael P. Lesser
University of New Hampshire, Department of Molecular, Cellular and Biomedical Sciences, Durham, New Hampshire, United States of America

David M. Baker
Geophysical Laboratory, Carnegie Institute of Washington, Washington, District of Columbia, United States of America

Index

www.ingramcontent.com/pod-product-compliance
Lightning Source LLC
Chambersburg PA
CBHW080529200326
41458CB00012B/4383